CW01072756

Heat Recovery Systems

A directory of equipment and techniques

Heat Recovery Systems

A directory of equipment and techniques

D. A. Reay

Business Development Manager,
International Research and Development Co. Ltd

London
E. & F. N. Spon

First published 1979
by E. and F. N. Spon Ltd.
11 New Fetter Lane London EC4P 4EE

© *1979 D. A. Reay*

Typeset by C. Josée Utteridge-Faivre and
printed in Great Britain by
William Clowes and Sons Ltd.

ISBN 0 419 11400 9

All rights reserved. No part of this book may be
reprinted, or reproduced or utilized, in any form
or by any electronic, mechanical or other means,
now known or hereafter invented, including photocopying
and recording, or in any information storage and
retrieval system, without permission in writing
from the publisher.

Distributed in the USA
by Halsted Press, a Division
of John Wiley and Sons, Inc., New York

Library of Congress Cataloging in Publication Data
Reay, David Anthony.
Heat recovery systems.

Bibliography: p.
Includes index.
1. Heat recovery – Equipment and suppliers – Directories.
I. Title.
TJ260.R4 621.4′02 79-10877
ISBN 0-470-26728-3

Contents

Acknowledgements

The author would like to thank those companies who supplied data for inclusion in this Directory, particularly those who completed questionnaires concerning their product range.

Preface

Energy conservation, be it practised in industry, transport, public services or the home, has during the past few years become a major feature of the economies and planning policies of most industrialized nations. Governments have been stimulated into activity for a number of reasons. Continuing increases in the cost of primary fuels – the outstanding example being the rise in the price of oil, in 1973 – have led to large import bills for countries unfortunate in not possessing sufficient indigenous energy resources to meet their current needs, and a rise in the 'energy-content' cost of most goods and services, thus resulting in inflation. A realization that fossil fuels, wherever they may exist, are not in infinite supply and must be conserved for purposes which they alone can satisfy, has spurred conservation, as well as the development of alternative energy sources such as wave power and solar energy.

The implementation of energy-conservation measures is generally carried out in three stages. Immediate remedies, coming under the general heading of 'good housekeeping', include thermal insulation, repair of steam leaks, and other practices which can be implemented rapidly and at little cost in terms of capital expenditure. On a much longer timescale, domestic heating systems may be changed, or manufacturing processes may be replaced with more energy-efficient systems, or ones which rely on alternative energy sources. An interim measure, which may be implemented in months but provides benefits for many years, is to install waste heat recovery equipment.

The recovery of waste heat – be it in an industrial process, from a prime mover or in air conditioning systems – is in many cases the most beneficial single energy-conservation technique which can be implemented, and because of the very significant cost reductions which can be achieved it has long been common practice in the more energy-intensive industries such as steelmaking and petrochemicals. In my first book on energy conservation (*Industrial Energy Conservation – a Handbook for Engineers and Managers,* Pergamon Press, 1979), I stressed the importance of the lessons and techniques which could be learned by one industrial sector from the experiences of another, and in my close association with waste heat recovery in the UK during the past few years, I have consistently sought to emphasize this fact. The opportunity to write this Directory, with its worldwide appeal, has, I hope, resulted in a book which will be of considerable assistance to existing and future users of waste heat recovery equipment in all industries.

Users of this Directory may like to speculate as to the form such a book might take in 20 years time – in many cases the need for waste heat recovery is an indication of process inefficiency which ultimately will necessitate changing the process itself. If there are as many manufacturers of these types of heat exchanger in the year 1998 as there are today, our future will indeed be bleak.

Introduction

The recovery of waste heat has been practised for many decades, particularly in power generation and the energy-intensive industries. Many common heat recovery devices have been in production for 50 years, although of course performance and manufacturing techniques have in most instances been greatly improved over the intervening period. During the past few years the interest in energy conservation has attracted an increasing number of products on to the market, and the potential user of such equipment has been set an enormous task in identifying the correct type of system to meet his requirements. The purpose of this Directory is to make his task considerably easier.

In order to do this, it is necessary to present to the potential user a considerable amount of data. As the Directory is aimed at the international market, the opportunity is afforded to present data on equipment specifications and applications which the individual user in any one country may otherwise never be aware of – for instance, a reader in the USA will probably find a sufficiently wide range of products detailed here which are manufactured within his own country, but a particular type of Swiss heat recovery unit may also attract his attention. As well as equipment-users, those marketing equipment will also find the Directory of value in identifying new products and competitors.

The format of the Directory is directed primarily at enabling the potential heat recovery equipment-user to identify the several types of heat recovery systems which could meet his particular requirements. This is done by describing, sometimes briefly, in a number of chapters, the main types of heat recovery equipment and their principal applications. These include:

Gas-to-gas heat recovery equipment;
Gas-to-liquid systems;
Liquid–liquid systems;
Prime movers as users and sources of waste heat;
Waste incineration with heat recovery;
Heat pumps; and heat recovery luminaires.

Within the gas-to-gas heat recovery equipment classification (for example) will include: heat pipe heat exchangers, rotating regenerators, run-around coils, plate heat exchangers, convection and radiation recuperators and recuperative gas burners. Basic descriptions of each type of heat exchanger will be given, with its primary advantages and disadvantages. Main application areas will be discussed, as well as the type of performance and payback period to be anticipated.

The major part of the Directory is devoted to data on heat recovery equipment obtained from the manufacturers. This data has been collected by a detailed questionnaire sent to several hundred companies located world wide, and by presenting performance figures, where available, for each item of heat recovery equipment, it is hoped that the potential user will find the task of selecting equipment best suited to his needs much easier. As with the preceding chapters, the data on the equipment is presented in seven major sections. Data for gas-to-gas heat recovery units includes: manufacturer; model number; size; air volume; duty; operating temperature range; pressure drop; efficiency; materials of construction; weight.

For each type of unit, a number of specific questions were also put to the manufacturer, concerning cleaning of the heat exchanger, 'optional extras', who carries out installation, and case histories.

For completeness, a number of reference sections are appended. Following the detailed manufacturers' data in Chapters 8–14, there is an alphabetical list of the names, addresses and telephone numbers of these manufacturers, with their product categories. Appendices 2 and 3 give financial analysis and conversion factors respectively. There is a comprehensive Bibliography, a Subject Index, and an Index to Manufacturers.

1 Gas-to-gas heat recovery equipment

1.1 Introduction

It is in the field of heat recovery from exhaust gas and air streams that in most cases the largest benefits from investment in energy-conservation equipment can be realized. There are a considerable number of uses to which this waste heat can be put, and these depend to a large extent on the temperature and condition of the exhaust gases, the heat recovery equipment used, and the economic assessment of the overall system performance. With regard to the uses which determine the type of heat exchanger to be used, we may identify three main areas, these being the heating of liquid, steam raising, and air heating. This chapter is concerned with the last of these categories, the recovery of heat from exhaust gas or air streams for preheating of other air or gas streams.

Within this category it is possible to identify three main application areas for the waste heat recovery equipment:

(i) Use of process waste heat for preheating process supply air.
(ii) Use of process waste heat for space heating and air conditioning.
(iii) Recovery of exhaust heat from an airconditioning system for preheating supply air. (Note that in summer such a heat exchanger may also be used for precooling incoming air, effecting savings on the refrigeration load.)

One can also identify a few more specialized areas of application involving, for example, pollution control, recovery of heat from prime movers, and the use of incinerator waste heat. Where this equipment differs significantly from gas-to-gas systems used extensively in process and air conditioning heat recovery, full descriptions are reserved for inclusion in later chapters.

Most items of equipment described in this chapter can be used when the waste heat is at a sufficiently high temperature not to require 'upgrading' for reuse. However, heat pumps which can use outside air or process air as a heat source, are also available. The heat pump, which is able to upgrade waste heat, may also be effectively used in some drying processes, and is dealt with in detail in Chapter 6.

A point often overlooked when selecting heat recovery equipment – whatever the type – is that when installed in a new building or process, it can generally result in a reduction in size of, say, boiler or refrigeration plant. However, when retrofitted, the equipment, while relieving the load on steam-raising plant, for example, may as a result cause it to operate at a lower efficiency. Such changes should be taken into account when carrying out an economic assessment, but, of course, such a loss in efficiency is unlikely to result in any dramatic reduction in the effectiveness of heat recovery equipment as an energy-saving technique. (In the same way, of course, extra fan power needed to overcome the pressure drop through a waste heat recovery unit should also be included in an operating-cost balance sheet.)

Three other points of importance should be taken into account when considering the installation of waste heat recovery equipment, particularly applicable to heat recovery from exhaust gas streams. First, while recovering heat obviously saves energy, it may result in such a reduction in fuel consumption that the burners used in the process to which the heat exchanger is being applied may not have a sufficiently large operating range to cope with reduced fuel requirements. In such a case it will of course be necessary to replace the existing burners, and this may have a serious adverse effect on the economic analysis. Obviously this will only happen if the recovered heat is being used to preheat the combustion air, and if burner turndown range is insufficient, and alternative use may be found for the waste heat (i.e. for space heating or replacing calorifiers).

Second, the use to which the waste heat is put can have a significant effect on the total installed cost of the heat recovery installation. Taking the example above, where preheating of combustion air may be required, a complex oven used for food baking may have several burners along a length of oven approaching 50 m. Such an oven will also have possibly four exhaust gas flues situated along its length. Ducting of preheated air to each burner (depending upon the type of gas–gas heat recovery system used) can be an expensive exercise, and in any heat recovery installation under study, it is extremely important to look in detail at all the associated costs, such as ductwork, fans, controls and additional thermal insulation, in arriving at a realistic cost. These financial features are dealt with in more detail in Appendix 1, but it is worth pointing out here that the installed cost of a gas–gas heat recovery system can be anything between 1·5 and 4 times the cost of the basic heat exchanger itself, depending upon the system complexity and operating temperature. (Installation costs increase with operating temperature, because of the need to use more expensive materials, etc.) Further variables which must be taken into account in equipment selection are discussed at the beginning of Chapter 8.

1.2 Types of gas–gas heat recovery equipment

Of the techniques for waste heat recovery, it is in the area of gas–gas heat exchange that the widest variety exists. However, commercially available systems may be broadly grouped into two classifications: recuperators and regenerators. The recuperator functions in such a way that the heat flows steadily and continuously from one fluid to another through a separating wall. In a regenerator, however, the flow of heat is intermittent, and is typified by rotary systems such as the heat wheel. In this unit a matrix of metal comes into contact alternately with the hot and cold fluid, first absorbing heat and then rejecting it. The run-around coil (or twin-coil) unit, which incorporates a pumped liquid circuit carrying heat between two gas–liquid heat exchangers, can also be classified as a regenerator.

From the point of view of the user of the heat recovery equipment, the distinction between recuperative and regenerative heat exchange may be regarded as largely academic, as each type of heat exchanger in both categories has its own merits and drawbacks, discussed later. A much more important distinction in the selection of heat recovery equipment is the operating temperature range. Temperature of course can have adverse effects on heat exchanger materials at both ends of the scale. If temperatures are too low, brought about by the removal of too much heat from the exhaust gases, the dew point may be reached and condensation can result in corrosive products affecting heat exchanger materials. Freezing in air conditioning heat recovery systems can also be a serious problem, and it will be noted that defrosting systems are offered as optional equipment on many heat recovery devices. The life of heat exchangers used at high temperatures can be seriously reduced if incorrect materials are selected. As discussed in more detail in later sections of this chapter, the fluids used in heat recovery systems such as heat pipes and run-around coils may also be subject to thermal degradation if temperature limits are exceeded.

1.3 Heat pipe heat exchangers

The heat pipe heat exchanger used for gas-to-gas heat recovery is essentially a bundle of fined heat pipes assembled like a conventional air-cooled heat exchanger. (In duties involving heat transfer to a liquid, it is generally unnecessary to fin the surfaces in contact with the liquid, as discussed in Chapter 2.) Because the heat pipe is a comparatively recent development in the waste heat recovery field, it will be discussed in some detail in this context. It is in many ways, however, similar to the Perkins tube, which was in regular use until the 1960s as a heat transfer device in bread ovens, and which was invented in the nineteenth century.

1.3.1 *The heat pipe*

A heat pipe (Dunn and Reay, 1978) is basically a sealed container, normally in the form of a tube, containing a wick lining the inside wall. The purpose of the wick is to transport a working fluid, contained within the heat pipe, from one end to the other by capillary action. The full operating cycle may be described with reference to Fig. 1.1. Heat applied externally to the evaporator section of the heat pipe causes the working fluid contained within the wick to evaporate, and the increase in pressure causes the

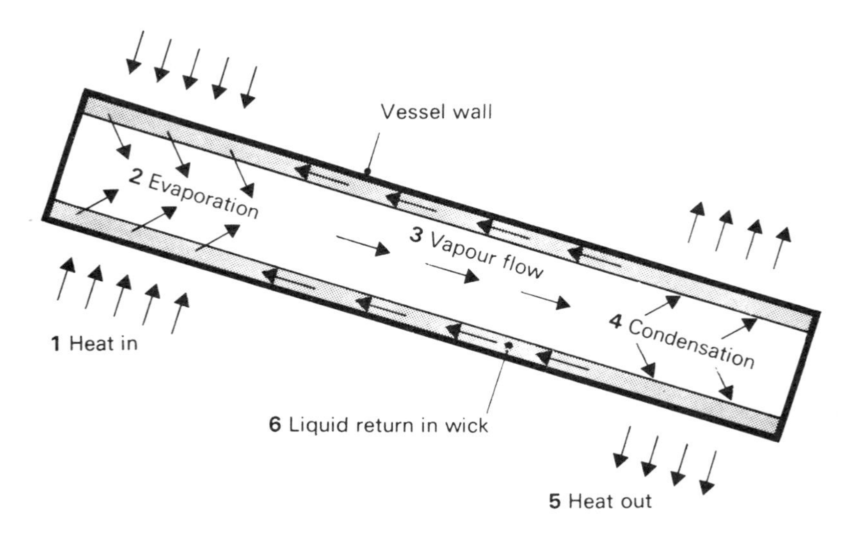

Fig. 1.1 Heat pipe operation.

vapour to flow along the central vapour space to the slightly cooler condenser section, where it condenses, giving up its latent heat of condensation. This heat is then rejected from the outer surface of the condenser, and the condensate is pumped back to the evaporator by the capillary action in the wick. Because of the reliance on capillary action to return the condensate to the heat-input section, the heat pipe is particularly sensitive to the effects of gravity, and hence its inclination to the horizontal. Depending of course on the type of wick used and its pore size, a heat pipe operating with the evaporator below the condenser may be capable of transporting several times as much as one having the evaporator above the condenser. This is particularly important in long heat pipes (more than 500 mm in length), and its implications with respect to heat pipe heat exchangers are discussed below.

In cases where gravity aids return of the condensate to the evaporator section, it is possible to omit the wick, either in whole or in part. The device then becomes a simple thermosyphon. It should be noted here that some manufacturers use the term 'thermosyphon' to describe the basis of their heat recovery equipment, while others use 'heat pipe'.

1.3.2 *General description of the units*

In a gas-to-gas heat pipe heat exchanger the evaporator sections of the heat pipes span the duct carrying the hot exhaust gas, and the condensers are located in the duct through which the air requiring preheating is passing, as shown in Fig. 1.2.

As mentioned above, heat pipes are influenced in their performance by the angle of operation. In the heat pipe heat exchanger, the tube bundle may be horizontal, or tilted with the evaporator sections below the condensers. Because of this sensitivity,

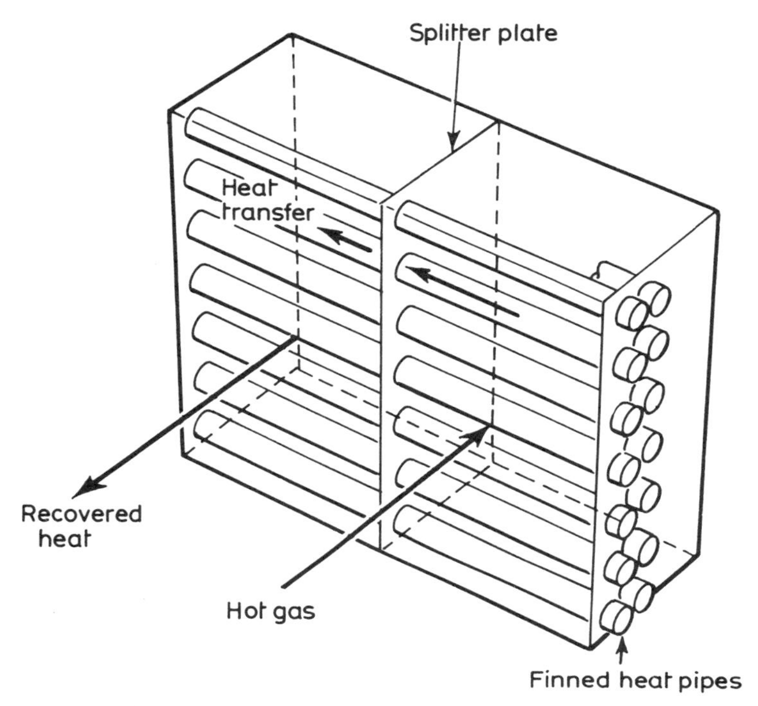

Fig. 1.2 Layout of a heat pipe heat exchanger.

the angle of the heat pipes may be adjusted *in situ* as a means of controlling the heat transport.

This is a useful feature in air conditioning applications, and the effect of the angle of inclination on performance is shown in Fig. 1.3. A number of proprietary units incorporate tilt-control mechanisms which, either manually or automatically, can be adjusted to cater for changes in heat transport requirements (Reay, 1977).

The basic finned heat pipe heat exchanger externally resembles an air-cooled condenser coil, complete with flanged casing and covers protecting the ends of the tubes. The main external difference is the incorporation of a splitter plate (as shown in Fig. 1.2), which is used partly to support the heat pipes, which can be several metres in lengths, but primarily to prevent cross-flow between the two air streams, effectively sealing them from one another. In common with other air-cooled heat exchangers, the finning may be applied individually to each tube, using integral or helically wound fins, or may be in the form of plates into which the tubes are expanded. In the latter case the contact between the tube and fin results in a higher thermal resistance between them, but the cost of such tube bundles is generally lower. Tubes are normally staggered, and the number of tube rows in the direction of flow is typically between 4 and 10.

Material selection for the heat pipes depends upon the working fluid contained within them, as well as the external environment. Working fluids used in heat pipe heat exchangers range from fluorocarbons and water to high-temperature organic fluids and,

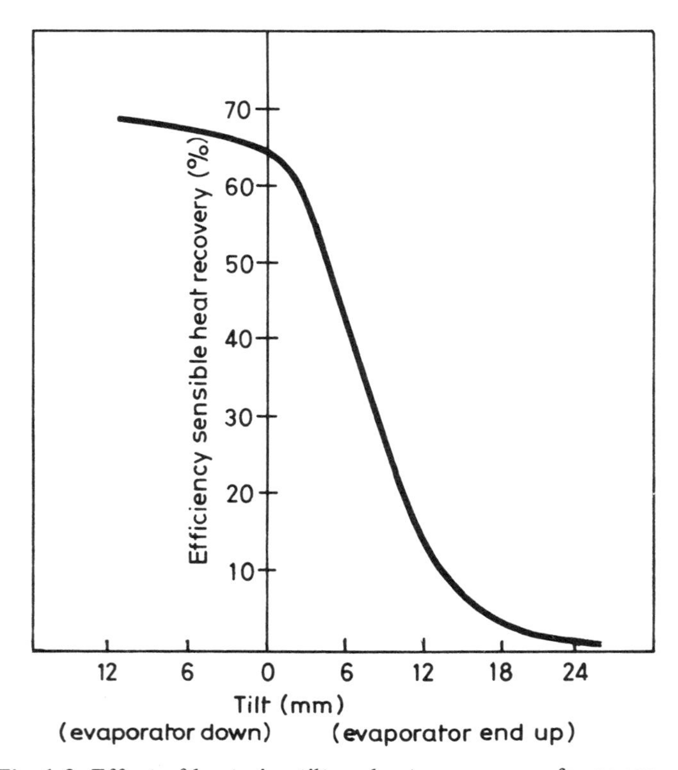

Fig. 1.3 Effect of heat pipe tilt on heat recovery performance.

for special applications, liquid metals such as mercury and sodium. Depending upon the fluid used, aluminium, copper and stainless steel are suitable container materials, and identical materials are used for the fins (although in some units aluminium fins may be applied on copper heat pipes). Environmental factors affecting the selection of the tube and fin material are common to all other types of heat exchangers – temperature, corrosion, erosion, etc. Similarly, fin pitching and the shape of the fin may be dictated by pressure drop or fouling considerations.

Unit size varies with the air flow, a velocity of about 2–4 m/s being generally accepted to keep the pressure drop through the tube bundle to a reasonable level. Small units having a face size of 0·6 m (length) x 0·3 m (height) are available, and the largest single units employ heat pipes having lengths of up to 5 m. A wider fin pitch may be used on the evaporator sections, where contaminated exhaust gas dictates easy cleaning, while retaining a pitch of 2 mm or less on the condensers located in the supply air duct. Some manufacturers offer this option.

In the case of a gas-to-liquid heat exchanger, the evaporator sections remain identical to those in gas-to-gas systems, but because of the significantly higher external heat transfer coefficients obtained on the external liquid-side of the condensers, the latter need not be finned externally and can be much shorter in length than the evaporators. While the gas-side generally incorporates only one pass, the liquid-side is commonly a multipass arrangement.

1.3.3 *Application areas*

Although the heat pipe heat exchanger has been in production for less than a decade, a review of manufacturers' literature and published papers on their development and use indicate a very wide range of applications in industry and commercial and municipal buildings (Basiulis and Plost, 1975; Feldman, 1976; Reay, 1977).

The general application areas for heat pipe heat recovery units are, in common with most other gas–gas systems, covered by the three categories listed at the beginning of the chapter. Heat pipe heat exchangers have lower efficiencies than some other gas–gas heat recovery systems, notably the heat wheel, but in common with tubular recuperators and plate have the advantages of zero cross-contamination, brought about by the presence of a splitter plate which effectively seals the inlet and exhaust ducts, and the fact that there are no moving parts, including pumps. Performance modulation and the full reversibility of the unit, the latter feature being of benefit in air conditioning systems, are other 'selling points'.

One of the main limitations of heat pipe heat exchangers, in common with many other types, is the maximum operating temperature, discussed in more detail later. Historically, heat pipe heat exchangers were developed initially to meet the requirements of the heating, ventilating and air conditioning (HVAC) industry, where temperatures are of course little in excess of ambient. During the past few years, however, they have become increasingly popular in process-to-space heating and process-to-process applications, where conditions are normally much more arduous, both in terms of operating temperature and fouling. Processes in which heat pipe heat exchangers are now routinely used include:

Air-dryer recuperation
Automotive paint-drying ovens
Brick-kiln heat recovery
Biscuit and bread ovens
Laundries
Paper dryers
Pharmaceuticals
Pollution control plus heat recovery
Spray-drying
Textile dryers
Welding-booth heat recovery

It is fair to say that heat pipe heat exchangers (including thermosyphon systems, a terminology used by some manufacturers) may be used with confidence in most industrial processes where the heat wheel, run-around coil and gas–gas plate heat exchanger are competitors. A typical installation of a heat pipe heat exchanger is shown in Fig. 1.4, and a photograph of a unit under test is given in Fig. 1.5.

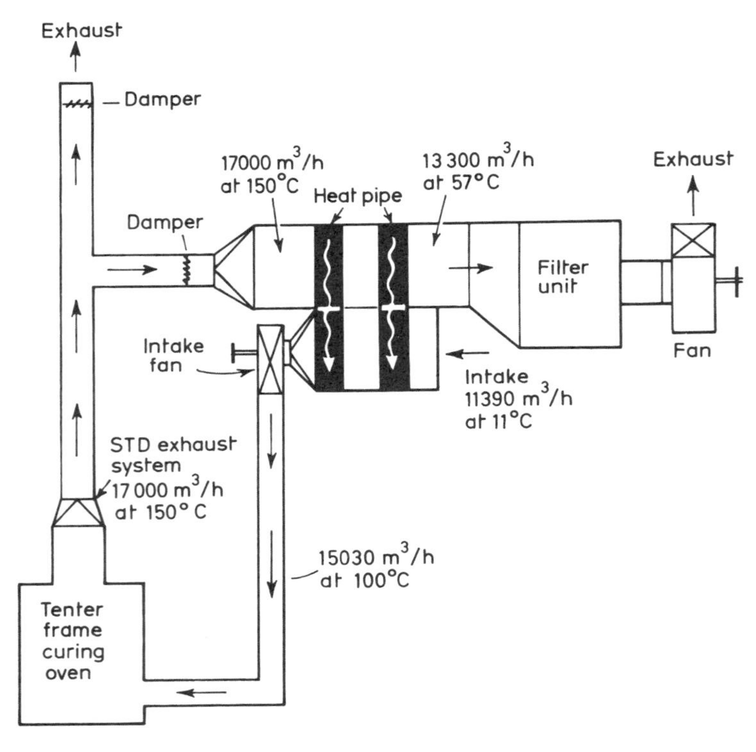

Fig. 1.4 Typical heat pipe heat exchanger installation.

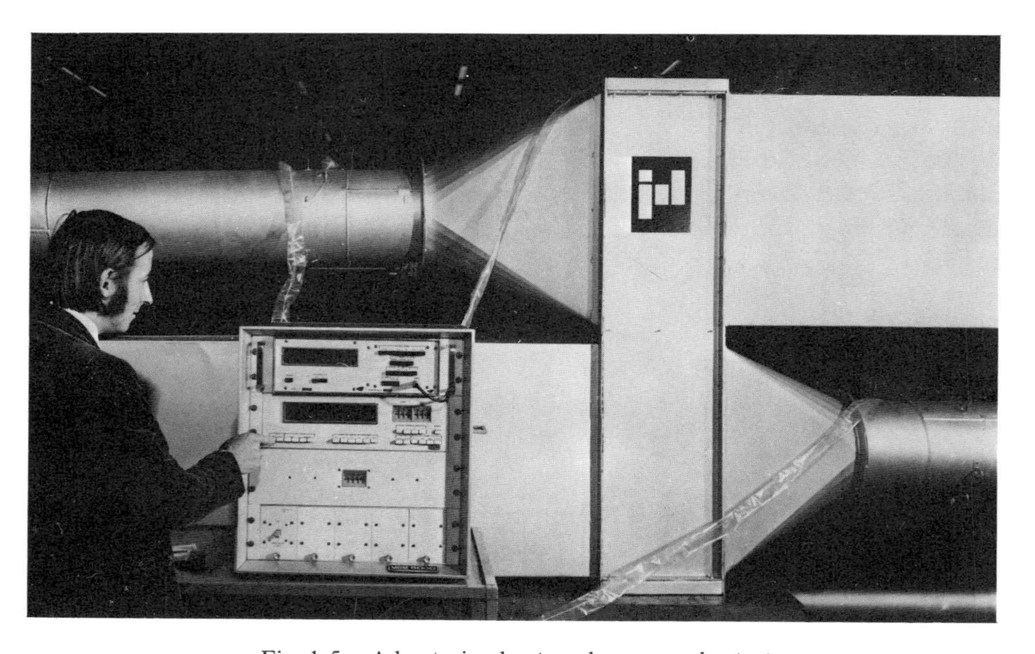

Fig. 1.5 A heat pipe heat exchanger under test.

1.3.4 *Specific design criteria*

Brief mention has already been made of a number of the factors which affect the performance of heat pipes, and hence the heat transport capability of heat pipe heat exchangers. It is proposed here to itemize these factors, and to discuss each in more depth so that the designer and user of such heat recovery devices may be aware of ways in which the requirements of his application may best be met.

Heat pipe length and orientation

The total (evaporator plus condenser) length of heat pipes used in heat pipe heat exchangers can be up to 5 m, but is more typically 1–2 m. Because of the reliance on the capillary action generated within the heat pipe wick to return the condensate to the evaporator section, wick design and performance becomes much more critical as heat pipe length increases. Even in comparatively short heat pipes, the inability of simple wire-screen wicks to support flow against gravity can be very limiting.

An illustration of the effect that the angle of inclination of the heat pipe has on its performance even if the length is only 0·2 m, can be obtained from the heat transport data obtained on a 12 mm diameter copper/water heat pipe operating at 120°C:

Evaporator vertical above condenser	60 W
Heat pipe horizontal	190 W
Condenser vertical above evaporator	600 W

Obviously more sophisticated wick structures may be designed to accommodate requirements for a high pumping capability against gravity, but in general for heat pipe heat exchanger applications they are impractical and would not be cost-effective. There are, however, techniques for achieving this aim which do not rely on capillary action, and one of these is discussed later.

Current heat pipe heat exchangers are almost always required by their manufacturers to operate with the heat pipes either vertical (with the hot exhaust gas stream passing below the inlet gas stream – i.e. evaporator down) or with a slight tilt to the horizontal, again with the evaporator slightly below the level of the condenser. Q-Dot Corporation, for example, recommend a tilt angle of 5·7 degrees from the horizontal for their heat exchangers, and quote a heat exchanger capacity 'figure of merit' as a function of tilt angle (evaporator below condenser) as follows:

Tilt angle	Figure of merit (equivalent)
5·7°	1·0
8·6°	1·3
90°	2·6

If the condensate return is being aided by gravity, one may query whether a wick is needed at all. However, as well as providing axial pumping, the wick structure fulfils three other important functions. First, it is responsible for circumferential liquid distribution, to ensure that in the evaporator section of the heat pipe, all of the available surface is wetted and can support the required radial heat flux. Second, the wick itself, particularly if it is integral with the wall of the heat pipe, in the form of grooves, can permit higher radial heat fluxes to be accepted than can a plain surface. Finally, the wick can inhibit entrainment, another feature which can restrict the heat transport capability of a heat pipe. (Entrainment occurs when the shear force between the counter-current liquid and vapour flows along the heat pipe is sufficient for the vapour to entrain droplets of liquid and carry them back to the condenser. The nature of the liquid/vapour interface is critical to entrainment, and if the liquid is contained within a wick, entrainment is inhibited.) Some heat pipe heat exchanger manufacturers incorporate a physical barrier between the liquid and vapour flow paths to eliminate this problem.

Mention has already been made of the fact that means other than capillary action can be used to transport condensate. In applications where plant layout prohibits installation of the heat exchanger in any way other than with the evaporator *above* the condenser, the artery-header system proposed by the National Engineering Laboratory, East Kilbride, offers promise (Chisholm, 1978). Because this artery does not rely on capillary action, its bore can be large and hence higher liquid throughputs can be obtained. Other work at NEL and in the USA on vapour bubble-pump anti-gravity thermosyphons suggest that units having a capability in the kilowatt range will be possible (Basiulis, 1977).

Heat pipe diameter

Reference has already been made to the need to ensure that the wick can support sufficient capillary action to circumferentially distribute liquid around the heat pipe wall. This is important, as it ensures that all of the evaporator surface can be used for heat transfer.

In heat pipe heat exchangers, the tube diameter is typically between 15 and 25 mm, the larger diameters being preferred. Heat pipes operating with some gravity assistance are able to support radial heat fluxes of 20 W/cm^2 (water at 100°C–150°C) and enhanced heat transfer surfaces such as integral circumferential grooves may be able to support fluxes of up to 150 W/cm^2 (Feldman and Berger, 1973). Thus one of the criteria for the selection of the correct heat pipe diameter, assuming that one satisfies the circumferential wicking requirement, is associated with the outside heat transfer coefficient – the diameter must be sufficient to accommodate enough finning to make maximum use of the ability to cope with high internal heat fluxes. A heat pipe heat exchanger currently operating on a dryer in the UK, for example, using 16 mm diameter tubes, incorporating 144 of these tubes and having a duty of 216 kW is running with a radial heat flux of only 5 W/cm^2. This suggests that there is substantial room for improvements in the heat transport capability to be gained if external heat transfer can be improved, and this will be discussed in more detail later.

Operating temperature

One of the most critical exercises in the design of a heat pipe heat exchanger is the selection of the correct working fluid. This is governed by the likely operating temperature.

Because the heat pipe is a completely sealed system, and provision for purging would make it economically unattractive, it is important that the working fluid used in the heat pipe is compatible with the wall and wick materials and also should not be liable to breakdown. To date, this has tended to limit the applications of heat pipe heat exchangers to installations where exhaust gas temperatures are less than 400°C, or where higher exhaust gas temperatures can be used only if the vapour temperature in the heat pipe is kept below 300°C by putting a much greater mass flow of cold air over the condensers than the quantity of hot exhaust flowing over the evaporators. Using this method, exhaust gases of 700°C can be dealt with, provided that the air to be preheated is at least 3·5 times the volume flow of the exhaust.

The most common working fluids in use in heat pipe heat exchangers are Freons, used in aluminium at up 50°C, water, used at vapour temperatures from 50°C up to 200°C in copper tubes, and a diphenyl-based fluid, commonly used in high temperature liquid heat transfer loops, which can be used at vapour temperatures approaching 300°C in steel containers.

Mercury can be used as a heat pipe working fluid, and has good compatibility with stainless steel. Sulphur has also been proposed. This, when mixed with iodine to reduce its viscosity, can be used at temperatures up to 600°C. Mercury, however, is toxic, expensive and, in common with sulphur, has a low latent heat. Insufficient data is available at present to comment further on the use of sulphur. There is considerable pressure on manufacturers to develop higher-temperature organic fluids with high latent heats for heat pipes, organic conversion cycles and solar applications, and this is one path likely to produce results.

Liquid-metal heat pipes have, of course, been successfully used in a number of applications for many years, and in the field of heat pipe heat exchangers have been suggested for gas turbine recuperators, where sodium would be used as the working fluid. It is probable that these will be developed for commercial heat exchangers during the next few years (Ranken and Lundberg, 1978). One such programme is described below.

Ceramic heat pipe heat exchangers

Moving close to the top of the industrial-process exhaust gas temperature range, work is progressing at Los Alamos Laboratory, New Mexico, on heat pipe heat exchangers for use in exhausts of up to 1500°C. Unlike high-temperature applications for use in space, where cost constraints are not so severe, enabling refractory metals such as niobium or tantalum to be used, the adoption of such systems for process heat recovery necessitates close examination of the unit cost-effectiveness (Ranken and Lundberg, 1978).

An alternative to the refractory metals is to construct the heat pipes from ceramic tubing. Ceramics such as silicone carbide and alumina have excellent corrosion and erosion resistance at these high temperatures, and are not excessively expensive.

One major problem exists with the selection of a ceramic as the container material, however. Of the two working fluids available for use in the heat pipes at these temperatures, namely sodium and lithium, the latter is particularly reactive with ceramics, and if used in conjunction with a ceramic tube, some form of protective coating must be applied inside the tube. The technique for overcoming this problem selected at Los Alamos is to deposit a thin layer of refractory metal on the inside of the tube by a method known as chemical-vapour deposition. This will provide an impervious barrier between the working fluid and the ceramic wall.

The principal application areas for heat recovery using ceramic heat pipe heat exchangers is in industrial furnaces. In the USA, for example, it has been estimated by Essenhigh (1975) (cited in Ranken and Lundberg, 1978) that these account for 12 per cent of the gross national energy usage, or about 9×10^{18} J per annum. Approximately 20 per cent of this energy is lost in the form of heat, in the stack gases. A potential annual saving, in financial terms, of 4×10^{9} J is therefore possible.

While ceramic recuperators and regenerators are not new, more conventional systems using this material have suffered badly in the past from thermal stresses and vibration, the large numbers of joints being unable to withstand excess stress cycling. The heat pipe unit, having individual tubes assembled as shown in Fig. 1.6 is not so susceptible to such wear.

Economic analyses have been carried out on ceramic heat pipe heat exchangers of this

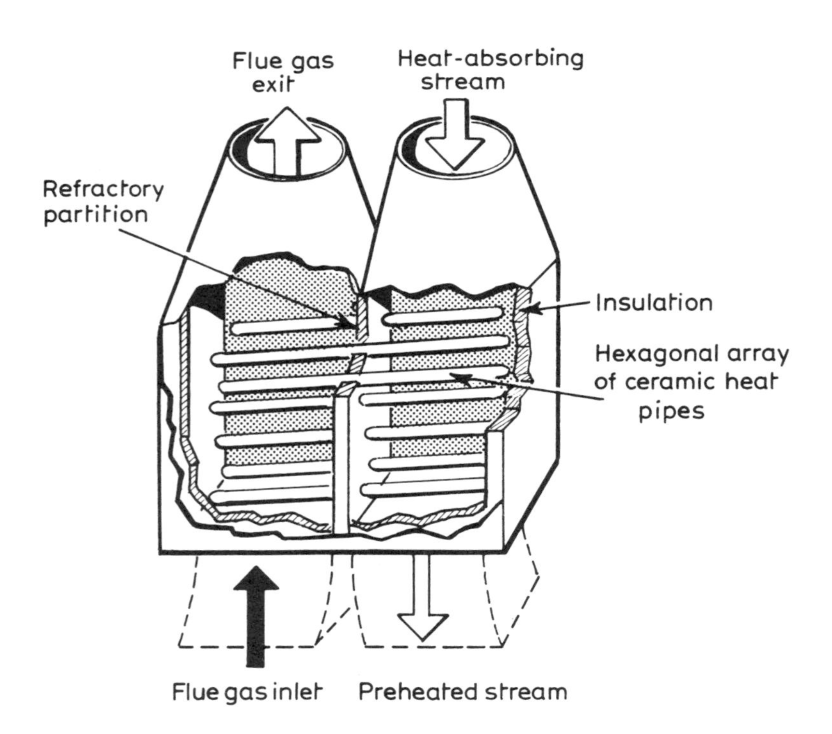

Fig. 1.6 Possible configuration of a heat pipe ceramic recuperator.

type. Based on plain tubes, paybacks of about 5 years using silicone carbide and 2 years using mullite may be possible, assuming that the ratio of the cost of each finished heat pipe to the tube material is 1·5 and the cost ratio of the finished recuperator to that of the heat pipes is 1·33, Finning of the ceramic tubes (with ceramic fins) would reduce heat exchanger size, although the cost reduction is less easy to predict. An alternative technique for raising the heat transfer would be to locate the unit in a fluidized bed.

Safety

One aspect of heat pipe technology which must be borne in mind in each application is safety. Heat pipes are obviously designed to withstand internal pressures likely to be encountered during operation and while failure, if it does occur, can take place by rupture of the filling tube seal, it is rarely associated with an explosion risk. Use of any toxic or liquid-metal working fluids can of course produce environmental problems, whatever the failure mode.

The legislation introduced as a result of the 1974 Health and Safety at Work Act in the UK, for example, has some bearing on the use of heat pipes in industry. Considering only one aspect, heat pipes are, in legal terms, pressure vessels and a water heat pipe comes under the UK Factories Act definition of a steam boiler. Regulations for pressurized systems now produced indicate that heat pipes may attract the requirement for systematic examination and testing, both by the manufacturer, which is common practice, and by the user at intervals – often a much more difficult standard to meet.

1.3.5 *Performance data*

Heat pipe heat exchangers have been in use for many years, companies in the USA pioneering their development in the late 1960s. Over this period steady improvements in performance have been achieved, and reported in the literature. However, much of the data from the USA is presented in papers written specifically for workers in the field of heat pipes, and is often overlooked by energy-conservationists in industry. Some of the performance data given here is from such sources (Ruch and Grover, 1976; Feldman and Lu, 1976; Bedrossian and Lee, 1978) and complements the information on overall heat exchanger performance presented in the tabulated data from manufacturers in Chapter 8.

The type of performance required from a heat pipe heat exchanger is qualified well by Ruch and Grover, who contrast this system with a plate fin exchanger with a multiplicity of walls common to both gas streams. The heat pipe units involve heat transfer resistances into and out of the heat pipe, instead of across a single wall, and the coefficients of evaporation and condensation must therefore be as high as possible to make the units economically attractive.

As the ratio of extended surface area to that of the bare tube is typically between 20:1 and 30:1, evaporation and condensation coefficients must be at least two orders of magnitude greater than the air-side heat transfer coefficients. In typical heat pipes, internal coefficients of the order of 10 $kW/m^2 K$ are readily achievable (compared to external forced-air coefficients of 50 W/m² K), so this criterion can be met. Using circumferentially grooved wick structures of the type proposed by Moritz (1969) and Feldman and Berger (1973), coefficients approaching 80 kW/m² K are claimed with water as the heat pipe working fluid. Thus the potential for increasing the internal heat transfer coefficient exists, and in order to improve on the performance of existing heat pipe heat exchangers, optimization of the external surfaces is of considerable importance.

In terms of axial heat transport per heat pipe, one is ideally looking for something in excess of 2 kW for a 25 mm diameter unit, having a 1-m long evaporator and a condenser of identical length. As discussed in Section 1.3.4, the main parameter affecting the magnitude of axial heat transport, assuming that one is not operating the heat pipe against a gravity head, is entrainment of the liquid by the countercurrent vapour. This can be minimized by shielding the liquid return path.

Feldman and Lu (1976) examined quantitatively the effect of varying the many geometrical parameters of a heat pipe heat exchanger, using a computer program. It will be useful to discuss some of their findings qualitatively, as this will give the reader an idea of the effect of changing geometrical parameters on performance. (The authors presented performance data in terms of heat exchanger effectiveness, defined as:

$$E = \frac{T_1 - T_3}{T_1 - T_2}$$

where
T_1 = hot-gas inlet temperature
T_2 = cold-gas inlet temperature
T_3 = hot-gas outlet temperature
T_4 = cold-gas outlet temperature.)

All external heat transfer coefficients were for Escoa Turb-X serrated fin tube with a diameter of 12 mm. The effectiveness of the heat exchanger varied as anticipated with increasing number of fins, tube rows, stagger/in line arrangements, etc.

With regard to heat pipe length, effectiveness tended to increase with increasing length between 1 m and 6 m, and when heat pipe diameter increased above 12 mm (up to19 mm) a 9 per cent increase in effectiveness was predicted. Feldman and Lu also showed that increases in the evaporator and condenser heat transfer coefficients from 4 kW/m² K to about 10 kW/m² K did not have a marked effect on overall performance, increasing effectiveness by 8 per cent. This indicates that the internal thermal resistance are fairly small, and large increases in external heat transfer coefficients would be more desirable.

To this end, AERE Harwell and IRD are working in the UK on a joint development and demonstration programme aimed at producing low-cost modular heat pipe heat exchangers for industrial-process heat recovery. A major part of the development programme is concerned with improvements in extended-surface performance, so that the full potential of the heat pipe can be exploited. With emphasis on new corrugated fins and on optimized internal heat pipe structure developed with cheapness as well as performance in mind, it is hoped that these units will be superior in cost-effectiveness than competing systems. Demonstration plans are proceeding in parallel with the development programme, and industries where it is proposed to install prototype units include paper, glass, ceramic, wool, laundry, food and malting plant. This programme is supported by the Department of Energy in the UK and the European Economic Community.

With regard to specific application performance data, Ruch and Grover (1976) cite several examples of Q-Dot installations for gas–gas duties. Using 3 bundles of heat pipe heat exchangers, each with 4 tubes rows in the direction of flow, and with external fins pitched at 3·1 mm, a thermal recovery of 71 per cent was achieved on a textile-drying oven. (Fin pitch was comparatively large because of contamination, including plasticizers, which were removed downstream of the heat exchanger by pollution-control equipment.) In this installation the exhaust air (1·58 m^3/s) entered the exchanger at 140°C and was cooled to 60°C. An identical volume flow of inlet air was preheated to 112°C from ambient, recovering 162 kW. A 500 kW unit, totalling 8 rows of tubes for use on the exhaust of a paint-drying oven operates with an efficiency of 65 per cent, preheating 2·83 m^3/s of inlet air using exhaust at 204°C.

1.3.6 *Detailed cost-effectiveness analysis*

Serious attempts at cost-effectiveness studies on any type of waste heat recovery equipment are very difficult to find. In many instances the cost data obtained through a manufacturer's literature, when it is available, relates solely to the capital cost of the heat exchanger and, particularly in the case of high-temperature heat recovery systems, bears little relationship to the installed cost, which may be up to four times as high as the basic unit price.

It is possible, although it can be an extensive exercise, to approach as many manufacturers as possible with requests for quotations for a range of heat exchangers. However, analysis of the data obtained, taking into account differences in pressure drop (hence operating costs), etc. will be the most time-consuming component.

Possibly because the heat pipe heat exchanger has grown up in an environment where serious academic interest is present (being the subject of much research and development in universities and other research laboratories), the development of computer optimization programs, both technical and economic, as an aid to unit selection, has become almost routine. One such study carried out at the University of New Mexico was recently published by Lu and Feldman (1977). The results were based on data supplied by the major heat pipe heat exchanger manufacturer in the USA, Q-Dot Corporation.

Cost data for a variety of heat pipe heat exchangers, covering operating temperature ranges from HVAC to process at up to 400°C, is given in Fig. 1.7. The initial equipment cost includes materials, labour and overheads, and makes allowance for the fact that the units will be part of a production run, rather than 'specials'. Installation cost includes ducting, controls and labour. The results are presented in such a way as to show the relationship between the cost per unit surface area of the heat exchanger, and

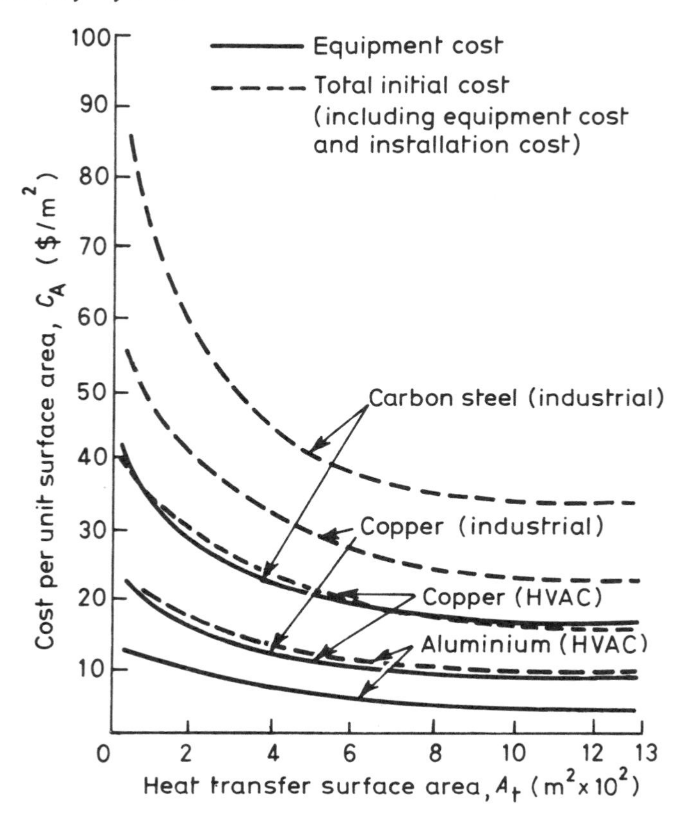

Fig. 1.7 Approximate cost per unit heat transfer surface area for aluminium, copper and steel heat pipe heat exchangers.

the total size of the heat exchanger. As one would expect, the cost per square metre of surface decreases as the heat pipe heat exchanger increases in size.

Other interesting points may be noted from the graph (Fig. 1.7). First the aluminium heat pipe heat exchanger has the lowest capital cost, followed by copper, then carbon steel. Second, the installation cost of a copper unit in a HVAC application is considerably less than the cost needed to install an identical heat exchanger in a process. Economies of scale are particularly noticeable on the carbon steel high-temperature heat exchangers.

Based on the cost data, the staff of the University of New Mexico designed a heat pipe heat exchanger to recover heat from the exhaust flue of the site boilers for air preheating. With an exhaust gas temperature of 316°C (flow of 8534 m^3/min), a carbon steel unit with Dowtherm A as the working fluid was selected.

Recovering approximately 1·5 MW, the unit would save fuel valued at $ 45 000 per annum, at a cost per annum of $ 8000 (amortized over 10 years at 10 per cent interest). Included in the costs were allowances for fan power and maintenance, totalling $ 1700 per annum. The life of the unit was estimated to be 15 years. Costing was based on operation for 24 hours per day, 355 days per year, and the effectiveness

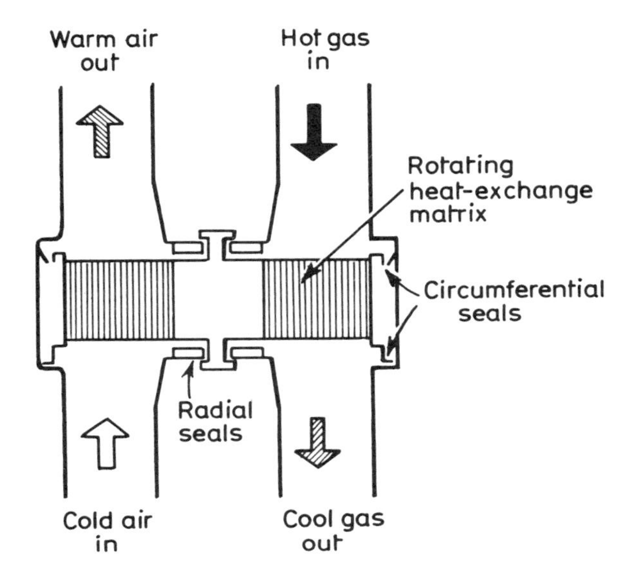

Fig. 1.8 Operation of the rotating regenerator.

of the heat exchanger was 60 per cent.

1.3.7 *Conclusions*

Heat pipe heat exchangers are established for gas–gas waste heat recovery and can be used for gas-liquid duties where economizer-type heat exchangers are unacceptable. Current development and demonstration programmes in the UK will enable much detailed data to be obtained on their performance in a number of industrial applications, and it is anticipated that these units will be even more cost-effective than current models. There are a considerable number of variables which can affect heat pipe performance, and limitations exist at present on maximum operating temperature.

1.4 Rotating regenerators

The rotating regenerator variously known as the heat wheel, rotary air preheater, Munter wheel, or Ljungstrom wheel after its Danish inventor[1], has been used over a period of about 50 years for heat recovery in large power-plant combustion processes

[1] Some power-generating plants use a second type of rotating regenerator, the Rothemuhle design, which incorporates a stationary matrix with rotating hoods to distribute the gas and air flows, as shown in Fig. 1.9. However, only Hitachi Zosen offer these commercially for process-plant application (Chojnowski and Chew, 1978)..

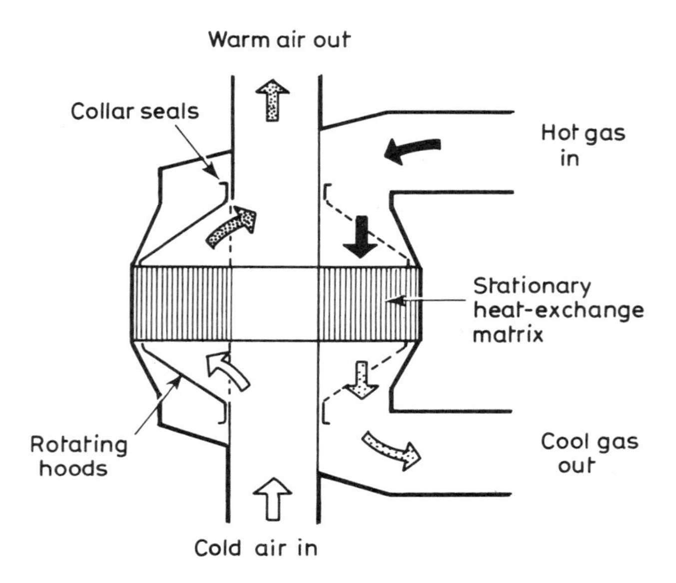

Fig. 1.9 Operation of the less common Rothemuhle regenerator.

(Applegate, 1970). It has also been widely used in airconditioning and a variety of industrial-process heat recovery applications – in 1975, it was estimated that in Europe alone upwards of 15 000 rotating regnerators were in use (Sohlberg, 1976).

The operation of the rotating regnerator is shown in Fig. 1.8. Like the heat pipe heat exchanger (Section 1.3), the regenerator wheel spans two adjacent ducts, one carrying exhaust gas and the other containing the gas flow which it is required to heat. The gas flows are counter-current. As the wheel rotates, it absorbs heat from the hot gas passing through it and transfers the heat to the cooler gas flow. A later development, the hygroscopic wheel, is able to transfer moisture, as well as sensible heat, between the two ducts.

Rotating regnerators, in common with many other heat recovery systems, can be used in hot climates for precooling air used for conditioning large buildings, and the wheel works effectively in applications where the temperature differences between hot and cold air streams are too low for effective use of the heat pipe heat exchanger.

Many manufacturers produce different types of rotating regenerator. The most common form utilizes a wheel made up from a knitted aluminium or stainless-steel wire matrix. This matrix is cheap, and the heat transfer efficiency is high as the air flow is exposed to a large amount of surface as it passes through the wheel. However, the pressure drop through this type of matrix can be relatively high, and the fouling of the matrix tends to be more severe than on other types.

Development of laminar flow wheels, in which the matrix is corrugated, resembling a small-pore honeycomb, has alleviated the pressure drop and fouling problems of the mesh matrix, and this type of wheel is easier to clean. In terms of thermal efficiency the performance of a metallic corrugated matrix wheel is similar to that of the mesh

Fig. 1.10 Fläkt 'Regoterm' rotating regenerator which uses an aluminium core.

type. An example is shown in Fig. 1.10.

A third form of matrix used in rotating regenerators is non-metallic. Known as the hygroscopic wheel, this type can transfer moisture as well as sensible heat, and is particularly useful in heating and ventilating applications. The structure is similar to that of the metallic laminar flow wheel. While the hygroscopic wheel is likely to be up to 35 per cent more expensive than the metallic type, Wing claim that the increased capital cost is generally more than offset by the increased heat transfer attributable to latent heat recovery. The latent heat content can vary considerably from one application to another, and should be carefully assessed before settling for a particular unit.

The use of a stainless-steel matrix in the regenerator permits operation in exhaust gases at temperatures in excess of 800°C. In some processes regeneration is required at even higher temperatures (for example, in gas turbines and in steelmaking plant) and details are given in Chapter 8 of a ceramic core used in a regenerator. Glass-ceramics and silicone nitrides have been used as core materials (Mortimer, 1973). During the last two or three years, however, difficulties have been encountered with material degeneration and cracking in some ceramic regnerators, and potential users are advised to consult closely with manufacturers when proposing to operate units at very high temperatures.

1.4.1 *Performance and operation*

In determining the performance and operating characteristics of a rotating regenerator, a number of factors must be taken into account, and these include the following:

Operating temperature
As discussed above, the type of matrix used, be it metallic or ceramic, depends upon the operating temperature range likely to be encountered in the application.

Operating pressure
In general it is desirable to operate this type of regenerator in situations where the exhaust and supply gas streams have similar pressures. If a knitted mesh is used for the matrix, there is a potentially large flow path available through the mesh if pressures are not equalized. In wheels of the laminar flow type, where the gas is restricted to movement in the axial direction, some pressure differential can be supported if suitable seals are fitted to the unit. To minimize carryover of contamination, it is preferable to operate the exhaust stream at a marginally lower pressure than the supply stream.

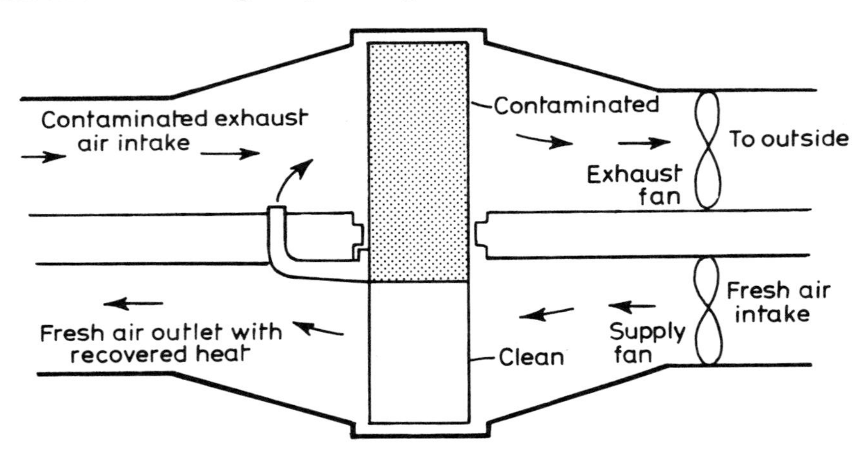

Fig. 1.11 Operation of the purge section.

Cross-contamination
The possibility that contamination of clean supply gas could occur with this type of regenerator has led to the incorporation by manufacturers of purge sections, the operation of which is shown in Fig. 1.11.

Here a proportion of the supply air is used to scavenge the matrix section leaving the exhaust duct, and the contamination or residual exhaust gas is blown back into the exhaust duct. Proper purge-section operation depends on correct fan location so that the pressure on the supply side is higher than on the exhaust side, and in cases where this requirement can be met, cross-contamination can be as low as 0·04 per cent by volume, and particle carryover less than 2 per cent. Some manufacturers specify a

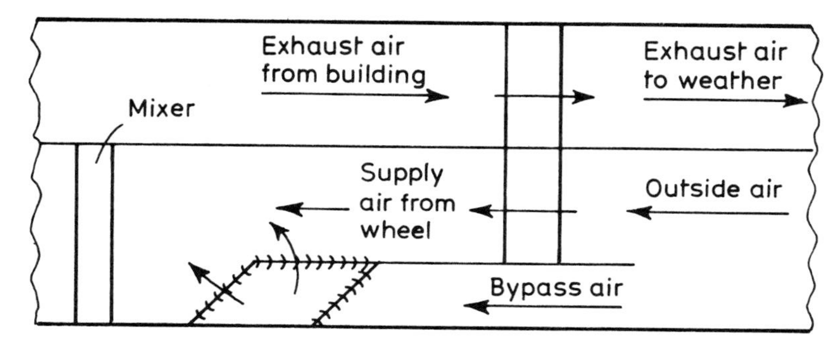

Fig. 1.12 Compensation for unequal flow rates in a rotating regenerator (by courtesy of Wing International Ltd).

carryover of less than 0·1 per cent by volume in their literature, and where this factor is critical, their advice should be sought (Applegate, 1974).

Purging is not obtained without sacrificing efficiency. The sizing of system fans may be increased by up to 10 per cent of rated volume to handle the greater gas flow requirements, and if the correct fan arrangement cannot be installed because of practical difficulties, seals can be incorporated at each radial partition, but these are not nearly as effective, cross-contamination rates possibly then approaching 8 per cent by volume.

On some of the higher-temperature metallic regenerators, a separate fan specifically for purging may be fitted. These may use up to 4 kW in electrical energy and, together with the much lower motor power for rotating the heat exchanger, should be accounted for in any cost analysis.

Very high-temperature ceramic regenerators also suffer from cross-contamination, but in the case of gas turbine units, for a 4:1 engine compression ratio transverse flow accounts for only about 0·3 per cent of engine air flow at full speed conditions.

Unequal flow rates

Potential users will find that most manufacturers quote performance figures based on equal supply and exhaust flow rates. If the flows are unequal, correction factors must be applied. It is normal practice to size the generator for the maximum flow rate and to select the efficiency on the basis of the lower flow (see example in Section 1.4.2).

In situations where the inlet-gas flow rate is much larger than the exhaust flow, say, by a factor of 4, the rotating regnerator may be selected on the basis of the exhaust flow, and used to heat only a proportion of the supply air, say, 25 per cent. The arrangement for this, shown in Fig. 1.12, permits the balance of the supply air to bypass the regenerator, after which it is mixed with the preheated air.

Pressure drop

The pressure drop through the heat exchanger is a function of gas velocity and matrix design. Some manufacturers offer two types of matrix, one designed for optimum heat transfer capability which, because of its greater amount of surface, has a pressure drop which may be unacceptable in some applications. By compromising on thermal efficiency, a lower pressure drop unit having larger pores may be adopted. Typically, a high efficiency unit (81 per cent at a velocity of 4 m/s) will have a pressure drop of approximately 35 N/m^2, whereas the corresponding low-pressure drop system will operate at 76 per cent efficiency at the same velocity, resulting in a pressure loss of only 17 N/m^2 (Wing Co. figures).

Control

Most rotating regenerators (excluding high-temperature ceramic types) are driven by electric motors, the power of which on the largest wheels (4 m diameter) approaches 0·5 kW. For optimum heat transfer, rotational speed is typically 20 rev/min. As this speed reduces, the thermal efficiency is lowered, and hence speed control can be used to meet reduced heat-load duties. Alternatively, the supply air may be regulated using dampers and/or a bypass system. In most industrial processes where a constant duty is required, any form of modulation will be unnecessary, and a constant speed motor is recommended.

Details of the performance of a typical rotating regenerator for sensible heat recovery are given in Tables 1.1 and 1.2. The unit selected is the Wing Cormed metallic regenerator, designed for operation at temperatures between 150°C and 400°C. It is available in a variety of sizes, wheel diameter varying between 1·1 and 3·2 m. Maximum flow rate through the largest size of unit is about 70 000 m^3/h, corresponding to a velocity of 7 m/s. The performance data presented is for the model WCM-1400, representing the middle of the range of this type.

Similar tables are presented by manufacturers giving correction factors for unequal flow rates and the effect of temperature on pressure drop. Thus, the selection procedure is quite simple and the user has no need to enter into elaborate thermal calculations. (Typical data for hygroscopic wheels is given later.) Table 1.2 is used to correct for efficiencies calculated at temperatures other than 150°C.

The selection procedure is as follows:

Select a regenerator to handle 35 000 m^3/h of dry exhaust gas at 260°C with an equal supply air flow of outside air at 10°C for return to the process:

(i) Determine exhaust air flow at NTP:

$$\frac{273 + 20}{273 + 260} = \frac{293}{533} = 0{\cdot}54 \times 35\,000$$

Flow at NTP = 18 900 m^3/h.

(ii) Determine average wheel temperature:

$\frac{1}{2}(260 + 20) = 140°C$

(iii) Select wheel size and base efficiency from Table 1.2, WCM-1400 has efficiency of 0·65 for nearest flow (19 796 m^3/h).

(iv) Correct base efficiency for average wheel temperature, Table 1.2. Negligible correction required.

Table 1.1 Performance of Wing WCM-1400

Efficiency (based on average wheel temperature of 150°C)												
0·83	0·82	0·79	0·76	0·73	0·70	0·68	0·65	0·63	0·61	0·60	0·58	0·57
Velocity (m/s)												
1·0	1·5	2·0	2·5	3·0	3·5	4·0	4·5	5·0	5·5	6·0	6·5	7·0
Pressure drop (N/m^2)												
21	30	41	51	61	71	81	91	102	112	122	132	142
Flow rate (m^3/h)												
4399	6599	8798	10998	13198	15397	17597	19796	21996	24196	26395	28595	30794

Table 1.2 Temperature correction factors

AVG temperature (°C)	Base efficiencies (AVG temperature = 150°C)							
	0·55	0·60	0·65	0·70	0·75	0·80	0·85	0·90
	Temperature-corrected efficiencies							
65	0·51	0·56	0·61	0·66	0·72	0·77	0·82	0·88
95	0·52	0·57	0·62	0·67	0·73	0·78	0·83	0·89
120	0·54	0·59	0·64	0·69	0·74	0·79	0·84	0·89
150	0·55	0·60	0·65	0·70	0·75	0·80	0·85	0·90
175	0·56	0·61	0·66	0·71	0·76	0·81	0·86	0·91
205	0·58	0·63	0·67	0·72	0·78	0·81	0·86	0·91
230	0·59	0·64	0·68	0·73	0·78	0·82	0·87	0·92

Average wheel temperature = exhaust temperature + supply temperature divided by 2. (Courtesy: Wing Co. Ltd.)

(v) Determine supply air leaving temperature:

$20 + (260 - 20) \times 0{\cdot}65 = 176°C$.

(vi) Determine heat recovered:
Flow x Sensible heat factor x Temperature rise

$18\,900 \times 0{\cdot}288 \times 176 = 958\,000$ kcal/h.

1.4.2 *Applications and economic assessments*

In common with the heat pipe heat exchanger and many other types of gas–gas heat recovery systems, the rotating regenerator is equally applicable in heating, ventilating and air conditioning (HVAC) and process heat recovery, the main differences being in the type of wheel used (hygroscopic or sensible heat only).

HVAC

The performance of a rotating regenerator in a HVAC application is shown in Fig. 1.13. As can be seen on the figure, it is able to precool incoming air to a building in summer, as well as preheating this air using recovered heat in the winter months. As with most other air conditioning exhaust heat recovery systems, the rotating regenerator can only function if the supply and exhaust ducts are adjacent to one another where the heat exchanger is cited, and this may affect 'retrofitting' of the system (Friedlander, 1973).

The hygroscopic wheels used in HVAC vary considerably in their matrix material. Until recently asbestos sheet, made up into a honeycomb core and then impregnated with a substance such as lithium chloride, was very popular[1]. It is now possible to obtain metallic matrix material for HVAC regenerators, the core of one such unit being of aluminium foil, again formed into a honeycomb. The hygroscopic properties in this instance are ensured by treating the wheel after fabrication so that a microporous layer of aluminium oxide is formed on the surface. This layer has the same properties as a dry desiccant, adsorbing water vapour into the pores.

The need to use a hygroscopic wheel, in spite of the fact that it may be 25 per cent more expensive on first cost than a simple sensible heat metallic wheel, can be seen from the fact that the latent heat content of air exhausted from buildings can reach almost the same magnitude as the sensible heat contribution. In addition to the recovery of the heat, however, an exchange of humidity itself is required in order either to dehumidify the supply air in summer, or to increase the humidity in winter (Kruse and Vauth, 1976). Recuperative heat recovery systems, by virtue of the fact that a solid wall separates the two air streams, are unable to transfer moisture, and can only affect the temperature of the incoming air.

[1] While there has been some concern over the possibility of asbestos particles detaching themselves from the regenerator matrix, thereby creating an asbestosis hazard, a study undertaken by Professor Schlipkoter of Düsseldorf indicated that the hygiene risk was negligible (Field, 1976). This conclusion was based on examination of room conditions in a number of thermal wheel installations of different ages.

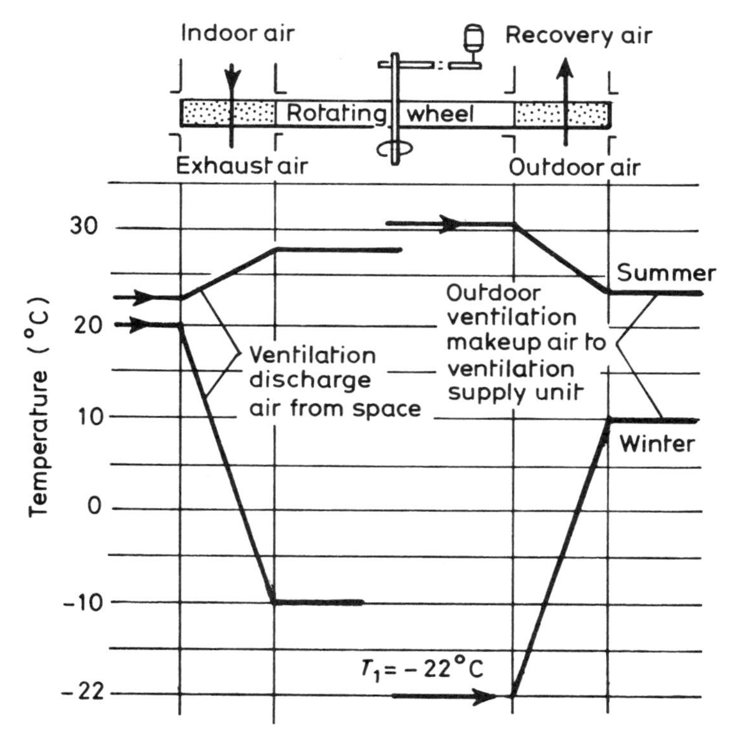

Fig. 1.13 Performance of a rotating regenerator used to heat or cool building makeup air.

Kruse and Vauth, following on the research of a number of other workers, including Spahn and Gnielinski (1971), Ruth *et al.* (1975) compared the performance of an asbestos/lithium chloride (hygroscopic) wheel and an aluminium knitted-wire (non-hygroscopic) wheel under climatic conditions appropriate to HVAC in Germany. Of particular interest was their work on the formation of frost on the matrices under adverse ambient conditions. Their results indicate that hygroscopic wheels ice up at lower temperatures (typically −20°C ambient in their tests) than metallic sensible heat wheels (icing commenced at −10°C). This was attributed to the fact that in the latter case, all moisture was in liquid form and invariably forms frost when the mean effective rotor temperature falls below the ice point. They also recommended that the performance of hygroscopic wheels should be presented by manufacturers in terms of both temperature and humidity.

An example of sizing procedures, where temperature and humidity efficiencies are given, is taken from the Fläkt literature and presented below. The design parameters for sizing the unit are the supply and exhaust air flows, the temperature and humidifying efficiencies and the pressure drop.

Equal supply and exhaust air flows

When the supply air and exhaust air flows are equal, the size of the heat exchanger for a specified air flow can be selected from Fig. 1.14 or Table 1.3. The efficiencies, pressure drop and air velocity can then be read off. The temperature efficiency η_t applies to both types of rotor, while the humidifying efficiency η_x applies only to the hygroscopic rotor.

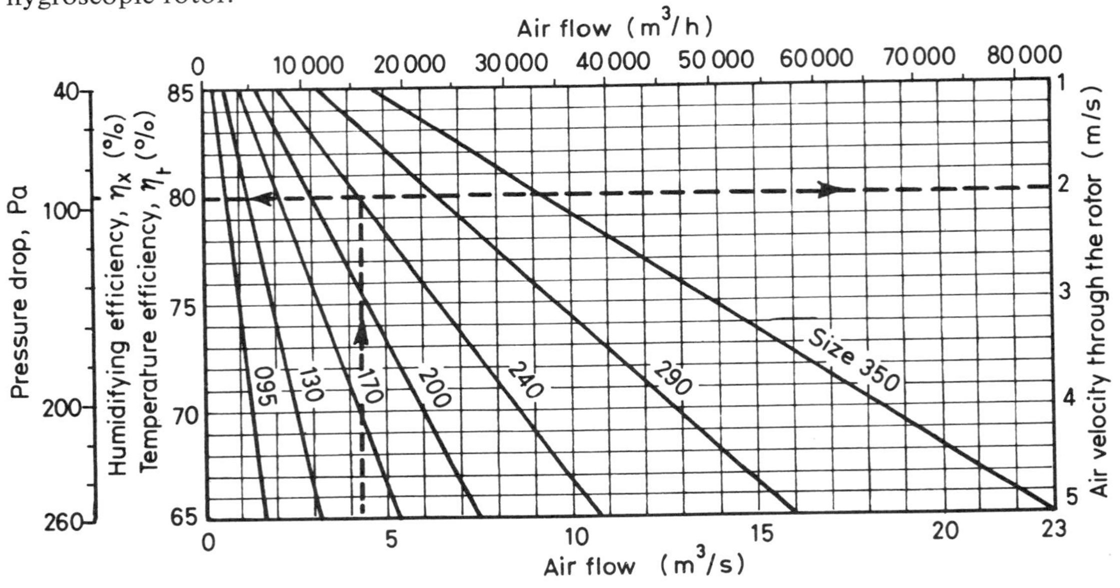

Fig. 1.14 Selection chart for Fläkt regenerator. The graph is valid for air at a density of 1·2 kg/m³ and for maximum rotor speed (approx. 10 rev/min).

Unequal supply and exhaust air flows

When the supply and exhaust air flows are unequal, the temperature and humidifying efficiencies for the supply air flow may first be read from Fig. 1.14. The value of the efficiency obtained from Fig. 1.14 is marked off on the horizontal axis in Fig. 1.15 and a vertical line is drawn to the curve corresponding to the exhaust/supply air ratio q_f/q_t. The point of intersection is transferred to the vertical axis and the temperature and humidifying efficiencies of the supply air read off. The temperature efficiency η_t applies to both types of rotor, while the humidifying efficinecy η_x applies only to the hygroscopic rotor.

The pressure drops corresponding to the various air flows are obtained from Table 1.3.

Example of sizing given:
Horizontal air flow
Supply air flow q_s = 4·3 m³/s
Minimum required temperature and humidifying efficiencies (η_{tt} η_{xx}) for supply air flow: 82 per cent.

Table 1.3 Equal supply and exhaust air flows

Temperature efficiency η_t (%)	Size code suffix aaa							Pressure drop Δp
	095	130	170	200	240	290	350	
Humidifying efficiency η_x (%)	Air flow (m^3/s)							(Pa)
65	1·67	3·21	5·44	7·66	10·82	16·10	22·91	260
66	1·60	3·07	5·30	7·39	10·41	15·25	21·95	250
67	1·53	2·93	5·02	7·12	9·90	14·72	20·99	240
68	1·46	2·81	4·74	6·69	9·48	14·16	20·03	225
69	1·39	2·66	4·60	6·41	9·07	13·36	19·06	215
70	1·32	2·56	4·32	6·14	8·65	12·78	18·10	205
71	1·26	2·42	4·13	5·86	8·23	12·21	17·28	195
72	1·19	2·29	3·91	5·58	7·81	11·38	16·32	185
73	1·12	2·16	3·72	5·16	7·39	10·82	15·36	175
74	1·06	2·03	3·49	4·89	6·83	10·22	14·54	165
75	0·99	1·89	3·27	4·61	6·41	9·62	13·58	150
76	0·92	1·77	3·04	4·32	6·00	8·92	12·62	140
77	0·86	1·64	2·83	3·96	5·58	8·22	11·80	130
78	0·79	1·52	2·58	3·63	5·16	7·67	10·84	120
79	0·72	1·39	2·36	3·32	4·74	6·97	9·88	110
80	0·64	1·25	2·17	3·02	4·32	6·42	8·78	95
81	0·58	1·12	1·93	2·69	3·91	5·72	7·96	90
82	0·52	1·00	1·71	2·38	3·41	5·02	7·13	90
83	0·46	0·87	1·49	2·08	2·96	4·46	6·31	65
84	0·39	0·74	1·26	1·78	2·50	3·74	5·35	55
85	0·32	0·61	1·04	1·47	2·08	3·07	4·39	40

This table is valid for air at a density of 1·2 kg/m^3 and for maximum rotor speed (approximately 10 rev/min).

Solution:
Assume a size 240 unit
From Fig. 1.14, temperature and humidifying efficiencies η_t and η_x = 80 per cent
Ratio q_e/q_s = 4·8/4·3 = 1·12
From Fig. 1.15, $\eta_{tt} = \eta_{tx}$ = 83%
Select a PABA-240 unit.

Unlike the majority of process-plant heat recovery operations, where, once set up, the process operates continuously with little change in exhaust gas conditions, heat exchangers used in air conditioning plant require, in general, some form of modulation. As mentioned in the previous section, this may be implemented on a heat pipe heat

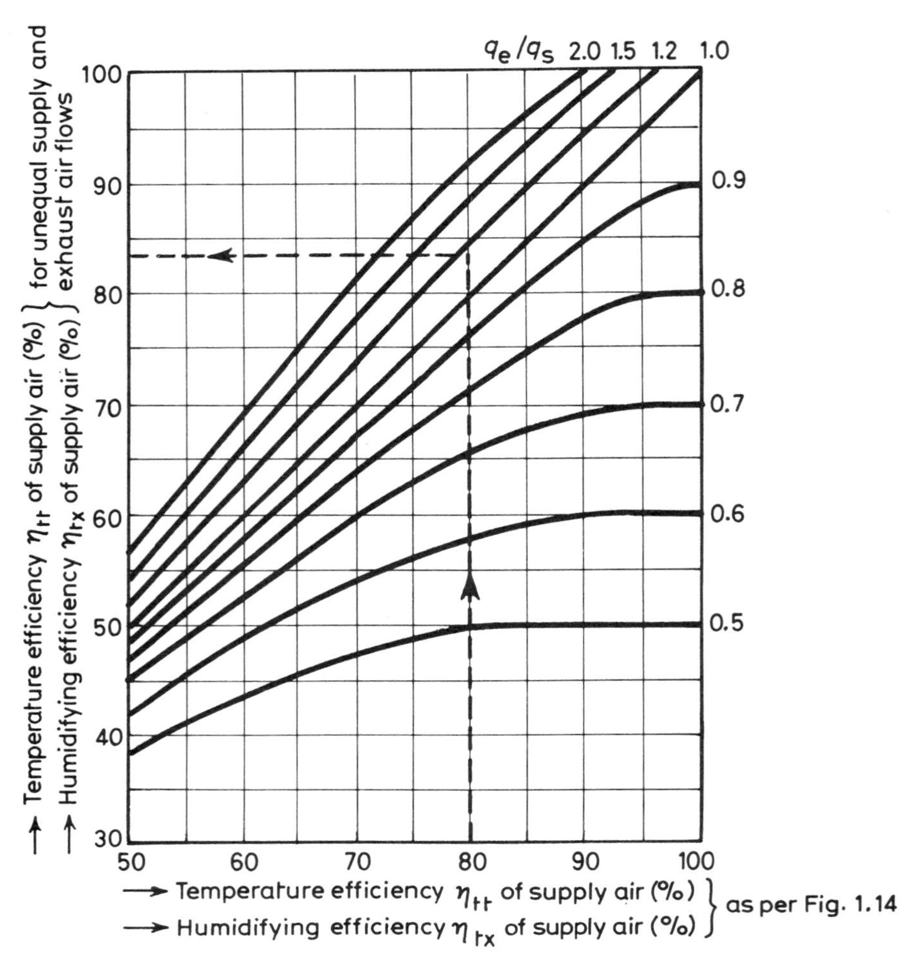

Fig. 1.15 Temperature and humidifying efficiencies of a Fläkt regenerator when handling unequal supply and exhaust flows. Exhaust air temperature and humidifying efficiencies:

$$\eta_{tf} = \frac{\eta_{tt}}{q_e/q_s}\,;\ \eta_{xf} = \frac{\eta_{xt}}{q_e/q_s}.$$

exchanger by tilting the heat pipes at varying angles to the horizontal. On rotating regenerators the primary control technique is to vary the speed of rotation, and Fig. 1.16 shows how a typical regnerator is affected in terms of efficiency as the speed of rotation is altered.

One system of control widely used relies on input signals supplied to the speed controller of the heat exchanger from a control unit or a proportional thermostat. In this particular system, a relay unit is used to over-ride the normal control circuit when purging, ice removal or other functions are required at varying intervals.

A further complication arises in air conditioning plants in which the outdoor air flow may be reduced, recirculated air then being mixed with the supply air through a system of dampers. The system adopted by Fläkt to meet this requirement is illustrated in Fig. 1.17.

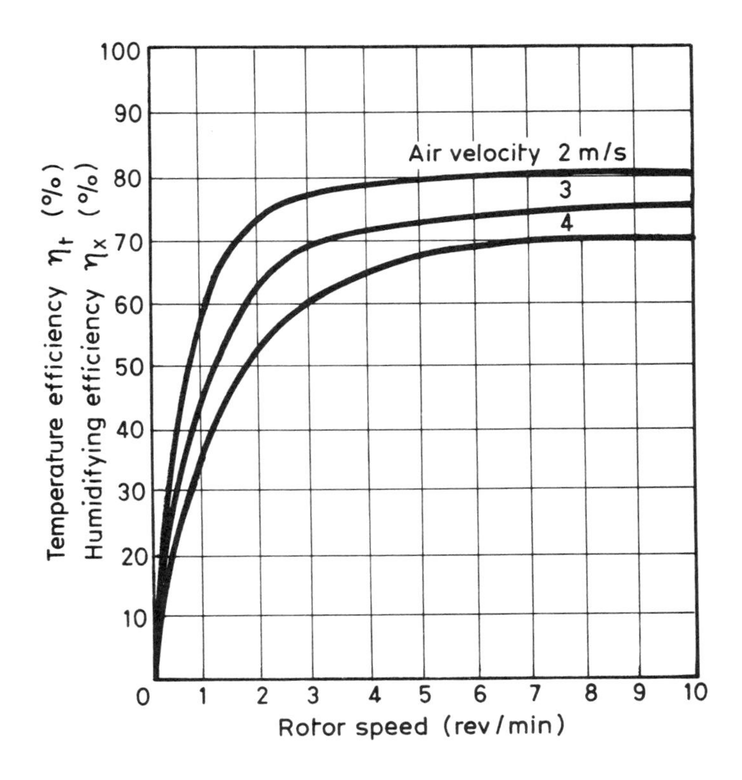

Fig. 1.16 The effect of speed of rotation of a regenerator on performance.

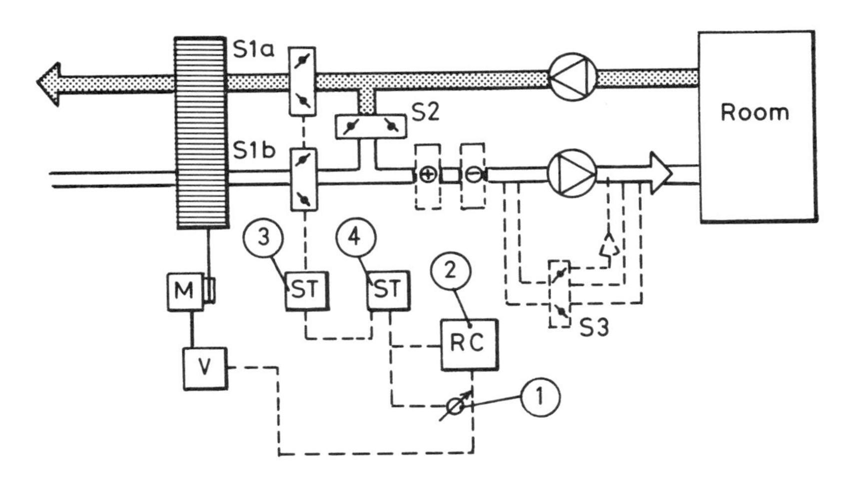

Fig. 1.17 Control circuit for a rotating regenerator application where mixed outdoor air and recirculated indoor air may be used.

Example
System using mixed outdoor air and recirculated air. In air conditioning plants in which the outdoor air flow may be reduced, recirculated air may be mixed with the supply air through a system of dampers. These dampers are then included in the control sequence.

To achieve a constant-supply air flow, the supply-air fan should be designed for the total pressure drop in the supply and exhaust systems when operating with recirculated air. For the same reason, the pressure drop across damper S1b in the open position and other pressure drop on the outdoor air-side should correspond to the total pressure drop in the exhaust-air system. S1b may also be replaced by a pressure-controlled bypass damper S3.

When there is a risk of freezing, the fans are stopped, whereupon dampers S1a and S1b close, and S2 opens. In cooling recovery operation, the heat exchanger rotor will run at maximum speed, while dampers S1a and S1b will move to minimum opening if the outdoor air temperature is higher than the recirculated air temperature.

Operating sequence from minimum to maximum heat content of supply air:

Initial conditions:
(a) Cooling valve (or valves) open
(b) Dampers S1a and S1b open, damper S2 closed
(c) Heat exchanger rotor running at minimum speed
(d) Makeup heating shutoff

Operating sequence:
(a) Cooling valve (or valves) close
(b) Dampers S1a and S1b move to minimum opening, damper S2 opens
(c) Heat exchanger rotor speed increases
(d) Makeup heating increases
e.g. Damper settings fixed by means of potentiometer 1.

Equipment needed (independent of the rest of the system):

Item	*Description*
1	Potentiometer
3	Damper control motor
4	Damper control motor

e.g. Damper control using control unit RC2. (This system may be used in combination with a sequential controller.)

Equipment needed:

Item	*Description*
2	Control unit
3	Damper control motor
4	Damper control motor

An example of a comprehensive rotating regenerator 'package' in an airconditioning system in a major hospital in the USA gives an insight into the many aspects of control,

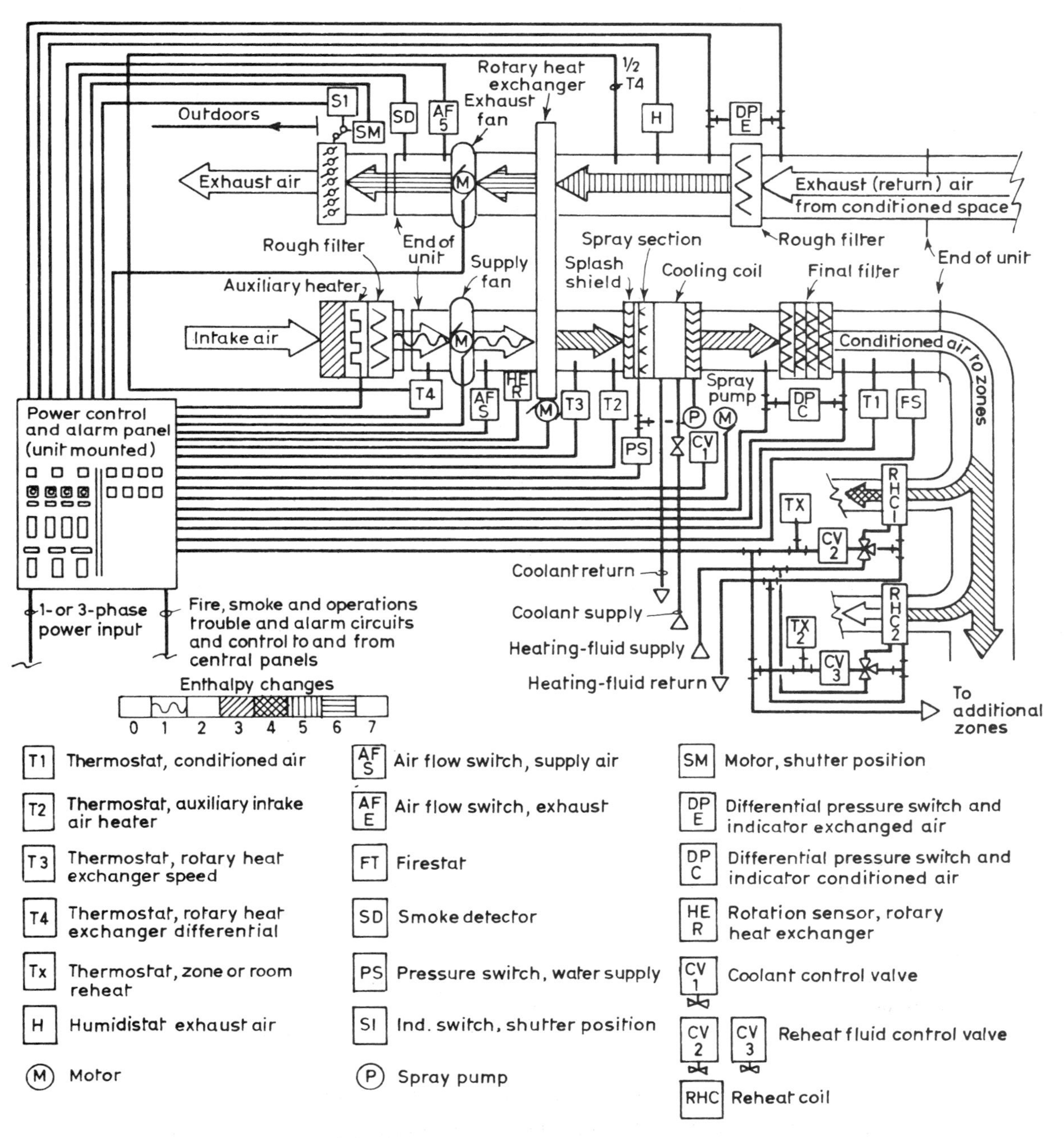

Fig. 1.18 Air flow and control diagram for a heat recovery module, using rotating regenerators in a US hospital complex. (Note: All fluid functions shown may be electric/electronic enthalpy changes.)

as well as the potential energy savings possible using heat recovery systems. This system, incorporated in the Boca Raton Community Hospital in Florida (Hayet and Maxwell, 1975), was selected to meet the requirements of aseptic environmental control, with a recommended number of air changes per hour and proper temperature and humidity control. The use of 100 per cent outside makeup air was selected to provide these conditions. Previous experience with rotating regenerators led to the concept of modular units.

The major components in the packaged heat recovery units, illustrated in Fig. 1.18, were the regenerator, intake and exhaust ducting, chilled-water spray and coil section, supply-air filters, plenums, exhaust-air filters, control panel and structural base.

The efficiency of the regenerator, which was of the 'total-energy' type, varied between 60 and 90 per cent, but it was found that the optimum efficiency, taking into account the most economical installed and running costs, was about 75 per cent for the following conditions:

Outside	Summer	32°C d.b. 27°C w.b.
	Winter	2°C d.b. 0°C w.b.
Inside (general)	Summer	24°C d.b. 50% r.h.
	Winter	24°C d.b.
Intensive-care unit	Year-round	21–24°C d.b. 50–60% r.h.

A total of five regenerator modules, handling 47 m^3/s air at a 370 tonne cooling load was required. Comparing this with a conventional system, where 670 tonne cooling would be required, a 45 per cent saving in installed chiller capacity resulted. With an installed total cost of $1876 per tonne, the rotating regnerator system saved over $500 000 on capital expenditure, compared with a conventional system. (This neglects ancillary services needed to support a much larger chiller system, thus giving a pessimistic picture of the true savings.) Energy savings per annum, including a small amount for boiler fuel, totalled $24 230.

The financial picture above brings to the foreground an important point which may often be overlooked in assessing the benefits of heat recovery equipment. While the rotating regenerators themselves would form a comparatively small part of the total installed cost of the system, and would have an attractive 'payback period', the economic decisions have not been taken solely based on that consideration. The major motivation behind the project has been associated with the capital-cost savings possible by using the heat recovery system. Of course, it is not always possible to make such a major impact – existing buildings where chiller capacity is based on conventional air conditioning systems would not show such marked improvements, in many cases having to rely on a more modest outlay for heat recovery units alone where the return on the investment is weighed against annual energy savings. However, the opportunity to build heat recovery equipment into hospitals and other structures, as well as into process plant such as dryers, can result in capital savings which can be much more significant than annual energy savings, in terms of finance.

Process heat recovery

Process heat recovery applications of rotating regenerators are many and varied, ranging from small units of 0·5 m to 1 m in diameter on dryers, to the large wheels in 500 MW power-station boilers, where diameters of 10 m are typical. Operating temperature ranges vary from just above ambient to 1000°C, and a wide range of materials are available to meet corrosion and thermal-stress requirements in these processes.

At the top end of the temperature scale, the British Steel Corporation has been working on a ceramic rotating regenerator for use on large reheat furnaces. Flexible seals are used to minimize gas leakage between the two flow regimes (Dugwell, 1977). The regenerator will provide a constant air preheat without any of the thermal downgrading or switching loss of static regenerators (such as the type used in glass melting furnaces) and will achieve efficiencies in the region of 90 per cent. Because the level of refractory utilization is very high (about 2000 W/kg compared with less than 100 W/kg for a static regenerator), the unit will be much more compact than the equivalent static unit.

It has been estimated by the British Steel Corporation that if applied to a 150 tonne/h reheat furnace, a rotating regenerator of this type could save £500 000 per annum by raising the air preheat temperature from 400°C to 900°C.

Howden, who produce a range or rotating regenerators for power-station boiler preheating, also manufacture smaller units for typical industrial applications and the fuel savings which can be effected by recovering heat from a boiler, economizer or process plant are illustrated graphically in Fig. 1.19.

With the temperature of the gas leaving the boiler, economizer or process as a base, plot the desired gas temperatures through the air preheater on graph A. Follow the grid lines through graph B, which relates temperature drop to gas flow – a measure of the heat saved. A scale for evaporation rate is included, permitting either gas quantity or boiler capacity to be used. The final graph C converts the heat saved into annual money savings based on 6000 h operation. To obtain the most realistic figure, the fuel cost used should be the mean value during the economical life of the plant.

Another area of application, where a hygroscopic wheel can be used is in the printing industry. In this instance a hygroscopic rotating regenerator was installed in the exhaust from coating and gravure machines at Harrison and Sons Ltd. These machines use large quantities of hot air for evaporating off solvents in the inks and coating materials. Heat is frequently wasted by directly exhausting the air to atmosphere following passage through the printing machine, although in some cases solvent recovery is practised (Butter, 1975).

In this particular factory, the air supply to the coating and gravure installation is heated by circulating heat transfer oil through finned tubes over which the air flows. The maximum heater output is 12 700 MJ/h, at which rate consumption is 0·39 m^3/h of 3500 s fuel oil. By recovering 70 per cent of the heat in the exhaust air, the fuel bill can be reduced by £600 per week, based on current fuel-oil prices. Two regenerators are used in the plant, and the location in one machine is shown in Fig. 1.20. They have diameters of 2 m and are 280 mm thick. The matrix is asbestos fibre sheet impregnated with lithium chloride. The hygroscopic properties attract the moisture, which is

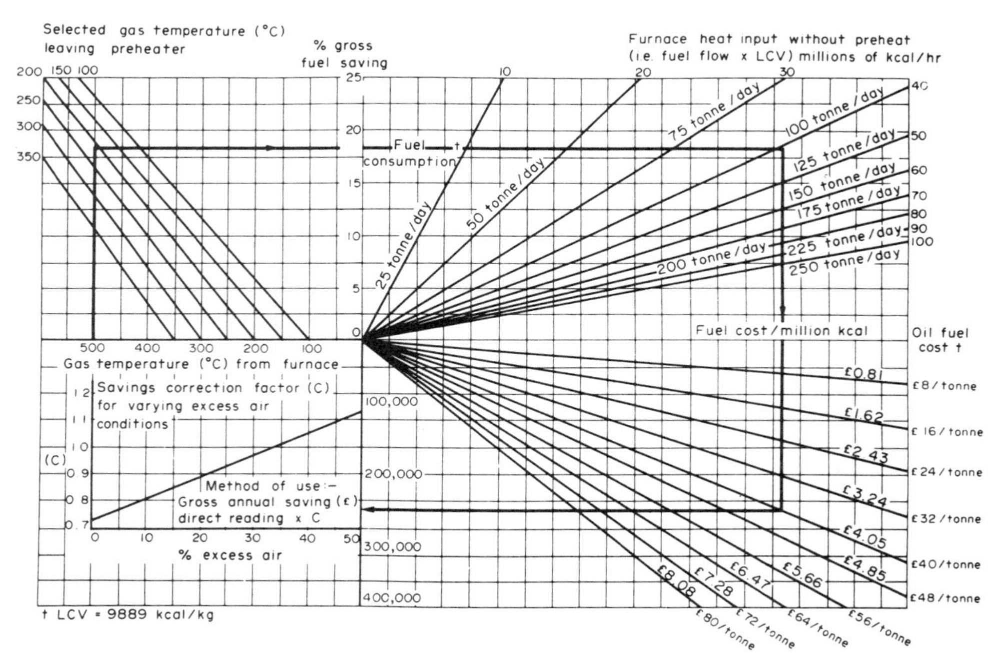

Fig. 1.19 Prediction of fuel savings, using a graph based on rotating regenerator performance.

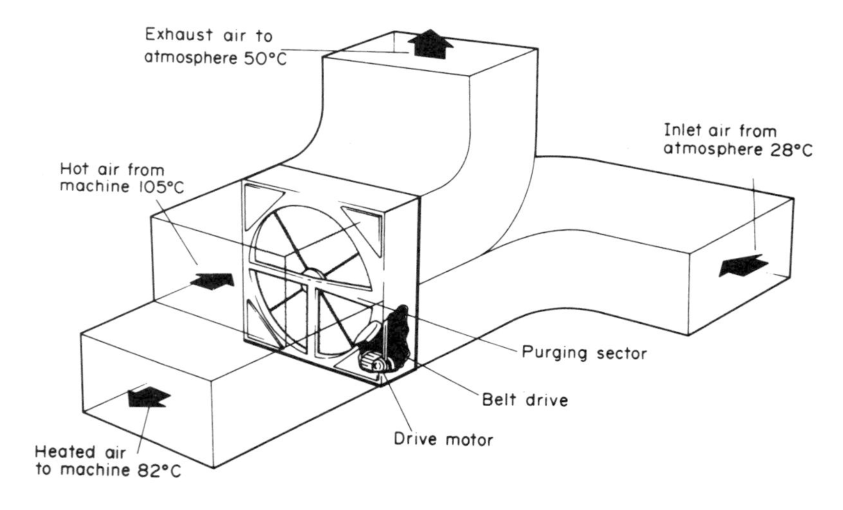

Fig. 1.20 Rotating regenerator used to recover heat from a coating and gravure plant in a printing works (by courtesy of *The Engineer*).

transferred into the preheating duct, and a purge section is fitted.

The total installation cost was £42 000, of which £11 000 is attributable to thermal wheel costs. Much of the ductwork used, which is included in the above cost, would be necessary to remove exhaust even if heat recovery was not used. In this case the payback period on the investment is substantially less than two years.

1.5 Run-around coils

1.5.1 *General description*

The run-around coil system, sometimes referred to in technical literature as liquid-coupled indirect heat exchanger network, can be an effective and readily assembled technique for waste heat recovery, particularly where it is desirable to install a heat recovery system without rerouting of ductwork. In its simplest form, used for HVAC and low-temperature process waste heat recovery, it consists of standard extended-surface finned tube water coils, used with a circulating pump. As illustrated in Fig. 1.21, one coil is located in the exhaust gas stream, and the other is located in the duct through which the air to be preheated is flowing. (As illustrated later, a multiplicity of coils may be used to serve several heat sources and sinks.) The pump is used to circulate water, an antifreeze solution or a higher-temperature heat transfer fluid such as Dowtherm through the two coils. The liquid picks up heat from the exhaust gas as it passes through the coil in that duct, and subsequently rejects the heat to the incoming air stream before being pumped back to the exhaust gas coil.

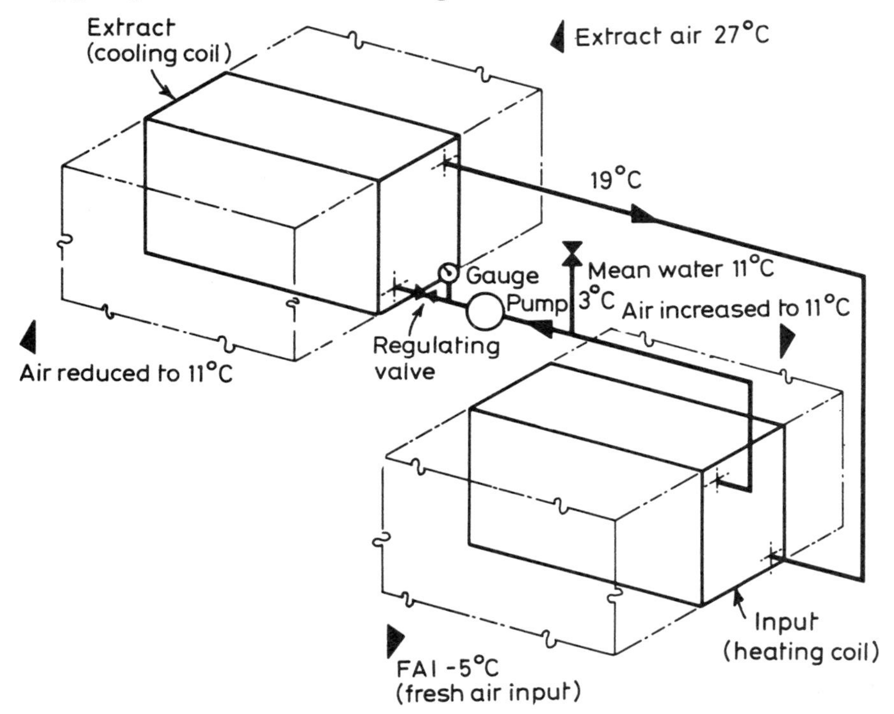

Fig. 1.21 Twin-coil run-around heat recovery system layout.

Run-around coils cater for air precooling in summer in HVAC applications and, in common with other types of air-air recovery systems, can be incorporated by many manufacturers in a 'package', together with fans, filters, control circuitry, etc.

One of the principal advantages of the runaround coil system is the ability to operate effectively without having to reroute ductwork. Because the intermediate heat transfer medium is a liquid, the coils in the exhaust and inlet air ducts can be many meters apart, even on different floors of a building, and the interconnecting pipework takes up little space and is much cheaper to install than air ducting. This has obvious benefits when the system is to be 'retrofitted' to a HVAC installation or item of process plant, but can also lead to cost savings when incorporated at the design stage.

Run-around coils rely on sensible heat recovery, at least in their basic form, although the occurrence of condensation on the coil in the hot exhaust duct may result in the capture of some latent heat content also. Maximum efficiencies are about 70 per cent, but operation in the efficiency range of 40 to 60 per cent is more typical.

Two other merits of the run-around coil may be cited. Because an intermediate heat transfer fluid is used in a sealed system, no cross-contamination between the two gas streams can occur. Also, the coils used in the ducts are widely used on their own as heaters, evaporators and condensers in most HVAC plant, and are therefore available from a large number of manufacturers and at very competitive prices.

The pump may introduce a reliability factor into the system, being the only moving part, but for water/glycol and water intermediate heat transfer fluids problems should occur very infrequently. Pumps for high-temperature heat transfer fluids can, however, be expensive. The theory of the run-around coil is dealt with extensively by Strindehag (1975) and the *ASHRAE Guide and Data Book* (1975), and Strindehag and Astrom (1974) give further data on system engineering.

1.5.2 *Applications and economics*

The applications of run-around coils are similar to those discussed for the items of equipment described in previous sections. However, the ability to have remote location of the gas streams, even at the points where the coils are situated, can be of benefit where the economics or practical difficulties of installing heat pipe or rotating regenerator systems prove prohibitive.

HVAC applications

Use of the run-around coil in HVAC heat recovery and precooling applications is extremely well developed, and its performance in such a situation is illustrated graphically in Fig. 1.22. (For a comparison with a rotating regenerator, refer back to Fig. 1.13.) For such low-temperature applications, the coils will be in the form of copper tubes, either expanded into aluminium-plate fins, or finned externally with spirally wound copper fins. Both of these construction methods are commonly used on tube bundles for heat pipe heat exchangers.

Control circuitry is particularly important in HVAC applications, and a typical system, used by Curwen & Newbery, is illustrated in Fig. 1.23. The amount of heat

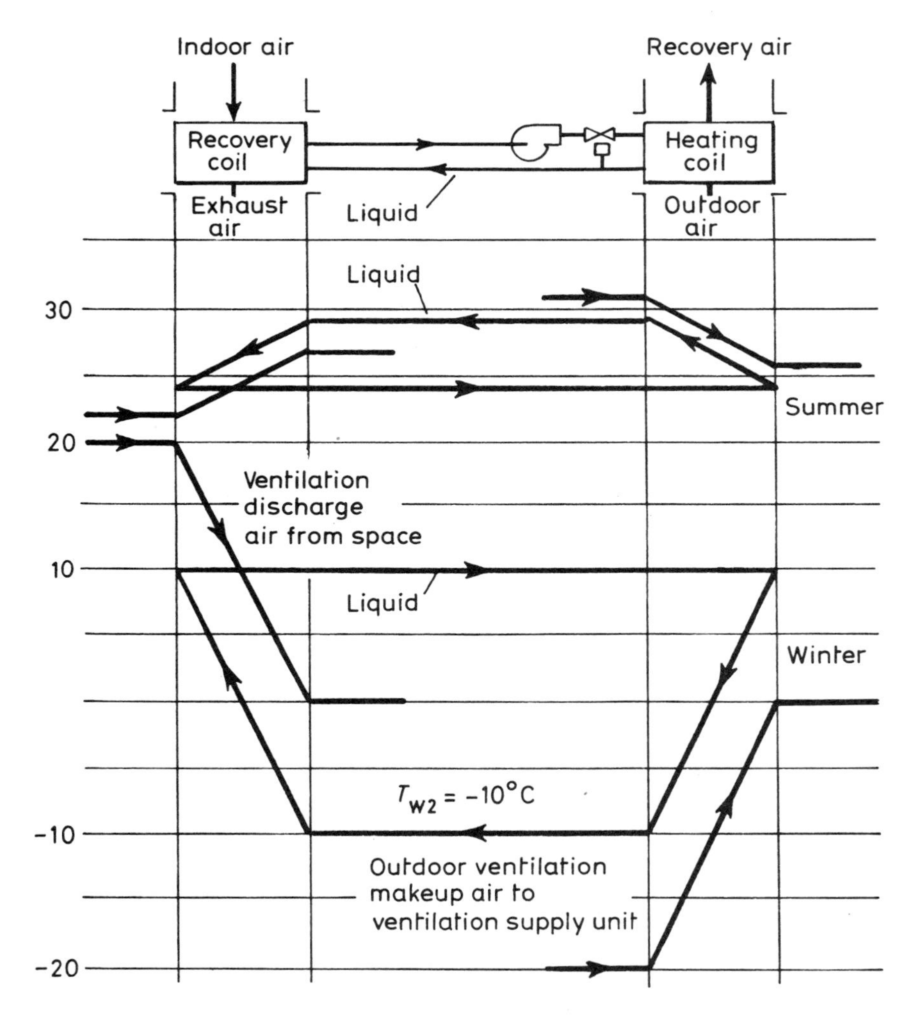

Fig. 1.22 A run-around coil used to transfer heat between air streams. Performance is inferior to the regenerator shown in Fig. 1.13.

transferred is controlled by a preset thermostat to maintain a constant supply air temperature.

The heat is absorbed during the winter heating cycle from the exhaust air stream by the heat exchanger A and is transferred to the incoming air stream via the closed pipe circuit and heat exchanger B. The thermostat C controls the heat transfer via the motorized three way valve I, sequentially controlling additional heating from the heater battery V via the three way valve II, in order to maintain the air temperature t_i constant at a preset level. If the demand for heat falls, V cuts out before B is reduced. The regulation works in a similar manner when the coils are used for a cooling duty in summer.

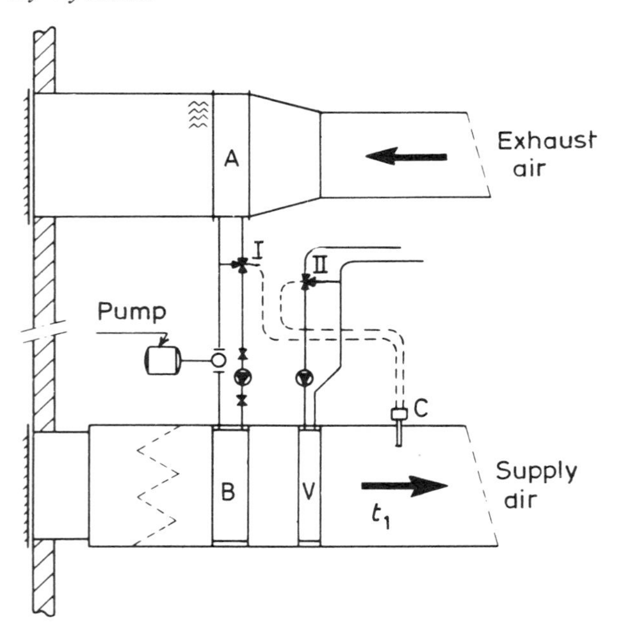

Fig. 1.23 Control system on a HVAC run-around coil installation.

Although glycol solutions may be used in HVAC run-around coils, giving some protection against freezing, this does not necessarily protect the outside of the exhaust recovery coil, as liquid leaving the supply-air coil may be at a very low temperature. The system offered by Air Energy Recovery Company in the USA, described below, caters for such an eventuality.

Along with the pump/controls assembly, two thermostats and one sensor are furnished along with the coils and pump/controls assembly. The two thermostats are field-installed in the fresh-air duct upstream from the recovery coil and will start the pump when the supply-air temperature is cold enough or hot enough to permit energy recovery. At temperatures at which energy cannot be recovered, the pump is stopped. The one temperature sensor provided for duct installation immediately downstream of the supply-air recovery coil prevents the system from furnishing too much heat during mild weather by signalling the three-way valve to bypass liquid around the supply coil(s).

During extremely cold weather, the liquid out of the supply recovery coil can become cold enough to cause ice to form on the exhaust recovery coil. When this happens, the liquid temperature sensor within the pump/control cabinet will again bypass liquid around the supply coil and will maintain a constant minimum liquid temperature to the exhaust recovery coil. The results are beneficial in two ways. First, a constant *maximum* energy recovery rate is maintained for all fresh-air temperatures colder than that at which ice starts to form on the exhaust coil. Second, the ice or frost is prevented from forming on the exhaust coil.

This manufacturer has also proposed the use of a liquid-liquid heat exchanger in the

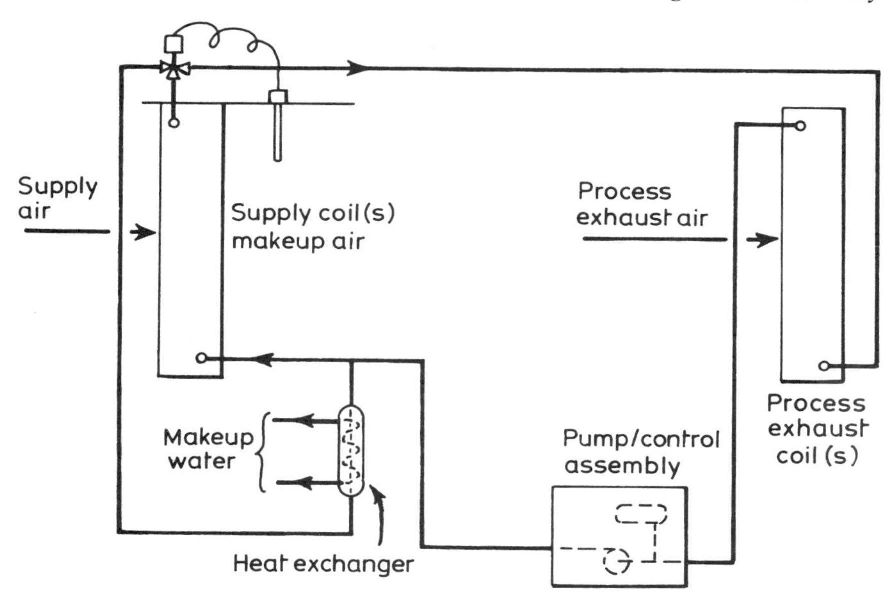

Fig. 1.24 Use of a liquid–liquid heat exchanger to boost the performance of a run-around coil system.

circuit, as shown in Fig. 1.24. This uses the heat contained in makeup water to boost the temperature of the fluid being circulated in the run-around coil before it reaches the supply-air coil. It is claimed that the use of such a heat exchanger can significantly reduce the payback period.

An economic assessment of the use of the run-around coil system in a HVAC application is given below (from Fläkt literature). In order to assess the economics, it is assumed that heating is provided using an oil-fired boiler, giving an equivalent energy cost of 0·46 p/kWh. The energy supplied to the pump and to the supply-air fan is considered as additional heat during the heating season, but fan power is classed as a loss during the remainder of the year. Capital-cost calculation is based on an annuity factor of 0·163 (10 years at an interest rate of 10 per cent) for all plant except the buildings, where a factor of 0·08 is taken.

The ventilation system has the following specification:

Supply-air flow	11 m^3/s
Exhaust-air flow	11 m^3/s
Exhaust relative humidity	55%
Supply-air temperature required	18°C
Operating time	8760 h/year.

Using an Ecoterm coil, the reduced heat demand amounts to 57 050 degree-hours which at the appropriate air flow and oil price represents a reduction of the fuel cost of £3456 per year. When the heat recovery system is installed, the boiler output can

Table 1.4 Cost exercise on an Ecoterm coil

Investments	Installation cost for the recovery system (coils, pumps, pipework, filter)	£4180
	Cost of increased building volume (25/m^3)	£1000
	Reduction in investment for boiler plant	£1830
	Net increase in investment	£3350
Capital and operating costs	Capital costs for the increased investment	£464/year
	Energy and power costs (pumps and fans)	£271/year
	Service and maintenance	£100/year
	Reduced fuel cost	£3456/year
	Net saving	£2621/year

Note: The operating costs take into account pump and fan energy costs and maintenance.

be reduced by 183 kW. Depending on the size of the boiler installation, the resulting reduction in investment costs varies substantially from case to case. In the installation being considered, the reduction in investment costs is assumed to amount to £10 per kilowatt, i.e. a total of £1830.

The investment and operating costs for installing the heat recovery system are summarized in Table 1.4. According to the results given in this table, a net saving of £2621 is obtained for an increased investment of £3350, which means that the payback period (increased investment/net saving) is 1·28 years. If similar calculations are carried out for systems with 6 and 10 row coils, one finds according to Table 1.5 that the payback periods are 1·41 and 1·54 years, respectively. Accordingly, 8 row coils seem to be the optimum size in this case.

Table 1.5 Coil optimization

Number of rows	System efficiency (%)	Increase in investment (£)	Net saving (£)	Payback period (years)
6	47	2910	2064	1·41
8	60	3350	2621	1·28
10	65	4150	2702	1·54

Industrial applications

Industrial use of run-around coils has not yet reached the scale of use in HVAC installations, as their efficiency is not as high as some competitive systems, and when

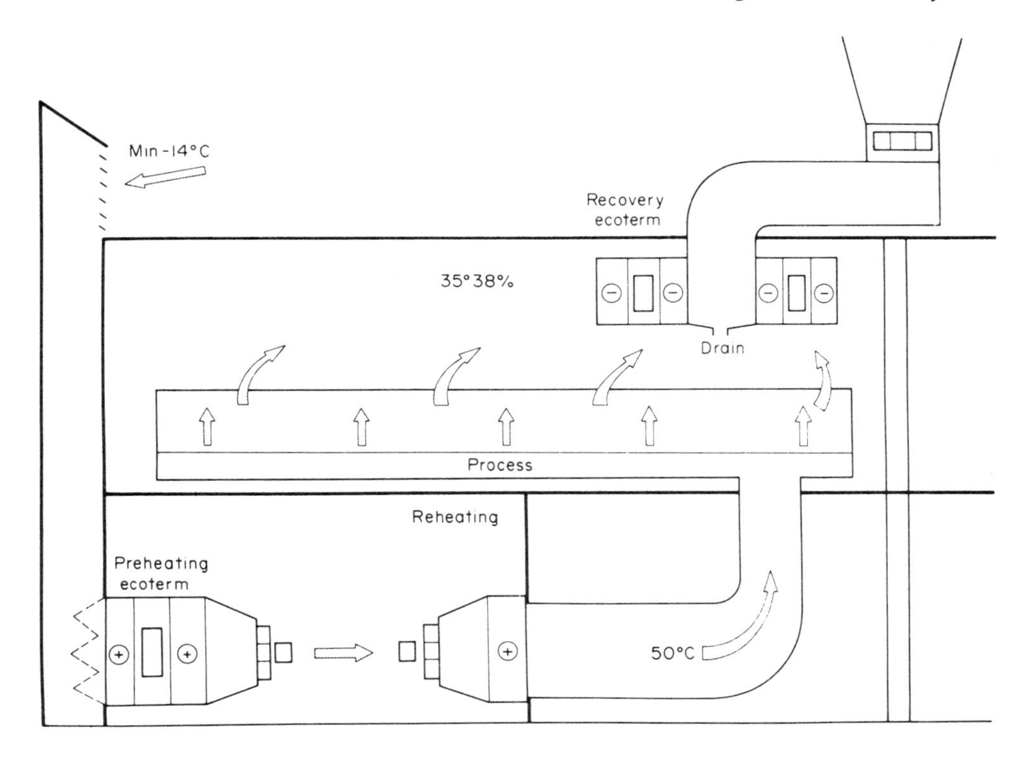

Fig. 1.25 A run-around coil system used to recover heat in a sausage-skin factory.

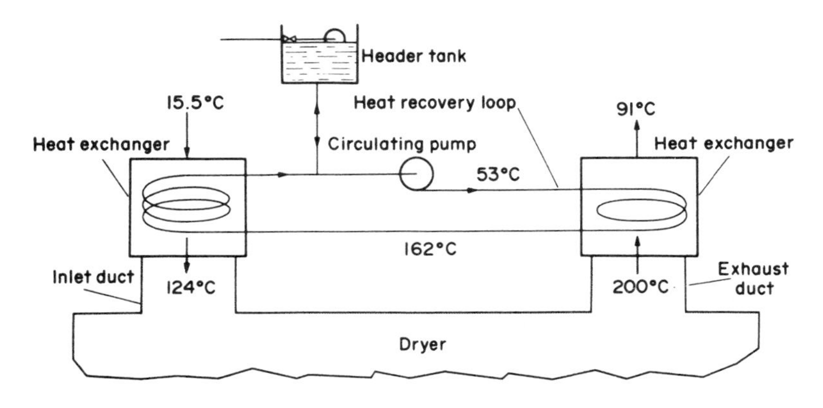

Fig. 1.26 A dryer incorporating a run-around coil heat recovery system.

used to preheat air on a single-process line or piece of plant, the cost advantages of liquid pipework over air ducts are not so great. Also, limitations on temperature (even when using high-temperature organic heat transfer fluids, some degradation can occur over prolonged periods, necessitating recharging) have proved a drawback to their use.

However, a number of successful installations have been demonstrated. Fig. 1.25

shows a complex coil system in a sausage-skin factory in Sweden. Five identical coil systems were used to recover heat from the process exhaust, the heat then being used to preheat process incoming air. The total investment of £70 000 was recovered in less than one year in this case. Should space heating be required, there is no reason why the runaround coil assembly should not be used to recover process heat for this purpose, also.

In most dryers, the inlet and exhaust ducts are at opposite ends of the plant, and therefore the run-around coil offers a convenient way for transferring heat from one end to the other without the expense of additional ductwork. An example of this heat exchanger applied to a continuous dryer is illustrated in Fig. 1.26. The coil in the inlet and exhaust duct take the form of extended surface heat exchangers, and in this particular example, the duct dimensions were 1·2 m x 0·75 m. The mean air velocity in the air inlet was 5 m/s and approximately 5 kg/s of air was passing to exhaust. Savings of up to 60 per cent in the heat needed to be provided by inlet burners were possible using this method. As in other applications of run-around coils, the circulating fluid may be water in most instances, but high-temperature organic heat transfer fluids may be used if the operating temperature is high (Mills, 1972).

The runaround system, by permitting remote location of the inlet and exhaust duct, is also worth considering for preheating dryer air using waste heat from other processes. It is possible to recover sufficient heat in this manner to provide all the requirements of the dryer, depending, of course, upon the dryer size and load schedule. Runaround coils have recently found popularity in malting plant, where drying of grain involves vast quantities of heat and air flow volumes are very high.

1.6 Plate heat exchangers

1.6.1 *General description*

The gas-to-gas plate heat exchanger, not to be confused with the units of identical name used for liquid–liquid heat transfer and described in Chapter 3, is one of the simplest recuperative heat exchangers, and is finding increasing popularity on HVAC and, more recently, process heat recovery applications. As will be seen from the data listed in Chapter 8, there are now a large number of manufacturers of equipment of this type.

Plate heat exchangers, two examples of which are shown in Figs. 1.27 and 1.28, consist of a framework containing a number of thin plates, which are normally of metal but, as in the unit in Fig. 1.27, can be of glass. These plates are located so that each one is a small distance from the adjacent plate. The two air flows pass either cross-wise or in a counter-flow mode between the plates, as illustrated in the exposed unit in Fig. 1.29. As in the liquid-liquid plate heat exchangers, the hot and cold gases pass adjacent to one another, separated only by one wall, through which conduction occurs, taking the heat from the hot to the cold stream.

In the case of the unit shown in Fig. 1.29, made by Munters, the plates are of corrugated aluminium foil, and the pack of plates is removable for cleaning. In common with the systems produced by some other manufacturers, the unit is sold in module form, and the casing has removable endcovers on each side to permit connection to other modules to be made when dealing with high air flows. Provision is normally made

Fig. 1.27 A conventional plate heat exchanger module.

Fig. 1.28 A plate heat exchanger in which galss plates are used, giving excellent corrosion resistance and aiding cleaning.

for condensate collection in these systems, and in the case of the Munters unit, a plastic coating may be applied to the plates to inhibit corrosion.

Some plate heat exchangers, notably that manufactured by Beltran & Cooper use extended surfaces in the form of corrugated fins on each plate to assist heat transfer from and to the gases.

In HVAC applications, certain manufacturers offer freeze-protection mechanisms.

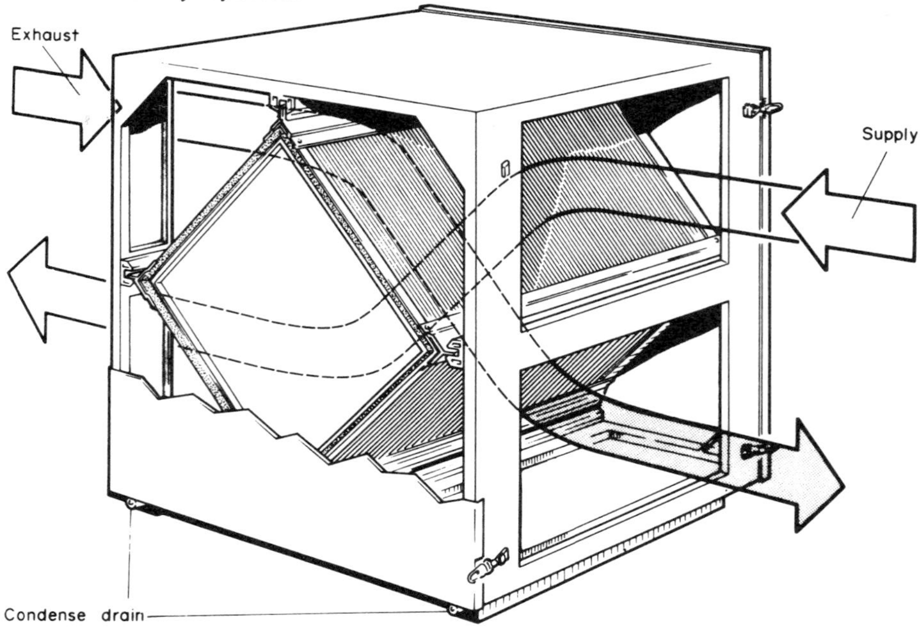

Exhaust
Supply
Condense drain

Fig. 1.29 The inside of a typical plate heat exchanger, showing the air flow paths.

In the case of the Munters Econovent model range, this is in the form of a sliding lid which moves to block off some of the passages, thus inhibiting the inflow of cold fresh air. In this way the plates of the heat exchanger are permitted to warm up, preventing icing.

The plate heat exchanger (with the exception of the above defrost mechanism) has no moving parts, and is designed for long life and ease of maintenance. By its nature, cross-contamination between the two air streams is not possible, and by using coated metals, or giving even better protection, by using glass plates as in the Air Froehlich unit, corrosion resistance is very high.

Capacities of plate heat exchangers vary considerably, modules being available having capacities of up to about 60 000 m^3/h air; but, as mentioned above, they may be mounted in parallel to cater for higher flows.

Some units are designed primarily for HVAC applications, and are limited in their maximum temperature. Most manufacturers, however, include in their range systems capable of operating at up to 200°C, and the unit manufactured by Beltran & Cooper is available with chrome-steel surfaces, enabling it to operate at higher temperatures (see Chapter 8).

The efficiency of plate heat exchangers is determined by a number of factors, including gas velocity, plate pitch, number and dimensions of the plates and the gas temperatures and humidity. (Of course, if condensation occurs in the exhaust gas stream as a result of cooling, some of this heat will be recovered, in addition to the sensible heat.)

Typical of the performance and selection charts available from the manufacturers is that for the Air Froehlich glass plate heat exchanger, and this is reproduced in

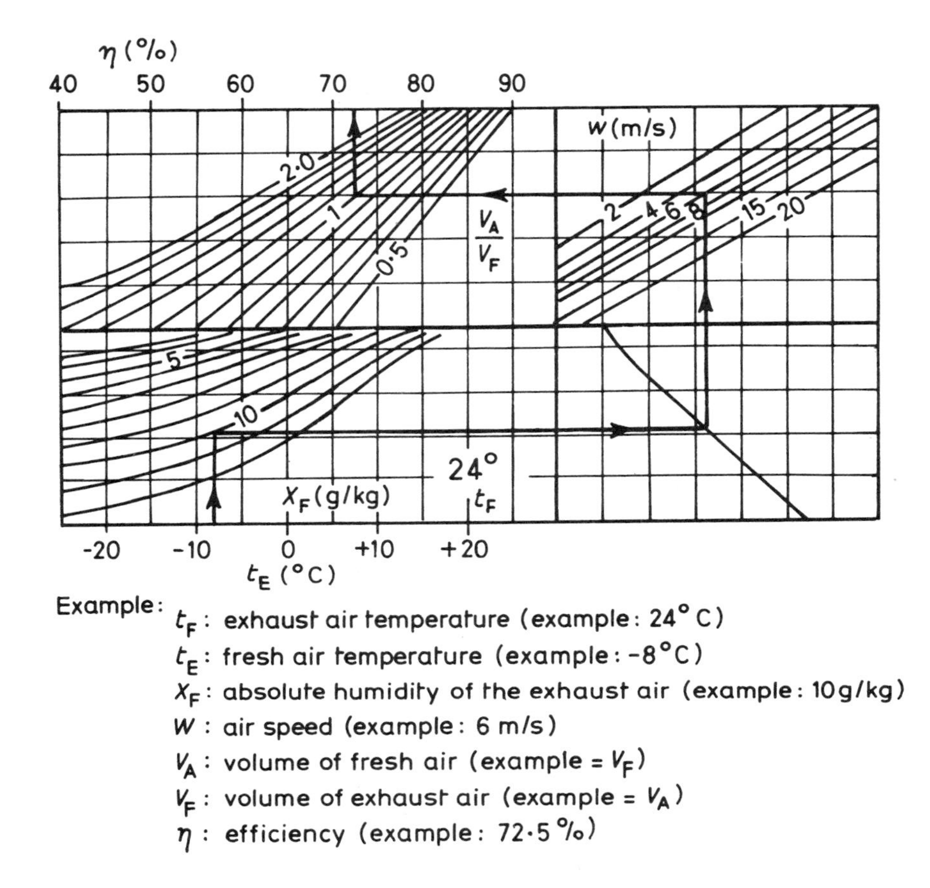

Fig. 1.30 Efficiency graph for an Air Froehlich plate heat exchanger.

Fig. 1.30. Using the values of the major parameters given in the table below the figure, it is possible, starting with a known fresh-air temperature (–8°C) and exhaust temperature (24°C), to read off the efficiency of the unit.

This type of heat recovery unit tends to occupy more space than a rotating regenerator, but competes on cost with heat pipe heat exchangers and is basically very simple. They are not yet available generally for very high temperature processes, but, as illustrated in Section 1.6.2, have proved attractive in several application areas.

1.6.2 *Application and economics*

The areas of application of the plate heat exchanger differ little from those of the heat pipe heat exchanger and runaround coil. One area opened up by the ability to use glass plates is the recovery of heat from contaminated and/or corrosive exhaust gases (see also Section 1.7, where glass tube recuperators are described).

HVAC applications are many, including commercial buildings, hospitals and swimming-pools. An example of the economics of a glass plate heat exchanger is given by the Air Froehlich data for a unit in a swimming-pool:

Problem	Design ventilation equipment for the indoor swimming-bath with two pools in Switzerland	
Technical data	Maximum number of swimmers	370
	Room temperature	30°C
	Room humidity	60%
	Water temperature	26°C
	Total surface of water	427 m²
Solution to the problem	Supply ventilation equipment with heat recovery using glass plate heat exchanger	
Quantity of air	Intake air/exhaust air	32 800/21 000 m³/h
	Fresh air/exhaust air	variable (depending on the humidity of the fresh air)
Design methods	Covered by the ventilation equipment	55 000 kcal/h
	Covered by local heating surfaces	16 000 kcal/h

Thanks to the glass plate heat exchanger, the following equipment can be reduced in capacity: the heating boiler; the pipe connections for the heating battery; and the heating battery in the air conditioning equipment.
This amounts to a reduction of 120 000 kcal/h in the heating requirement. The reduction in running costs amounts to Swiss Francs 17 700 a year at a fuel price of approximately Swiss Francs 18 per 100 kg (prices in 1973).

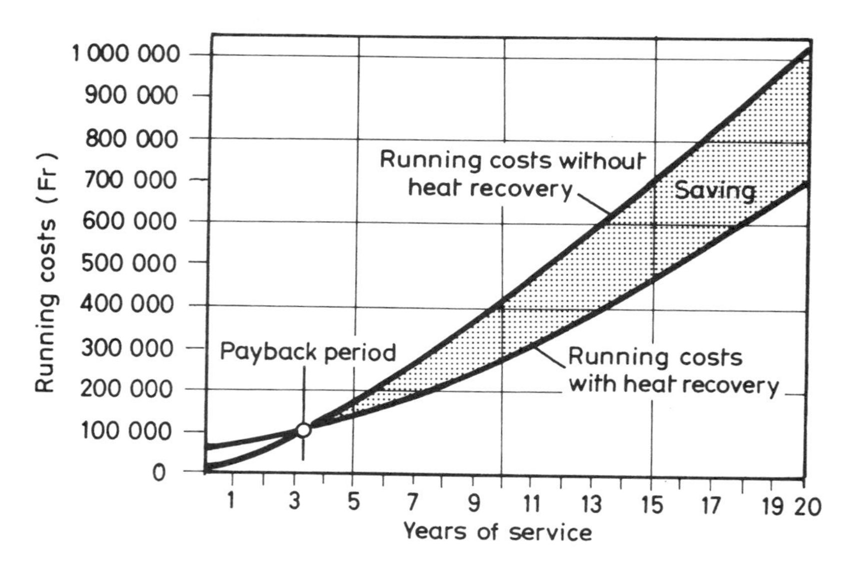

Fig. 1.31 Effect of a plate heat exchanger on the energy costs of a swimming-pool HVAC system.

The payback period of the additional heat recovery equipment would be approximately 3·3 years. The savings over the life of the pool are illustrated in Fig. 1.31.

Industrial applications of plate heat exchangers have concentrated largely on drying plant, partly because of limitations on the maximum temperature of some of the units. However, process-to-process heat recovery applications are becoming more widespread.

The recovery of process heat for space heating is also a viable proposition, as can be shown by the following example of a plant recently installed in a heat-treatment furnace:

Objective Recover heat from the exhaust stack of a continuous heat-treatment conveyor tunnel furnace and use for industrial warm-air space heating, to provide basic background heating over a wide area, thereby gaining maximum utilization of the plant.

Factors
(1) Stack temperature max. 95°C, min. 45°C. *Average* 65°C (from recorder).
(2) Stack gas flow rate average 3·14 m^3/s (at temperature).
(3) Workshop temperature 20°C.
(4) Utilization: furnace runs for 120 h/week: heated areas occupied for 80 h/week.

System
(1) Total stack loss based on an internal temperature of 20°C = 145 kW.
(2) Using 70 per cent heat recovery efficiency = 101 kW heating input.
(3) Based on current fuel costs and overall plant efficiency of 75 per cent cost to generate 1 kW = 0·76 p/h.
Cost saving = 76·5 p/h.
(4) Based on 80 h/week utilization.
Cost return rate = £61·00/week.
(5) Heat exchanger laminaire S2–336.
Ex-works cost £2000. Approximate cost returned in 33 weeks'use.

This data was prepared by Laminaire Products. Note that the heat exchanger cost is for the basic unit, and in order to do a more realistic payback analysis, installation costs should be known. In this case the operating temperature is not sufficiently high to inflate installation costs, and the only unknown is the availability of ductwork to transfer heat from the furnace flue to the areas to be space-heated.

1.7 Convection (tubular) recuperators

The tubular convection recuperator comes in many forms, depending upon the operating temperature range, plant size and the type of fouling in the exhaust gas stream. In this section a number of different types are described and applications listed. It is difficult to generalize on the rate of return following investment in these units because of the wide variety of applications, but data from Air Froehlich, tabulated later in this section, gives an idea of the potential for their particular unit in several process industries.

1.7.1 *Low-temperature units*

As with any type of heat recovery unit, the technological problems to be overcome in the design and operation of tubular recuperators increase with increasing process exhaust-gas temperatures, and at temperatures below about 250°C, several interesting types of tubular recuperator exist which bear little resemblance to their higher-temperature counterparts.

Fig. 1.32 A glass tube recuperator, ideal for use in highly corrosive or fouled gas streams.

The most interesting of these is the glass tube recuperator, an example of which, manufactured by Air Froehlich, is shown in Fig. 1.32. A glass tubular heat exchanger (see also the glass plate heat exchanger described in Section 1.6.2) is very easy to clean and has excellent resistance to corrosion. In common with all other tubular recuperators, one gas stream flows across the tubes, which in this case are not finned, and the other stream passes through the tubes, effecting a cross-flow arrangement. Of course, the two gas streams are completely sealed from one another, thus no cross-contamination can occur.

In some tubular recuperators, a number of passes on the tube-side are used, and this is illustrated later in the case of a high-temperature unit. It is also possible, although not common, to put the dirty gas stream through the insides of the tubes, which may, with some types of deposits, be easier to clean than the outside of the tubes.

Table 1.6 gives data on the heat recovery capacity and the resulting payback period for glass tube recuperators in a number of industrial processes. As with other heat exchangers, the heat recovery capability depends on the temperature difference between the hot and cold gas streams, and this is to some extent reflected in the payback figures, when related to the exhaust temperatures in the third column. In most cases these units are designed to recover between 60 and 70 per cent of the heat in the exhaust gas stream.

As an example of a plant in operation for several years, consider an installation in a textile factory. In a textile-finishing machine with cylinder dryers, fumes and vapours are extracted through hoods over the dryer section and then exhausted to the atmosphere. This exhaust air has to be replaced by outside air preheated to a temperature of 20°C to 25°C for working-space ventilation. As illustrated in Fig. 1.33, complete heating of

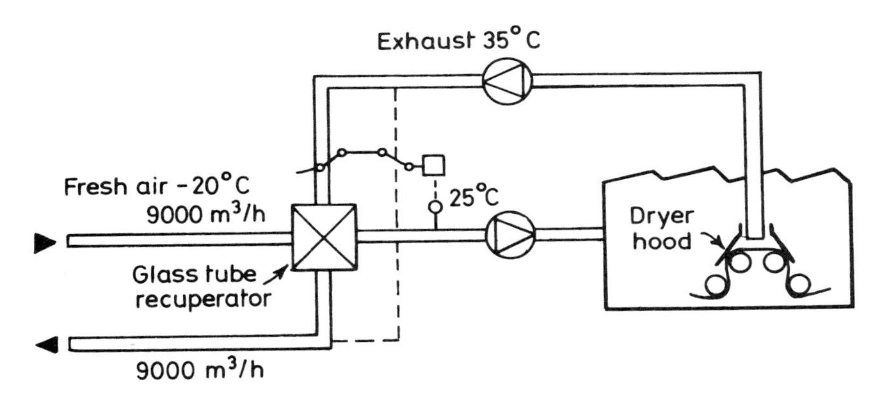

Fig. 1.33 Schematic layout of heat recovery plant using heat from a dryer exhaust for space heating.

outside air is accomplished without any external energy, i.e. no heating coils are installed in the ventilation plant. Even if the outside-air temperature is as low as –20°C, all energy necessary to raise the temperature to the required level is recovered from the exhaust air which is at 30°C to 35°C. In order not to overheat the textile plant during the warmer periods of the year, supply-air temperature is controlled by a bypass in the exhaust air duct.

In this installation, the fouling in the exhaust air was the primary reason for selecting a glass heat exchanger. This consists mainly of duct, fibres and sticky components from the finishing process, and settles on the inlet-side of the heat exchanger. From time to time the residues are flushed away with a permanently installed sprinkler nozzle. No depositions could be found in the unit itself, mainly due to the fine surface structure of glass and the high air velocity along the surface. Operating savings on this unit amounted to about 6000 Swiss Francs per annum.

Table 1.6 Application and economic data on glass tube recuperators

Industry	Process description	Exhaust air temperature (°C)	Average air volume (kg/h)	Exhaust air impurities	Average yearly operating time (h)	Recoverable energy per year with 10 000 kg/h air volume (Gcal)	Payback period for additional investment (years)
Textile finishing	High-temperature air drying of printed textiles	70–160	5 000–20 000	Dyestuff urea fibres	2500	400	1·1
	High-temperature thermofixation	150–250	up to 100 000	Fibres resin	4000	1100	0·4
Ceramics	Drying and granulating with spray dryers	50–150	5000–50 000	Ceramic powder combustion impurities	6000	780	0·6
Food	Powderization of milk, coffee, potatoes, etc. with spray dryers	60–140	up to 200 000	Powder	8400	1080	0·4
Chemical	Powderization of colouring agents	100	20 000	Powder	4000	520	0·85
Brewing	Kiln-drying of malt	20–70	30 000–250 000	Dust, combustion impurities	4000	170	2·2
Pulp and paper	Pulp and paper drying	60–100	40 000–200 000	Pulping fibres	5000	510	0·9
Fertilizer	Drying of animal dung	90–110	5000–20 000	Dust, partially moist	6000	540	0·8
Fodder	Grass drying with cylinder	90–100	5000–50 000	Dust graphite	3000	360	1·2
Wood	Pressboard chip drying	120–150	20 000–50 000	Wood-dust resin	4000	690	0·6

A number of simple tubular recuperators using metal tubes are also available for use at low temperatures. If fouling is not a problem, the use of external (and in some cases preferably with internal) tube finning is advantageous in helping to keep the size of the heat exchanger within reasonable proportions.

1.7.2 *Higher-temperature recuperators*

Metallic convection recuperators can be used where gas approach temperatures are less than 1000°C, although higher temperatures can be permitted if special materials and construction techniques are used. Originally all recuperators were made from ceramic materials, but these units suffered from serious leakage problems, and have today been largely superseded by metallic recuperators, except in some special cases (see below).

There are two basic types of convection recuperator, those which use cast tubes and those using drawn tubes which are assembled in bundles, in common with many other types of heat exchanger. The use of cast tubes is normally recommended for low-pressure applications, where leakage is unlikely to be a significant problem. It is most common to bolt each tube to the header boxes, and thus tube replacement is relatively easy (Kay, 1973).

The tubes used are available either plain or with a wide variety of extended-surface configurations. Wide pitching of surface projections is used when the exhaust gas flow may be heavily contaminated. In some cases, the outside of the tubes may be left completely bare. However, the overall heat transfer through a tube of this type may be maintained at an acceptable level by retaining the extended surfaces on the inside of the tube, through which the air to be heated is passed.

Two different designs of tubular recuperator for use at high temperatures are illustrated in Figs. 1.34 and 1.35. The unit in Fig. 1.34, manufactured by Harris Thermal Transfer Products Inc., utilizes closely pitched plain tubes, whereas the heat exchanger illustrated in Fig. 1.35 has a 'pin-fin' type of extended surface. This can accept exhaust gases at up to 950°C, preheating to 450°C–500°C. Cast tubes are used.

An alternative method of location used on cast tube recuperators involves the use of endflanges with integrally cast steel expansion joints. This permits the tubes to be welded together to form tube banks of any size. A four-pass horizontal flue-gas composite tube recuperator designed by Thermal Efficiency Ltd. is illustrated in Fig. 1.36. It can be seen that both bare and externally finned tubes are used in this installation. These rows of plain spun alloy steel tubes, containing a percentage of chromium and nickel, are arranged in front of the composite tubes to safeguard the latter from localized heating, created by non-uniform and excessive radiation. The chromium provides a resistance to oxidation at high temperatures, and the nickel content improved ductility in areas where high thermal stresses are likely to be encountered. Typical uses of these units include soaking-pit and reheat surface recuperation, where an additional requirement is for resistance to abrasive and sintered dust-laden gases. Although tube replacement is not as conveniently carried out as with bolted units, each tube can be removed from the bundle once the weld beads have been

Fig. 1.34 A high-temperature tubular recuperator made by Harris Thermal Transfer Products, Inc.

ground off.

Composite tube recuperators are used exclusively as convection-type heat exchangers with waste gas temperatures of up to 950°C. Using the plain spun-tube system described above, the temperature range can be slightly extended. Drawn-steel tube recuperators are available in many forms. Each tube bundle is attached to header boxes, and the construction technique used allows the tubes, and individual tube bundles, to expand relative to one another. In some systems the tubes are also bent at their mid-point to minimize stresses arising from thermal expansion. These recuperators are often used where it is required to recover a considerable proportion of the radiation heat, and the tubes are generally not finned. However, conduction through the wall is enhanced by the fact that whereas cast recuperator tubes have wall thicknesses of the order of 8 mm, drawn tubes may have a thickness of only 3 mm.

It has been stated above that metallic recuperators have largely superseded the

Fig. 1.35 An extended surface cast tube recuperator made by Thermal Efficiency Ltd.

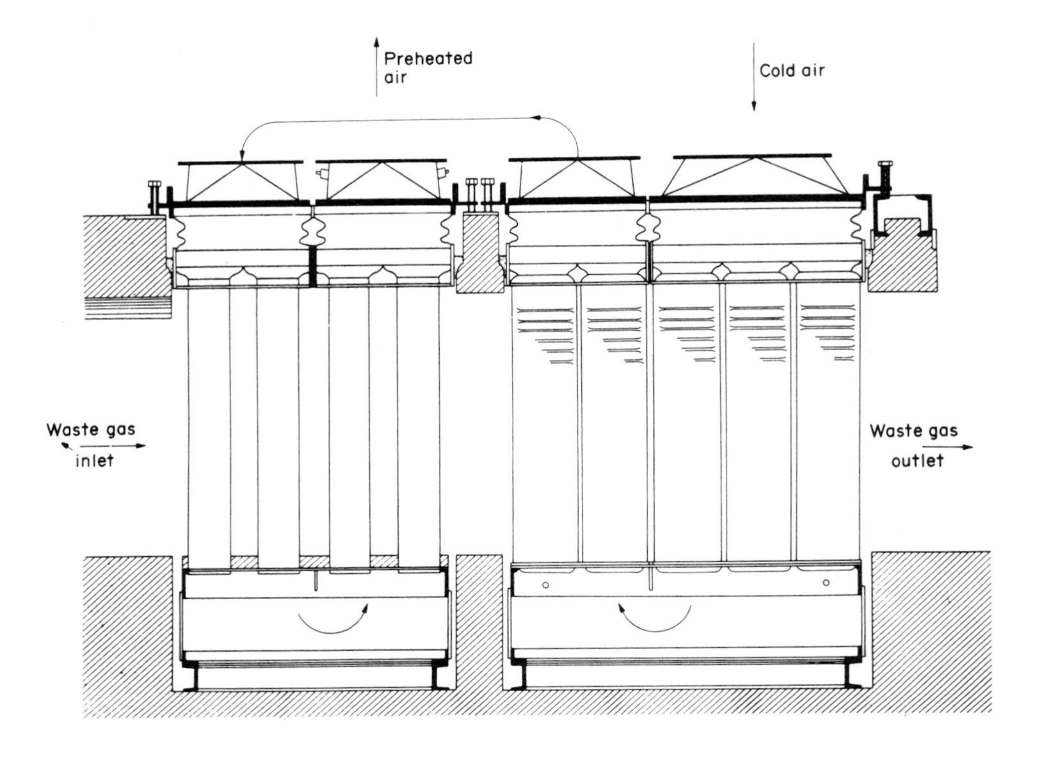

Fig. 1.36 A four-pass cast composite tube recuperator installation (by courtesy of Thermal Efficiency Ltd.)

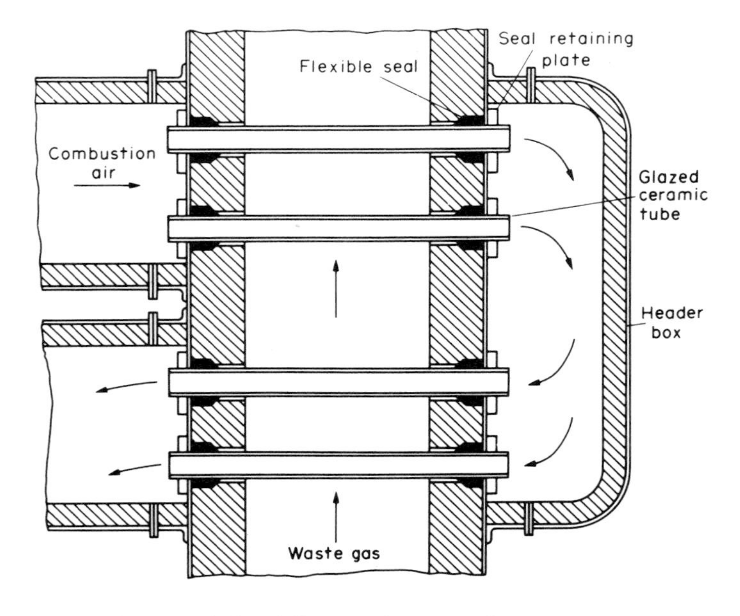

Fig. 1.37 British Steel Corporation ceramic recuperator.

refractory type. However, high-pressure/high-temperature ceramic recuperators have been developed for some specialized application. A number of manufacturers are able to supply ceramic recuperator tubes capable of operating at exhaust gas temperatures of up to 1800° C. It is claimed that the sealing problem on this type has been overcome using a ceramic-fibre packing based on aluminium silicate. An example of such a recuperator, with tubes of 'Carbofrax' and 'Refrax' produced by the Carborundum Company, connected directly to the outlet of a high-temperature kiln, is designed to accept 0·86 m^3/s of gas at 1800° C, giving an air preheat temperature of 1200° C and a heat transfer rate of 104 kW.

The Corporate Engineering Laboratories of the British Steel Corporation (BSC) worked for a number of years on ceramic high-temperature recuperators and regenerators for steel-plant applications. The ceramic recuperator, which can preheat combustion air to 650° C, is illustrated in Fig. 1.37. A prototype system commenced operation in November 1973, and its performance is considerably better than metallic recuperators, particularly as far as restrictions on operating temperature are concerned. Leakage problems are also minimized by the use of flexible ceramic seals.

1.8 Radiation recuperators

Radiation recuperators take two basic forms. They may consist of two concentric cylinders, the air to be heated normally flowing through the outer annulus while the exhaust gas flow through the central duct. Alternatively, the unit may be built up with tubes between two headers.

Compared with the convection recuperator, the radiation-type offers very low resistance to gas flow and in most instances never needs cleaning. The dirtiest of exhaust gases can be permitted through it, and by its nature this type of recuperator can also act as part of the chimney or flue.

The size of radiation recuperator can vary considerably, the largest units being about 50 m long and 3 m diameter. Radiation recuperators can be constructed using tubes to separate the exhaust gas and air, rather than a single annulus. These are used in instances when preheat temperatures in excess of 600° C are required. It is preferable to use parallel flow of air and gas in these, as the tubes tend to be subjected to near-equal temperatures along their whole length, thus keeping temperature-induced stresses to a minimum. The radiation recuperator is generally regarded as being the most reliable of the two main types available, and has the longest life.

As well as applications involving heat recovery from boiler and furnace exhausts, the radiation recuperator may be used in conjunction with a radiant tube heater, forming a self-contained radiant tube heater and recuperator unit, as shown in Fig. 1.38. The recuperator replaces the normal exhaust stack on individual radiant tubes, absorbing heat which is then used to preheat combustion air for the burner.

Figure 1.39 shows a number of ways in which a radiant tube recuperator may be mounted on a furnace, one example incorporating the burner type shown in Fig. 1.38.

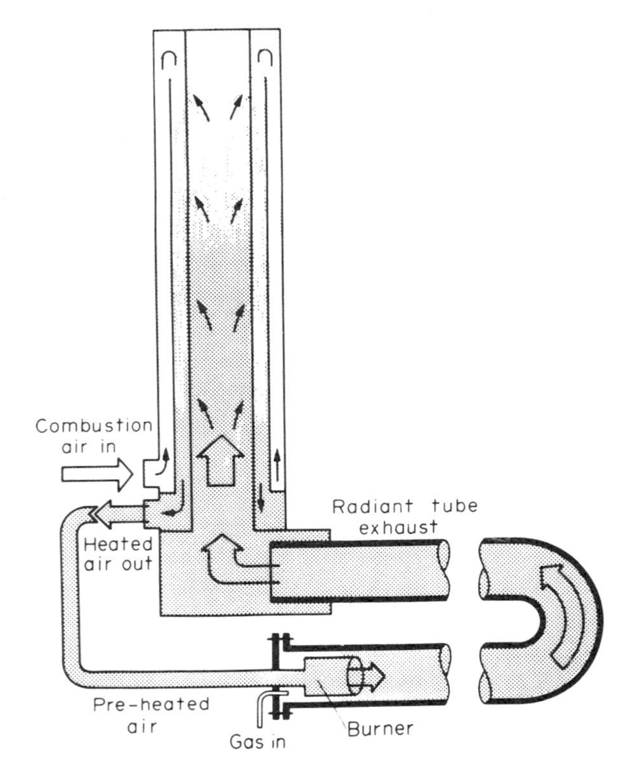

Fig. 1.38 A radiant tube recuperator linked to a burner.

One of the main application areas for tubular radiation recuperators is in glass-melting furnaces. Such a unit, manufactured by Johnson Construction Company AB (see Chapter 8), consists of a bundle of tubes hanging freely in a vertical refractory shaft. The bundle of tubes, which serves as the heating surface, is made of high-alloy heat-resistant steel and welded as one unit. The inlet and outlet are placed on two ring manifolds, between which the tubes are welded.

One of the main advantages of such a recuperator over conventional ceramic regenerators used on glass furnaces is the weight reduction. Metals in glass-furnace exhaust gas streams have suffered from corrosion and severe fouling, but with the correct selection of materials, and reliance on radiation as the mode of heat transfer, these difficulties can be overcome and lives equivalent to that of the furnace lining (5 to 10 years) can be achieved without repairs.

1.9 Recuperative burners

Descriptions have already been given of several types of recuperators, many of which are applied for the preheating of combustion air. These include the radiant tube

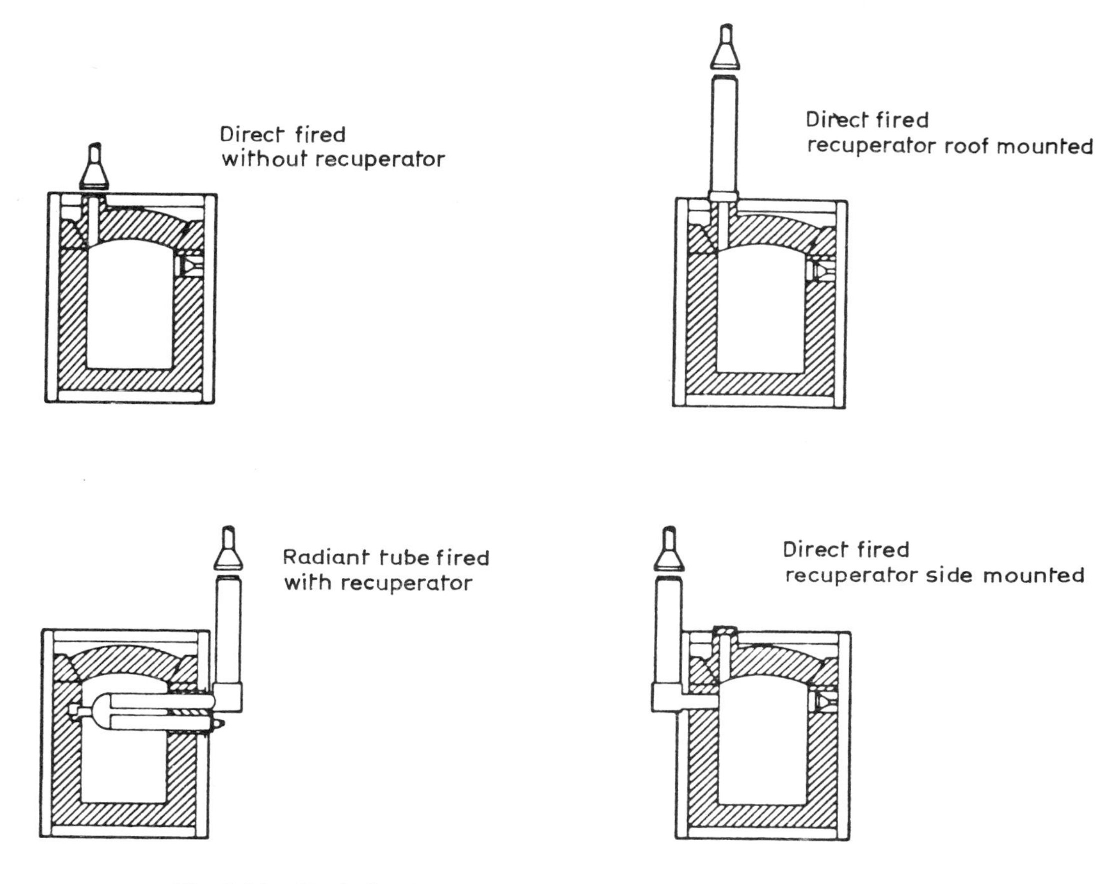

Fig. 1.39 Typical radiant rube recuperator installations on a furnace.

recuperator, described above, which may be directly linked to a burner to provide this combustion air preheat. However, it is also possible to integrate the recuperator within the body of the burner.

1.9.1 *Burner description*

One of the most successful recuperative burners is that developed by the Midlands Research Station of British Gas, and their design is illustrated in Fig. 1.40. The unit consists of a high-velocity nozzle-mixing burner, in itself an effective burner because of the good heat transfer obtained by directing high-velocity hot gases over the furnace load, surrounded by a counter-flow heat exchanger. This heat exchanger supplies hot combustion air to the burner nozzle. The heat exchanger consists of a series of concentric tubes which act as interfaces between the combustion gases and the air to be preheated, and is made from a heat-resisting steel.

In operation, the combustion air enters the burner at a manifold, which totally

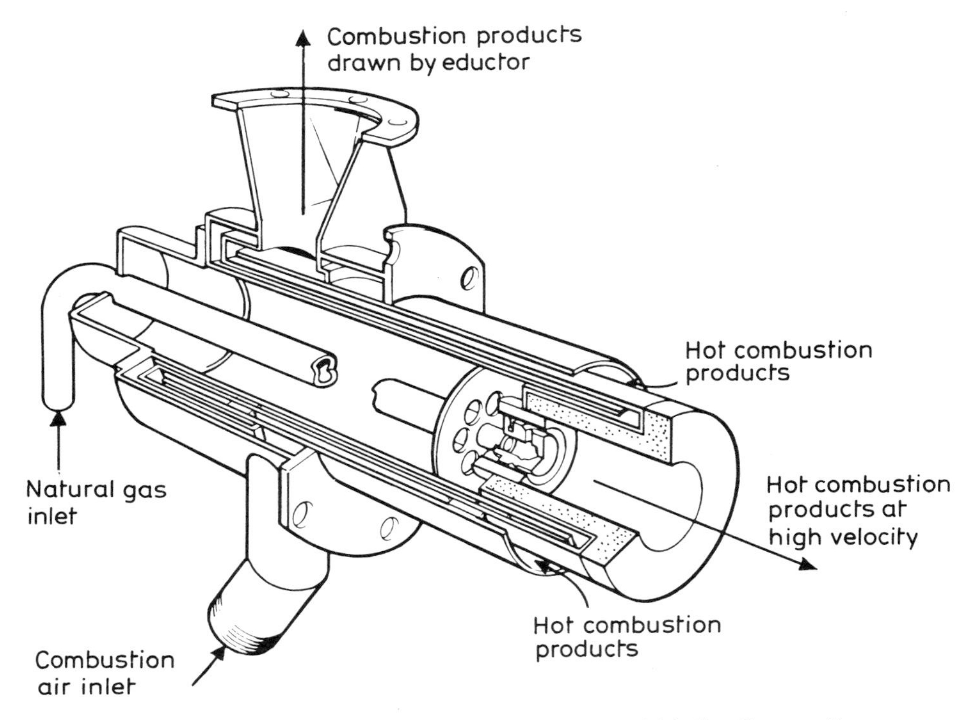

Fig. 1.40 Recuperative burner developed by the British Gas Corporation.

encloses the flue annuli. It then passes forward along the air annulus, thus keeping the external surfaces of the recuperator cool. Flue products are extracted in a counter-flow direction around the outside of the air annulus and thus preheat the air. At the front of the recuperator the air flow doubles back and enters the burner nozzle. The gas and preheated air mix at the nozzle, on the face of which the flame stabilizes, and combustion is essentially complete before the gases pass into the furnace chamber. Exit velocities from the burner tunnel considerably exceed 50 m/s. Normally all the combustion products are extracted through the recuperator, assisted by the use of an air-driven eductor mounted on the flue of the burner. By controlling the amount of eductor air the furnace pressure can be maintained at the desired level.

The series of burners now designed covers a range of thermal inputs from 100 kW (3·5 therm/h) to 900 kW (30 therm/h). The nozzles used give a stable flame, can operate if necessary with over 100 per cent excess air and have a turndown ratio of over 10:1.

The control system for a recuperative burner resembles that required for a conventional nozzle-mixing air blast burner, but it compensates for the increase in backpressure which occurs when the air for combustion becomes preheated. This is achieved by incorporating volumetric governing on both air and gas; a governor across the throughput control valve is backloaded in such a way that a constant pressure differential is maintained across it.

1.9.2 *Applications in production furnaces*

Batch-steel reheating furnace

The furnace is situated in a smith's shop where it is used to heat a variety of steel sections to 1300°C for open-die forging into links, shackles and other heavy-duty chain tackle. It was previously fired by two 350 kW (12 therm/h) oil burners and these were replaced by two dissimilar prototype 225 kW (7·5 therm/h) recuperative gas burners. The hearth area of the furnace is 2·7 m by 1·35 m and the chamber height is 1·1 m – half the original height prior to changeover. The fuel saving due to the use of recuperative burners was predicted to be over 30 per cent, in addition to that to be expected from the lowering of the roof.

Intermittent kilns

Intermittent kilns used for firing ceramicware are essentially refractory-lined chambers of rectangular section into which ware is usually charged by means of a wheeled car, on which it is stacked in an open arrangement to facilitate the penetration of heat. On single-truck kilns, burners usually fire from diagonally opposite corners in such a way that the gases are encouraged to flow around and through the stacked ware. The gases are flued through ports in the top of the car supporting the ware and these ports connect via insulated ducting with a flue at the rear of the kiln.

When processing refractories or kiln furniture the firing cycle usually starts with a period of drying, followed by firing to a high temperature, at which the ware is held, before being cooled to ambient conditions. The quality of the ware produced from such a kiln is sensitive to the amount of heat transferred to the load during various parts of the firing cycle. During the early, low-temperature phase the heat is required to drive off water and any other volatile constituents in the greenware. At the higher temperatures the ceramic bond is developed. In order to obtain consistent ware quality, it is important that each piece in the kiln has a similar thermal history. Thus, one of the prerequisites of a good firing system is even temperature distribution. Another requirement for this particular ware is that the kiln atmosphere must at all times remain oxidizing; furthermore, to assist in the rapid turnround of cars and promote efficient kiln utilization, it is desirable to cool the kiln after firing as quickly as the thermal capacity of the ware and kiln lining will allow, while avoiding too rapid chilling of walls or ware could lead to failure due to thermal shock.

A twin-truck kiln, of capacity 18·4 m^3, heats a charge of 3 tonnes of industrial refractories and abrasives to 1200°C. It was previously fired by 8 nozzle-mixing burners with a total capacity of 1·6 MW (56 therm/h), which fired down channels between the ware contained in saggars in 4 stacks. The kiln was controlled automatically to a set time/temperature cycle, except during the first 24 h, when the burners were operated under manual supervision. In the following 9 h the temperature was raised to 1200°C and a 5 h soak completed the firing. The saggars full of ware were carried in on two cars which also accommodated the flueways, as shown in Fig. 1.41.

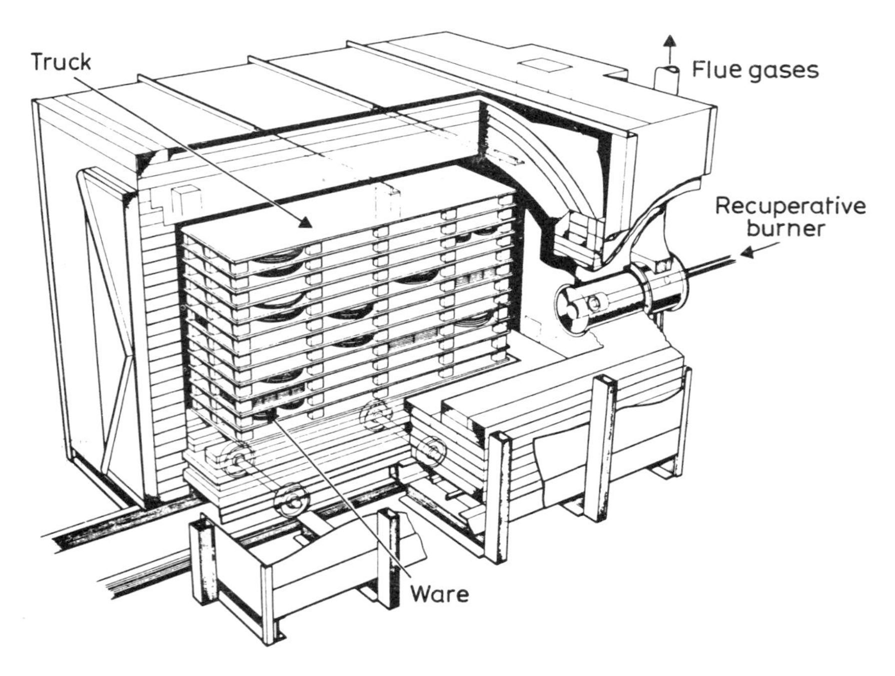

Fig. 1.41 A single-truck intermittent kiln fired by a recuperative burner.

It is common practice in this and other kilns to put sample pieces of a special clay in a selection of saggars. These samples, known as 'Bullers rings', contract according to the degree of thermochemical change in the clay. By measuring the size of a ring before and after firing, a gauge of the degree of firing of the neighbouring ware can be obtained. This method provides a convenient indication of temperature uniformity throughout the kiln and shows any corrections that may need to be made to the firing cycle.

The result of modelling studies and calculations indicated that two 450 kW (15 therm/h) recuperative burners, located as shown in Fig. 1.42 could be used during drying and that

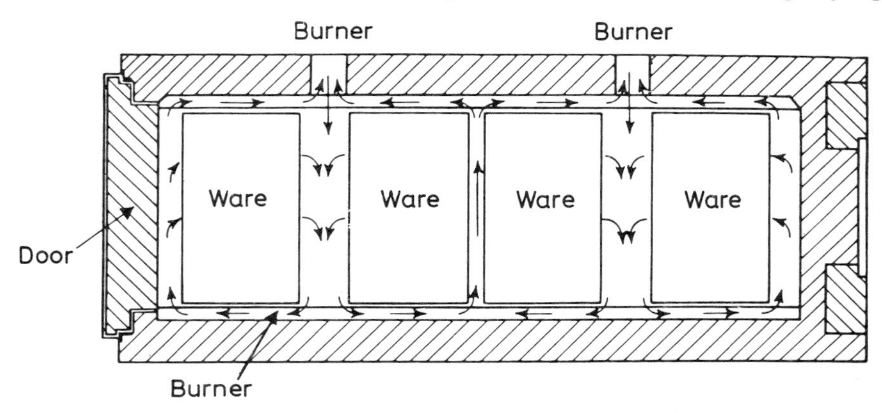

Fig. 1.42 Plan section of twin-truck intermittent kiln after modification, fired by two recuperative burners.

heat input should be reduced to 20 per cent at the higher temperatures. A fuel saving of over 40 per cent was predicted.

After the first two months' operation, attempts were made to utilize the improved recirculation and greater mass flow by shortening the overall cycle. The drying, firing and cooling periods were all shortened and the kiln turnround time was reduced from 50 to 35 h, thus enabling an extra firing to be carried out each week.

1.9.3 *Economics*

Examples of the fuel saving achieved with each of the two examples quoted are outlined. Because there is a significant amount of hidden expenditure involved in burner installation and application engineering, it is not easy to provide details of individual installation costs. It is, however, possible to state that the payback times for this development are 1 to 2 years. Current development is aimed at reducing capital costs of burner and controls so that these periods might be halved.

Batch-forge furnace

Results of the trials are shown in Table 1.7. It can be seen that overall fuel savings of over 50 per cent have been obtained, part of this being due to lowering the roof. A feature which is especially attractive to the furnace operator is the virtual elimination

Table 1.7 Comparison of recuperative and non-recuperative firing of a batch-reheating furnace

	Non-recuperative	Recuperative
Burner capacity,		
kW	2 x 350	2 x 225
(therm/h)	(2 x 12)	(2 x 7·5)
Time to reach working temperature, (h)	4	2·5
Fuel consumption to reach working temperature,		
MJ	10 000	4 100
(therm)	(96·0)	(37·5)
Fuel rate for holding at temperature,		
kW	533	293
(therm/h)	(18·2)	(10·0)
Fuel consumed in one shift,		
MJ	25 400	12 400
(therm)	(241)	(117·5)
Fuel saving on heating up (%)	–	61
Fuel saving on holding (%)	–	45
Overall fuel saving (%)	–	51

of the discharge of hot gases from the door, due to the balanced furnace pressure.

Twin-truck intermittent kiln

The results of the trials on the twin-truck intermittent kiln are given in Table 1.8.

Table 1.8 Comparison of recuperative and non-recuperative firing of a large intermittent kiln

	Non-recuperative	Recuperative
Burner type	Nozzle mixing	MRS recuperative, 450 kW
No. of burners	8	2
Total burner capacity	1·6 MW	0·9 MW
Fuel consumption over 1 cycle	72 800 MJ (690 therm)	42 200 MJ (400 therm)
Fuel saving (%)	–	42·0

Substantial fuel savings of 42 per cent have been achieved in this application. The operation of the kiln has been greatly simplified and as a final advance the necessity for manual supervision during the drying period has been eliminated by mechanizing the heating cycle according to the results obtained in the early firings. Acceptable ware is consistently produced and it has been possible to fire even the most difficult ware without shattering. A Bullers ring variation of less than ± 4 throughout the kiln is common. By the correct adjustment of the educator air, little or no fume is now allowed to enter the working area, as the products pass through flueways which discharge above the roof of the factory.

A more general illustration of the savings which may be achieved using a recuperative burner is given in Fig. 1.43, which is a nomogram relating the combustion air preheat temperature to the percentage fuel saving achievable.

The left-hand column shows flue gas exhaust temperatures on the left-hand side under stoichiometric conditions (0 per cent excess air), and at 20 per cent excess air on the right-hand side. The right-hand column shows combustion air preheat temperature. In the absence of actual temperature measurement these may normally be taken as approximately 50 per cent of the exhaust temperature. The middle column shows percentage fuel savings.

Example:

Assume a furnace exhaust temperature of 1200°C and a combustion air preheat temperature of 600°C. Strike a line across these two values, and at the point where it crosses the middle column the expected percentage fuel saving will be shown, in this case 32·5 per cent. Using similar temperatures at 20 per cent excess air, and the same process, a fuel saving of 42·5 per cent is indicated.

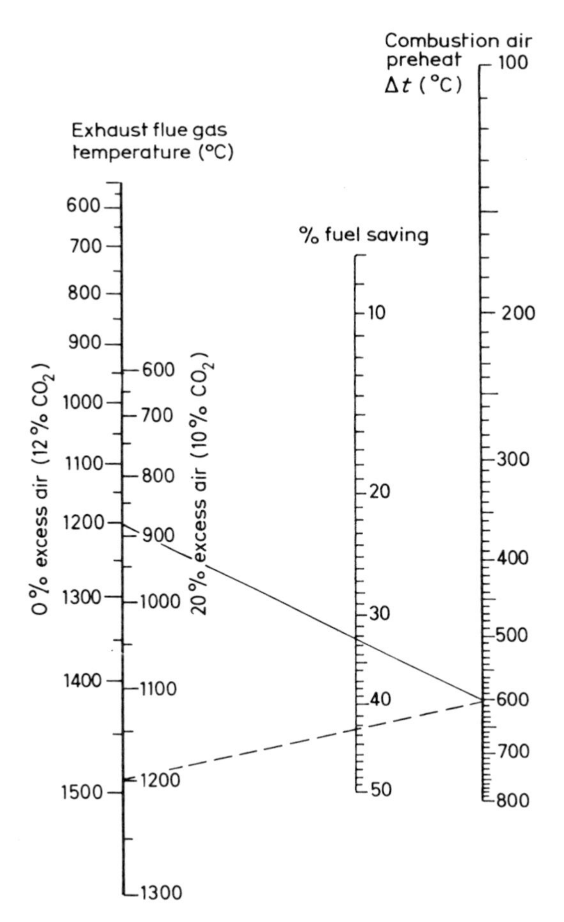

Fig. 1.43 Nomogram giving fuel savings possible with combustion air preheat, applicable to a recuperative burner.

References

Applegate, G. (1970), Heating a factory for nothing, *Heating Ventilating Eng.*, vol. 44, 21–24.

Applegate, G. (1974), Heat regeneration by thermal wheel, *Proc. Waste Heat Recovery Conf. Inst. Plant Eng., London, 25–26 September 1974.*

ASHRAE (1975), Air-air heat recovery equipment, *ASHRAE Guide and Data Book,* Equipment volume, chapter 34, 34.1–34.12.

Basiulis, A. (1977), Vapour bubble pump heat pipes for energy storage, *Proc. 11th IECEC,* Paper 769155.

Basiulis, A. and Plost, M. (1975), Waste heat utilization through the use of heat pipes, *Trans. ASME,* Paper 75-WA/HT-48, New York.

Bedrossian, A. and Lee, Y. (1978), The characteristics of heat exchangers using heat pipes or thermosyphons, *Int. J. Heat Mass Transfer* (in press).

Butler, P. (1975), Using the Munter wheel to recover your heat and make big savings, *Engineer,* 20 November, 20.

Chisholm, D. (1978), Improved heat pipe performance: Header and artery system, *Int. J. Heat Mass Transfer* (in press).

Chojnowski, B. and Chew, P. E. (1978), Getting the best out of rotary air heaters, *CEGB Research,* no. 7, 14–21.

Dugwell, D. R. (1977), Waste heat recovery and use in the British Steel Corporation, *Steel Times Annual Review,* 784–793.

Dunn, P. D. and Reay, D. A. (1978), *Heat Pipes,* 2nd ed., Pergamon Press, Oxford.

Feldman, K. T. (1976), Simplified design of heat pipe heat exchangers, *Proc. 2nd Int. Heat Pipe Conf.,* Bologna, Italy, 1976, ESA Report SP/112.

Feldman, K. T. and Berger, M. E. (1973), Analysis of a high heat flux water heat pipe evaporator, University of New Mexico Report ME-62(73), AD-765 803.

Feldman, K. T. and Lu, D. C. (1976), Preliminary design study of heat pipe heat exchangers, *Proc. 2nd Int. Heat Pipe Conf.,* Bologna, Italy, 1976, ESA Report SP/112, vol. 1.

Field, A. A. (1975), Heat recovery from air systems – what's new in Europe, *Heat Piping Air Conditioning,* vol. 47, pt 11, 73–78.

Friedlander, G. D. (1973), Energy conservation by redesign, *IEEE Spectrum,* vol. 10, no. 11, 36–43.

Hayet, L. and Maxwell, T. H. (1975), Heat recovery wheel saves energy in Florida facility, *Actual Spec. Engr.,* vol. 33, pt 5, 54–56.

Kay, H. (1973), Recuperators – their use and abuse, *Iron and Steel International,* June, 231–240.

Kruse, H. and Vauth, R. (1976), Application limits and performances of hygroscopic and non-hygroscopic regenerative rotary heat exchangers, *Sci. Tech. Froid,* vol. 1, 705–713.

Lu, D. C. and Feldman, K. T. (1977), Cost-effectiveness study of heat pipe heat exchangers, *Proc. ASME Winter Annual Meeting,* ASME Paper 77-WA/HT-5.

Mills, L. (1972), Recover waste heat systematically, *Power,* December, 36–37.

Moritz, K. (1969), Ein Warmerohr neuer Bauert, *Chemie Ingenieur Technik,* vol. 41, 37–40.

Mortimer, J. (1973), Exchanging heat to save fuel, *Engineer,* 2–9 August, 34–37.

Ranken, W. A. and Lundberg, L. B. (1978a), High temperature heat pipes for terrestrial applications, *Proc. 3rd Int. Heat Pipe Conf., Palo Alto, California, U.S.A., 1978,* AIAA Report CP 784.

Ranken, W. A. and Lundberg, L. B. (1978), High temperature heat pipes for terrestrial applications, *Proc. 3rd Int. Heat Pipe Conf., Palo Alto, California, U.S.A., 1978,* AIAA Paper 78–435.

Reay, D. A. (1977), The heat pipe heat exchanger, a technique for waste heat recovery, *Heating Ventilating Engr.,* vol. 50, no. 593.

Ruch, M. A. and Grover, G. M. (1976), Heat pipe thermal recovery applications, *Proc. 2nd Int. Heat Pipe Conf., Bologna, Italy, 1976,* ESA Report SP112, vol. 1.

Ruth, D. W., Fisher, D. W. and Gawley, H. N. (1975), Investigation of frosting in rotary air-to-air heat exchangers, *Trans. ASHRAE, vol. 81, pt 2, 410–417.*

Sohlberg, J. (1976), Klumatechnische Tendenzen in schwedischen Krankenhausbau, *Heizung Luftung Klumatechnik,* vol. 27, no. 3.

Spahn, H. and Gnielinski, V. (1971), Heat and mass transfer performance of a hygroscopic rotating regenerator under summer conditions, *Verfahrestechnik,* vol. 5, 143 *et seq.*

Strindehag, O. (1975), A liquid coupled system for heat recovery from exhaust gases, *Building Serv. Eng.,* vol. 43, 52–56.

Strindehag, O. and Astrom, L. (1974), Energy conservation by Ecoterm, *Fläkt Review,* Sweden.

2 Gas-to-liquid heat recovery

2.1 Introduction

Unlike the waste heat recovery equipment covered in Chapters 1 and 3, namely gas–gas and liquid–liquid heat exchangers, systems available for gas–liquid heat recovery seldom can be looked at in isolation, if justice is to be done to their potential applications. In addition, certain gas–liquid heat exchangers need no longer be regarded strictly as waste heat recovery units (for example, superheaters and economizers, particularly the former) as they are almost standard items of plant, at least as applied to boilers.

Gas–liquid heat recovery units, particularly waste heat boilers, are frequently used in conjunction with other major items of plant which are an integral part of the waste heat recovery system. One extensive area of application in this context is the recovery of heat for raising steam for use in a turbogenerator. Typified by systems such as the Hawthorn Leslie 'Seajoule' and Peter Brotherhood turbines, these applications described later in the chapter could equally well appear in Chapter 4, where heat recovery both from, and using, prime movers is discussed. In Chapter 5 the reader will also see examples of waste heat boilers used on incineration plant, some again in conjunction with a turbine.

Applications of waste heat boilers and economizers on reciprocating engines and gas turbines are also very frequent. This topic, more generally covered in Chapter 4, has attracted a number of specialized waste heat boiler manufacturers, whose products are detailed separately.

The purpose of this is to stress the wide and often specialized field of application of some of these heat exchangers. Rather than making excuses for the fact that three chapters cater for similar items of equipment, the fact that this is necessary serves to emphasize the complexity and diversity of the field. Lest the reader become confused, all manufacturers, and their fields of application, are cross-referenced in Appendix 1, and specific data on the products manufactured may be found in the following chapters:

Chapter 9 General gas–liquid heat recovery equipment
Chapter 11 Heat recovery equipment (including gas–liquid), for use on prime movers
Chapter 12 Heat recovery equipment (including gas–liquid) for use on incinerators

2.2 Waste heat boilers

2.2.1 *General description*

Waste heat boilers are boilers which use heat that is produced without firing the boiler itself. In the case of direct-fired boilers, the generation of heat for the production of steam or hot water is the primary object, and takes place in the boiler itself by burning solid, liquid or gaseous fuel. Waste heat boilers, however, are supplied with heat generated externally, in the form of hot exhaust produced during chemical or metallurgical processes (or others), or from the exhaust of prime movers and incinerators. As a rule chemical reactions do not occur in waste heat boilers.

Waste heat boilers with supplementary firing are exceptions. The purposes of supplementary firing can be for capacity increase of the boiler or compensation of load variations due to changes of the chemical or metallurgical processes or continuance of steam production if the preceding process should fail.

These supplementary-fired boilers belong to the waste heat boiler group as they have little in common with direct-fired boilers. Problems of waste heat boilers are of a much greater variety than is the case with direct-fired boilers. Dust loads, hazard of erosion, proneness to fouling, possibilities of cleaning, resistance times, hazard of corrosion and adaptation to the preceding process, to name but a few, are all design and construction criteria of waste heat boilers.

Considering the nature of the gas entering the waste heat boiler, there are three classes of processes to be distinguished requiring different applications of waste heat boilers:

(1) The waste heat gas undergoes no further processing and is led off into the atmosphere. The continuance of the process requires no cooling. Boilers behind copper-melting furnaces, copper-refining furnaces, openhearth furnaces, glass-melting furnaces and gas turbines afford some examples. The purpose of the waste heat boilers is to reduce waste heat losses, which, in turn, increases the operational economy of the whole plant. (Cooling, however, may become necessary in this case if government regulations require the temperature of the gas escaping into the atmosphere to be brought below a maximum permissible level).

(2) The waste heat gas undergoes further processing. The object of the boiler is to cool down the gas until its temperature makes it suitable for the subsequent processes. Some examples are roasting plants for pyrite and zinc blende, sulphur-combustion plants, hydrogen sulphide combustion plants and residual-acid cracking plants.

(3) The waste heat gas undergoes further processing. Its cooling should be performed within a given period. In this case, the boiler's purpose is also to secure the stability of the preceding process and to prevent further chemical reactions. The best examples of this are the cracked-gas coolers of ethylene plants.

In each case, the heat obtained as a byproduct of the processes is transformed into steam. The steam produced can be used for the generation of electricity for heating, drying and driving machinery, which contributes to eliminate additional sources of energy. Thus waste heat boilers increase the efficiency of total plants, and so contribute to more rapid investment return on such plants. Especially in these days, where major investments are readily accepted provided that greater economy can be achieved, the exploitation of waste heat by waste heat boilers has assumed great importance.

The use of waste heat in exhaust gas streams to generate steam has a number of advantages compared to, say, gas-to-gas heat recovery, and these are listed below:

(i) As the boiling process involves very high rates of heat transfer in terms of heat flux per unit area, the waste heat boiler is one of the most compact types of heat recovery unit.
(ii) Installations are in most instances lower on capital cost than other heat recovery systems of comparable duty.
(iii) Waste heat boilers can in general withstand high-temperature exhausts without incurring problems of materials selection and life because the high heat transfer coefficients maintain the tubes at a comparatively low temperature, close to that of the fluid being boiled.
(iv) Response rate is high, and the duty may be conveniently varied by adjusting operating pressure on the steam-side.
(v) Precise control of gas and water flow rates is not demanded.

The advantages above must be weighed against some of the drawbacks of steam-raising. Assuming that a use exists in the plant for high-grade steam, the need for good-quality boiler feedwater implies the use of some water-treatment process, which must be accounted for in capital and operating costs. It may also be found that the waste heat boiler alone will not recover sufficient heat to reduce the exhaust-gas temperature as much as desired, and additional heat recovery equipment may be needed downstream of the boiler.

There are a number of different types of waste heat boiler, each of which can be used singly or in conjunction with the other types. Typified by the range of one of the major waste heat boiler manufacturers, the four major systems are natural- and forced-circulation boilers, which may be either water tube or shell and tube designs.

Natural circulation

The prime requirement of any boiler, particularly one used for waste heat recovery, is that it should be reliable and require the minimum amount of attention. Natural-circulation boilers are best employed when large changes of density occur as the result

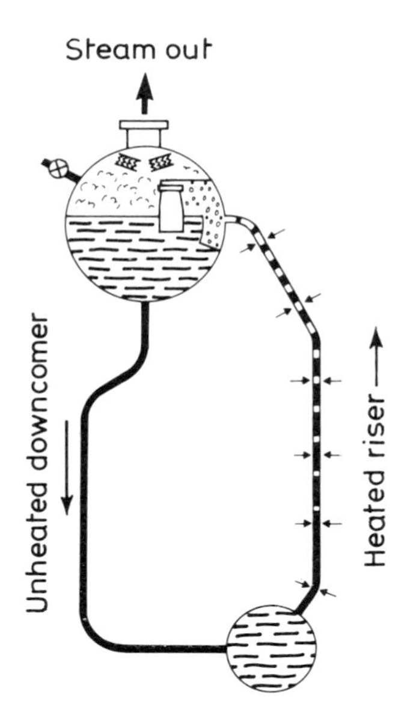

Fig. 2.1 Natural circulation boiler.

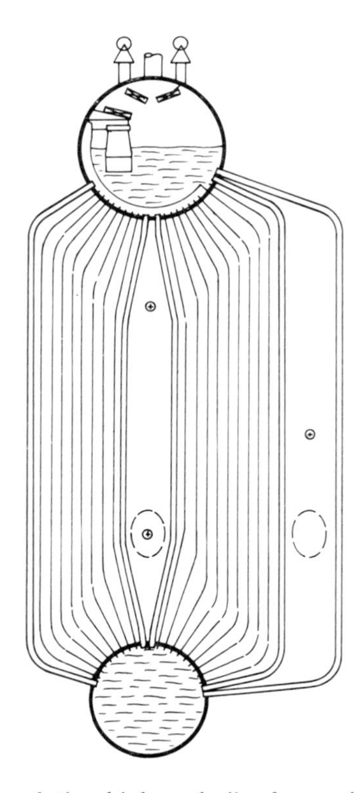

Fig. 2.2 Natural circulation bi-drum boiler for use in dust-laiden exhausts.

of heat absorption, as illustrated in Fig. 2.1. Under varying load, the density differential between the water in the downcomer and the steam/water mixture in the riser creates a simple self-adjusting circulation. Apart from responding to the rapid process changes, a natural-circulation boiler can accommodate a gradual change of load resulting from higher or lower gas conditions. The strength, simplicity and ruggedness of a natural-circulation boiler owing to the large-diameter, thickwalled tubes has made it popular and successful in many applications.

Figure 2.2 illustrates a typical natural-circulation design having excellent dust-shedding capability. In this unit, internal baffles are included in the upper steam drum, and cyclones and steam separators give high steam purity. Sootblower locations are indicated on the figure, with manhole access doors for inspection and tube replacement.

Natural-circulation boilers are independent of any external energy source to provide circulation under all conditions, and the flow, as mentioned above, is self-regulating.

Forced circulation

In general forced circulation increases the rate of heat transmission on the water-side enabling smaller diameter and lighter tubes to be assembled in the same envelope as a natural-circulation boiler, as shown in Fig. 2.3.

Forced-circulation designs are, therefore, popular for high-heat flux application and give greater geometrical freedom. This flexibility is important where weight and wind loading are major considerations and in clean gas conditions where performance can be

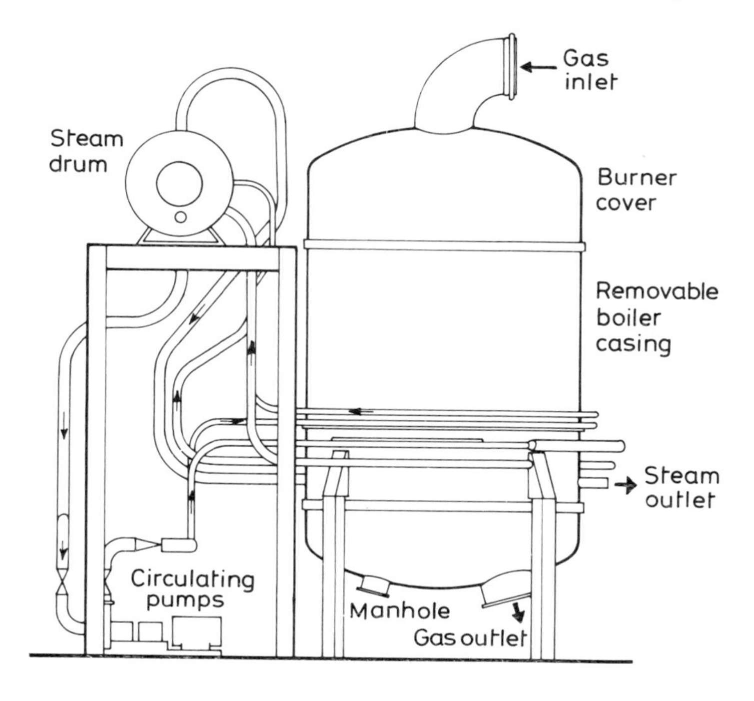

Fig. 2.3 Forced circulation heat recovery boiler.

enhanced by using extended surface.

As the smaller-diameter tubes of a forced-circulation boiler can accept higher stresses than a larger tube, normal carbon steel can be used more widely with a greater factor of safety and a cost benefit. The smaller tube is easier to manipulate and weld onsite, and because a bank of them present a more plane surface, they will shed slag deposits more easily and facilitate the application of the wall backing and insulation. The smaller water quantity circulating not only reduces the weight, but reduces the startup time, and cooling time for routine inspections, which helps reduce maintenance. The unit illustrated in Fig. 2.3 is supplied as a complete 'package', including steam drum, circulation pumps, piping and valves.

Natural-circulation shell and tube boilers

Shell and tube boilers are attractive because of the simple gas flow pattern and ease of installation with their compact, low profile. These boilers are usually of the natural-circulation type with a steam drum able to absorb the variation in heat load during startup without the forced circulation of unnecessarily large quantities of cooling water. Depending on the gas characteristics, the inlet cones and tube sheet can be lined with refractory to protect the metal, and the tube inlets protected by ceramic or metallic ferrules.

Internal bypass may be incorporated in these boilers to permit small and varying quantities of high-temperature gas to be introduced during startup conditions. A natural-circulation shell and tube boiler for use on reformed gas plant is illustrated in Fig. 2.4. This comprises two shell and tube boilers in series and one shift converter boiler with a common steam drum mounted above. An internal bypass is incorporated to control the gas outlet temperature at low load.

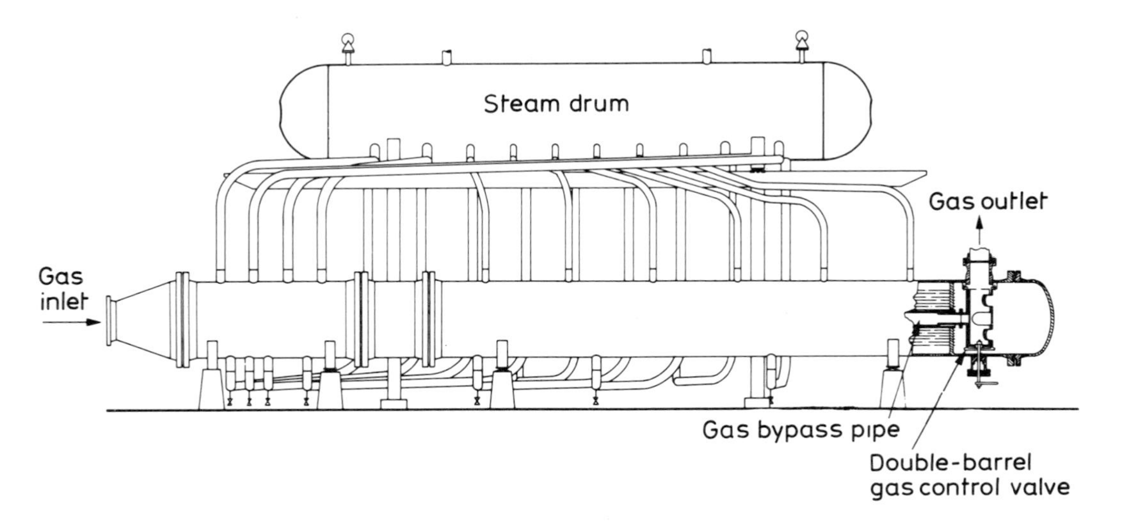

Fig. 2.4 Natural circulation reformed gas boiler comprising two shell and tube boilers in series.

Waste heat boilers can be economically used for very low duties (e.g. 500 kg to 1000 kg steam per hour), and there is no real upper limit to size. The shell and tube boiler is generally selected, depending upon the manufacturer's recommendations, for ratings of up to 20 000 kg steam per hour, and for operating pressures of up to about 20 bar. Water tube boilers are suitable for much larger ratings, both in terms of steam-raising capability and operating pressure, pressures of 60 bar being readily achievable. Other relative merits of shell and tube and water tube boilers have been well summarized by Fanaritis and Streich (1973) (see below).

Guidelines for shell and tube units

(1) Elevated-temperature streams being cooled may be liquid or gas.
(2) Adaptable to cooling of elevated-temperature gas streams operating under pressure. Cooling of streams at pressures of 30 bar and higher is not uncommon.
(3) Most efficient when process-gas streams having a reasonable heat-transfer film coefficient are being cooled.
(4) Generally limited to process-stream flow rates that can be handled by shop-assembled units.
(5) Can handle clean or highly fouling elevated-temperature streams. Generally less susceptible to fouling and easier to clean on the high-temperature side than a water tube design.
(6) A fire tube unit will generally be less expensive in applications where either fire tube or water tube design may be used.
(7) In high-temperature service (500°C to 1000°C inlet-gas temperatures), the inlet or hot tube sheet of a fire tube unit is highly vulnerable.
(8) Generally requires a higher pressure drop on the high-temperature stream side than a water tube design.
(9) Natural-circulation and forced-circulation designs are available.

Guidelines for water tube units

(1) Can be designed for any steam pressure, including supercritical.
(2) Wider range of designs is available than in fire tube units, including the use of extended-surface tubes as well as bare tubes.
(3) Ability to use extended heat transfer surface makes this design more efficient than a fire tube unit when cooling gas streams that have poor heat transfer characteristics or low allowable pressure drop.
(4) Designs are available for handling low-pressure, elevated-temperature gas streams at much higher flow rates, since water tube designs may be field-assembled in large sizes.
(5) Water tube designs do not lend themselves to cooling elevated-temperature gas streams with highly fouling characteristics. Even high-density, bare tube designs do not lend themselves to efficient sootblowing.
(6) Water tube designs lend themselves more readily to supplementary firing, where such additional heat input is desired, than fire tube units.

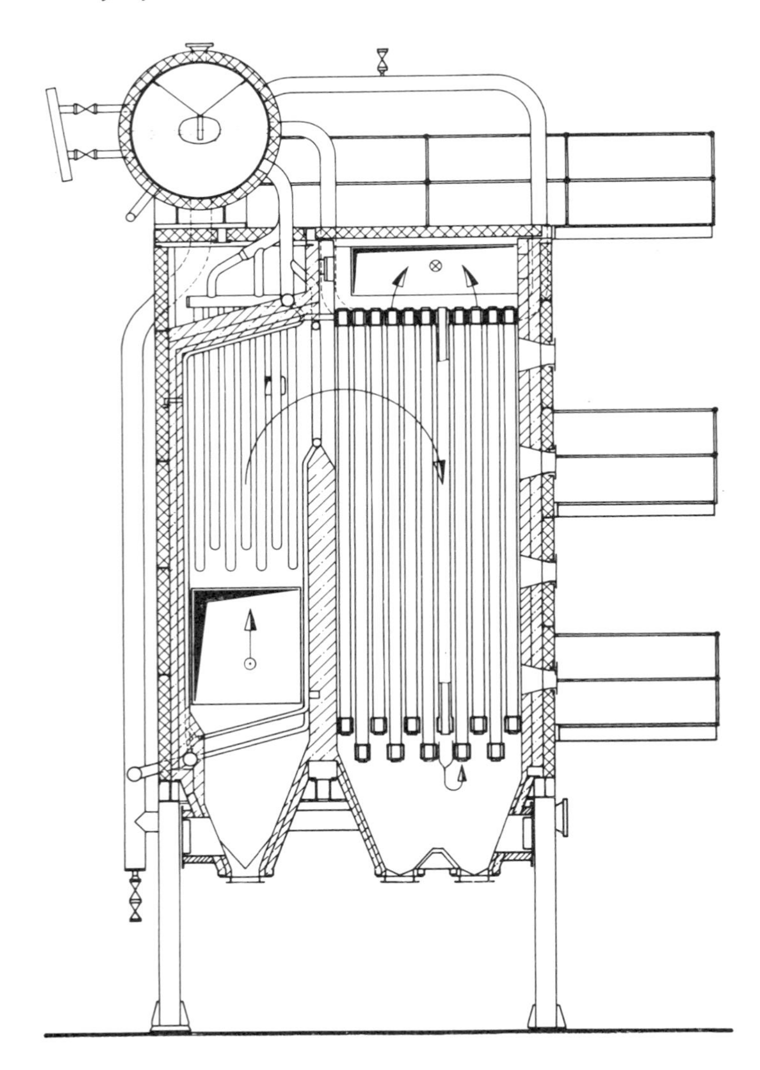

Fig. 2.5 Waste heat boiler behind anode furnace.

(7) Water tube units can be designed for low friction losses on the high-temperature side, while handling high flue-gas flows, and still provide efficient heat transfer.
(8) Water tube units are generally less susceptible to mechanical failure as a result of malfunction than fire tube units.
(9) Natural-circulation and forced-circulation designs are available.
(10) Certain water tube designs may incorporate steam superheater and other auxiliary service coils beyond the primary function of steam generation. Included in a single housing can be superheater, generator and economizer coils to provide a high-efficiency heat recovery unit.

An example of an application of a waste heat boiler in a copper works, in this case behind an anode furnace, illustrates how a waste heat boiler can be linked with other heat recovery components such as superheaters, economizers, air preheaters and the use of supplementary firing. It also permits elaboration of some cleaning methods. Fig. 2.5 shows a waste heat boiler in this application, the unit being part of a system constructed by Schmidtsche Heissdampf-Gesellschaft mbH. From the furnace the gas passes through a brick-lined duct into the radiant chamber of the boiler. The chamber is lined with steel tubes welded to headers at either side. The space between the tubes and the outer sheeting is brick-linked. Since the boiler is continuously run at low vacuum, there is no need for it to be gastight. The radiant chamber is necessary for different reasons. The great variety of impurities entrained in the gas would cause a convection heating surface to foul and become restricted very quickly, since most impurities leaving the furnace in liquid form solidify in the boiler. On the other hand, a radiant chamber is also preferable to a contact heating surface for economical reasons. Efficient heat dissipation by radiation is ensured due to the gas-inlet temperature varying between 1100°C and 1400°C, the H_2O content approximating 6 to 10 per cent and the CO_2 share varying between 6 and 16 per cent. Furthermore, the radiation heating surface is highly suited for absorbing the temperature peaks occuring. These temperature peaks arise as a result of the different stages of the metallurgical process for copper extraction.

Temperature peaks and gas quantity peaks rarely coincide so that the variations are considerably lessened at the radiant chamber outlet, where on an average the gas temperature should not be in excess of 1000°C. This temperature permits further heat transfer in the convection section without introducing the hazard of faulty operation due to solidified impurities being deposited on the tubes – provided, however, that the heating surfaces are cleaned periodically. The surface consists of double-tube registers. In the upper part, at the side, the gas enters the space between the outer tubes. Flowing along the outer tubes from top to bottom, it returns to the dust hopper. Then, entering the inner tubes from the bottom, it is collected above these inner tubes in the outlet chamber. The upper headers of the double-tube registers are welded to each other by sealing angles preventing the gas from passing directly from the radiant chamber into the gas-outlet chamber.

All evaporators operate on a common steam drum under natural circulation. In order to supply the heating surfaces with water from the steam drum, it is advisable to arrange a ring piping at the bottom end of the boiler, to which downcomers as well as coils can be connected. Separate risers have to be installed, in order to connect the different evaporators to the drum. Any joining of tubes, which might produce faulty water circulation, should be avoided. Special cooling devices at the furnace, such as cooled door frames, can be connected to the boiler drum without difficulty, using natural circulation or forced circulation.

Cleaning requirements in boilers of this type vary, but can be quite considerable. The radiant chamber need not, as a rule, be cleaned during operation. Nevertheless, it is advisable to provide for cleaning ports for removing, by blowing or with the aid of bars during the furnace shutdown, any heavy accumulation of solid slag on the walls from outside. The outer tubes are cleaned during operation with the aid of handlances, using compressed air or steam as a blowing medium. For this purpose, there are blow-off openings at two adjacent walls of the double-tube section. These openings are arranged in front of all lanes between the double-tube rows. The vertical clearance between two openings is approximately 1·5 m.

The inner tubes of the double-tube system can be cleaned by a band cleaner or a series of driven rods. In the case of band cleaning, a steel band is permanently suspended in each inner tube throughout its full length. Each band is connected to a vertical bearing above the tube via a hook, and when the bearing is driven by an electric motor, the bands scrape and strike the tube surface, cleaning them. Automatic timing of the operation of these devices may be used.

A second cleaning system, sometimes known as the 'four-fold' cleaning device, consists of four driven rods, each end of which carries a cutter head with three cutters of hard-wearing material. Above each tube there is a cover in the boiler roof. For cleaning, four covers open at a time, and the four rods are introduced simultaneously through the boiler roof into the tubes. When the cutter heads reach the tube-end, the rods start rotating. At the same time, the entire device is lifted at very slow speed in such a manner that the scraping cutters, pushed against the tube wall under the effect of centrifugal force, scrape the tubes clean during their upward travel. This device requires an operator's services. Whereas the band cleaning method is recommended where the bands are not exposed to corrosion and deposits can be removed by striking, the four-fold device is decidedly preferable in corrosive atmosphere and with glutinous and greasy deposits. Such a boiler may also be equipped with superheaters, feedwater preheaters, air preheaters and supplementary firing.

Superheating

The boiler represented in Fig. 2.5 produces saturated steam. Its feedwater is pumped direct into the drum by the feedpumps. For the production of superheated steam, the radiant chamber and the double-tube part are separated. A tubular bundle superheater is suspended in the connecting duct from the top. Its design requires in any case the consideration of an adequate control range and the provision of intermediate cooling, otherwise the temperature of the superheated steam is hard to control at the different gas loads. This is essential if the steam is used for driving turbines and if great value is set on the availability of superheated steam at constant temperature.

The desuperheater is generally of the surface-cooler type since the fully demineralized water required for an injection cooler is rarely available in metallurgical works. The superheater is of the two-stage type. In the first stage, approximately two-thirds of the heat including the control range is transferred, in the final stage approximately one-third. This provides ample temperature difference for desuperheating, and the steam temperature at the superheater outlet can be efficiently limited.

Feedwater preheating

On the boiler shown, the gas-outlet temperature is approximately 80°C above the temperature of the saturated steam. For cooling the gas still further, the provision of an economizer (see Section 2.3) is required. Its tube bundles are cleaned in the same way as the outer tubes of the double-tube system. Note that water-inlet temperature must exceed the gas dew point, it should reach approximately 160°C. Consequently it is best to preheat the water by other means, preferably by tube coils placed in the drum. Steam heating is also possible in a separate preheater.

Air preheating

For saving fuel on the furnace burners at the expense of steam production, it is possible to connect at the outlet side a furnace combustion air preheater, replacing in whole or in part the convection evaporator. Such a preheater consists of vertical tubes typical of types described in Chapter 1. The waste heat gas is led through the tubes, the air circulating in cross-flow around the tubes in multipass. Dew point and maximum tube wall temperature should be taken into consideration for the design. The air preheater is cleaned in the same way as the inner tubes of the double-tube system.

Supplementary firing

Boilers behind copper furnaces and, of course, in many other applications are highly suited for being equipped with supplementary firing. The additional expenses are extremely low, a radiant chamber being available anyhow. The oil or gas burner is installed in a side wall. For this purpose, the wall is bent in such a manner that causes the flame to point obliquely downwards for mixing the waste heat gases with the burner waste gas. The burner is simply connected to the fuel feed system of the furnace burners. If the boiler is equipped with an air preheater, it is advisable also to preheat the combustion air of the supplementary burner. The load variations of the waste heat system due to the process can be fully compensated for by the supplementary burner. The steam quantity to be maintained constant is measured by means of an orifice which is connected to an automatic fuel controller. It is advisable to allow the burner to operate at minimum load, even during maximum waste heat operation, as frequent cutting in and off can be detrimental.

The determination of the burner output should be based on the assumption that throughout the whole process the steam quantity produced reaches a level corresponding to the maximum waste heat quantity plus minimum burner output, or that alternatively full steaming capacity is reached also with the furnace shut down.

A further example of a fully integrated system incorporating several heat recovery devices centred around a waste heat boiler is shown in Fig. 2.6. Constructed by Fives-Cail Babcock in France, this unit cools flue gases resulting from the heating of petrochemical reforming furnaces from 1000°C to 200°C. Recovered heat in a 1000 tonne per day ammonia plant, using such a system, can reach 40×10^6 kcal/h, operating without auxiliary burners.

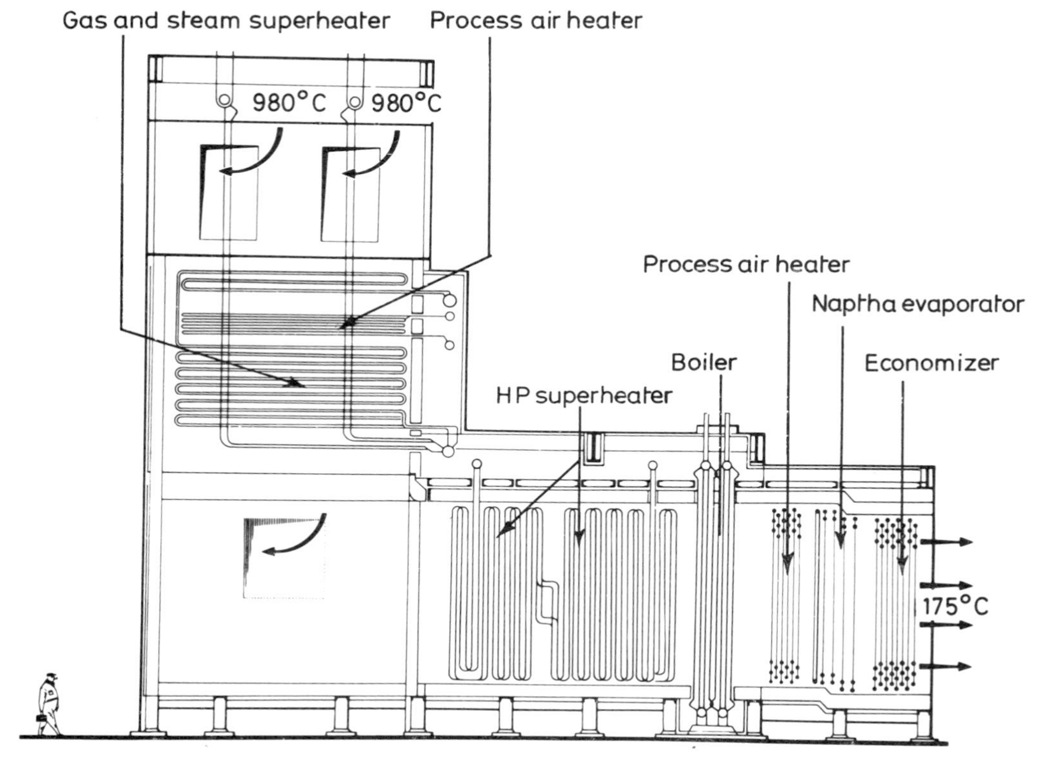

Fig. 2.6 A waste heat recovery boiler and ancillary heat recovery equipment, installed by Fives-Cail Babcock on an ammonia plant (AGQ-Plant at Grand-Quevilly, France, 1000 tonne/day ammonia).

2.2.2 *Applications and economics*

Some applications of waste heat boilers have already been mentioned, and one has been discussed in some detail as it affects the boiler design and operation. Broad application areas, in terms of the type of cooling duty required in a process, and the uses to which the steam might be put, have been given in Section 2.2.1. Plant on which waste heat boilers may be applied include many chemical and metallurgical processes, kilns, furnaces, internal-combustion engines and breweries and distilleries (case histories of which are given later).

One of the most interesting application areas, discussed also in Chapter 5, is the recovery of heat from internal-combustion engines. One example where this heat is subsequently used in the form of steam to drive a turbogenerator is discussed below.

The Seajoule waste heat conversion 'package'

'Seajoule', manufactured by Hawthorn Leslie (Engineers) Ltd. is based on systems developed for marine use, but is now equally applicable to land-based applications.

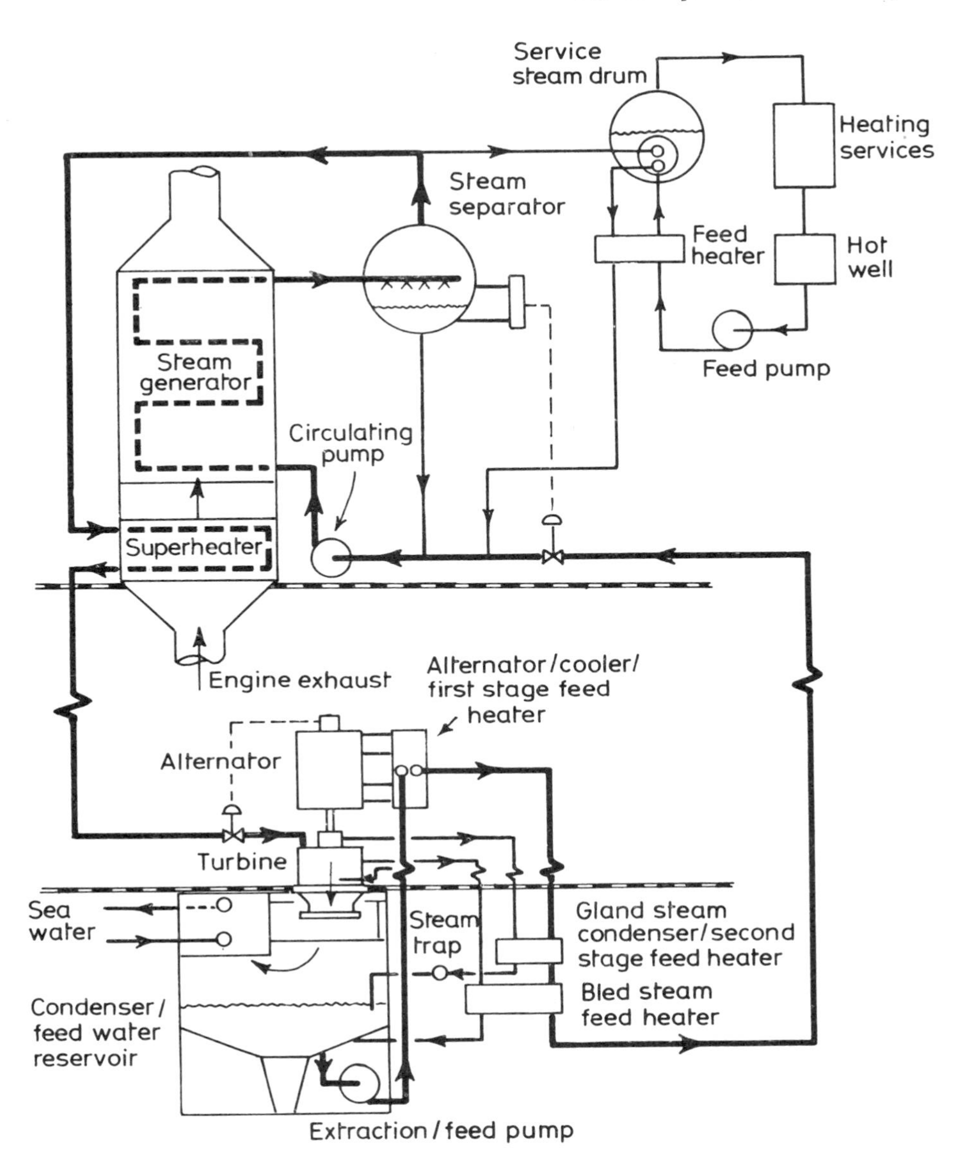

Fig. 2.7 Flow diagram of the Seajoule concept incorporating a waste heat boiler and turbogenerator.

The basic function of the system, a flow diagram for which is given in Fig. 2.7, is to recover heat from diesel engine or gas turbine exhausts to generate low-pressure steam. A portion of all of this steam may then be used to generate electrical energy. When installed in a power station, the 'Seajoule' increases the maximum power output of the station. In a marine installation, the 'Seajoule' output may substantially exceed the normal electric load. In such cases equipment can be provided for turning the surplus electrical load into additional propulsive power. Supplementary oil firing can also be provided should power be required when the main engine is not operating.

The waste heat boiler, A in Fig. 2.7, extracts approximately half the heat in the exhaust gas to generate superheated steam. The operating pressure is normally chosen so that the maximum quantity of electricity can be produced. The temperature of the boiling water at this pressure is sufficiently high to ensure that no condensation of combustion products can take place, permitting mild steel to be used in construction of the heat exchanger.

The superheated steam is used to drive a turboalternator in which it expands down to a vacuum condition corresponding to the temperature of the coldest cooling medium available. A direct-drive, vertical-axis turbine has been adopted to minimize the plan area of the plant, eliminate gearing and flexible couplings and to achieve the highest possible efficiencies without having recourse to highly stressed, tapered and twisted blading. The hollow turbine rotor is made up of separate stainless-iron discs in which the blade profiles are formed by an electrochemical process, a manufacturing technique which results in a rotor of the highest possible standard.

To avoid the complication and space requirements of an exhaust-end gland, the combined turbine journal and thrust bearing is water-lubricated. At the upper end the turbine shaft is solidly coupled to the alternator shaft which runs in grease-lubricated roller bearings. The absence of oil-lubricated bearings eliminates the need for a lubricating oil system. As speed-governing is effected by electronic means, the turboalternator has only one moving part.

The turboalternator can be installed directly on the condensing module or, if more convenient, remote, in which case the two units are interconnected by an eduction pipe about 1 m in diameter. Exhaust steam is led to the underside of the condenser tube stack, thereby avoiding as far as possible any undercooling of the condensate. Reserve feed is stored within the enlarged condenser well, thereby avoiding the use of a separate tank and minimizing the risk of contamination by oxygen and other undesirable substances.

For a unit employing a considerable amount of capital equipment, the payback period is comparatively short. In the case of power generation schemes, the calculation is relatively simple. The net investment is the cost of the basic 'Seajoule' modules less the cost of the additional prime generating plant that would otherwise be required (due account being taken of reduced service ratings). This net cost will probably vary between £100 and £250 per kilowatt, depending on whether the prime generating plant consists of heavy-duty, base-load diesel equipment suitable for burning high-viscosity fuel at one extreme, and aero-type gas turbines burning gas oil at the other.

The tabulation (Table 2.1) gives fuel and lubricant costs, discounted at 10 per cent, for two notional generating plants. Plant A relates to heavy-duty diesel engines burning high-viscosity fuel at a cost of £46 per tonne and running for 5200 h/year on base load. Lubricating oil is assumed to cost 5 per cent of the fuel cost and the specific consumption of the plant is taken to be 220 g/kWh. Plant B relates to gas turbines burning fuel at £76 per tonne with a specific consumption of 300 g/kWh running 50 per cent total time.

Table 2.1 Accumulated fuel and lubricant costs in £-Sterling per kilowatt generated

	1	2	3	4	5 (years)
Plant A	55	104	148	189	225
Plant B	100	190	270	344	410

It will thus be seen that, even ignoring lower maintenance costs, the payback period will be under three years, even with modest utilization. Another way of looking at the earning potential of a 'Seajoule' unit as an adjunct of an existing generating station is to evaluate the selling price of 'units' sent out. At 1·5 p per unit, the annual net revenue of a 'Seajoule' 1000 unit running continuously at full load would be £122 286.

The marine case is complicated by the fact that the optional 'Seajoule' modules should be taken into account, but in no case is the payback period likely to exceed three years unless the time spent in harbour exceeds 100 days per annum. Of course, when export credit facilities are applicable, the annual savings in operating costs completely offset the annual repayments, leaving the operator with a substantial profit.

The heat in the exhaust of a diesel is typically equivalent to 90 per cent of its mechanical power output – in a gas turbine it can be as high as 400 per cent. For practical and commercial reasons, it is only possible to extract about half of this heat, and for thermodynamic reasons it is impossible to convert more than about one-fifth of such low-grade heat into mechanical work.

Thus, in the case of a diesel engine, the maximum theoretical recovery is equivalent to about 18 per cent of its own power output, but no practical plant can expect to approach this theoretical performance and no claim is made for the 'Seajoule' that it represents the last word in energy recovery. What is claimed is that it does present an extremely attractive economic case, recovering about half the theoretical possible. Table 2.2 gives the recovery that can be guaranteed from a representative selection of prime movers.

Waste heat boilers in distilleries

Two case studies presented in the Department of Energy newspaper in the UK (1978) relate to the use of waste heat for steam-raising in distilleries. The first of these, at William Teachers & Sons Ltd. is saving approximately £57 000 per annum. By far the major part of these savings (£49 000 per annum) is directly attributed to using the heat extracted from the hot flue gases to raise steam – previously steam was generated by an oil-fired boiler (now virtually redundant). A further £8000 a year is being saved in reduced coal burn – effected by introducing a system for controlling the air draught to the fires.

The basis of the scheme is that exhaust gases from the 8 coal-fired stills are fed into 2 ducts. Extraction is provided by natural draught from a chimney.

The gases at 555°C are extracted from the main ducts and passed through a cantilevered waste heat boiler, situated immediately over the flues to reduce resistance, where they give up 50 per cent of their heat to produce steam. The gases, now at 250°C, are then

Table 2.2 Energy recovery by the 'Seajoule' system for various prime movers

Prime mover type	Service output		Exhaust at service power		Gross output from 'Seajoule' (kW)	Gross output from 'Seajoule' as a percentage of prime mover output
	As mechanical power (BHP)	As electricity (MW)	Mass flow (kg/s)	Temperature (°C)		
Slow-speed, 2-stroke loop scavenge diesel	30 000	21·5	60·0	295	1350	6·0
Slow-speed, 2-stroke uniflow scavenge diesel	20 000	14·4	40·06	340	1285	8·6
Medium-speed, 4-stroke diesel	15 000	10·7	24·2	385	990	8·8
Medium-speed, 4-stroke diesel	8 000	5·7	13·5	415	635	10·6
High-efficiency industrial-type gas turbine	18 000	12·7	56·0	520	3830	28·5
Aero-type gas turbine	8 500	6·0	33·3	475	1970	31·0

fed through a vortex grit arrester – to remove grit – and hence via an induced-draught fan having a head of 200 mm w.g. and which can be modulated automatically by a damper, to the chimney.

The scheme is based on a waste-gas mass flow rate of 32 600 kg/h, an inlet temperature of 555°C, and an exit temperature of 250°C. This gives an average output of 1·35 MW and a maximum output of 2·4 MW.

The cost of fuel for producing 1·35 MW in a residual oil-fired boiler – which this waste gas boiler has replaced – works out at £9.90 per hour. Allowing for additional electricity to drive the draught fans, costing £1 per hour, the net savings are £8.90 per hour. Over an operational year of 5500 h the savings are, therefore, £48 950.

The effect of having a controlled, rather than random, draught on the coal fires is an imrovement in combustion efficiency. Currently savings in coal, directly attributable to this feature, are running at 230 tonnes per year – worth £8050 per year.

At present the steam from the waste heat boiler is used for the mashing process in whisky production, for cleaning, and for drying. It is proposed in future to use spare steam capacity to drive a single-stage 80 kW turbine, saving a further £10 000 per annum.

A second installation, also engineered by Integrated Energy Systems, is at the William Grant plant. There are two distilleries adjacent to one another, the Glenfiddich and Balvenie distilleries. A waste heat boiler generating slightly over 2000 kg/h steam was installed in 1976, and in 1977 it was decided to install a second unit, using waste heat from the stills, as in the case of Teachers, which were butane-fired.

It was recognized that the combined outputs of this boiler and the existing waste heat boiler would produce, on average, more steam than required by Glenfiddich Distillery and the decision was taken to pipe this steam to the neighbouring Balvenie Distillery, a distance of 700 m. The average steam sent to Balvenie is 6800 kg/h.

Steam is used at Glenfiddich for: heating mashing water; cleaning mashing vessels and pipelines; cooperage (shaping timber for barrel-making); space heating in the bottling hall and cask shed; and at Balvenie, for: directly heating some of the stills; heating mashing water; and cleaning mashing vessels and pipelines.

Initial designs showed that this pipeline must transport waste heat steam to Balvenie from Glenfiddich and also be capable of bringing steam from Balvenie to Glenfiddich at certain points of the production cycle. This was achieved with loop and non-return valeves, the steam flow in both directions being metered to provide accurate fuel costings for both distilleries. Heat is recovered from two main flues by pulling the hot gases at 350°C through extended-surface water tube heat exchangers by induced-draught fans. Pressurized hot water is circulated through the heat exchangers from a remote steam drum positioned in the boiler house some 20 m away.

The advantages of the system are that no civil-engineering work was necessary, intrusion on the stillhouse is minimal and the hot-gas ductwork is short. Additionally, using an extended-surface heat exchanger allows feedwater to be preheated so achieving optimum heat recovery. The output of the waste heat boiler is cyclic, depending on the distillation process, and in practice averages some 1150 kg/h. The total output of steam from the waste heat boilers is 3200 kg/h, and operational experience over the first four months of 1978 has shown that annual savings will be £180 000 for a capital investment of £230 000.

2.3 Economizers

2.3.1 *General description*

An economizer is commonly regarded as a waste heat recovery unit unique to boilers, although now its application has spread to other processes where hot flue gas could usefully be harnessed to heat water or raise steam. The majority of economizers have in the past been constructed for boiler applications, where they are used to preheat the boiler feedwater using the flue gases. In a boiler the heat transfer from the hot gas to the boiler tubes is comparatively poor, and as a result there is a very large temperature difference between the boiler shell and tube sides. This necessitates high flue-gas temperatures, and by installing an economizer in the flue, some of the heat in this gas can be recovered. Where one conomizer is insufficient to recover all the useful heat, an additional unit, or a subeconomizer, may be incorporated downstream of the main unit.

Economizers are normally constructed of steel, although originally cast iron was used as tube material. Most economizer tubes are finned, the steel fins being welded to the tubes. Their construction is much more substantial than on many other types of recuperative heat exchangers; tubes having outside diameters of 50 mm with approximately 1 m^2 of extended surface per metre length weighing over 10 kg/m. Economizers used on large power stations can contain 100 000 m of tube, and are able to operate in conditions where flue gas temperatures approach 900°C, although in large power plant the temperatures are considerably lower than this.

The two main points to be borne in mind when using economizers are associated with the highest temperatures met on the tube-side, and the lowest encountered on the gas-side. 'Water hammer' can occur at the hot end of the economizer, created by the formation of steam pockets. Alternatively, if too much heat is removed from the flue gas, and condensation on the gas-side occurs, this can lead to severe corrosion, which is more detrimental to the steel units than older cast-iron economizers. It can lead to uneconomic operation if these restrictions are permitted to dictate the operating conditions of the economizer, and it is worth remembering that a 6°C rise in boiler feedwater temperature can save 1 per cent on fuel consumption for the same boiler output. Alternatively, it may be stated that an increase in water temperature of 1°C via the economizer lowers the flue gas temperature by 2°C (Lyall, 1956).

Techniques for improving economizer performance, and at the same time overcoming the above problems, are illustrated in Figs. 2.8 and 2.9. When the economizer was raising the feedwater to too high a temperature, the feedpump was kept at full capacity and some of the hot water from the economizer was allowed to flash into a 34 kN/m^2

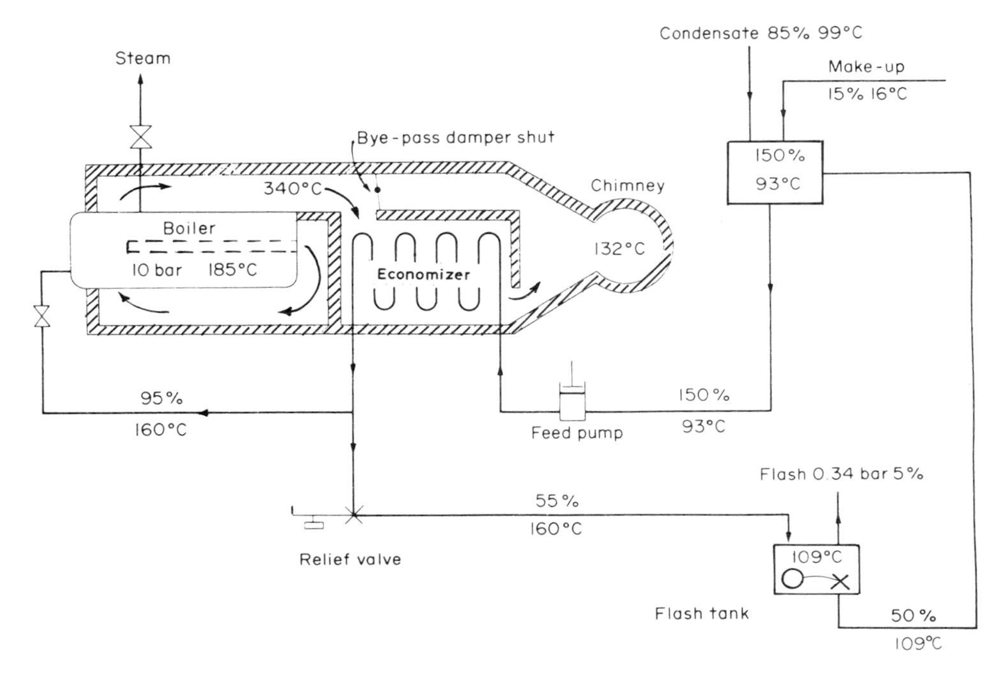

Fig. 2.8 An economizer installation with control of maximum feedwater temperature (by courtesy of HMSO).

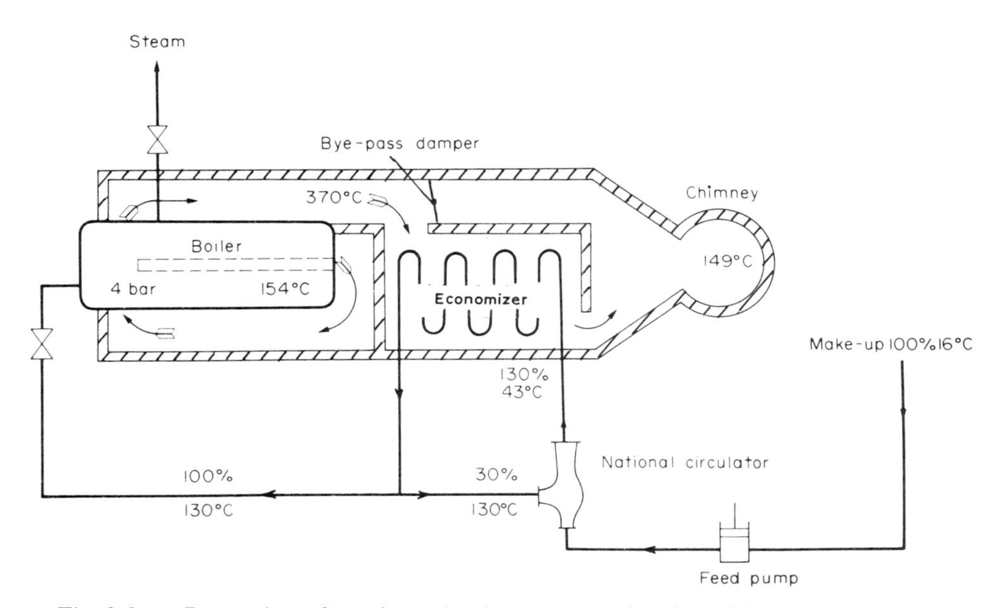

Fig. 2.9 Prevention of condensation in an economizer by raising the operating temperature (by courtesy of HMSO).

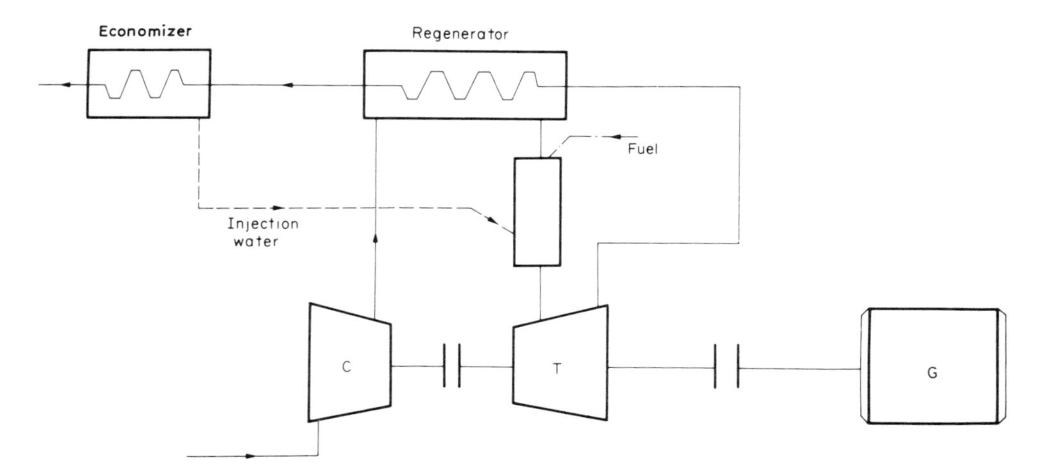

Fig. 2.10 Use of an economizer on a gas turbine to preheat injection water.

process main (Fig. 2.8). A relief valve at the economizer outlet controlled the flow to the flash tank, where the high-temperature water was immediately flashed down to a lower temperature, resulting in about 10 per cent evaporation. Approximately 5 per cent of the total steam requirement was generated in this way, and additional economies of 15 per cent resulted from return of extra condensate (Bunton, 1974).

Where condensation was occurring on the economizer tubes, a proportion of the output from the economizer was returned to the inlet to preheat the makeup before it entered the economizer. With a faster circulation rate, the stack temperature was reduced sufficiently to effect an additional 3·5 per cent recovery without a return to conditions leading to condensation.

In large power-generating plant, economizers operate at high temperatures and pressures, heating feedwater to temperature of 200°C to 300°C. However, in smaller boiler plant, where normal feedwater temperatures of only 40°C or thereabouts may be common, substantial economies can still be obtained using these heat exchangers. Take as an example a boiler operating at 690 kN/m^2 without a superheater, using feedwater at an inlet temperature of 49°C. The total heat in the steam above that of the feedwater will be 2558 kJ/kg. An economizer installed in this boiler, heating the feedwater from 49°C to 127°C, would add approximately 424 kJ/kg, resulting in a saving in fuel of the order of 12·5 per cent (Lock, 1972).

As mentioned in the introduction to this section, the economizer, while normally associated with boiler heat recovery, can be used in any suitable exhaust gas stream as a process water heater, superheater or heater for liquids other than water, including heat transfer fluids. Economizers are being increasingly used in waste incineration plant, and furnace flue gases are used to generate process heat via economizers.

In gas turbines, the injection of water leads to a very significant reduction in the amount of air to be compressed, with a resulting reduction in its sensitivity to pressure-ratio changes. The water also controls the maximum temperature, and is in effect part of the working fluid. By using an economizer on a gas turbine, as shown in Fig. 2.10, this injection water can be preheated, and the economizer remains effective at lower temperatures than the gas-to-gas regenerators (Wood, 1969).

2.3.2 *Case history*

To illustrate the potential savings obtained using an economizer system available for retrofitting to industrial boilers, the 'Airaqua' unit manufactured by Beltran & Cooper has been selected.

The 'Airaqua' economizer was designed to offer high-efficiency heat recovery in a compact unit which can easily be removed from the exhaust duct for inspection or cleaning. The layout of the system is shown in Fig. 2.11. In this unit spirally wound finned tubes are used to transfer heat from the exhaust air to the water, which may be heated to well above 100°C using the pressure generated by the feedwater pump.

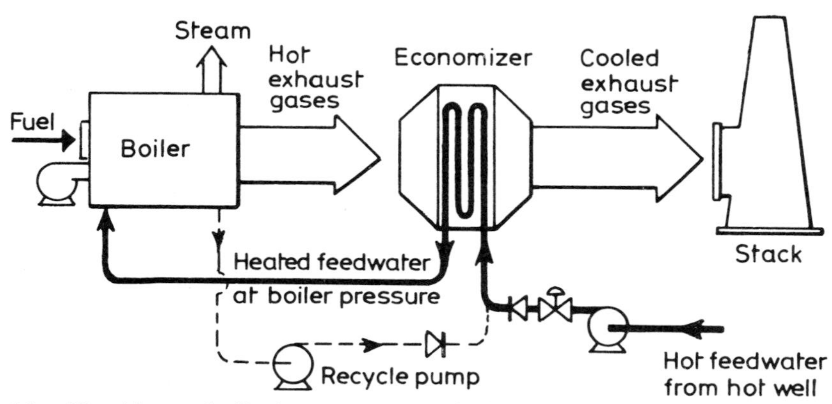

Fig. 2.11 The Airaqua boiler heat recovery unit. Section shown by broken line is only used when firing heavy fuel oil to prevent corrosion of the coil bank by sulphur oxides.

The heat exchanger can measure as little as 600 mm across flanges in the direction of gas flow, and gas-side pressure drops of 6 mm w.g. to 12 mm w.g. are typical. This obviates the need for an induced-draught fan in the exhaust duct, as the existing combustion air fan can normally cater for this slight increase in flow resistance. Similarly the water-side pressure drop is designed to be within the limits of the feedwater pump.

Instead of utilizing a bypass duct with an automatic damper to control the heat recovered, the system is controlled by mixing a varying quantity of boiler water (at the boiler saturation temperature) with the incoming feedwater, which has the effect of reducing the heat transferred in the economizer coil. The coil design is usually optimized for operation on the fuel with least sulphur (when a multifuel burner is fitted to the boiler) at 100 per cent of the boiler MCR (maximum continuous rating) with no recycle of boiler water. The recycle flow is then controlled at a suitable valve for operation on the other fuel, or at reduced steam outputs, when a reduced heat output must be tolerated to prevent corrosion of the back tubes of the coil (if the coil surface

Fig. 2.12 The tube bundle of a Beltran & Cooper economizer subsequently installed in a carpet-factory boiler flue.

temperature falls below the gas dew point).

On steam boilers which average at least 50 per cent MCR throughout the year, this system will generally be economically attractive. Typical payback times of between 12 and 24 months are achieved.

Ideally the system should be applied to boilers running at consistently high outputs; but the use of steam storage vessels can give reasonably constant operation even on traditionally unpredictable loads such as dyehouses.

The installation cost of such a system rules out its addition to boilers smaller than 5000 kg/h (rated output of steam, from and at 100°C), unless they are run at better than 80 per cent MCR. However, if a unit is installed at the same time as the boiler, the overall cost is much reduced, and the economics are thus improved.

One of the Beltran & Cooper economizers was installed in the boiler flue of a

Fig. 2.13 The Beltran & Cooper economizer located in a horizontal boiler flue. Payback is 10 months.

factory manufacturing carpets. The system fitted to the boiler, which operates mainly on natural gas with 3500 s 2 per cent sulphur heavy oil as the standby fuel, consists of a steel finned economizer as illustrated in Fig. 2.12, this showing the tubes prior to installation in ducting. The exchanger, rates at 600 kW, measures 2·5 m square and 300 mm deep in the direction of gas flow. It is installed in a horizontal flue, shown in Fig. 2.13, and precautions were taken to ensure that when the boiler is operating on sulphur fuel oil, the economizer metal temperatures are kept above the dew point to prevent back-end corrosion. With a total installed cost of £17 000, the payback period was only 10 months.

Other data appertaining to this application includes:

Exhaust gas temperature	232°C
After economizer	160°C
Exhaust flow	10·6 m^3/s
Water temperature in	71°C
Water temperature out	95°C
Water flow	5·7 kg/s
Efficiency	31%

2.4 Thermal fluid heaters

As well as using steam or high-pressure hot water for transferring heat around a plant to the various consumers, this heat may be generated in and supplied by a thermal fluid heater. These heaters operate in a similar manner to a domestic warm-water central-heating system, but instead of circulating water, a heat transfer fluid, commonly an organic with a boiling point above that of water, is used. This allows the fluid to be circulated at high temperature (organic heat transfer fluids can be used at temperatures between −50° C and 400° C) without the need for a pressurized system, as would be the case with steam. Typical of these fluids is 'Thermex', a diphenyl-diphenyl oxide eutectic having a boiling point of approximately 260° C. The main disadvantage of these fluids is the fact that they tend to degrade or 'crack' at high temperatures, generally well in excess of their boiling point. They are also susceptible to oxidation.

There are three aspects of thermal fluid heaters which are of interest in energy conservation. These are their ability to use the combustion of waste products for firing the burners, the use of these systems to heat and cool plant in the same building, and their application in heat recovery.

Thermal fluid heaters may be fired, or, as economizers, which they essentially are when using waste heat, be installed in a hot exhaust gas stream. Treated in depth by Boyen (1976), the major economic advantages of using thermal fluid heaters instead of steam or high-pressure hot water are associated with the generally lower installation and operating costs.

Discussion of the application of thermal fluid heat transfer loops associated with heat recovery from prime movers is included in Chapter 4 (see Figs. 4.13 and 4.15), and little detail will be given here. However, of the components of the system, the heat transfer element in the hot exhaust gas stream is the most critical, as it is in this region that the fluid will see the highest temperatures, and thus design here will have the strongest influence on the life of the heat transfer medium.

Fluid heaters, as with waste heat boilers, come in two types – fire tube or liquid tube. Both systems may be fired or may use waste heat sources such as exhausts from

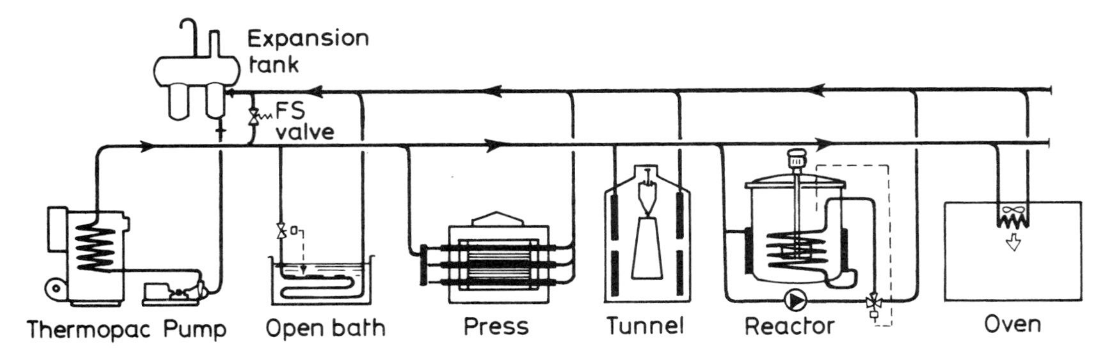

Fig. 2.14 Diagrammatic thermal fluid heating system showing five typical process plants.

process plant, gas turbines or incinerators. In fire tube heaters, the fluid flows through the shell of the heat exchanger, the exhaust gases passing through the tubes. In the liquid-tube the reverse occurs. It is more practicable to have controlled velocities in the liquid-tube type, as in the fire tube units the flow of liquid in the shell can vary greatly throughout the unit, possibly leading to local hot spots and fluid degradation.

One can envisage the use of thermal fluids in 'run-around' coils (see Chapter 1). Here a high-temperature system could be designed, recovering heat via a finned coil in one exhaust duct,and transferring it to a second duct via a single-phase organic fluid operating at high temperature. Thermal fluids can also be used in their vapour phase.

Figure 2.14, while showing a fired thermal fluid heating system manufactured by the Wanson Company, indicates the possible uses of a single loop serving a number of processes.

2.5 Gas–liquid HVAC systems

The 'run-around' coil or liquid-coupled heat recovery system, described in Chapter 1, may be regarded as two gas–liquid heat exchangers, but as the ultimate heat sources are gases, as in the case of the conventional heat pipe heat exchanger, it has been described in that context. However, an individual element of a 'run-around' coil may be used as a gas–liquid heat exchanger in HVAC applications, and one such unit, which differs considerably from the conventional coil used in such systems, is described below.

Developed by Studsvik AB Atomenergi, Sweden, and manufactured by Arex, the 'Retherm A' plastic heat exchanger is illustrated in Fig. 2.15. Thin plastic tubes are wound in even spirals with intermediate spacers. The coil 'package' is located between two plastic endplates, and at the internal and external peripheries the spiral tube ends are connected to an inner and outer tube, which form the headers. Air is pumped through the central inlet duct and flows radially outwards between the spiral tubes. The liquid is pumped from the outer distribution header through the spiral tubing to the inner collection header. In this way, a counter-flow type of heat exchanger is obtained as far as temperature differences are concerned.

The unit is made exclusively of plastics. Tubes are of polyethylene, and to compensate for the low thermal conductivity of plastics, the heat transfer surface available is made larger than that in the equivalent metal unit. Most of the other components are made of PVC. This selection of materials enables the unit to operate in corrosive environments, and the minimum use of joints helps to give the unit resistance to blockage in dust-laiden atmospheres. In fact, dust deposition on the tubes can, claim the manufacturers, actually stabilize the thermal performance, as although the thermal resistance may be increased, the deposits bring about a compensatory increase in the total heat transfer surface.

More common applications for this device in HVAC involve linking of the liquid side to a second, generally conventional, heat exchanger coil for preheating makeup air. In this case, the unit would be classified as a gas–gas system under the definition

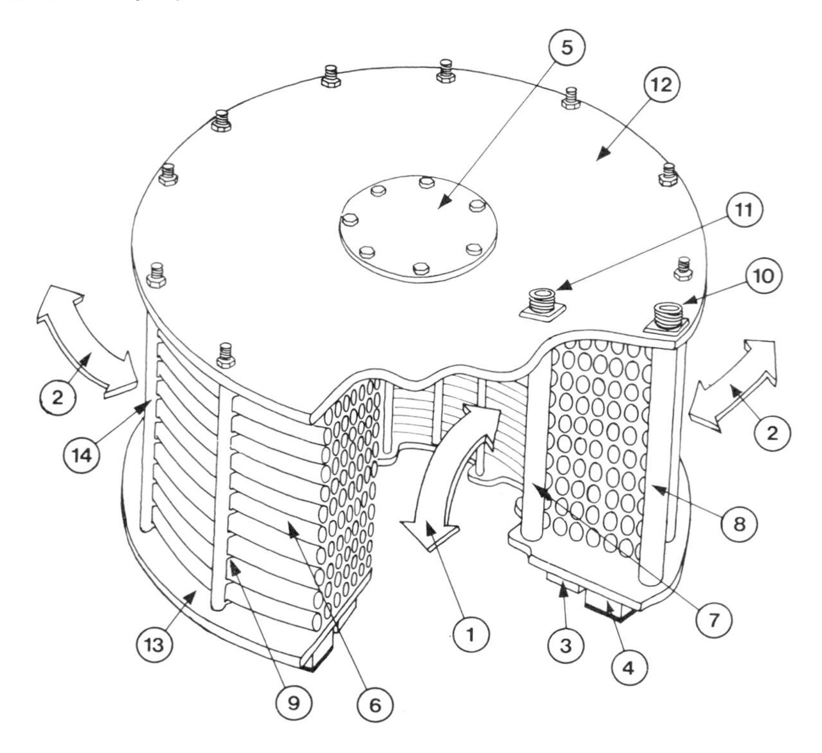

1. Inlet (outlet)
2. Outlet (inlet)
3. Connection flange
4. Lay-up frame
5. Clean-out opening
6. Coiled tubing
7. Inner distribution (or collection) header
8. Outer distribution (or collection) header
9. Spacer
10. Connection, inlet of (water/glycol mixture)
11. Connection, outlet of (water/glycol mixture)
12. Upper front plate
13. Lower front plate
14. Retaining plastic rods with nut lockings at both ends

Fig. 2.15 The Retherma gas-water heat recovery system, using a plastic heat exchanger.

used in this Directory.

2.6 Fluidized bed heat exchangers

2.6.1 *The fluidized bed*

As with many other techniques which tend to arouse much more interest at times when their attractiveness becomes evident as a result of a new emphasis in a particular field, in this case energy conservation, fluidized bed technology is anything but new. Leva (1959) cites 15 years of development in his study of fluidization; and the technique was used by the Romans as a means for purifying ore.

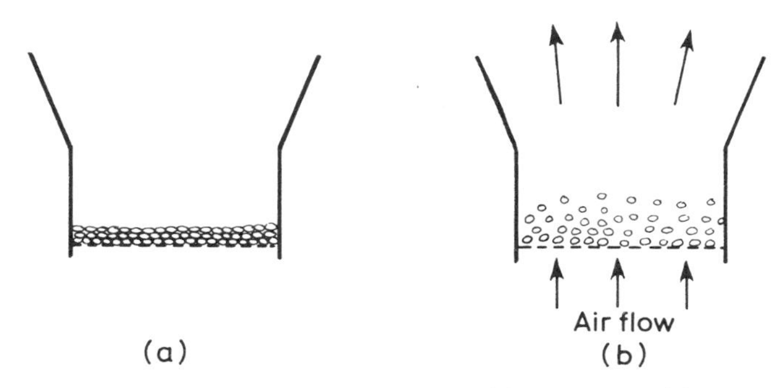

Fig. 2.16 The phenomenon of fluidization: (a) static bed; (b) fluidized bed.

The phenomenon of fluidization relates to a particular mode of contacting granular solids with fluids, either liquid or gaseous. Consider a bed of sand particles resting in a vessel with porous bottom, as shown in Fig. 2.16(a). If air is then passed upwards through the porous base, there will be a particular air flow rate at which the sand particles will be moved slightly away from one another and become suspended in the flow, to take the form shown in Fig. 2.16(b). The bed of sand particles will then resemble a high-viscosity liquid, and the particles can be moved around with the expenditure of much less energy than when they were tightly packed together. The sand is then said to form a 'fluidized bed'. There are a number of states of fluidization. A bed in which the onset of fluidization has just occurred is known as a 'quiescent fluidized bed', whereas at much higher air flows the particle movement is aptly described as a 'turbulent fluidized bed'. When the flow becomes sufficiently high to entrain the particles and carry them upwards by some distance, the bed becomes dispersed. Most fluidized bed applications require operation in the quiescent or turbulent region. The bed may operate at atmospheric pressure (normal for heat recovery units), or be pressurized (Patterson, 1978).

Solids in a fluidized bed are perfectly mixed because they can more in any direction, horizontally or vertically, within the confines of the bed. Gas passing through the bed experiences a pressure drop which is a hydrostatic head, independent of gas flow once fluidization occurs. The gas flow needed to create fluidization is a function of particle size, weight and shape, the size being the most important. Typically sizes of particles range from 0·5 mm up to several millimetres in diameter. In most cases, it is necessary for a fluidized bed to accommodate a wide variety of particles, sizes, and the air flow necessary to fluidize the largest particles may cause carryover, dispersing the smaller particles. In dryers using this technique, filters will be provided downstream of the bed to collect these particles.

The main area of research associated with fluidized beds has been as a technique for highly efficient combustion, particularly on incinerators and boilers, and in coal-gasification plant, where heat pipe heat exchangers immersed in the bed are used to effect heat removal. However, the fluidized bed may also be used as a heat recovery unit.

2.6.2 Fluidized bed heat recovery

A typical fluidized bed heat recovery installation is shown in Fig. 2.17. This is the Stone-Platt Crawley 'Fluidfire' unit, and is used primarily for gas–liquid heat recovery. In this unit the flue gases, which may, when used in conjunction with a fluidized bed, be dirty, are diverted from the main stack through the heat exchanger. The basic heat exchanger itself consists of a shallow fluidized bed above a distributor plate, with finned tubes through which the secondary heat transfer fluid is passed. (Plain tubes may also be

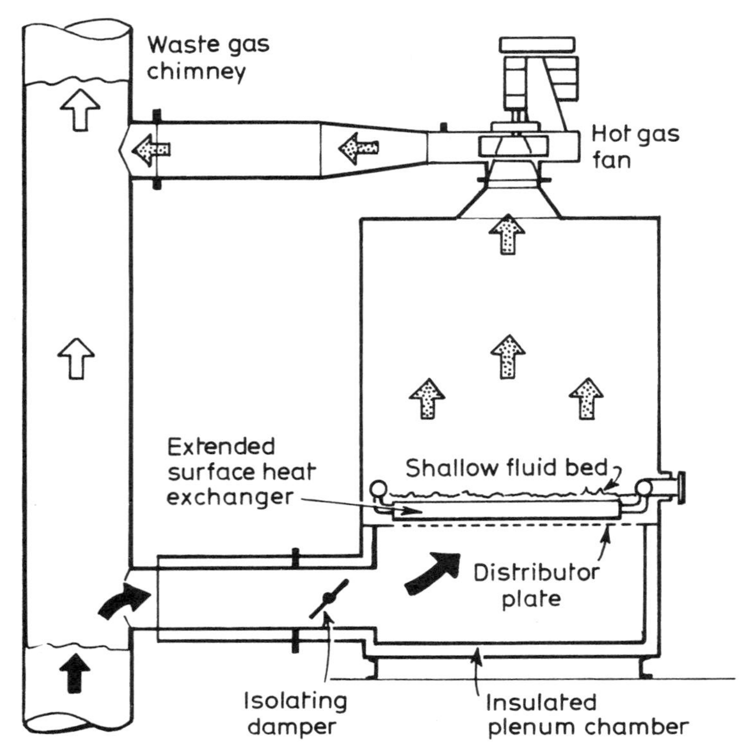

Fig. 2.17 Schematic diagram of the Stone-Platt fluid-fire waste heat recovery unit.

effectively used in fluidized beds, as the outside heat transfer coefficient due to fluidization is much higher than that achievable in a conventional forced-convection heat exchanger. For example, the external heat transfer coefficient for gases flowing across a tube is typically 50 W/m^2 K, whereas for a gas flowing through a fluidized bed, the coefficient rises to between 150 and 500 W/m^2 K, depending upon the diameter of the particles in the bed). Having passed through the shallow fluidized bed, the exhaust gases are returned to the flue. This particular package system comes complete with waste gas fan and all controls necessary to automatically provide specific outlet conditions for the secondary fluid. The fluid to be heated can be water, high-pressure hot water, steam or a thermal heat transfer fluid.

Applications
The Stone-Platt unit, in common with other types of fluid bed heat exchangers, is suitable for use in exhaust gases at temperatures up to 1000°C. This particular system is available with outputs of between 75 kW and 1500 kW, and multiple units are available for higher duties. It is generally applicable to heat recovery from most gaseous combustion products, and can be used in the following application areas:

Furnaces
Kilns
Ovens
Boilers
Incinerators
Reciprocating prime movers
Gas turbines

Typical duties for the recovered heat include heating boiler feedwater, process tanks, washing machines, dyeing and printing-plant fluids and space heating plus, of course, domestic hot-water services.

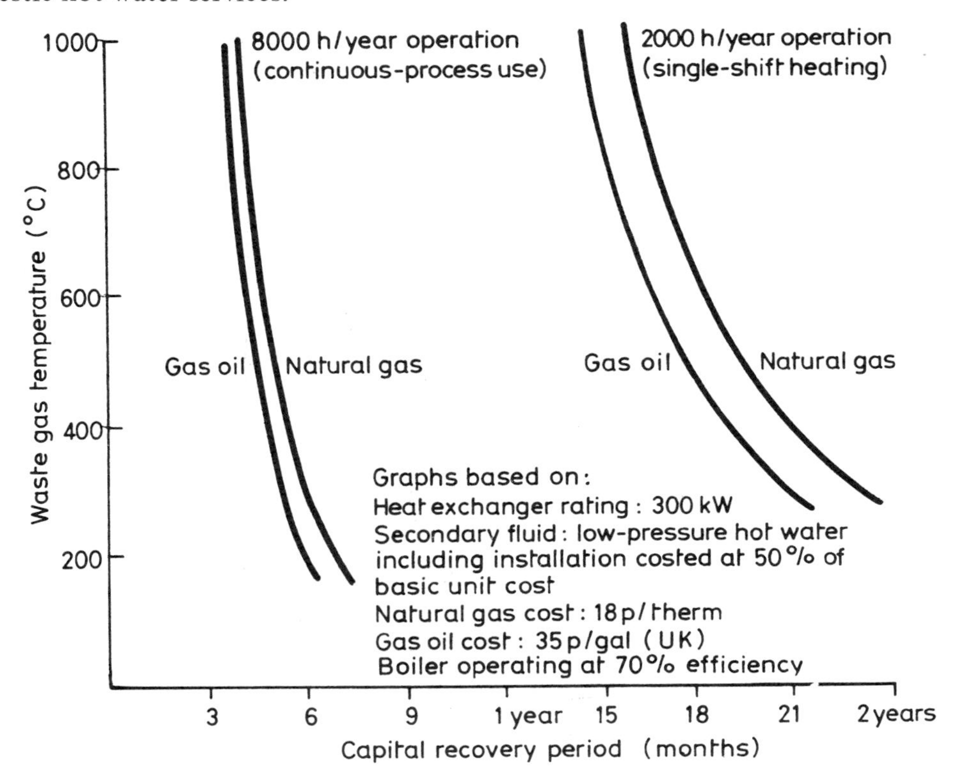

Fig. 2.18 Capital recovery periods for fluidized-bed heat recovery units (from Stone-Platt literature).

Economics

Figure 2.18 gives an indication of the payback to be expected from a fluidized bed heat recovery unit, as a function of the exhaust gas temperature (Stone-Platt figures). The graph is based on the assumption that the alternative heat source and running costs are for a conventionally fired boiler plant. It can be seen that, particularly in the case of continuous 24 hour per day process use, the payback period is dramatically short.

Fig. 2.19 A gas–liquid heat pipe heat exchanger.

2.7 Gas–liquid heat pipe heat exchangers

Worthy of brief mention is the fact that the heat pipe heat exchanger, described in detail in its gas–gas form in Chapter 1, may also be used for gas–liquid heat recovery.

Although its application areas are somewhat limited – normally a conventional economizer will fulfill the requirements of most gas–liquid heat recovery applications not involving a change of phase – the system does have some attraction. Illustrated in Fig. 2.19, the gas–liquid heat pipe heat exchanger may be used to transfer heat effectively between a gas stream and a liquid in cases where it is undesirable to have only a single interface between the gas and liquid, or where a high degree of safety is necessary to prevent gas–liquid mixing.

In a gas–liquid heat pipe unit, as illustrated, the heat is taken out of the gas stream via the finned heat pipe (or thermosyphon) evaporators, and the evaporation-condensation process within the heat pipe results in transport of this heat to the plain (finning is unnecessary because of the higher heat transfer coefficients) condenser sections. These are immersed in a remote jacket carrying the liquid to be heated. Thus, there are two solid interfaces between the gas and liquid, and failure of a heat pipe does not necessitate shutdown of the system.

In the unit illustrated, the heat pipes transfer 33 kW from exhaust gases at 250°C to heat high-pressure hot water from 120°C to 150°C. Equally applicable to other gas–liquid heat recovery systems, the use designated for the high-pressure hot water in this unit during the summer is to heat the generator of an absorption-cycle refrigerator. Thus, the recovered heat can be used for heating in winter and cooling in summer.

References

Boyen, J. L. (1976), *Practical Heat Recovery,* John Wiley, New York, USA.

Bunton, J. F. (1974), Recovery of low grade heat, *Proc. Inst. Plant Eng., Waste Heat Recovery Conf., London, 25–26 September, 1974.*

Department of Energy (1978), Whisky makers invest in waste heat recovery, *Energy Management,* June, 6.

Fanaritis, J. P. and Streich, H. J. (1973), Heat recovery in process plant, *Chemical Engineering,* 28 May.

Leva, M. (1959), *Fluidization,* McGraw-Hill, New York.

Lock, A. E. (1972), Boiler economics, Part 3, *Industrial Process Heating,* no. 12, 32–33.

Lyall, O. (1956), *The Efficient Use of Steam,* HMSO, London.

Patterson, W. (1978), To bed betimes, *New Scientist,* 20 July, 180–181.

Wood, B. D. (1969), *Applications of Thermodynamics,* Addison-Wesley, Reading, Mass., USA.

3 Liquid-to-liquid heat recovery

3.1 Introduction

It is not proposed to dwell at length on the subject of liquid–liquid heat recovery equipment, because most of the data on the heat exchangers, in common with much of that for equipment described in Chapter 2, will be already familiar to engineers involved in plant design and specification. The liquid–liquid heat exchanger, be it of the shell and tube type or plate form, is used in most processes involving liquid flow, and its application to waste heat recovery follows similar guidelines to those laid down for solving any liquid–liquid heat exchange problem. (The exception to this, the liquid–liquid heat pump for process heat recovery, is discussed in detail in Chapter 6.)

The industrial processes, rather than commercial buildings, are the major application areas for liquid–liquid waste-heat recovery of the type described in this chapter, although in both areas the heating of boiler feedwater and the recovery of boiler blow-down heat can be very beneficial. Systems for this are described later.

Manufacturers of equipment of the type described in this chapter are listed following Chapter 14, together with their addresses. More complete data on the product range of each company can be found in Chapter 10, where information is also given on several detailed facets of the heat exchangers, including cleaning, standards met, operating temperature ranges and pressures.

3.2 Types of equipment available

Two basic types of heat exchanger, albeit with several variations, are used for liquid–liquid heat recovery. They are the shell and tube heat exchanger and the plate heat exchanger.

3.2.1 *Shell and tube heat exchangers*

The standard shell and tube heat exchanger is one of the oldest and most commonly

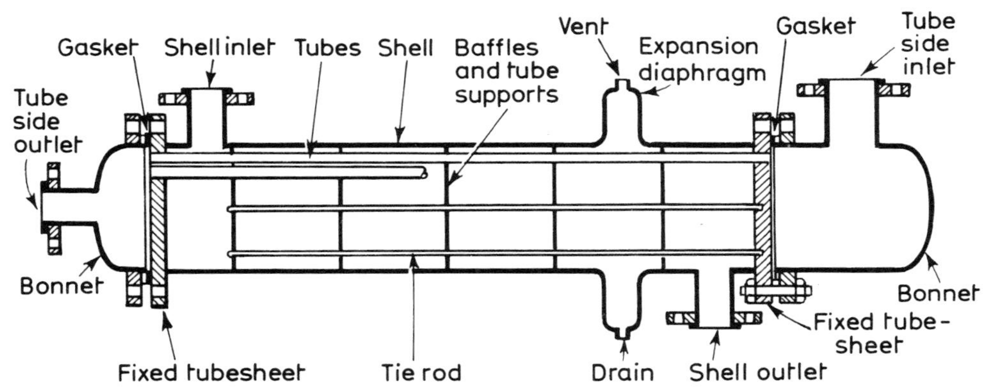

Fig. 3.1 Shell and tube heat exchanger with fixed tube sheet and non-removable tube bundle.

used heat exchangers in industry. The heating surface consists of a number of tubes, spaced apart, with one fluid flowing through the tubes and the other fluid outside the tubes. The ends of each tube are joined to corresponding holes in two tube sheets, being rolled or welded to the tube sheets. The tubes are generally kept in position on the outside by cross-baffles.

Figure 3.1 shows a one-pass, fixed tube sheet heat exchanger. The periphery of the sheet is welded to the shell at both ends. Endcovers are flanged to the shell at both ends. This type can be built in almost any size.

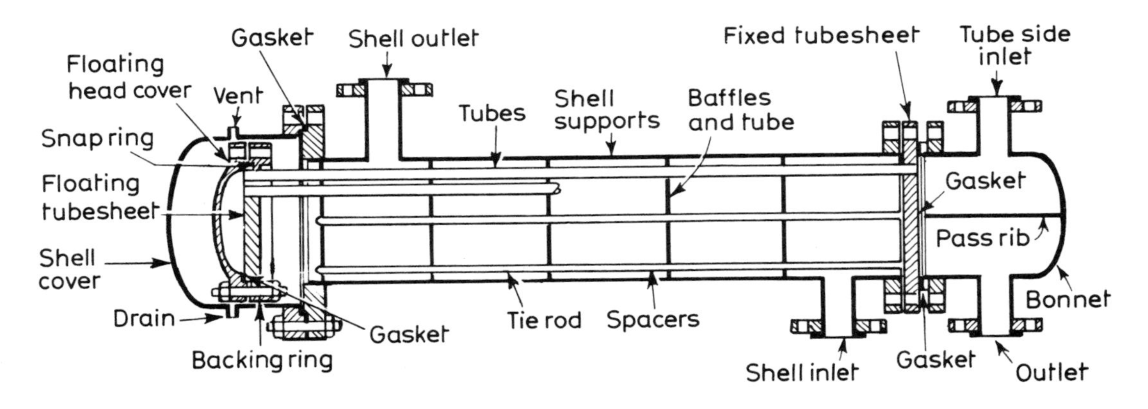

Fig. 3.2 Shell and tube unit with internal floating head and removable tube bundle.

Figure 3.2 shows a typical floating heat exchanger with retractible tube bundles, giving one shell-side pass and two tube side-passes. The floating tube sheet is clamped between the floating head flange and its backing device, and by removing these flanges after opening up the shell flanges the tube bundle can be taken out.

A third configuration of the shell and tube heat exchanger is illustrated in Fig. 3.3. In this case U-tubes are employed, permitting the elimination of one of the tube plates.

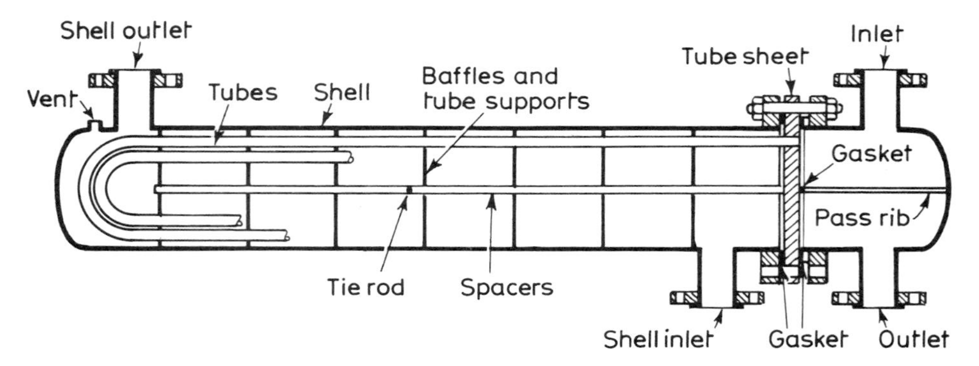

Fig. 3.3 Shell and tube heat exchanger employing a U-tube bundle.

The conventional shell and tube heat exchanger is easy to produce in large quantities. It is made of readily available standard materials and the manufacturing technique is well known. Because of this and its simple design it is generally inexpensive in carbon steel and copper alloys, in spite of being larger and heavier than other types of heat exchangers for specific applications.

Shell and tube heat exchangers are also available with tubes manufactured in Teflon, primarily for handling acids. A number of manufacturers also offer this type of heat exchanger in graphite, again for highly corrosive liquid handling.

However, one of the principal advantages of the shell and tube heat exchanger, when compared to most other types of liquid–liquid heat recovery units, is its ability to be constructed to withstand very high pressures.

Shell and tube heat exchangers are generally built for cross-flow outside the tubes. The number of tubes of a certain diameter, in parallel on the tube-side, is decided by the flow on the tube-side. The number of passes, with a certain tube length, is decided by the heat transfer area required. The calculated cross-flow velocity on the shell-side will then give the spacing required for the flow guiding cross-baffles. Most companies have computer-aided design facilities for such heat exchangers.

Counter-current flow is seldom obtained for liquid-to-liquid duties and the efficiency will therefore be reduced, due to the reduction in effective temperature difference. The calculated velocities are difficult to obtain on the shell-side because of stagnation behind baffles and baffle leakage.

The shell and tube heat exchanger is easy to repair in most cases. If a tube is damaged, it can be plugged or exchanged. A shell and tube heat exchanger can be manually cleaned on the tube-side and, if not too big, units with retractible tube bundles can also be cleaned on the shell-side. These procedures are, however, rather time-consuming and the costs are not negligible, especially if the costs for a plant shutdown have to be taken into consideration. It is not very suitable for chemical cleaning on the shell-side, or on the tube-side if a tube is completely blocked. There are a considerable number of variations on the basic shell and tube heat exchanger, and these are described below. Some are unique to one particular manufacturer (see Chapter 10).

The close tube heat exchanger

The close tube heat exchanger, CTHE, is an all-welded compact shell and tube heat exchanger with longitudinal flow in one pass on both sides. Fig. 3.4 shows a version of the CTHE, especially arranged for high-temperature service, to be mounted in a vertical position.

Fig. 3.4 A close-tube heat exchanger of the type produced by Alfa-Laval.

The tube bundle is connected to the tube channel shroud which hangs in upper nozzles, and at the lower end it is connected to the shell via a bellows. The tube bundle is thus hanging free from all external stresses and any difference in thermal expansion is taken by the bellows at the cold end of the heat exchanger. The tube bundle is surrounded by a shroud of thin sheet, from which the hot fluid enters directly into the channel outside the tubes without being in contact with any pressure-taking parts (shell, flanges and nozzles).

The hot fluid is cooled evenly, by the fluid in the tubes, during its longitudinal flow inside the shroud, and then after leaving the tube bundle at the cold end it passes upwards between the shroud and the shell to the cooled-fluid outlet nozzle in the upper end. All other parts under pressure are subjected to an even temperature close to the cooled-fluid outlet temperature, which releases them from thermal stresses, caused by high temperature, and the loss in material strength.

The tubes are closely pitched and welded to the tubes sheets, which are made of ferrules in a honeycomb pattern or thin sheet depending on the pitch. The thickness of the tube sheet material is chosen as close to the wall thickness of the tubes as possible, to ensure stress-free weld joints, and to avoid a differential temperature between the tube sheet and tube-ends during temperature cycling. Because of the close tube pitch and the pressure balance in the tube bundle, the tube sheet can be made of very thin material, since the strain it has to take is caused only by its weight and the forces from the bellows.

The tube joints are completely protected from temperature attack from the hot fluid, and generally the tube sheet is protected on the shell-side by a full-sized cross-baffle close to the tube sheet with stagnant fluid in between. The bellows which has to take the differential pressure at the cold end, is cooled by the cold fluid going in and has cooled hot fluid on the shell-side, stagnant in a pocket.

In order to support the tubes on the shell-side and still permit a well-distributed flow along the tubes, spiral wires of a diameter close to the distance between the tube walls are wound around the tubes in a steep pitch. The design on the CTHE may seem costly, but when the cost of the materials involved is the deciding factor it can, in spite of the exclusive design, be less expensive than a conventional shell and tube heat exchanger, partly because of the lower heat transfer area needed and partly because of the lower total weight due to its compact design.

For some high-temperature applications, the pressure-vessel authorities have approved the calculation of stresses on the covers, shell and flanges for a temperature close to the maximum outlet temperature of the cooled hot fluid. Since the external dimensions are smaller than a conventional tubular unit, a great saving in material is achieved and the cost is correspondingly lower.

The CTHE is easily arranged for true counter-current flow and is thus better able to utilize a given temperature difference, and therefore requires less heat transfer area. It is also possible to design a greater or smaller longitudinal flow section on the shell-side than on the tube-side, which facilitates the right choice of velocities on both sides. Also, since the flow distribution is good, the heat transfer coefficients are generally the highest possible in a tubular exchanger. The heat exchanger is also manufactured in a less exclusive design as a medium-temperature liquid-to-liquid, or hot liquid-heated reboiler, in high-alloy materials.

The close tube design has certain advantages regarding resistance to vibration and resonance phenomena. A tube firmly supported on the outside along a spiral line will not have a resonant frequence of its own. Since the tube bundle is assembled and completed separately before it is inserted in the shell, the tie bands around the tube bundle can be tightened successively, while hammer tests for vibrations are made, until the bundle is one fixed unit without any possibility of the separate tubes vibrating.

A tube bundle manufactured in this way shows complete resistance to vibration during operation. Other types of heat exchangers, failing because of vibrations, are therefore sometimes substituted by CTHE.

The repair of a CTHE is accomplished in the same way as for a conventional tubular unit. The tube joints have to be drilled away. Each tube replacement, however, has to be completed before the next tube can be withdrawn by pushing the failed tube out with the next tube. The welding in of the replaced tubes can be made in one continuous operation. If required the whole tube bundle can also be taken out by grinding away two welds.

The tubes are not so easy to clean manually because of the small tube diameter, 10 mm to 25 mm, generally used. However, because of high turbulence and well-distributed flow with the highest possible velocities on both sides the CTHE does not foul up easily. For the same reasons, it is well suited for chemical cleaning.

The lamella heat exchanger

The lamella heat exchanger can be considered as a variant of a shell and tube heat exchanger, but designed for longitudinal flow on both sides and having a tube bundle of flattened tubes as shown in Fig. 3.5. These flattened tubes, called lamellas, are made

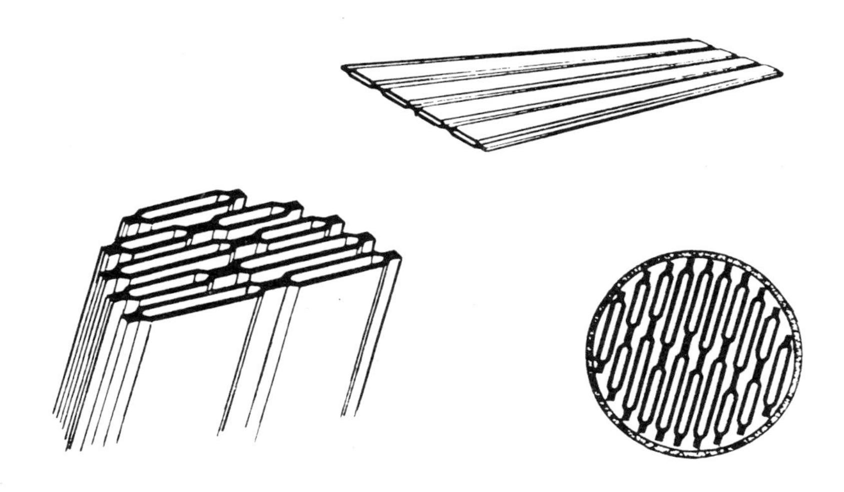

Fig. 3.5 A lamella heat exchanger.

up of two strips of plates, formed and spot-or seam-welded together in a continuous operation. The forming of the strips creates the space inside the lamellas, and dimples on the outside act as spacers for the flow sections outside the lamellas on the shell-side.

Instead of tube sheets the lamellas are welded together at both ends either directly at bent-out end edges, or by joining the ends with steel bars in between, depending on the space required between the lamellas. The lamella bundle is joined to the channel cover at both ends by peripheral welds, which at the outer end is welded to the inlet or outlet nozzles. The tube-side is, thus, completely sealed in by welds. At the fixed end the channel cover is equipped on the outside with a seal ring, which is clamped between the shell flange and a backup flange, for sealing up the shell side in the fixed end. In the flotation end the channel nozzle pipe is inserted in a special Teflon-sealed box for sealing up the shell-side. The end connection here is designed with a removable flange. By removing this flange and loosening the fixed-end shell flanges, the lamella bundle can be pulled out of the shell.

In some executions, especially large units, the tube bundle has a more conventional seal ring, clamped between the shell flange and the flange of a loose endcover, sealing both sides with gaskets on both sides of the ring. The units can be built in sizes up to

1000 m^2 and has a lighter weight than conventional tubular heat exchangers, because of its design and smaller diameter.

Because there is a comparatively free choice of flow sections on both sides, the channels can be sized to utilize the permissible pressure drops in the most efficient way. Therefore, the highest possible heat transfer with a smooth surface can be achieved. The smallest suitable spacing and even flow distribution also improve heat transfer.

The difference of expansion between the heating surface and the shell is well taken care of by the box in the floating end, and this design improves the reliability and protects the lamella bundle against failure because of thermal stresses and strain from external forces. The completely welded channel side gives full safety against gasket leakage, when such a possibility must be avoided.

The lamella heat exchanger is generally easy to repair, because of the accessibility of all parts and welds. A single lamella element cannot, however, be replaced completely without cutting the bundle apart. The lamellas can be manually cleaned only on the shell-side. This is done by pulling out the bundle and bending the elements apart, one by one, since they are generally long and flexible. Because of the high turbulence, uniformly distributed flow and smooth surfaces the units do not foul up easily. They are well suited to chemical cleaning. Cleaning fluids which usually attack gaskets can be used in the all-welded channel system inside the lamellas, e.g. hot nitric acid. This is also valid for the shell-side when Teflon seals are used in the fixed-end shell flange.

Spiral tube heat exchangers

A type of shell and tube heat exchanger which is becoming increasingly popular because of its simplicity, particularly with regard to the way it solves the thermal expansion problem, is the spiral tube heat exchanger, illustrated in Fig. 3.6 with the tube bundle partly exposed. The spiral tubes have good heat transfer characteristics in counter-flow configurations, and this leads to a compact unit for a given duty. As with other shell and tube units, a wide range of materials is available. The tube bundle is easily removable for external cleaning of the surfaces. An example of the application of these systems to heat recovery is given in Section 3.3.1.

3.2.2 *Plate heat exchangers*

The original idea for the plate heat exchanger was patented in the second half of the nineteenth century, and the earliest commercial units, introduced by the APV Company in the 1920s, used cast-gunmetal as the plate material. In the 1930s, however, pressed stainless-steel plates were introduced, and this material is now standard on the plate heat exchangers produced by most manufacturers. (A wide range of alternative materials is available.)

Most plate heat exchangers consist of a frame in which closely spaced metal plates are clamped between a 'head' and 'follower'. The plates, which are corrugated to achieve maximum heat transfer by promoting turbulence, have corner ports and are sealed by gaskets around the ports and along the edges. The plates are grouped into

Fig. 3.6 The spiral tube heat exchanger, partly opened to show the form of the tube bundle.

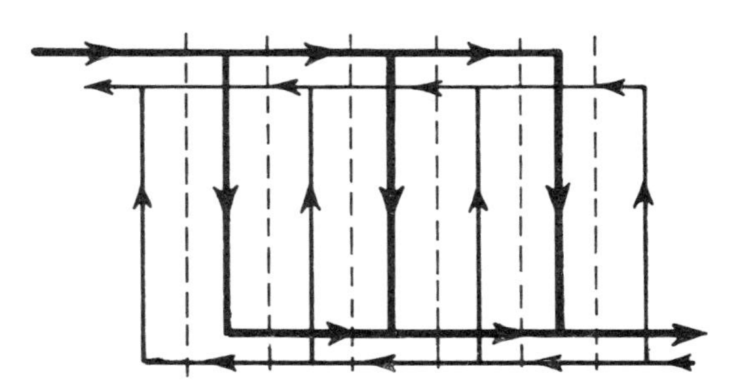

Fig. 3.7 Single-pass operation of a plate heat exchanger.

passes with each fluid being directed evenly between the paralleled passages in each pass. Normally a single-pass counter-flow system is used, as illustrated in Fig. 3.7.

One of the critical features of plate heat exchangers is the material used for the gaskets. Table 3.1 shows a selection of gasket materials used by the APV Company, covering operating temperature ranges up to 260°C.

The heat transfer coefficients obtained with plate heat exchangers vary between

Table 3.1 Gasket materials for plate heat exchangers

Gasket material	Approx. maximum operating temperature (°C)	Application
Paracril (medium nitrile)	135	Resistant to fatty materials
EPDM	150	High-temperature resistance for a wide range of chemicals
Paratherm (resin cured butyl)	150	Aldehydes, ketones and some esters
Paradur (fluorocarbon rubber base)	177	Mineral oils, fuels, vegetable and animal oils
Paracaf (compressed asbestos fibre)	260	Organic solvents such as chlorinated hydrocarbons

2500 kcal/m^2 /h/°C and 6000 kcal/m^2 /h/°C, and plates have typical surface areas of 1 m^2. Heat losses are negligible, as only the edge of the plate is exposed to the air, and no thermal insulation is necessary.

There are some thermal limitations to the plate heat exchanger. For single-phase liquid–liquid duties, the plate heat exchanger can be designed for moderately low pressure loss. However, if the pressure loss across any plate passage which has liquid flowing downward is lower than the available liquid static head, the plate will not run full and performance therefore will be reduced. This is termed 'low plate rate'. Use of a plate below the minimum plate rate is inadvisable, since it causes a wastage of surface area and results in unreliable operation. It is, however, possible to function below the minimum plate rate in a single-pass arrangement by making sure that the low plate rate is operated with a climbing liquid flow. These problems are not so severe with a tubular exchanger and, therefore, operation at a moderately lower available pressure loss is possible.

A second form of plate heat exchanger is known as the spiral heat exchanger. The spiral heat exchanger, sometimes called the spiral plate heat exchanger, was originally designed to solve various kinds of heat exchange problem in the cellulose industry, which were often cases of severe fouling or corrosion. It was one of the first heat exchanger made of stainless-steel plate. The heating surface consists of two relatively long strips of plate, spaced apart and wound around an open, split centre to form a pair of concentric spiral passages. Space studs, flash-welded to the two strips, maintain a uniform spacing between these strips.

Depending on the type, the edges of the strips can be bent and seam-welded on one or both sides of the channel, or only machined against the covers to prevent bypassing. In type 1 (Fig. 3.8(b), the edges are formed and seam-welded alternatively, so that one channel is open on one side and the other channel on the other side when the covers on each side are removed. Type 1 is mainly used for heat exchange in true counter-current manner between liquids or other fluids of limited volume.

In type 2 (Fig. 3.8(c), one channel is completely closed by welds both in the upper and lower side of the channel, and the other channel is open at both sides to allow vertical cross-flow through the spiral. Type 2 is generally made for handling large volumes on the open cross-flow side, the other fluid flowing in the close-welded side in a spiral-flow pattern.

In type 3 in Fig.3.8(d), one channel is open upwards and the other downwards as in type 1. Part of the upper cover, however, is cut away in the centre and equipped with a distribution device (in the figure shown as a cone), so that a large volume can enter in cross-flow, being distributed to the main part of the inner channel simultaneously. When the volume has been reduced by partial condensation, the remaining volume is cooled in true counter-current spiral flow and exhausted at the periphery. The distribution device can be calculated and executed so that a spiral flow of even, high velocity is accomplished even in the middle section. The steam, vapour or fumes are fed into each turn of the channel in quantities just sufficient to replace the successively condensed volume and maintain the required velocity in spiral flow in the centre section.

As the units offer such a wide variety of channel arrangements it generally has a high performance, when properly calculated and custom-made to suit a certain application. The fouling-up tendency is low, since good flow distribution and turbulence are obtainable in a single, long pass without bypassing or stagnation. Type 1 achieves, for all practical purposes, true counter- or co-current flow.

Because of its compactness, and since the cold end of the cooling-fluid channel is generally in the outer spiral turn, the radiation losses are small and no insulation is generally required. The spiral heat exchanger is not suitable for high-pressure service. The upper pressure limit depends on the size, but 15 atm can be considered as an average. The design pressure is generally less than 10 atm.

Repair of corroded channel seam welds can generally be done without difficulties, since most of them are easily accessible. Corrosion damage on the inner parts of the strips are more complicated to repair, and are often difficult to do well onsite. In most cases parts of the strip have to be cut away, turn by turn, until the damaged spot is reached and can be repaired. The cut-away parts then have to be welded back again, one by one, working outwards out until the shell is reached. Fortunately repairs are seldom required if the right material for the application has been chosen.

Manual cleaning cannot be done easily on the standard spiral with spacer studs in the channels. If such cleaning is required, a special execution without studs must be chosen which will require thicker plate and is generally uneconomic in high-alloy materials. Carbon steel SHEs are, however, sometimes specially made without studs and with easily removable covers for manual cleaning. These units will be easy to clean manually. It is well suited for cleaning in place, because of its good flow distribution and the one single channel for each fluid.

3.2.3 *Comparison of plate heat exchanger and shell and tube heat exchangers*

A number of general statements may be made concerning the desirability of both of these liquid–liquid heat exchangers:

(1) For liquid–liquid duties, the plate heat exchanger will usually give a higher overall heat transfer coefficient and in many cases the required pressure loss will be no higher (see Table 3.2).
(2) The effective mean temperature difference will usually be higher with the plate heat exchanger.
(3) Although the tube is the best shape of flow conduit for withstanding pressure, it is the wrong shape for optimum heat transfer performance since it has the smallest surface area per unit of cross-sectional flow area. However, the spiral tube unit offers advantages over other shell and tube units in this respect.
(4) Because of the restrictions in the flow area of the ports on plate units, it is usually difficult, unless a moderate pressure loss is available, to produce economic designs when it is necessary to handle large quantities of low-density fluids such as vapours and gases.
(5) A plate heat exchanger will usually occupy considerably less floor space than a tubular for the same duty.
(6) From a mechanical viewpoint, the plate passage is not the optimum and gasketed plate units are not made to withstand operating pressure much in excess of 20 bar.
(7) For most materials of construction, sheet metal for plates is less expensive per unit area than tube of the same thickness.
(8) When materials other than mild steel are required, the plate will usually be more economical than the tube for the application.
(9) When mild-steel construction is acceptable and when a close temperature approach is not required, the tubular heat exchanger will often be the most economic solution since the plate heat exchanger is rarely made in mild steel.
(10) Plate heat exchangers are limited by the necessity that the gasket be elastomeric. Even compressed asbestos-fibre gaskets contain about 6 per cent rubber. The maximum operating temperature, therefore, is usually limited to 260°C.

3.3 Applications and economics

3.3.1 *Shell and tube heat exchangers*

The application range of shell and tube heat exchangers (and, indeed, plate heat exchangers) in heat recovery is enormous, and they have a place in every industry where liquid effluent is at a temperature in some cases only a few degrees above the supply water which needs to be heated.

It is proposed to give here two illustrations of how heat recovery can be effected using these heat exchangers.

Table 3.2 Comparison of plate and tubular heat exchanger surfaces and pressure drops

Fluid A (kg/s)	Fluid B (kg/s)	Duty (°C)		Tubular pressure drop (kN/m^2)			Plate pressure drop (kN/m^2)	
			(m^2)	(A)	(B)	(m^2)	(A)	(B)
1·69 (hydrocarbon)	6·72 (water)	160A49 49B33	28	2·8	20·6	8·4	2·1	15·2
109 (water)	9·84 (seawater)	42A28 37B22	1130	131	131	330	74	61
7·42 (petrol)	2·52 (water)	105A35 35B26	84	22	31	26	9·7	31
2·2 (solvent)	24·3 (water)	60A40 35B26	170	20	31	41	21	24
18·8 (effluent)	59·5 (salt water)	106A38 41B20	140	72	55	36	31	69

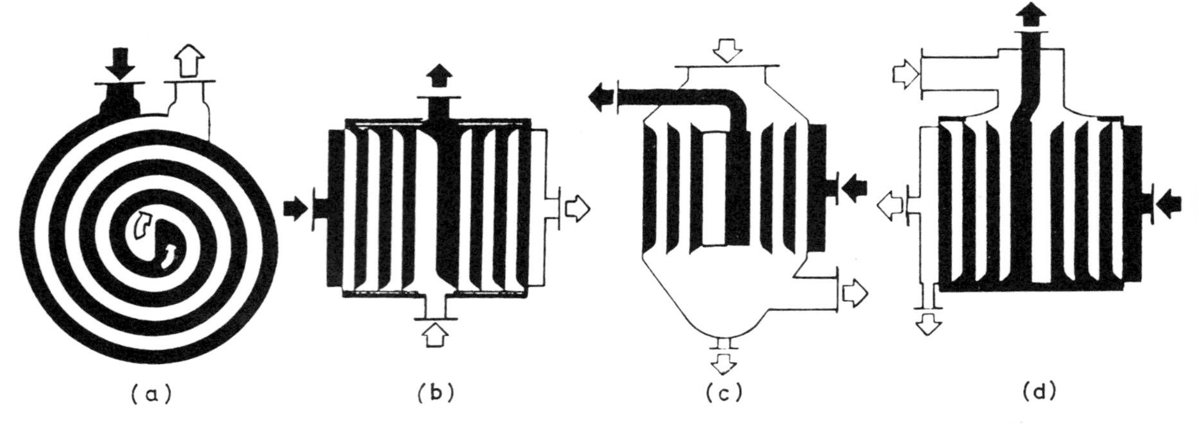

Fig. 3.8 A spiral heat exchanger: (a) principle of SHE; (b) principle of type 1; (c) principle of type 2; (d) principle of type 3.

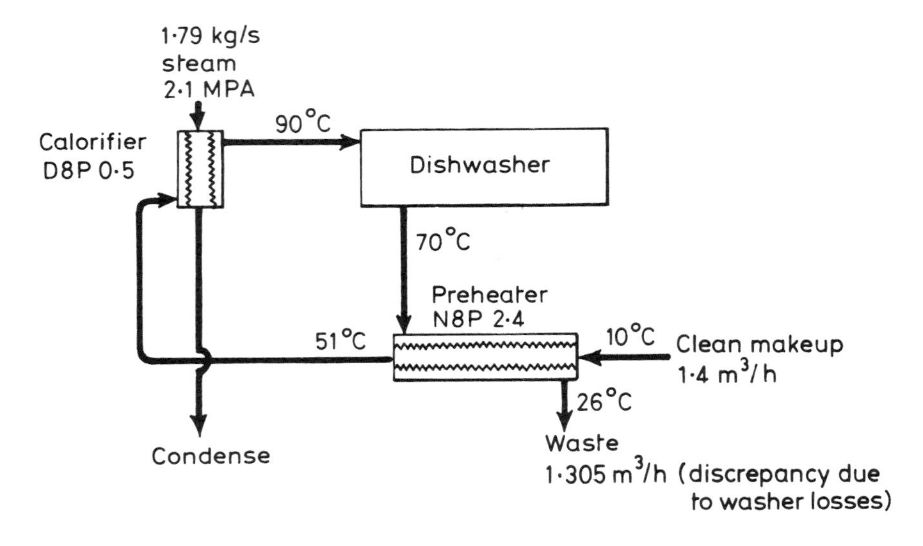

Fig. 3.9 Spiral tube heat exchangers used as calorifier and heat recovery unit in a dishwasher (by courtesy of D.J. Neil and Co. Ltd).

Heat recovery using spiral tube unit

A spiral tube heat exchanger has been installed downstream of the dishwasher in a large industrial canteen (serving 600 staff). The dishwasher is in use over 2500 h per year, and its water consumption is 1·4 m^3 /h, the temperature of which has to be raised from 10° C to 90° C using a steam calorifier.

The data below shows the effect of the addition of a simple spiral tube heat exchanger on the unit economics, this being implemented as shown in Fig. 3.9, using the type of heat exchanger illustrated in Fig. 3.6.

With calorifier only

Q/h = 1400 kg/h x 80°C = 112 000 kcal/h = 130 kW
Steam consumption 3·66 kg/s
Fuel cost £1854 per annum
Cost of calorifier type D8P1, £285

With heat recovery

(a) Preheat 1·4 m^3/h, 10°C to 51°C by waste water 70°C to 26°C (NB: Due to losses in the machine, waste water discharge is reduced to about 1·305 m^3/h)
Q/h = 1400 kg/h x 41°C = 57 500 kcal/h = 57 kW = *GAIN*
(b) Heat 1·4 m^3/h, 51°C to 90°C by steam at 200 kPa
Q/h = 1400 kg/h x 39°C = 54 600 kcal/h = 63 kW
Steam consumption 1·79 kg/s
Fuel cost, £907 per annum

Cost of preheater, type N8P2.4	£ 580.00
Cost of calorifier, type D8P0.5	210.00
Total	£ 790.00
Less cost of original calorifier D8P1	285.00
Additional capital cost of heat recovery	505.00
Fuel cost, calorifier only	1854.00 p.a.
Fuel cost, heat recovery	907.00 p.a.
Saving in fuel cost	£ 947.00 p.a.

As can be seen from the cost figures, the original calorifier may be replaced by a smaller unit, and the payback period, taking into account what should be comparatively low installation costs, will be of the order of 1 year, a highly acceptable return. The reader may also be interested in the successful application of heat pumps to dishwashers, (described in Chapter 6).

Recovery of boiler blowdown

It is necessary regularly to clean out a boiler, even though water treatment is normally carried out, because most feedwater will contain suspended or dissolved solids which will be deposited in the boiler as steam is generated. The cleaning procedure is either intermittent or continuous, and involves removing water which has become highly concentrated with impurities from the boiler, and replacing it with clean feedwater. The criterion for determining the amount of water removal, or 'blowdown', required is based on the measure of the total dissolved solids (TDS) level. The TDS level in a modern packaged boiler is generally recommended to be in the range 2000 to 3500 ppm (Urbani, 1975).

It has been common practice on most boilers to 'blowdown' the unit intermittently. This enables the cleaning process to be carried out at the most convenient time, when TDS level is high and generation rates are low. It also ensures that precipitate near the blowdown outlet is carried away with the sudden water flow.

However, intermittent blowdown is becoming less desirable for a number of reasons. In the UK the grid-water distribution system has led to a reduction in quality of feed-water, meaning a higher TDS and a greater blowdown requirement. Also in modern packaged boilers intermittent blowdown, with the associated thermal shocks and complete falloff in pressure, can create problems. Local authority regulations concerning maximum temperatures at which waste water can be dumped into the sewerage system also work against intermittent blowdown.

The alternative is continuous blowdown of the boiler, which has the attraction of considerable potential for waste heat recovery. This system can also be fully automated, obviating the possibility of neglect which could lead to corrosion associated with intermittent, manually directed operations.

It has been recommended that continuous blowdown be applied if the average boiler requirement for blowdown exceeds 10^{-2} kg/s. Equipment is available for implementing this and it is claimed that the use of a continuous blowdown system does not interfere with the boiler steam-raising rate or pressure.

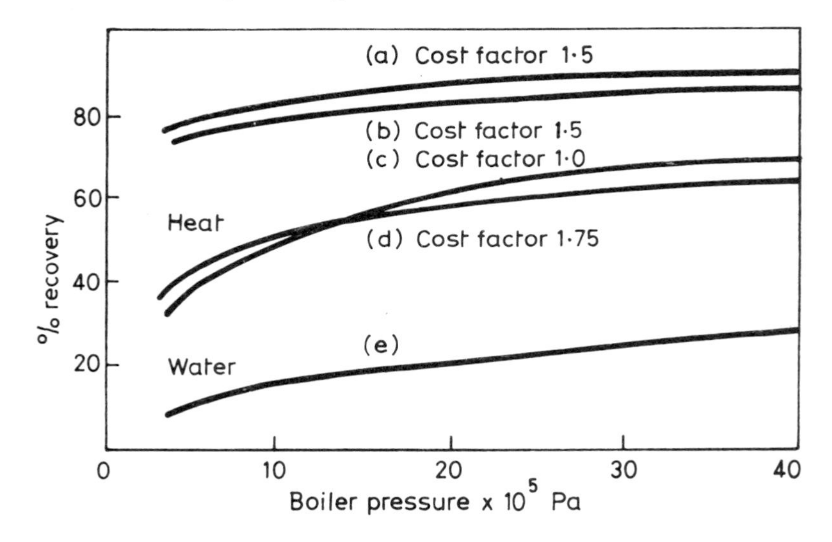

Fig. 3.10 Heat and water content recovery from boiler blowdown (by courtesy of Gestra (UK) Ltd.)

Continuous blowdown, by its name, implies that a continuous supply of hot water is being discharged from the boiler. The heat contained in this water may be recovered using a flash vessel or a heat exchanger, the recovered heat being used for feedwater heating, or preheating of fuel oil.

The quantity of heat which can be recovered from the blowdown depends largely upon the use to which it is put, and this is illustrated in Fig. 3.10. The most efficient recovery processes, and the most expensive, are represented by the two upper curves

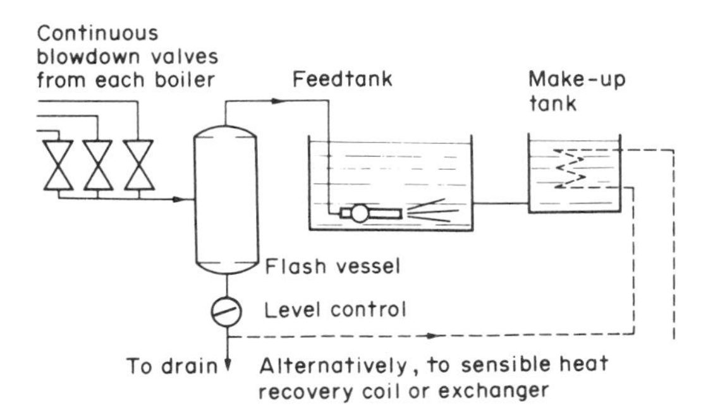

Fig. 3.11 Flash steam recovery and sensible heat recovery for feedwater heating (by courtesy of Gestra (UK) Ltd.).

in this figure. These involve flash steam recovery by directly injecting the water into the feedwater. This is combined with sensible heat recovery into cold makeup water using a conventional heat exchanger, and the layout is illustrated in Fig. 3.11. Alternatively, the flash vessel may be omitted, and all the sensible heat recovered used for makeup water heating. The lowest line on the graph illustrates the amount of water which can be recovered when a flash vessel is used. This is obviously becoming increasingly attractive in view of rising water costs.

With regard to capital cost of the equipment needed to implement a continuous blowdown system plus heat recovery, an outlay of £1000 to £5000, depending on the degree of automation and other 'optional extras', appears typical. The user may expect to recover this expenditure within one year, and thereafter to save several thousands of pounds every year. In addition, of course, the user benefits from a reliable blowdown system which will increase boiler life and protect associated plant.

3.3.2 *Plate heat exchangers*

One heat recovery application where plate heat exchangers have been successfully used is illustrated in Fig. 3.12. The APV Company supplied the plate heat exchanger for this installation, applied to large-scale evaporation plant at the Marchon Division of Albright and Wilson Ltd., at Whitehaven in the UK, Cumberland, (Anon, 1974).

In 1964 a 'Paraflow'-type R55 plate heat exchanger was supplied for the purpose of providing hot water for the boiler feed. The heat source is exhaust vapour from two forced-circulation evaporators which is ducted to a blower unit feeding a condenser. Hot condensate at 74°C is cycled through the 'Paraflow' to heat the boiler feedwater from 8°C to 65°C before returning to the condenser at 34°C. Overflow from the hot condensate is employed as hot process water.

The 'Paraflow' has 106 stainless-steel flow plates that give a total heat transfer area of 55 m^2, the whole unit occupying a space of only 3 m x 0·75 m x 2 m. Nevertheless, the division's services engineer estimated that it could supply approximately 750 000 l per day of hot water, or the equivalent of half a boiler. The 'Paraflow' has never needed

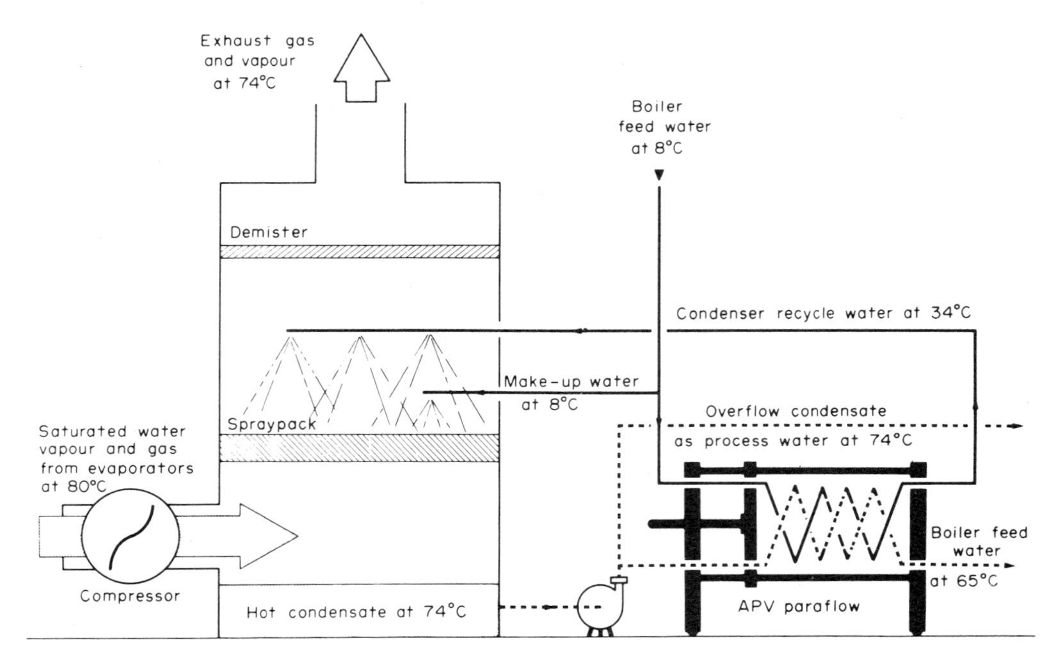

Fig. 3.12 An APV plate heat exchanger using waste heat to preheat boiler feedwater (APV Company diagram).

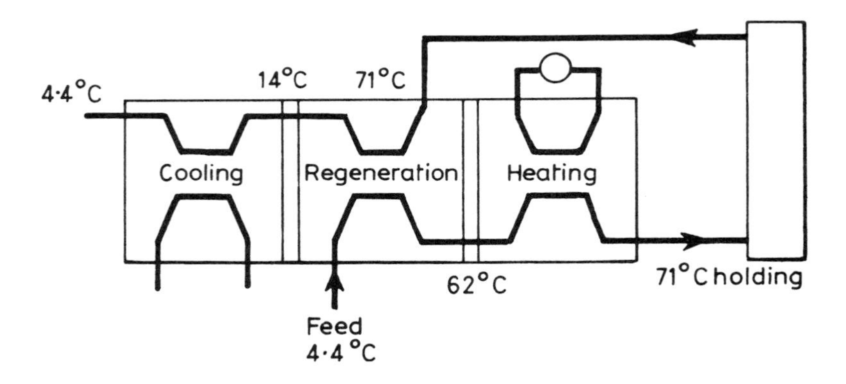

Fig. 3.13 The flow path in a dairy pasteurizing plant. % regeneration $= \frac{62-4\cdot4}{71-4\cdot4} \times 100 \triangleq 86\%$. (APV Company diagram).

to be opened up.

The original outlay on the installation, which included the 'Paraflow', its associated piping, the blower unit, the spray pack and tower, was £9200.

A more detailed quantification of the performance of plate heat exchangers as an aid to heat recovery can be made with reference to the dairy industry. Their use is widespread in a number of dairy processes, but it is in pasteurization of milk where they

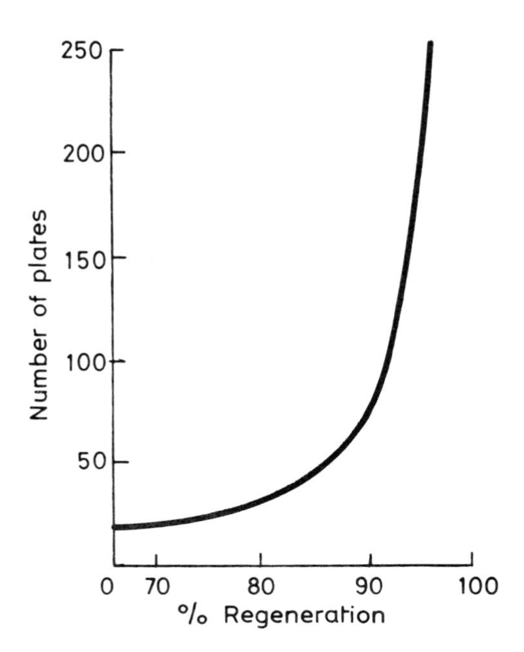

Fig. 3.14 Relationship between regeneration efficiency and number of plates (APV Company diagram).

have brought the greatest benefits. Fig. 3.13 shows the flow path in the pasteurizer. As the milk has to be cooled prior to being bottled, it is possible to preheat the milk entering the pasteurizer using heat from milk leaving the process for storage prior to bottling. In this way heat which would normally be rejected to cooling water is recovered and used effectively to reduce the energy expended in the pasteurzation process by almost 90 per cent. In the example shown, the quantity of regeneration is 86 per cent. (Starkie, 1975).

Obviously, by increasing the number of plates in a heat exchanger, one could theoretically approach a recovery efficiency of 100 per cent, but this would require an infinitely large heat exchanger owing to the lack of any temperature driving force available. This is illustrated in Fig. 3.14, which shows that while recovery of slightly over 90 per cent of the heat is possible while retaining a heat exchanger of sensible proportions, the number of plates needed for any further improvement makes further regeneration unrealistic. In order to assess the correct amount of regeneration in any particular process, account must be taken of the costs of energy, the capital cost of the plates and pumping power. The effect of increasing energy costs in pasteurization illustrates how this works. Until recently it has been standard practice to use regenerators having efficiencies of 85 per cent, but now it is worth the increased capital expenditure to install additional plates to give regenerator efficiencies in excess of 90 per cent. Fig. 3.15 shows how the operating costs of a pasteurization plant capable of treating 5·7 l/s of milk varies with the quantity of heat recovered. The degrees of regeneration

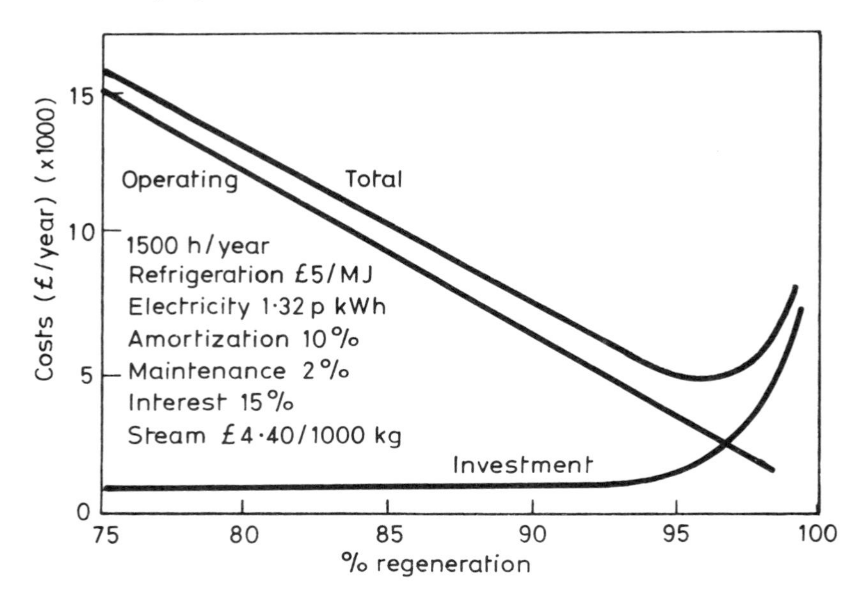

Fig. 3.15 Effect of heat recovery rate on the costs of pasteurizing plant (APV Company diagram).

for minimum operating cost, as indicated on Fig. 3.14 is not achievable in practice for reasons of increased pressure drop, etc., but a value of 92 per cent is feasible. Wherever heat can be usefully recovered from a liquid process stream and used without being upgraded (i.e. if the liquid temperature is higher than about 45°C), some form of heat exchange is worth considering, and the plate-type appears attractive, certainly from a first-cost point of view (Garrett and Reay, 1978). Plate heat exchangers have also been successfully used in conjunction with economizers (discussed in Chapter 2). A normal finned tube economizer unit is installed in the flue gas system and water, softened to ensure no scale deposition, is recirculated to pass through the economizer and one side of the 'Paraflow' (the APV Company's name for their plate heat exchanger), while the process liquor or primary water to be heated is passed through the other side.

A major advantage of this system is that scaling or fouling-process media can be heated. Deposits can easily be cleaned from the heat exchanger plates, whereas this could not be readily removed if the media were to be directly heated in the economizer. The system has particular appeal, therefore, in areas where hard untreated water is required to be heated for process use.

Typical installation

At John Smith's Tadcaster Brewery in the UK fresh cold water is heated by flue gases from the boiler to 77°C to 82°C for use in the cask-washing plant during the daytime. During the night when the cask washer is not in operation, the contents of a 290 000 l hot-liquor tank are heated by recirculation through the 'Paraflow' as shown in Fig. 3.16.

In such a system, the flue gases must not be cooled to below dew point because

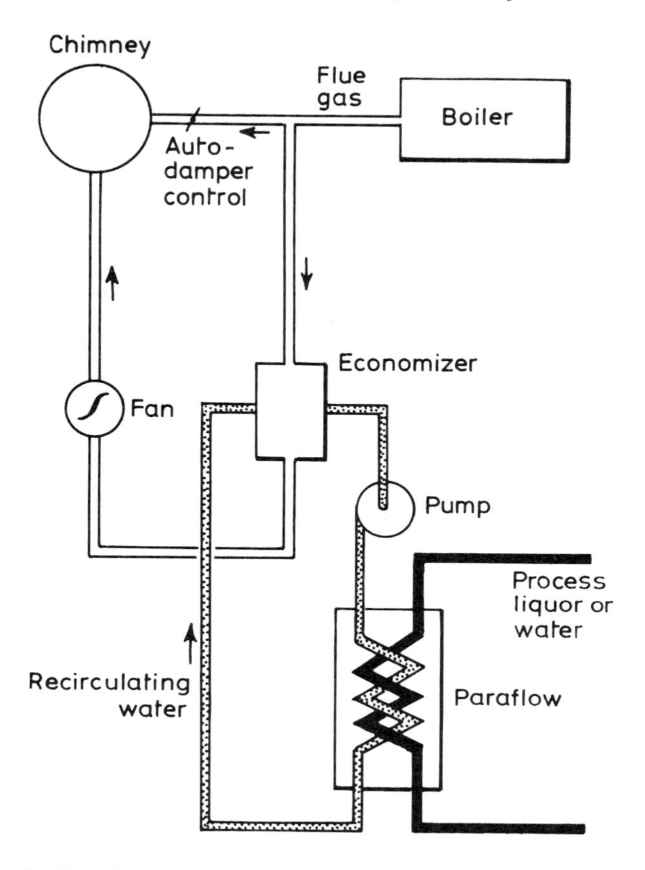

Fig. 3.16 Use of a 'Paraflow' plate heat exchanger in conjunction with an economizer.

severe corrosion can occur, as discussed earlier, in the chimney. The hot gases are therefore passed to the chimney via autocontrol dampers, to ensure that final gas temperature in the chimney is maintained above the dew point.

References

Anon. (1974), Regeneration and the energy crisis, *APV Spearhead – Engineering in the Process Industries,* APV no.3, June.

Garrett, N. and Reay, D. A. (1978), Energy conservation in the dairy industry, *Proc. Dairy Trades Conf., Eastbourne, 1978.*

Starkie, G. L. (1975), Some aspects of energy conservation in dairy process plant, *J. Soc. Dairy Technol.,* vol. 28, 121–129.

Urbani, A. C. (1975), Boiler blowdown and the economic viability of heat recovery, *Proc. Inst. Mech. Engrs. Conf. Energy Recovery in Process Plant, London, 1975,* Paper C.20/75.

4 Prime movers – sources and users of waste heat

4.1 Introduction

This chapter is concerned with two aspects of prime movers, in so far as they are linked to waste heat recovery. First, the prime mover, be it a gas or steam turbine or a reciprocating diesel or gas engine, is a source of waste heat. These heat sources include exhaust gases, condensate, cooling water and lubrication oil. The techniques for recovering this heat are many and varied, and include much of the equipment described in preceding chapters. Waste heat boilers, shell and tube heat exchangers, thermal fluid heaters, recuperators and regenerators are commonly used on prime movers, as are several other types of heat recovery unit.

Heat recovered from prime movers, particularly gas and steam turbines, may be used either for use remote to the prime mover generating the heat, or it may be used again in the prime mover itself. For example, to raise the combustion chamber inlet temperature, as discussed in Section 4.2.

The second major field of heat recovery associated with prime movers, normally linked to a gas turbine, but also available for steam turbines and process exhaust gases, is to use a turbine downstream of another turbine, or a process, utilizing the waste heat to generate power. Described later in the chapter, steam turbines may be located downstream of a gas turbine exhaust raising steam, via a waste heat boiler, which then drives the steam turbine. For lower exhaust-gas temperatures, a fluid such as Freon may be expanded through the turbine, as it can produce sufficiently high pressures at temperatures too low for steam to be effective.

Manufacturers making equipment specifically for prime mover waste heat recovery are listed in Appendix 1, and their equipment is fully described in Chapter 11, together with data on its selection. If the user is interested in gas–liquid or gas–gas heat recovery from prime movers, he will also find references in the alphabetical list of manufacturers to manufacturers in other sections of this book who include such equipment in their general range. Those with a particular interest in reciprocating engines are advised also to consult Chapter 6, where a section is devoted to the use of gas engines (with waste heat recovery) as drives for the heat pump compressor.

4.2 Gas turbines

Gas turbines are particularly convenient for onsite power generation, when compared to steam turbines, because, as internal combustion engines, they need no boiler and condenser. Also, the pressures achieved as much lower than those in steam systems.

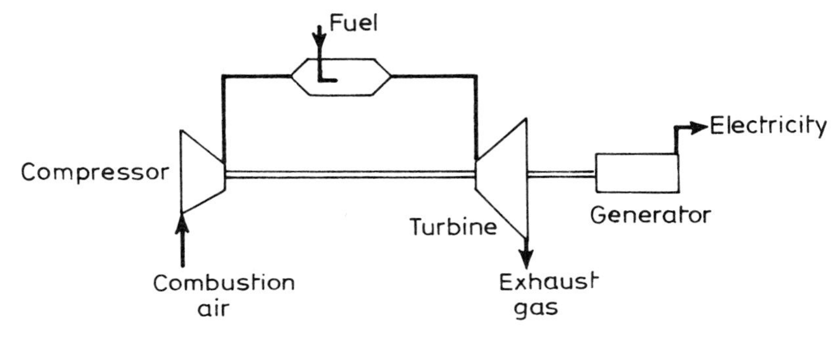

Fig. 4.1 Gas turbine cycle.

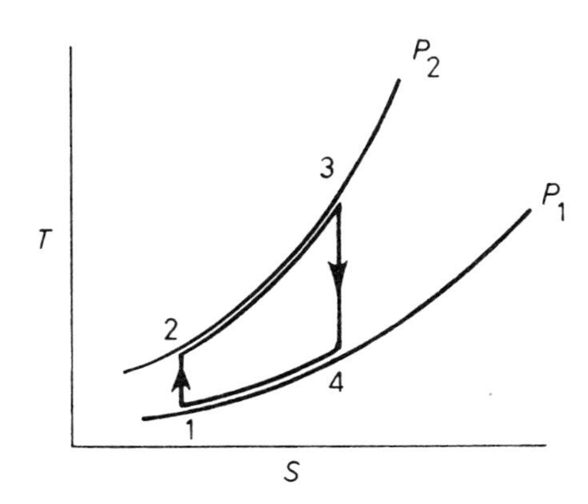

Fig. 4.2 Temperature-entropy diagram for gas turbine cycle.

A gas turbine cycle is illustrated in Fig. 4.1. The fuel, which may be one of a variety of gases or light oils, is burnt in a pressurized combustion chamber in the presence of air supplied by the compressor. This combustion gas is then expanded to atmospheric pressure through the turbine, which drives the compressor and also the electricity generator. Most gas turbines, particularly those used for industrial duties, are of the single-shaft type, with the air compressor and turbine mounted on a common shaft. However, it is possible to obtain split-shaft units which use two turbine stages, one driving the compressor and the other providing shaft output for generator drive (or the other rotating equipment).

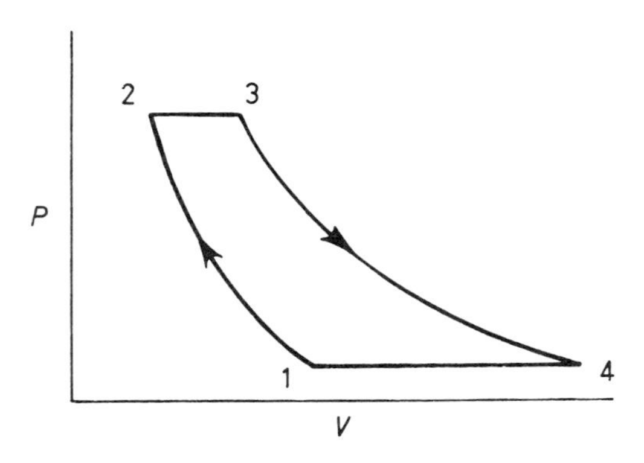

Fig. 4.3 Pressure-volume diagram for gas turbine cycle.

It is possible to examine the factors which affect the amount of shaft power that a gas turbine can produce with reference to the temperature/entropy diagram in Fig. 4.2, and the pressure/volume diagram in Fig. 4.3:

The compressor work is $W_{12} = C_p\ (T_2 - T_1)$
The turbine work is $W_{34} = C_p\ (T_3 - T_4)$
The heat supplied in the combustion chamber is given by $Q_{23} = C_p\ (T_3 - T_4)$
where T_1 = air-inlet temperature to compressor
T_2 = air-outlet temperature from compressor
T_3 = temperature of gas leaving combustion chamber
T_4 = turbine exhaust temperature
C_p = specific heat.

The cycle efficiency may therefore be written as the ratio of the net work done (that is, the useful work from the turbine) to the quantity of heat supplied,

$$\text{i.e.}\quad \eta = \frac{C_p\ (T_3 - T_4) - C_p\ (T_2 - T_1)}{C_p\ (T_3 - T_2)}\ .$$

Alternatively, the cycle efficiency of the gas turbine can be expressed in terms of the pressure ratio P_2/P_1, denoted by r_p.

For isentropic compression and expansion:

$$T_2 = T_1 r_p{}^{(\gamma-1)/\gamma}$$

$$T_3 = T_4 r_p{}^{(\gamma-1)/\gamma}\ .$$

where γ is the ratio of specific heats

$$\text{therefore } \eta = 1 - \left(\frac{1}{r_p}\right)^{(\gamma-1)/\gamma}$$

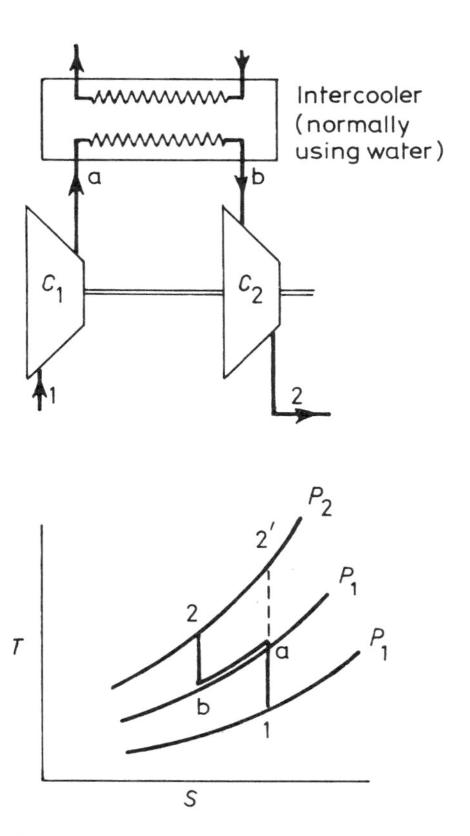

Fig. 4.4 A gas turbine system incorporating an intercooler between stages.

The work ratio r_w is the ratio of the net work output to the total turbine work, and is given by:

$$r_w = \frac{C_p\,(T_3 - T_4) - C_p\,(T_2 - T_1)}{C_p\,(T_3 - T_4)}$$

$$= 1 - \left(\frac{T_1}{T_3}\right) r_p^{\;(\gamma-1)/\gamma}\,.$$

Thus, while the ideal cycle efficiency (which neglects to take into account the isentropic efficiencies of the compressor and turbine) is a function of the pressure ratio r_p, the work ratio depends upon temperatures T_1 and T_3. In a gas turbine, for maximum work output it is therefore desirable to have the air-inlet temperature as low as possible, and the combustion-chamber gas temperature as high as permitted by gas turbine materials, etc. At present turbine-inlet temperatures are limited to 850°C to 950°C, giving overall thermal efficiencies approaching 30 per cent, with some industrial gas turbines operating with efficiencies below 20 per cent. The raising of the turbine-inlet temperature could lead to thermal efficiencies of over 40 per cent, eliminating the

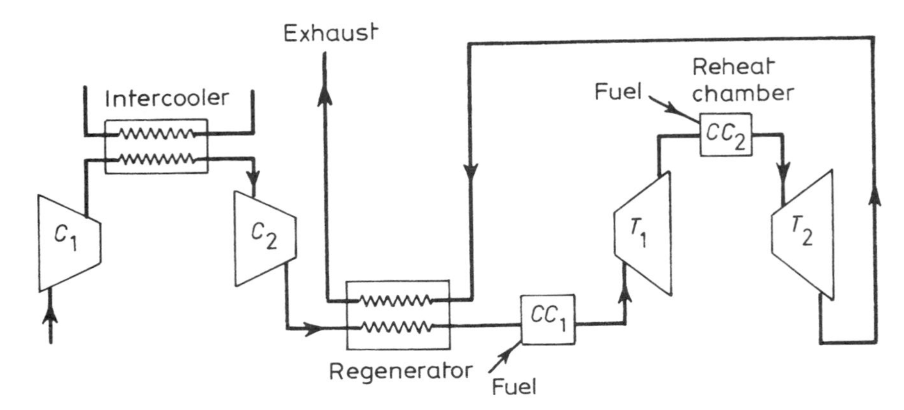

Fig. 4.5 A gas turbine incorporating intercoolers, reheat and regeneration.

major disadvantage of the gas turbine when compared with steam turbine plant. The efficiencies of the turbine and compressor are typically in the range 80 to 90 per cent, and this can also have a significant effect on plant efficiency.

Apart from increasing turbine inlet temperatures, which can only be carried out using blade cooling and/or new materials (this is the subject of many development programmes, temperatures of 1300°C to 1400°C being the aim), the use of intercoolers, reheaters and regenerators can assist.

The application of an intercooler necessitates a two-stage compressor, as shown in Fig. 4.4. With reference to the temperature/entropy diagram, it can be seen that the compressor work $C_p\ (T_2 - T_b) + C_p\ (T_a - T_1)$ will be less than that without intercooling, $C_p\ (T_2 - T_1)$.

Reheating involves at least two turbine stages, with a combustion chamber added between each stage. Again it can be shown that the turbine work output $C_p\ (T_3 - T_c) + C_p\ (T_d - T_4)$ is greater with reheat.

A third and most important technique for improving the efficiency of the gas turbine, and one which will depend on users' requirements for shaft power and heat, is to use exhaust heat to raise T_2, the combustion-chamber inlet temperature. Incorporation of all the above modification results in the cycle shown in Fig. 4.5 and discussed in more detail below.

Some users of industrial gas turbines have opted for combined-cycle units, in which the exhaust from the gas turbine is used to raise steam which then drives a steam turbine. Combined systems can operate at efficiencies in excess of 40 per cent, and are discussed in Section 4.4.

4.2.1 *Gas turbine regeneration*

There are several types of regnerators and recuperators available for use on gas turbines for preheating air for the combustion chamber. Plate regenerators have been widely applied to gas turbines, but their design has some shortcomings.

The plate regenerator is, in general, a series of hollow plates separated by corrugated grids. The design has many narrow gas passages where dirt and contamination can accumulate. While the design does provide an enormous heat transfer surface, its rigid construction requires very large numbers of welded and brazed joints, all of which are subject to expansion stresses and thermal distortions which can easily break welds and joints and cause leaks. Because the entire regenerator is made of carbon steel, it is also vulnerable to simple oxidation and corrosion, especially at welds and brazed joints. Normal oxidation and corrosion rates are accelerated by the condition that plate regenerators must run dirty, unless their system is fuelled with gas or clean oil. Both of these fuels are growing less attractive because of their cost and availability.

Owing to its complex structure, the plate regenerator tends to accumulate dirt rather than shed it, and the only way to clean such a structure is to shut it down and hose it out with very high-pressure water. Designers have attempted to cure some of these shortcomings by proposing the use of a low-level, 409 stainless and by specifying heavier components. These proposals so far have not been tested in any operating units.

One alternative design is that made by Farrier Products Division of Coaltech, who build U-tube-type heat exchangers in capacities suitable for gas turbine test stands. Farrier selected the U-tube configuration illustrated in Fig. 4.6 because seamless tubes, hung in clusters of U-shapes, where the best containers for air at extremely high temperatures and pressures. The strength-to-material ratio of the cylindrical shape was the most economical for containing high pressures. Without welds or brazing, this envelope had no weak points, and the pendant U-tubes could move and expand in all directions without causing any fractures or failures. The U-tubes are built in conformance with section VIII, ASME Code.

The U-tube design also permitted complete freedom in metal selection, allowing Farrier to extrude the tubes of 304 stainless, a steel with exceptionally high resistance to oxidation and corrosion. These seamless tubes have, on their exterior surfaces, cold-rolled fins which perform a dual function. By increasing the area of exposed surface, they improve heat transfer. Between cycles the tubes which have expanded during operation, cool, contract the fins and squeeze out the dirt. Where dirt accumulations grow too rapidly to be removed entirely by contraction, the tubes can be cleaned easily with sootblowers while the regenerator continues to operate, whereas plate regenerators must normally shut down for cleaning.

U-tube regnerators are less prone to breakdown and failure, are easier to inspect and repair and are more highly resistant to oxidation and corrosion than any other type of regenerator. With the only welds being between the tubes and the tube sheets, and these not subjected to any unusual stress, U-tubes have few points at which a failure could occur. The fins, being cold-rolled on the tubes, are impervious to damage.

Because the tubes hang free inside their housing, they are completely accessible and easily inspected. Where a leak occurs in a tube, that tube's individual bundle can be removed and replaced without disturbing the other bundles and without any need to dismantle the entire regenerator. Leaks in plate regenerators are difficult to find and impossible to repair.

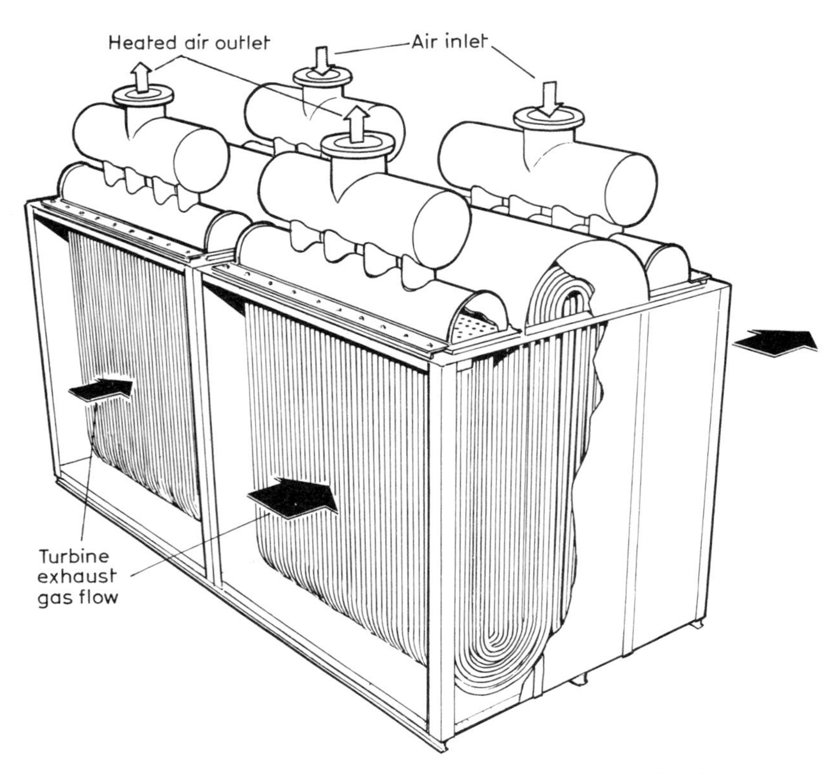

Fig. 4.6 A U-tube gas turbine regenerator manufactured by Farrier.

With regard to the performance of tube heat recovery units, Farrier claim that the percentage of available heat extracted is between 82 and 87 per cent, effectively reducing gas turbine fuel consumption by 20 to 30 per cent. They are also attractive in applications where cyclical operation (e.g. starting up of the gas turbine once or twice per day) is necessary.

GEA, in Germany, manufacture a tubular heat recovery unit for gas turbines, but this differs from the Farrier-type in that it resembles a shell and tube heat exchanger (as shown in Fig. 4.7). The exhaust gas flows through the tubes, while the air to be heated up is caused by adequately designed baffles to pass around the tubes in a multiple cross-counterflow pattern. Since the interior and exterior flow passages are free from tubes, each of the disc and doughnut-type baffles can support all and any of the tubes, and since the clamping conditions are consequently equal for all tubes, the vibration characteristics of the tubes can be calculated precisely at the design stage.

The tubes, normally sized to 25 mm o.d., are welded to the top and bottom tube plates. Carefully applied welds will minimize any leakage risks even under frequently varying loads. This careful fabrication ensures that any leakage, which might affect turbine efficiency, is virtually kept below the sensitivity level of commonly used leakage-test methods, e.g. pressure-holding tests, even after extended periods of operation

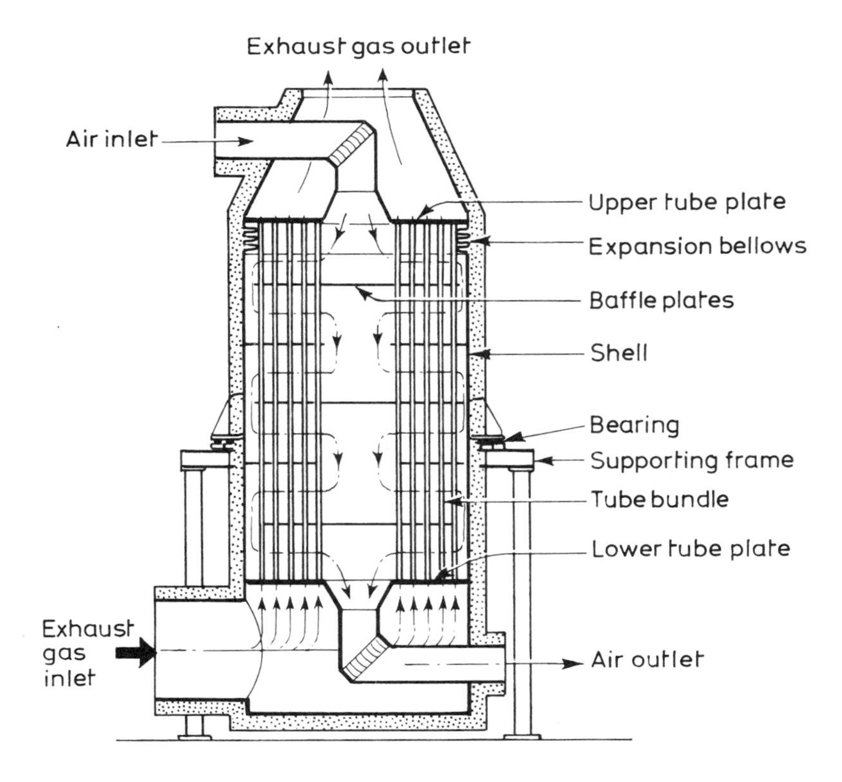

Fig. 4.7 GEA shell and tube gas turbine recuperator.

The major advantage of this regenerator design, which mainly becomes effective with plants subject to frequent starts and shutdown, lies in the completely rotational symmetry of the heat exchanger proper, especially of the pressure-exposed shell and of the interior flow scheme. As a result, the major items, such as tube bundle and other parts subject to pressure, show a completely symmetrical temperature distribution not only during stationary operation, but also under varying load conditions or at frequent starts and stops, precluding any substantial thermal stress.

The differential thermal expansion in the longitudinal direction of tube bundle and shell caused by different average wall temperatures, even during continuous and even more during instationary operation, is compensated by an expansion joint mounted in the shell which prevents any additional stress. In a North German thermal power station, for instance, a gas turbine with GEA regenerators has operated for 60 000 h and accomplished 4025 starts in 14 years of operation with no signs of leakage ever observed during regular leakage tests. The application of round tubes as heat exchanger elements is additionally advantageous in as much as the interior tube surface which is exposed to the exhaust gas is completely smooth, so that impurities, if any, can be easily removed.

The heat exchanger can be arranged either above the turbine with exhaust-gas inlet

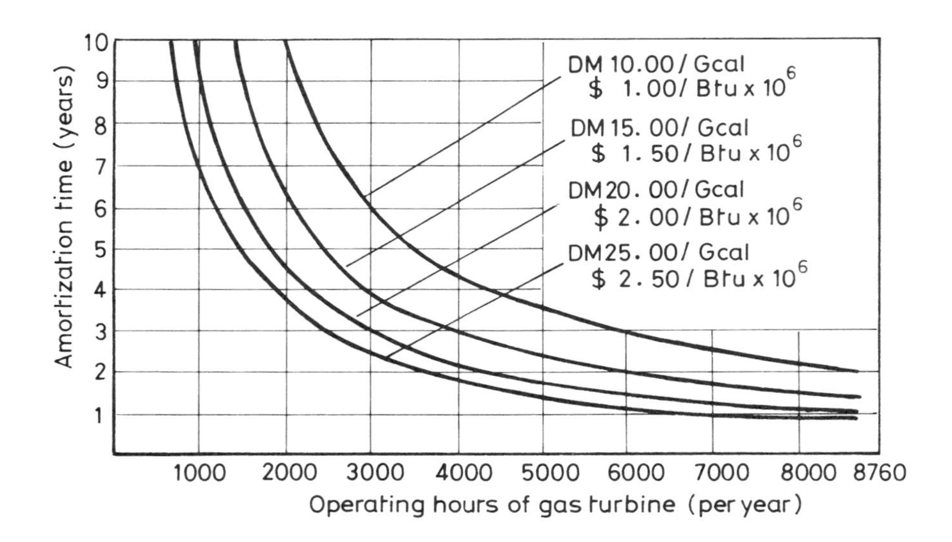

Fig. 4.8 Return on investment for a GEA gas turbine regnerator on a 15000 h.p. unit.

from below if the turbine-gas outlet flange is mounted vertically upwards, or outside the turbine hall with axial or lateral exhaust-gas inlet. Even a horizontal arrangement has been employed for a number of applications.

The decision to use a regenerator mainly depends on the actual fuel costs on the planned hours of operation of the gas turbine plant. Using a regenerator, the total investment cost will rise about 6 per cent with about 25 per cent fuel cost saved.

Considering the pressure losses the total efficiency of a gas turbine installation will improve by about 6 per cent points, hence attaining values which come very close to those of a modern steam-power station. As a function of fuel price to be paid and hours of operation involved, the additional investment cost required for a GEA regenerator will in most of the cases amortize after one or two years of operation (see Fig. 4.8).

Studies on the economy of the regenerators previously supplied by GEA have shown that the optimum effectivity is in the range of 80 per cent, with effectivity defined as follows:

$$E = \frac{\text{Air-outlet temperature} - \text{Air-inlet temperature}}{\text{Exhaust-gas inlet temperature} - \text{Air-inlet temperature}}$$

As a result of the favourable operating experience made with GEA regenerators even under severe operating conditions, additional GEA regenerators were ordered to supplement existing compressor stations of a European gas pipeline. This clearly shows that the economic advantages of GEA regenerators will still be maintained even with the extra costs required for their additional installation in existing plants.

Data on these and other gas turbine recuperators and regenerators are given in Chapter 1

4.2.2 *Gas turbine heat recovery for use in processes*

The exhaust of a gas turbine consists predominantly of air, regardless of the type of fuel used, because the turbine operates with large quantities of excess air. Typical exhaust-gas temperatures are in the range 350°C to 600°C, but it must be remembered that if recuperators or regenerators of the types described above are used, the final exhaust-gas temperature will be considerably reduced, possibly down to about 270°C. Although this leads to higher turbine efficiencies, the quantity of heat recoverable for use in processes will be significantly lower.

Factors such as altitude and ambient temperature affect exhaust-gas conditions, as well as influencing the performance of the gas turbine itself. Manufacturers are able to provide full data on the performance under such conditions, as well as part-load performance. Some typical gas turbine inputs and recoverable outputs are given in Table 4.1.

Because of its low thermal efficiency when used solely to provide power for driving generators, etc., and the fact that it does not in general offer cost advantages over steam turbines and other prime movers such as diesel engines, industrial use in process plant of the gas turbine as a generator drive for continuous duty should only be envisaged when the exhaust heat can serve a worthwhile function. Thus the industrial gas turbine should form part of a 'total energy' package, particularly where fuel costs are high. (In this context, the term 'industrial gas turbine' includes modified aeroengines although, as pointed out by manufacturers of industrial units designed from the outset as such, maintenance may be needed more regularly on aeroengine-derived units. The Ruston gas turbine, designed solely for industrial use, will run for 20 000 to 30 000 h between blade inspections and 80 000 –100 000 h between overhauls.)

In the area of industrial energy applications, gas turbines are used to power compressors, pumps, fans and electric generators. Their exhaust heat is used for steam generation, process energy, drying, space heating and air conditioning. The shaft power output of gas turbines ranges from 50 kW to 100 MW, with up to four times as much heat being available, depending upon the particular thermal efficiency of the machine.

The applications in industry may also be depicted as in the diagram supplied by Ruston gas turbines (Fig. 4.9). Ruston highlight two main areas where their 'total energy' units have been successfully used, and the comments appertaining to these applications are concisely described in their brochure (Stocks, undated).

It is of course necessary to have a requirement for both steam (heat) and electricity in a plant before considering a total energy set, and in the case of a gas turbine the low thermal efficiency as a prime mover, possibly as little as 17 per cent, suggests that the heat/power ratio should be of the order of 4:1 or higher. Thus in a gas turbine installation the heat generated is of considerable importance in making an economic case for installation. Quantifying heat and power outputs, a Centrax CS600-2 unit in its basic form requires a fuel input equivalent to 3100 kW to produce 500 kW of electrical power. Taking into account small losses, the waste heat in the turbine exhaust is 2500 kW, which could be directly employed in, for example, a drying process.

Table 4.1 Typical gas turbine inputs and recoverable outputs

	150 kW; C 3·4: 1; 30 000 rev/min; P 690 kN/m^2				200 kW; C 4·1: 1; 35 000 rev/min; P 690 kN/m^2			
		Outputs				Outputs		
Shaft load (%)	Fuel rate (kW)	S (kW)	H (kW)	S/H	Fuel rate (kW)	S (kW)	H (kW)	S/H
100	1690	150	820	1/5·5	2170	200	1080	1/5·4
90	1615	135	815	1/6·0	2040	180	985	1/5·5
80	1545	120	805	1/6·7	1890	160	895	1/5·6
70	1475	105	790	1/7·5	1760	140	800	1/5·7
60	1405	90	760	1/8·5	1640	120	710	1/5·9
50	1330	75	700	1/9·3	1530	100	620	1/6·2

Notes: C – compression ratio; S – shaft output; H – heat output; P – gas pressure.

Table 4.1 (*continued*)

Shaft load (%)	300 kW; C 6: 1; 43 500 rev/min; P 1030 kN/m^2				900 kW; C 4: 1; 6000 rev/min; P 690 kN/m^2			
		Outputs				Outputs		
	Fuel rate (kW)	S (kW)	H (kW)	S/H	Fuel rate (kW)	S (kW)	H (kW)	S/H
100	1850	300	760	1/2·5	6630	900	3290	1/3·7
90	1670	270	700	1/2·6	6220	810	3040	1/3·8
80	1600	240	615	1/2·6	5800	720	2790	1/3·9
70	1490	210	585	1/2·8	5420	630	2550	1/4·1
60	1330	180	530	1/2·9	5040	540	2310	1/4·4
50	1240	150	470	1/3·1	4630	450	2080	1/4·7

Notes: C – compression ratio; S – shaft output; H – heat output; P – gas pressure.

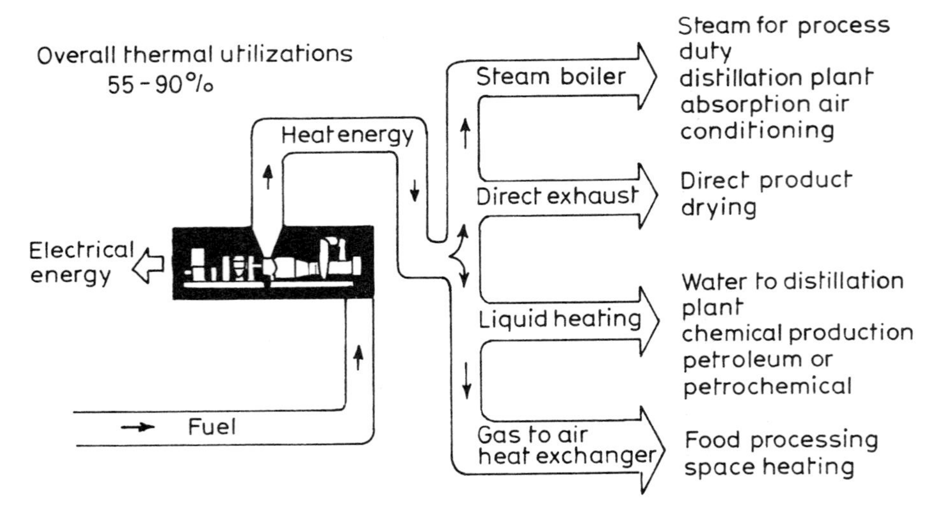

Fig. 4.9 Typical total energy applications using a Ruston type TA industrial gas turbine as the single power and heat source.

This would raise the overall efficiency to 65 per cent.

Alternatively, considerable quantities of steam may be raised. It is common practice to supplement the exhaust-heat capacity of the gas turbine, making use of the oxygen present, by supplementary firing. Applied to the Centrax unit, a supplementary-fired waste heat boiler could produce approximately 3·5 kg/s of steam, increasing the overall efficiency to almost 80 per cent. (This assumes a boiler efficiency of 80 per cent.) The heat input via auxiliary firing in this case is considerable, accounting for 75 per cent of the total heat input to the boiler. However, it does ensure optimum use of the waste heat produced by the gas turbine, although the total heat/power ratio is 16, possibly rather high for most industrial applications. If the gas turbine set is only providing a portion of the total electricity requirements of the plant, the balance can be more easily accommodated, however.

The economics of a system such as the above appear very favourable, Centrax estimating payback periods of between 12 300 and 12 400 h, even with allowances for maintenance costs. Leasing arrangements available on gas turbines ease the burden even further. (A point to consider for other heat recovery equipment?)

Ruston see the gas turbine set in industry as alleviating some of the problems associated with standby generators. Standby sets are most commonly based on diesel or gas reciprocating engines having a lower capital cost when compared with some industrial gas turbines. As utilization of standby generators is low, maintenance is also inexpensive and long-life reliability is not an important consideration. A gas turbine, however, is able to fulfil both continuous and standby duties. Even in industries where the power requirements are high, the relationship of the generation capacity needed under emergency conditions to the heat capacity under normal operating conditions could well fall within the range of power/heat ratios available from a total energy system

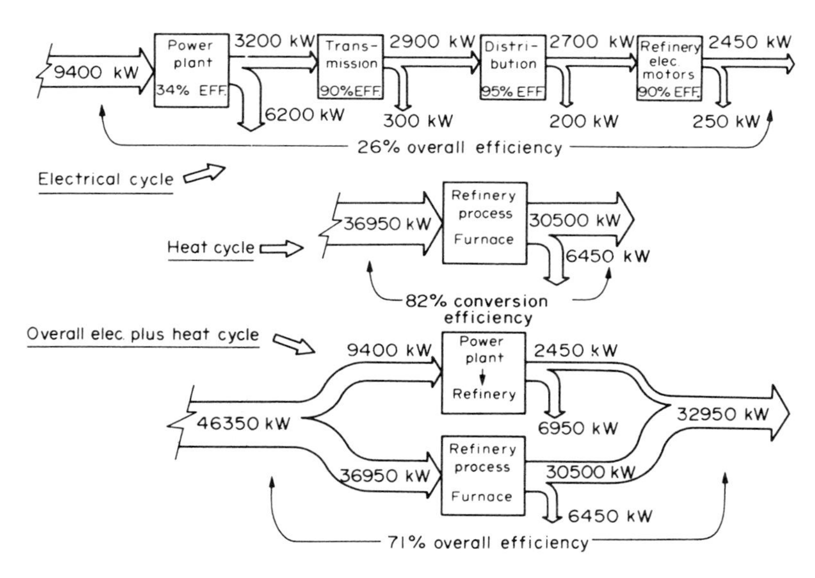

Fig. 4.10 Standard Oil Company electrical and steam cycles before installation of gas turbine plant.

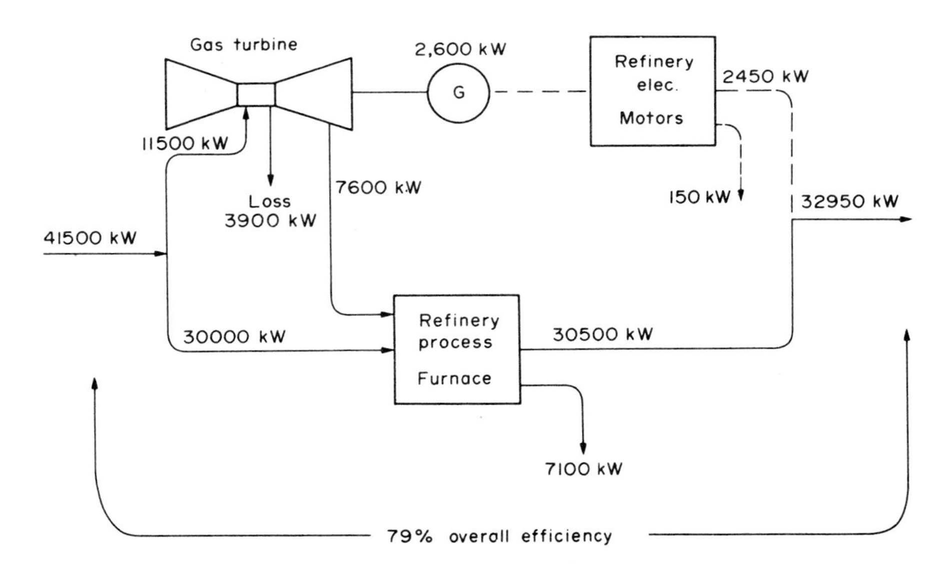

Fig. 4.11 Onsite generation of electricity using a gas turbine, waste heat being applied to furnace preheating.

based on the gas turbine. Thus gas turbine units sized for the emergency-power requirement can be operated to supply a proportion of the base-load electricity needs while also supplying the heat requirements, with or without auxiliary firing. In an emergency

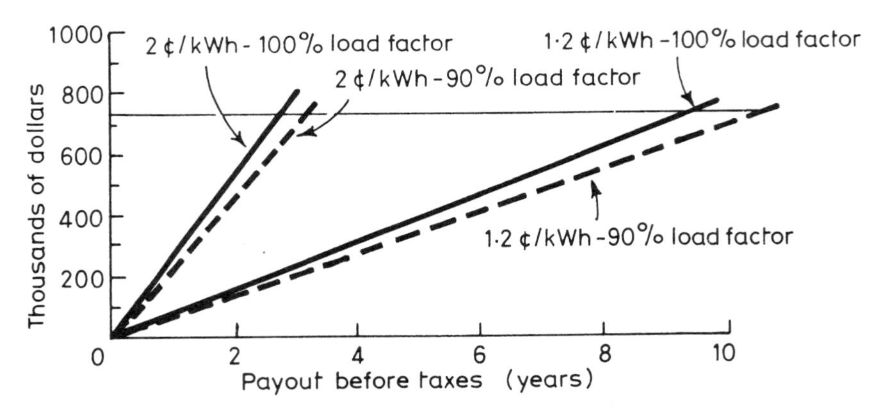

Fig. 4.12 Economics of a gas turbine generator with waste heat used for furnace preheating.

Cost of purchased power	Load factor	1·2¢/kWh	2¢/kWh
Incremental plant investment($)		724 000	724 000
Purchased power savings ($/year)	0·9	246 000	411 600
	1·0	271 000	452 000
Added fuel cost (for electrical generator)	0·9	164 000	164 000
$/year @ $ 1·40/kJ	1·0	180 200	180 200
Added annual maintenance and operation costs	0·9	12 000	12 000
$ /year	1·0	13 000	13 000
Net annual savings ($ /year)	0·9	70 000	235 600
	1·0	78 000	259 200
Payout before taxes (years)	0·9	10·3	3·1
	1·0	9·3	2·8

situation, the gas turbines would of course maintain their thermal duty.

The economics of total-energy gas turbine operation may also be examined with reference to well-established plant at the Standard Oil Company. This particular plant possessed a large furnace. When electricity purchased from utilities and national authorities was comparatively cheap, all plant electric motors were supplied with power from outside the refinery, and although the furnace efficiency was high, the combined conversion efficiency of furnace and power station → electric motor is low, as shown in Fig. 4.10 (Levers, 1975).

A more attractive scheme, selected for installation in a Standard oil plant in 1975, is shown in Fig. 4.11. In spite of the fact that the furnace was already installed, a gas turbine unit was conveniently fitted local to the furnace, ensuring only the minimum amount of ducting between the two units. This is attributed to the compactness of the gas turbine, and its low auxiliary system requirements. The gas turbine was selected so that its exhaust gas stream met the combustion-air requirements of the furnace. The relatively large generator eliminated the need for other steam turbine emergency generators that would otherwise have been required. In this case, the gas turbine generator supplies critical motors normally connected to the refinery emergency-power system. The utility-supplied electric power thus becomes the 'emergency' source, i.e. critical motors are transferred to the utility system during short periods when the gas turbine generator is inoperative. After adding the gas turbine generator, the overall cycle efficiency increased from the 71 per cent shown in Fig. 4.10 to 79 per cent shown in Fig. 4.11.

Economics applying to this example are shown in Fig. 4.12. The savings involved are modest compared to a large refinery system. On the other hand, there are a number of possibilities in any typical refinery, so in aggregate they can add up to significant savings.

This concept can be utilized to provide a considerable share of the refinery's electric power if several units are operated in parallel. One normally thinks in terms of centralized power generation, but with modern speed-control governors, fast-acting voltage regulators and reliable protective relaying, multiple units located strategically in various process plant are a real possibility.

Thermal fluid heaters

Most types of wate heat recovery equipment can be used to recover heat from the exhaust of gas turbines, as will be evident from a reading of Chapter 11. One system becoming increasingly popular is the use of this heat to raise the temperature of thermal heat transfer fluids such as Dowtherm, Thermex and Mobiltherm.

One of the major manufacturers of these heaters is the Eclipse Lookout Company, who produced packaged thermal fluid heat recovery units ranging in output from 450 kW

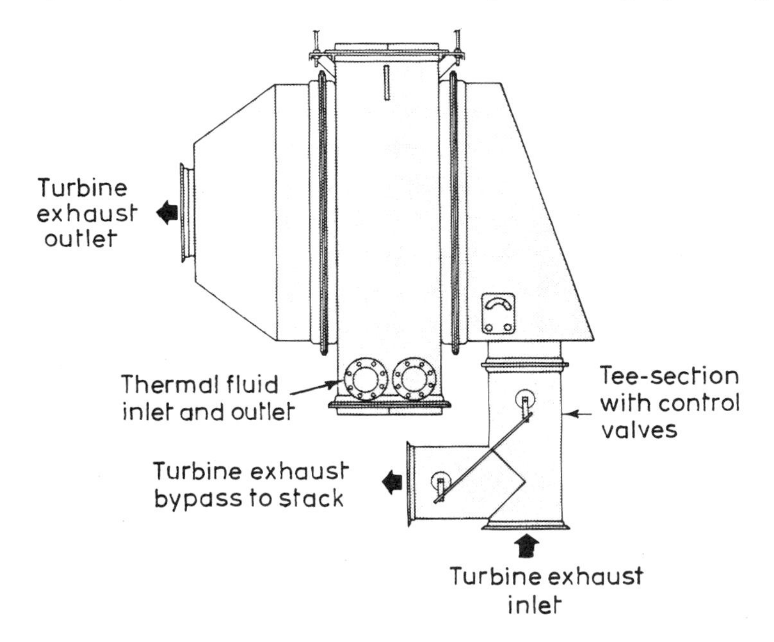

Fig. 4.13 An Eclipse Lookout thermal fluid heater for gas turbine applications, rated at 1·2 MW.

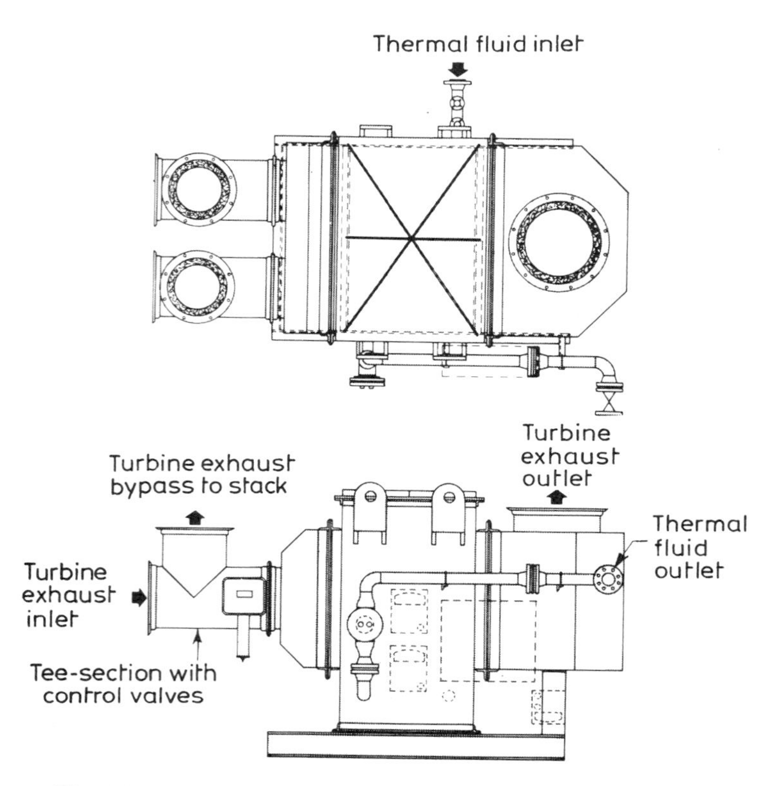

Fig. 4.14 A unit, also by Eclipse Lookout, rated at 600 kW.

to 5·5 MW. Two examples of these heaters are shown in Figs. 4.13 and 4.14. In these units, the turbine exhaust gases are passed over a finned tube heat exchanger through which the thermal fluid is pumped. Normally the thermal fluid remains in its liquid phase, being transferred to the processes where the heat is required. (It is important in the application of thermal fluid heaters to ensure that liquid-film temperatures are controlled. Some fluids can operate with film temperatures as high as 400°C, although continuous but predictable thermal degradation does occur, necessitating fluid treatment. Other fluids have lower operating maxima, and users will find the fluid manufacturers fully able to advise on their use.)

Note in the figures that both units incorporate an exhaust gas bypass, so that the tube bundle may be isolated. This may be used when cleaning is required. Gas turbine exhaust gases are normally relatively clean, particularly if natural gas is used as the fuel. Light oil-fired turbines have some pollutants in the exhaust, and these increasing during load changes. Although it is desirable to use finned tubes for the heat exchanger, whatever the fuel, some form of sootblowing or other cleaning method should be allowed for (Boyer, 1975).

The associated distribution system for a thermal fluid heated by gas turbine exhaust heat, as recommended by Eclipse Lookout, is illustrated in Fig. 4.15. Note

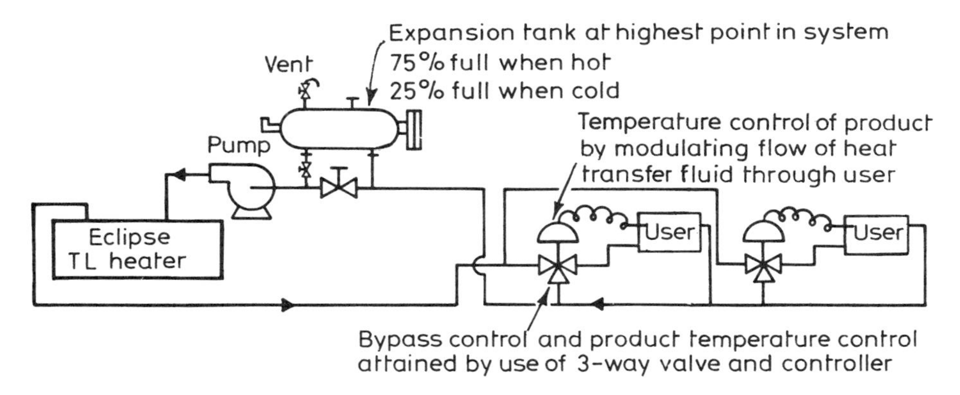

Fig. 4.15 Thermal fluid control circuit, including an expansion tank and process temperature control.

particularly the necessity to incorporate an expansion tank, and the provision of temperature control at the process itself by modulation of the flow of the thermal fluid.

Waste heat boilers

The waste heat boiler (see also Chapter 2) can be effectively used as a waste heat recovery unit on gas turbines. The efficiency of a gas turbine generator set by itself is very low. Hot-water heat recovery systems and low-pressure steam systems recover up to 75 per cent of the available heat in the turbine exhaust and can produce an overall efficiency of 80 per cent. The application of a high-pressure steam waste heat boiler results in recovery of up to 65 per cent of the available heat, giving an overall efficiency value of 72 per cent.

The potential heat recoverable from turbine exhaust gas is approximately 10 kW/kg/s per °C temperature drop in the exhaust gas. With an exhaust mass flow of 15·4 kg/s and a gas temperature drop through the heat exchanger from 554°C to 160°C, the recovery rate is approximately 6·1 MW. As a comparison, if the equivalent fuel is used in a fired boiler having an efficiency of 80 per cent, the heat of combustion is slightly greater than 7·9 MW. This heat release rate is equivalent to 14 l/min of fuel oil or 720 m^3/h of natural gas. The manufacture of the system meeting this specification estimates a payback period on the heat recovery equipment of about three years, taking into account operating and maintenance costs.

In the gas turbine waste heat boiler illustrated in Fig. 4.16(a), a supplementary-firing capability is provided. This releases the particular process for which the steam is being raised from its dependence on the performance of the gas turbine, and also allows boiler efficiency to be raised. (Normally the waste heat boiler for a gas turbine is sized on the basis of the exhaust flow rather than the required steam output, making the unit relatively expensive. Thus, any means whereby its output could be greatly increased with little additional expenditure is of interest.) As considerable amounts of oxygen

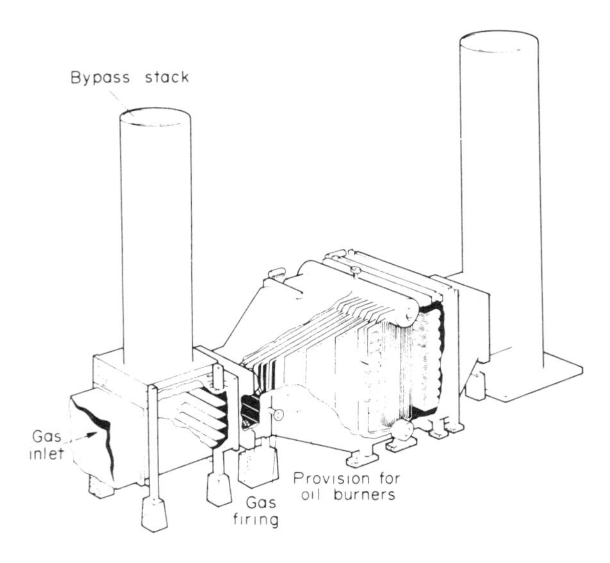

Fig. 4.16(a) A waste heat boiler used to recover heat from a gas turbine exhaust, aided by supplementary firing.

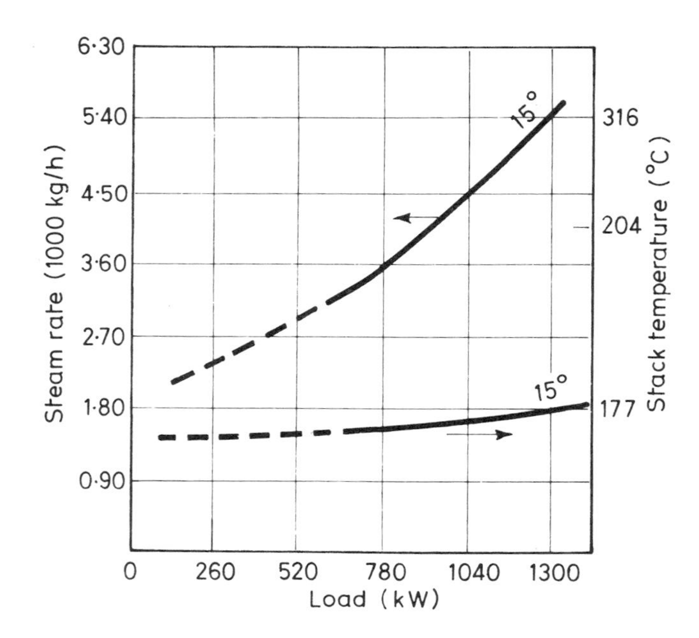

Fig. 4.16(b) Effect of part-load operation on the steam-raising rate of a Ruston TA1750 turbine using a Conseco waste heat boiler giving steam at 19·25 kg/cm^2.

still exist in the exhaust gases, this can be used to assist supplementary firing, and in some cases it is possible to increase the steam-raising capacity over and above that produced by the unfired boiler by 200 per cent. As a result a very large gain in boiler output is achieved at the expense of the installation of simple grid burners. (In the unit illustrated, the bypass stack is included to permit boiler operation at lower loads than those dictated by exhaust mass flow, if the need to reduce steam production arises, Bunton (1974). Fig. 4.16(b) illustrates the effect of part-load operation on the steam-raising rate of a gas turbine waste heat boiler – in this case a Conseco unit applied to a Ruston TA1750 turbine.

4.3 Reciprocating engines

4.3.1 *General characteristics*

In applications where it is advantageous to use a gas turbine or reciprocating engine as a prime mover, instead of an electric motor as discussed above, it is generally necessary on grounds of economy to recover a large proportion of the waste heat generated, rather than rejecting it to atmosphere or to cooling water.

Reciprocating engines have several advantages over the gas turbine when a substantial mechanical output is required. In general their shaft efficiency is greater than of the turbine, ranging from 27 to 35 per cent, and their cost is less, at least at the lower end of the output scale. On the debit side, the ancillary systems required for reciprocating engines tend to be more complex, particularly in view of the water cooling required, and oil consumption can be considerable. Units have a low power-to-weight ratio, and bad atmospheric pollution can occur, particularly if maintenance is neglected. Maintenance is one of the features which, certainly in the past, has held back wider use of reciprocating engines in industry, particularly in applications where very high utilization is required.

There are two basic types of reciprocating engine, dual-fuel and spark-ignition units. Dual fuel in this context applies to an engine which runs on the diesel cycle, using a small amount of diesel fuel as the igniter when using natural gas as the main fuel. This results in simultaneous combustion of atomized liquid fuel and gas, and the engine has similar economy and emission characteristics to the pure natural-gas engine. In instances where a company has an interruptable natural-gas supply, this type of engine can be switched immediately to 100 per cent diesel fuel, without switching off the power supply. The diesel fuel requirement to initiate combustion is typically 5 to 10 per cent of full-load diesel requirement.

Spark-ignition engines have in the past been restricted to use with petrol and diesel fuels, particularly in the UK. This was because the price of town gas, together with its high hydrogen content, restricted the performance of spark-ignition engines. Only very low compression ratios could be achieved. As a result, large spark-ignition gas engines have been manufactured mainly in the USA and on the European continent, most UK types being dual fuel.

In general, the use of natural gas as a fuel is preferred in industrial applications. It burns cleanly, hence reducing maintenance of the engine as no carbon is deposited.

Also, no oil dilution occurs and hence lubricant life is improved. As the main constituent of natural gas is methane, which has a high 'knock' resistance, high compression ratios (up to 12:1) can be obtained in spark-ignition engines, leading to higher efficiencies. The supply of gas, via a pipe, is also convenient, and storage facilities on the site are avoided.

Spark-ignition gas engines may be either purpose-built or converted from petrol or diesel engines. Conversion from a diesel engine requires modification of the pistons to correct the compression ratio, introduction of spark plugs into the cylinder head, and changing of the carburation system such that the new unit can homogeneously mix the gas and air in the necessary ratios. Petrol-engine conversions also require new carburation systems, and may require some adjustment to the compression ratio. The ignition timing on gas engines is advanced further than on petrol engines because the flame-propagation rates are lower for natural gas.

The power obtained from natural-gas engines is somewhat less than that obtained from the corresponding petrol engine because of the large volume occupied by the gas in the cylinder compared with atomized liquid petrol. Furthermore the intake manifold of a petrol engine is usually heated to produce partial evaporation of the liquid petrol. For gas-engine duty, this heating merely increases the specific volume of the gas and further reduces the available power.

Gas-engine manufacturers quote different figures for the derating of maximum brake horsepower for continuous duty. The American Petroleum Institute recommends derating the maximum BHP by 35 per cent for continuous duty. Gas-engine thermal efficiencies lie between 27 and 33 per cent.

Because many available engines are generally converted quite cheaply, from automotive or marine applications, it is considered that derating is important in order to extend engine life and reliability as much as possible. The best way to derate an engine is to slow it down under full throttle to the required horsepower; this will, if anything, slightly improve the efficiency over full load.

4.3.2 *Engine heat recovery*

The quantities of heat which can be recovered from natural-gas reciprocating engines are shown in Table 4.2 (compare with gas turbines which have significantly lower efficiencies in Table 4.1). This data is presented in a different manner in Table 4.3, itemizing the quantities extracted in the cooling jacket and engine exhaust (Anon, 1965).

A system based on a diesel generating set, using both exhaust and coolant-water heat, installed at the Petbow factory in the UK, is illustrated in Fig. 4.17. The diesel engines are 500 kVA Rolls-Royce units, one of the three installed always being kept on standby (Butler, 1975). The system for recovering heat centres round a 95 m^3 (21 000 gal) thermal storage vessel. Here water is kept near boiling point by extracting heat in heat exchangers through which the engine-radiator cooling water and exhaust gases are passed. Water from the storage vessel is then pumped to the factory areas to special heating units in each bay at roof level. Hot air is circulated round the bays by fans blowing cold air over hot pipes. The temperature in each bay is controlled

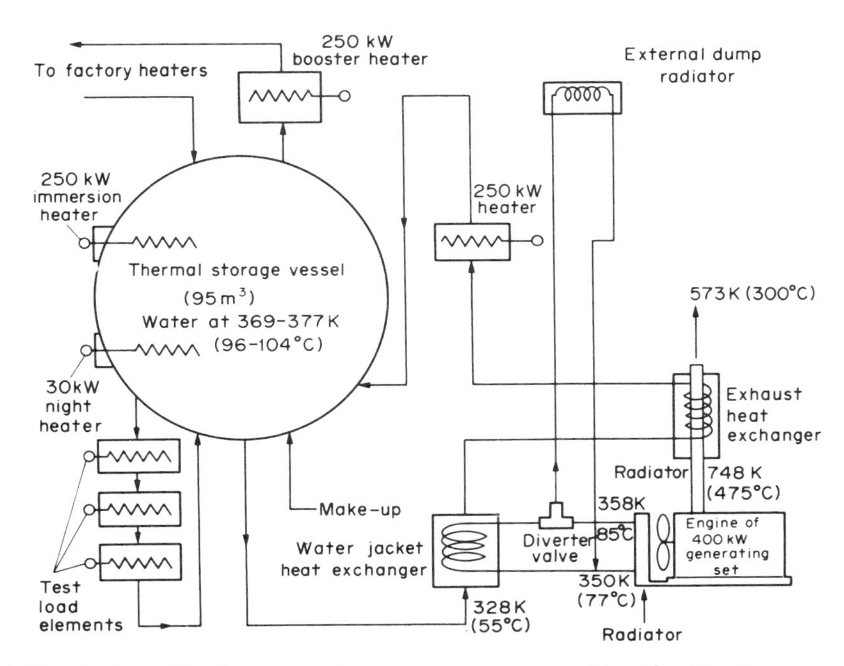

Fig. 4.17 Petbow diesel generator/waste heat storage combination (*The Engineer*, 23rd Oct. 1975).

individually by thermostats. Refinements include a 30 kW immersion heater to keep the water in the storage vessel hot overnight, 250 kW booster immersion heaters to provide quick startup early in the morning, and an external radiator to dump heat during exceptionally warm spells.

The company aims to use its system to gather data on how total energy works in practice. A full set of instruments will permit continuous monitoring of data such as fuel and power consumption, and air temperature.

The total cost of the system is reported to be £200 000. The diesel engines will only be used to generate electricity when there is a demand in the plant for the waste heat from the engines, and the use of this energy for space heating suggested that it would be most economical to generate electricity 'inhouse' during the day for eight months in the year, night and summer power requirements being obtained from the national grid. The electricity authorities were found to be most helpful in implementing this scheme.

There are many other applications where waste heat can be successfully utilized. As well as generating electricity, reciprocating engines can be used in air conditioning and refrigeration, and gas engine-driven chiller units are marketed on a wide scale. Also, where modulation may be applied easily when variations in load occur. Drives for centrifugal compressors, which may have rotational speeds up to six times those of the gas engine, are another application. It has been found that to obtain the best compromise between equipment first cost (engine, couplings and transmission) and maintenance cost, engine speeds of 900 rev/min are recommended.

Table 4.2 Typical reciprocating gas engine inputs and recoverable outputs

	75 kW; C 12:1; 1800 rev/min NA; P 11 mm h.g.				175 kW; C 10:1; 1200 rev/min TA; P 100 kN/m^2			
		Outputs				Outputs		
Shaft load (%)	Fuel rate (kW)	S (kW)	H (kW)	S/H	Fuel rate (kW)	S (kW)	H (kW)	S/H
100	293	75·0	148	1/2·0	638	175	247	1/1·4
90	272	67·5	134	1/2·0	595	158	222	1/1·4
80	253	60·0	121	1/2·0	545	140	198	1/1·4
70	233	52·5	109	1/2·1	504	122	177	1/1·4
60	214	45·0	97	1/2·2	457	105	154	1/1·5
50	194	37·5	87	1/2·3	408	88	132	1/1·5

Notes: C – compression ratio; NA – naturally aspirated; TA – turbocharged; S – shaft output; H – heat output; P – gas pressure.

Table 4.2 *(continued)*

	225 kW; *C* 10:1; 1200 rev/min *TA*; *P* 100 kN/m^2				450 kW; *C* 10:1; 1200 rev/min *TA*; *P* 100 kN/m^2			
		Outputs				Outputs		
Shaft load (%)	Fuel rate (kW)	*S* (kW)	*H* (kW)	*S/H*	Fuel rate (kW)	*S* (kW)	*H* (kW)	*S/H*
100	805	225	311	1/1·4	1610	450	624	1/1·4
90	735	203	276	1/1·4	1475	405	556	1/1·4
80	688	180	250	1/1·4	1370	360	495	1/1·4
70	624	158	218	1/1·4	1240	315	433	1/1·4
60	562	135	189	1/1·4	1110	270	375	1/1·4
50	500	113	162	1/1·4	980	225	319	1/1·4

Notes: *C* – compression ratio; *NA* – naturally aspirated; *TA* – turbocharged; *S* – shaft output; *H* – heat output; *P* – gas pressure.

Table 4.3 Typical heat recovery rates of various types of engines

(kcal/kWh) (based on 100 kN/m^2 steam pressure and 38°C ambient)

Type of engine	Fuel input at rated load	Heat recoverable at rated load and b.m.e.p. Jacket water: Air-cooled manifold	Water-cooled manifold	Exhaust unit
Two-cycle				
Mechanical supercharged gas	2800	590	750	410
Naturally aspirated gas	4100	1100	1300	510
Blower charged diesel	2800	560	660	370
Four-cycle				
Naturally aspirated gas	2900	650	790	420
Naturally aspirated diesel	2900	650	790	420
Turbo-charged diesel	2500	370	460	410
Turbo-charged dual-fuel	2200	320	370	340
Gas turbine				
Simple cycle	6200			4300
Regenerative cycle	4400			2500

One of the most interesting applications of the diesel/gas engine is in driving a heat pump compressor. In this case the ability to use the waste heat to full advantage makes the system more efficient on primary energy consumption than an electric system, and this is also discussed fully in Chapter 6.

Waste heat recovery equipment

Many reciprocating engines are available direct from the manufacturer with heat recovery units utilizing the engine water cooling and lubrication oil heat. Waste heat boilers or hot-water heaters served by the exhaust gas are available as separate units, or may be incorporated in a single 'package', as illustrated in Fig. 4.18. In common with many other exhaust-gas systems, this unit manufactured by Pott Industries also acts as an exhaust silencer. Units are available catering for engines ranging in output from 100 kW to 15 MW.

As shown in the sectional drawing of the 'Vaporphase' unit shown in Fig. 4.19, the exhaust gases pass through tubes immersed in the water before venting, this water being preheated by the heat in the water-jacket cooling water, which is subsequently returned cool to the engine. Provision is also made for blowdown.

Of major importance in the operation of any heat recovery installation connected to a prime mover whose load may vary, is the control system. As shown in Table 4.4,

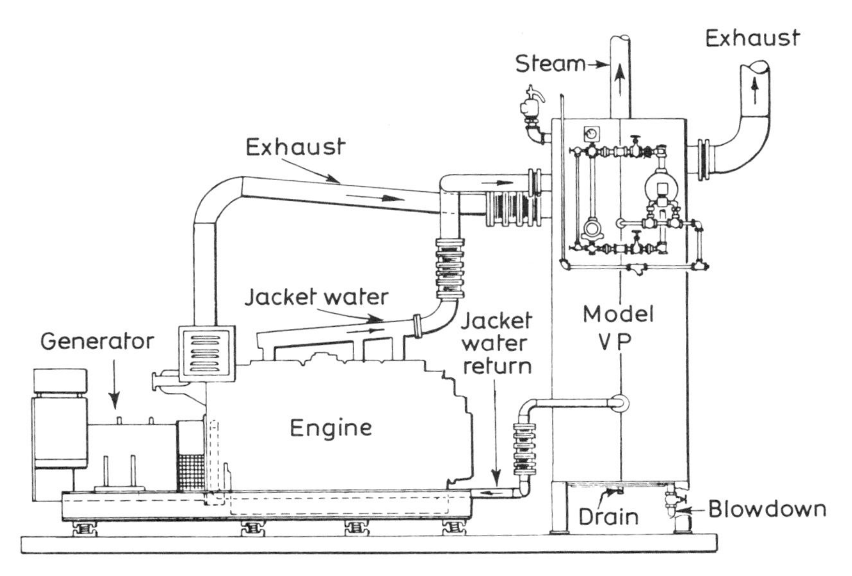

Fig. 4.18 Vaporphase model VP heat recovery system, integrating water jacket and exhaust heat sources.

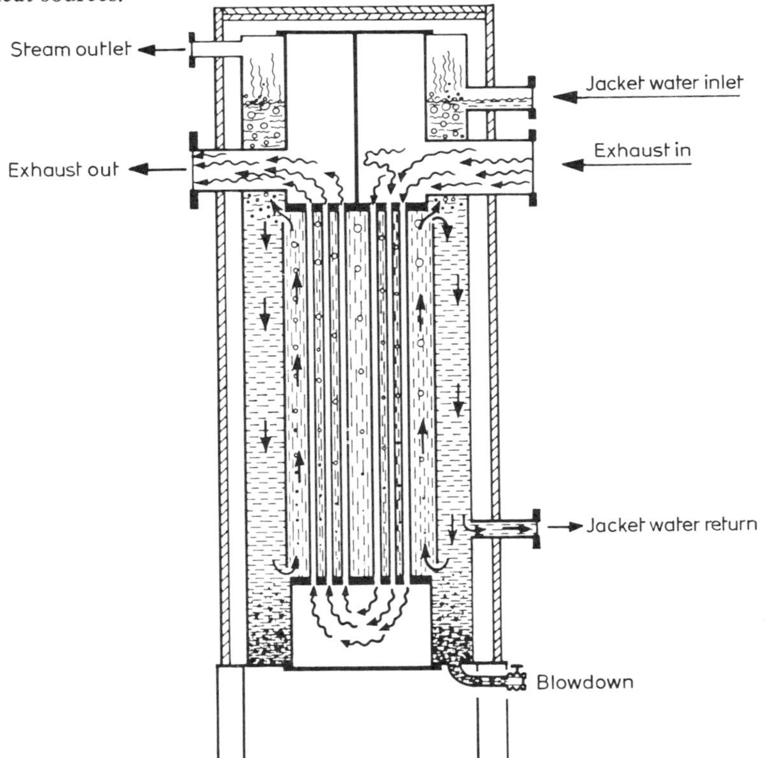

Fig. 4.19 Combined water jacket and exhaust-gas heat recovery boiler, vertical unit.

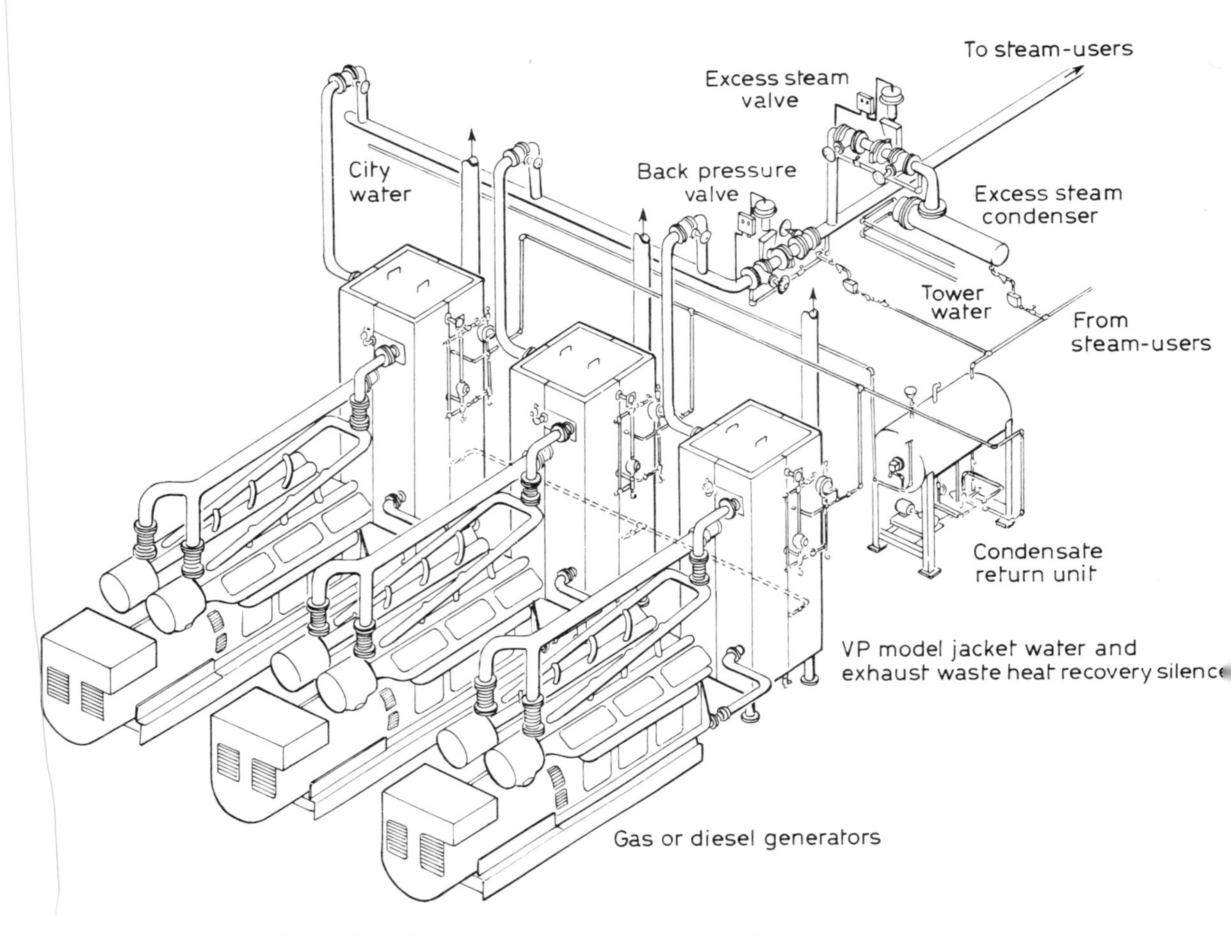

Fig. 4.20 Heat recovery system for a typical total energy plant.

where the standard and optional control features on the 'Vaporphase' unit are listed, comprehensive safety and control devices are available. A layout of a typical total energy plant employing reciprocating diesel or gas engine generators, shown in Fig. 4.20, illustrates the additional items of plant needed to serve the basic boilers.

Other factors to keep in mind when using water-cooled engines are as follows:

(i) Ebullient cooling, because it involves evaporation in the engine, requires special precautions as far as water treatment is concerned (see below).
(ii) In a multiengine installation, separate cooling systems to each machine should be used to avoid total shutdown should one coolant-system component fail.
(iii) Care should be taken to ensure provision for condensate removal, this forming in exhaust passages when the engine cools.
(iv) Water-level control of separate-engine cooling systems that produce steam must cater for the possibility of backflow through the system nozzle of units which are not running. Use of a steam check valve is recommended.

Table 4.4 Controls and safety features on a reciprocating engine heat recovery system

Standard controls and instrumentation

Control	Function	Features
Water-control valve	Maintains normal water level	Modulating, float operated for close level control Tight shutoff Union mounted with isolation valves for easy service Popular control insures availability of replacement, parts or service
Low-water shutoff switch	Stop engine in event of low water	Union-mounted for quick changeout or service Popular control insures availability of replacement, parts or service Properly located to prevent false shutdowns
3-Way bypass valve	Allows bypass of water-level control valve	Single-lever valve operation allows manual feed to unit in event of level control failure or normal service operation
Air-vent/vacuum breaker	Discharge air out of J.W. System and prevents vacuum	Thermostatically operated Provides reliable air elimination
Tubular-gauge glass	Allows visual check of water level	Covers full range of control column for constant knowledge of actual water level Provided with gauge guard for maximum protection against breakage Provided with automatic shutoff cocks for protection in event of glass breakage
Safety valve	Pressure-relieving device to protect system from overpressures	Sized for 100% production capacity of unit Quality valve provides tight shutoff ASME approved
Pressure gauge	Indicates operating pressure of system	4½ in dial gauge insures easy readability from distance 0-30 # range places normal operating pressure in middle of range with 2% accuracy Mounted with shutoff valve and syphon
Blowdown valves	Provide surface and mud drum blowoff for solids-concentration control	Surface blowoff facilitates removal of foaming agents, thereby eliminating priming and carryover of solids that can foul heat transfer surface of absorption equipment and other steam-users Blowdown located in mud drum (dead-water space) where solids formed by water treatment are precipitated out

(*This table is continued on the next page.*)

Table 4.4 Controls and safety features on a reciprocating engine heat recovery system (*continued*)

Optional controls

Control	Function	Features
High-water alarm or shutdown switch	Sounds alarm or shuts down engine in event of high-water level	Union-mounted for quick changeout or service Popular control insures availability of replacement, parts or service Properly located to prevent false shutdowns
Low-water alarm	Sounds alarm in event low-water level is occurring	Union-mounted for quick changeout or service Popular control insures availability of replacement, parts or service Properly located to prevent false-alarm signal
Low-water alarm and emergency feeder	Sounds alarm and feeds treated city water in event of failure of normal feed system	Union-mounted for quick changeout or service Popular control insures availability of replacement, parts or service Properly located to prevent unnecessary city water feed Alarm sounds at same time as city water feeds Allows operator to determine source of normal system failure and correct condition without shutdown
High-pressure switch	Sounds alarm or shuts down engine in event of high steam pressure	Mounted on control column Popular control insures availability of replacement, parts or service

There are several publications which give excellent accounts of the many factors to be taken into account when installing this type of plant, and a manual such as the *ASHRAE Systems Handbook* provides much useful data.

As an alternative to the conventional water-jacket cooling system on engines of this type, an 'ebullient' system may be used. This cooling system involves evaporative heat transfer, and operates with a low-temperature differential, peak temperatures being in the range 100°C to 125°C. As the coolant ascends in the engine jacket and riser leading to the steam separator, evaporation takes place and increasing amounts of steam are generated. The system is aided by natural circulation brought about by the changes in coolant density. In the separator, steam at about 100 kN/m^2 is removed from the coolant and transported to the process where it will act as the energy source. The water from the separator is returned to the engine, supplemented by the condensate returning from the process heated by steam.

When an ebullient system is specified for engine cooling, a waste heat boiler utilizing exhaust heat will provide additional steam. These can be obtained incorporating silencers. Should the exhaust heat be required to provide hot water, this can also be implemented. Engines capable of utilizing ebullient cooling are generally more expensive than those employing sensible heat removal methods.

Table 4.5 Operating costs on total heat recovery – 1 MW to 2 MW unit size of engine

Fuel type	Class A/B	Class F	Class G	Dual fuel
Cost p/therm	13·4	11·4	9·6	11·0
Engine speed rev/min	750	750–600	600–500	750–600
Capital cost £/kW				
(a) Elect/mech. equipment	110–124	120–170	150–190	120–180
(b) Switchgear and cables	12–20	12–20	12–20	12–20
(c) Civil work erection cooling system	20		20	20
(d) Total £/kW	142–164 (153)		182–230 (206)	152–220 (186)
Fixed annual charges				
Insurance, interest and depreciation on 1½ MW unit £/kW installed (av.)	1.5304	1.8107	2.0608	1.8607
Operating cost p/kWh				
(a) Fuel	1.292	1.004	0.850	1.100
(b) Lubricating oil	0.050	0.053	0.056	0.050
(c) Spares and labour	0.027	0.033	0.036	0.023
(d) Staffing	0.067	0.133	0.200	0.067
(e) Total p/kWh	1.436	1.223	1.142	1.240
Capitalization p/kWh				
(a) 50 per cent utilization (4380 h)	0.349	0.413	0.471	0.425
(b) 10 per cent utilization (876 h)	1.747	–	–	–
Gross generating costs p/kWh	3.183	1.636	1.613	1.665
Waste heat recovery				
(a) Exhaust only 0·45p/kWh				
(b) Total 0·95 p/kWh	0.950	0.950	0.950	0.950
Net generating costs p/kWh*	2.233	0.686	0.663	0.715

* Using total recovery figures for heat value.

Economics

Any economic analysis applied to heat recovery from reciprocating prime movers must take very careful account of maintenance and reliability costs. Engine maintenance costs are not as easily estimated as fuel consumption, recoverable heat, or initial cost;

they are, however, not entirely elusive. The increasing use of guaranteed maintenance and service contracts most common in the USA has eliminated much of the estimating formerly required in feasibility studies. Maintenance contracts vary from complete maintenance and service, including all parts, supplies and labour, to contracts that provide only a guaranteed cost for engine rebuild. For this reason, a maintenance cost figure is meaningless unless well defined. Complete maintenance costs are composed of three basic items:

(a) The miscellaneous maintenance and service cost, including service manual recommendations plus makeup oil (excluding labour to perform this routine duty).
(b) The overhaul maintenance cost, usually expressed in terms of cost per engine operating hour. This item should cover all labour and parts necessary to perform major and minor overhauls at the recommended intervals.
(c) The third item is the labour cost necessary to perform the miscellaneous service for item (a).

Items (a) and (b) will vary considerably with the severity of service the engine must perform. For onsite power installations the conditions under which the engines operate, the quality of fuel, the routine maintenance the engine receives, are all usually considered to be good. Item (c) will vary with labour costs and the location of the engine plant with respect to the point from which service personnel must be dispatched. Because of the many variables, it is difficult to provide realistic figures that would be useful for all applications. Maintenance costs should be based on past experiences in the area being considered.

Quantification of maintenance costs, and any other cost associated with the operation of plant, can be misleading because of the time factor involved, inflation making it desirable, where possible, to use the latest data available. Shearer (1976) presents figures based on 1976 costs for reciprocating engines running on gas oils (Class A/B), residual oils (Classes F, G) and natural gas (in dual fuel systems), and he takes a low average value for the recovered waste heat of 0·45 p/kWh for the exhaust heat and 0·95 p/kWh for total heat recovery (including water-jacket heat). The cost are given in Table 4.5 for reciprocating engines in the size range 1 MW to 2 MW. For further discussion on this topic, the reader is referred to a book dealing solely with total energy plant (Diamant, 1965).

4.4 Turbines as users of waste heat

There are three major areas for the use of turbines in waste heat recovery, converting waste heat into power. These are:

(i) Combined cycles
(ii) Steam turbines linked to waste heat boilers

(iii) Vapour turbines using low-grade waste heat

4.4.1 *Combined cycles*

If the requirements of an industrial plant demand considerable generation of electricity in preference to very large quantities of waste heat, the use of a gas turbine alone to drive the generator would be uneconomical. However, because of the high (= 600°C) exhaust temperatures of the gas turbine, its waste heat can be used to raise steam, which in turn can drive a second turbine. This combined cycle, involving gas and steam turbo-generator plant, is illustrated in Fig. 4.21. Use of this system can raise the thermal efficiency to 42 to 44 per cent, and it retains several advantages over the isolated steam turbine installation. It is possible to start up and shut down this combination rapidly, and if a number of gas turbines are used in conjunction with a single steam turbine for generating large powers, high efficiencies can be maintained even if the load is sometimes as low as 75 per cent of peak value.

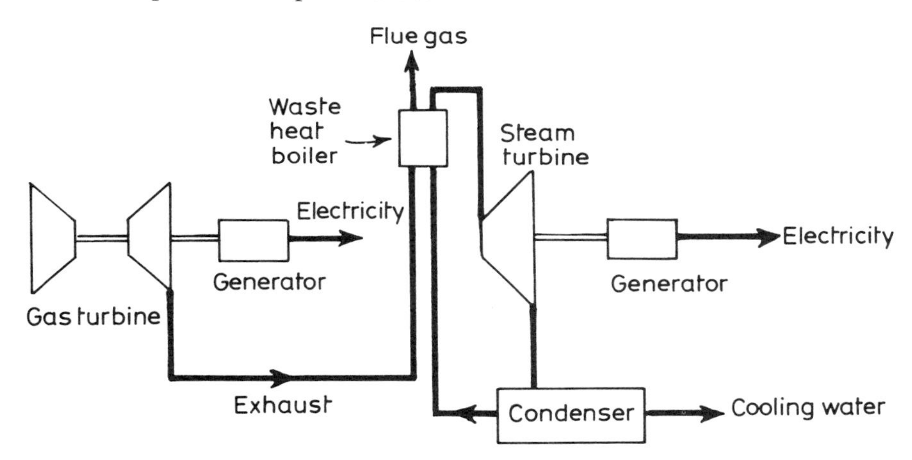

Fig. 4.21 Combined power cycle, using gas turbine exhaust to raise steam for driving a second turbogenerator set.

The potential of the combined cycle will be even greater when gas turbine inlet temperatures can be raised well above 1000°C. It has been predicted that efficiencies in excess of 50 per cent are achievable when high inlet-temperature units are combined with steam plant.

The type of steam turbine used in conjunction with the gas turbine depends upon the duty. If electricity generation is the most important consideration, a condensing steam turbine will be used. If, in addition, there is a need for process heat in the form of steam, a backpressure turbine, in which the steam exhausts into the condenser at a relatively high pressure, is used. Alternatively, an extraction (or passout) steam turbine can meet this second requirement; in this system steam for process use is extracted between turbine stages at the required condition.

The first such installation in Europe was commissioned in Germany in 1973. With a peak thermal efficiency of 39·8 per cent using coke-oven gas as fuel, and 42 per cent using natural gas, this system has a net output of approximately 35 MW, the emphasis in this instance being on power generation. Apart from the increased thermal efficiency compared with an isolated steam turbine, other advantages claimed are (Mitchell *et al.*, 1973):

(i) Lower cooling-water consumption, only about 33 per cent of that of a corresponding conventional steam turbine plant.
(ii) Lower atmospheric pollution (NO_x).
(iii) Lower cost of investment per kilowatt installed. In this case the installed cost has been quoted as just less than $100 per kilowatt.

Obviously it will be possible to 'ring the changes' when selecting a combined-cycle system, depending upon the power/heat requirements. At present operational experience in Europe is limited, but there are a number of equipment manufacturers producing both prime movers, and in many cases the combined cycle will prove more attractive than isolated steam or gas turbines.

One area where such a system has been successfully applied is in the paper industry. A paper machine having a capacity of 10 ton/h will use approximately 30 MW of heat and 3·5 MW of electricity. While much of the steam required for heating is at a low pressure, it is generally raised at comparatively high pressures, making backpressure power generation attractive. In a papermill which combines pulp production and paper manufacture, steam at 2400 kN/m^2 is used in the hardboard-mill and in the chemical digester, 760 kN/m^2 steam is used in the packaging papermill, while steam at only 172·5 kN/m^2 is passed through the drying cylinders of the machine manufacturing newsprint paper. Among the major uses of electricity are the log grinders the digester, and the drives for the cylinders in all the paper drying plant.

This would indicate that a total energy plant capable of meeting the demand for both steam and electricity would be attractive in a papermill. However, these figures relate to one specific mill, and it is very difficult to satisfy the ratio of steam:power consumption in all plant of this type. Owing to the differences in the grades of paper produced and the different production techniques used to manufacture these various grades, this ratio can vary considerably. Similarly, raw-material composition and product quality can also affect energy usage.

The rate of production also affects this balance. Demand for steam required per kilogram of product reduces with increasing machine speed, whereas the electricity consumption, calculated on the same basis, is essentially constant. Thus, a total-energy plant would be required to have characteristics which permit the steam output to be raised, while maintaining a constant electrical load (Frei, 1975).

A further influence on the relative heat and power requirement of the plant not previously mentioned is the number of machine stoppages. Most of these are likely to occur because of a break in the paper-web, and although the stoppages are often of fairly short duration, they are accompanied by a sudden drop in the steam consumption.

In spite of this, processes further upstream such as log preparation, and finishing processes downstream of the cylinders, may proceed as normal, leaving the demand for electricity comparatively unaffected.

Sulzer argue that the power coefficient S defined as the ratio between the specific power requirement (MJ/tonne of product) and the specific heat requirement (Gcal per tonne of product), may vary from 540 to 3600 MJ/Gcal, depending upon the state of the plant and the product. This variation in power coefficient may be difficult to meet with a conventional backpressure steam turbine total energy plant, mainly because the electrical output from the available steam rate is too low.

This leads to a surplus of steam and lack of power, which, with the trend towards higher power coefficients, makes the backpressure turbine alone less attractive. The unit is also less efficient when operating outside its optimum range, as may be required due to the various factors discussed above which can create changes in plant power coefficient.

The solution proposed by Sulzer utilizes a gas turbine/generator set, the waste heat

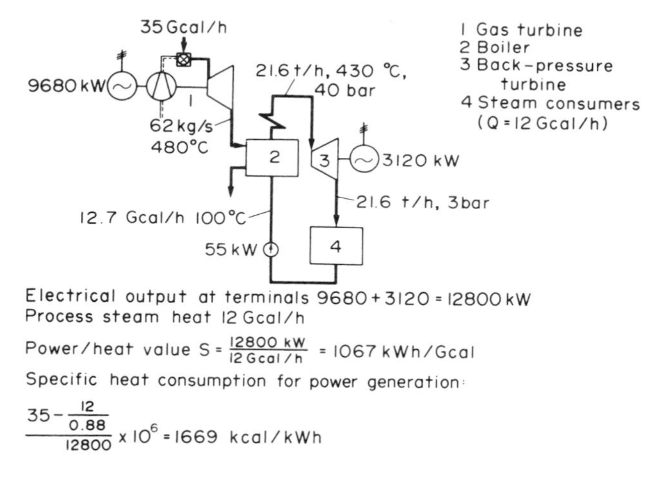

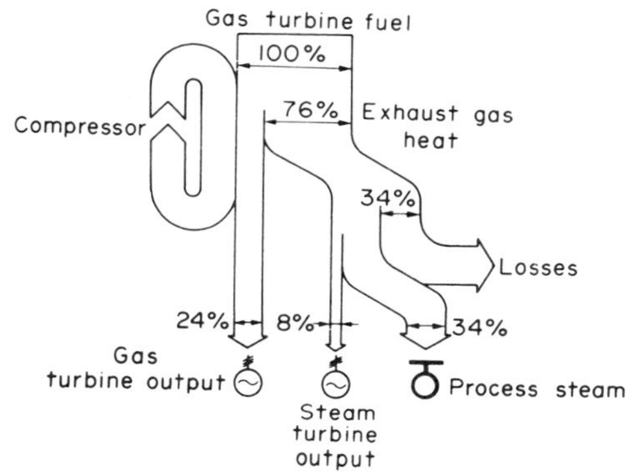

Fig. 4.22 Combined cycle heat and power plant based on a gas turbine/backpressure turbine package, with heat flows (by courtesy of *Sulzer Technical Review*).

from the gas turbine exhaust being used to raise steam which in turn is fed to a back-pressure turbine. This turbine generates a limited amount of electricity, and the steam, having expanded through the turbine, is then available for process use at a pressure of about 3 bar. The system is illustrated in Fig. 4.22.

Compared with a system relying solely on a backpressure turbine, this arrangement can give a peak value of S three times as great, while producing 25 per cent more steam. An energy balance carried out on the gas turbine/steam turbine plant reveals that 34 per cent of the energy input is lost as exhaust gas heat, and greater efficiencies might be attainable if a proportion of this heat could be used to warm air used for drying above the cylinders. This is already done in tissue plant, where the power coefficients are very high (Frei and Holik, 1973).

Mention should also be made here of a new waste heat recovery concept, which may be linked to any process steam system, and which has also been proposed as particularly applicable to papermills (Gaggioli *et al.*, 1977). In this system, a compression/expansion unit would be located between the outlet of a turbogenerator (providing both electricity and process steam) and the paper machine, which is the steam-user, as illustrated in Fig. 4.23. Basically a heat pump (see Chapter 6), the important feature of the layout is

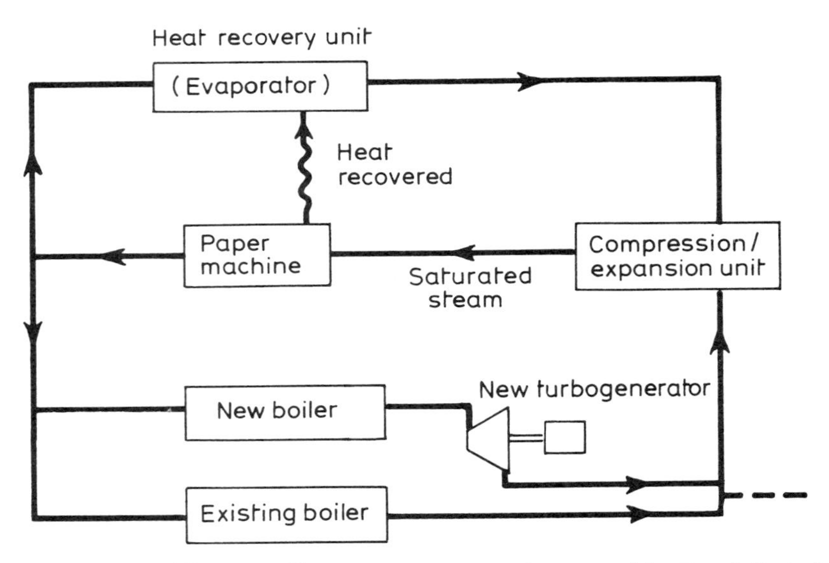

Fig. 4.23 Schematic diagram of heat recovery concept proposed by Gaggioli *et al.* for a papermill.

the appreciation that the paper machine, which acts as a condenser of steam, could also act as the condenser for a refrigeration loop – the upper loop in the figure. In this case the 'refrigerant' would be water, and the branch of the loop incorporating the paper machine carries vapour for both the heat pump circuit and the power cycle (the lower loop). The evaporator, as in any heat pump, operates at a lower temperature than the condenser, and can use process effluent or other heat sources from the papermill (e.g. hot air). A steam turbine-driven centrifugal compressor or, as an alternative to a heat pump, a steam ejector, would be used to convert the recovered heat into steam useful for the paper machine.

Many possible configurations were examined, one of the most promising economically resulting in a capital-cost saving of $ 4·1 million (compared to an original plant cost of $ 12·5 million) and a net annual saving of $ 350 000.

4.4.2 *Steam turbines linked to waste heat boilers*

This section briefly discusses one particular type of steam turbine which can use steam from a waste heat boiler or, for that matter, any other steam source, even at pressures of less than 4 bar. One company, the Terry Steam Turbine Company in the USA, has pioneered the commercial use of low-pressure turbines for recovery of energy available in low-pressure steam.

Low-pressure turbines are usually multistage units and because of the cost of the attendant condensing facilities, installations for powers less than 150 kW are rare and most of them are above 250 kW. However, ratings below 250 kW are becoming more economically feasible as fuel and utility costs continue to spiral up. Among this group of machine are drives for cooling-tower circulating water pumps, boiler feedpumps and refrigeration compressors, both gas and Freon units.

With low-pressure installations usually costing more than those for other turbine conditions, the incentive is that these units are practical when the steam cost is zero or else very cheap. This condition occurs when low-pressure steam would otherwise be vented to the atmosphere or to a barometric condenser. Thus it is the free, or exceedingly cheap fuel cost, that justifies the larger initial cost of low pressure turbines. It should also be mentioned that cooling water for the condenser will usually have to be conveniently at hand or else the already high initial cost becomes still higher.

The turbines range from single-stage to multistage units. Some users are interested in keeping the first cost down, whereas others are interested in economy of steam consumption. Of course, this brings up a point that must be resolved: steam rate against initial capital investment. One certainly cannot evaluate steam consumption on a low-pressure turbine only in terms of steam cost. Terry argue that the real answer should be based on how much one can afford to spend in order to produce power without fuel cost. Both surface and barometric condensers are used with this type of machine. The barometric condensers are cheaper; can attain slightly higher vacuum; are perhaps slightly more reliable; and have less maintenance cost. The surface-type condenser will reclaim the condensate and this generally is a chief reason for using them, because the value of the recovered condensate frequently is significant. This is particularly true in areas where preparation of raw water for boiler feed purposes is expensive.

In specifying turbines of this type, the manufacturers stress the following points:

(1) Care should be excercised in specifying the inlet pressure as accurately as possible.
(2) A steam/water separator is essential. This is because the starting of a non-condensing

machine upstream could easily dump slugs of condensate into the line that could reach the turbine and damage it. This separator will also help reduce the moisture content, where the low-pressure steam is quite wet, to satisfactory levels.
(3) Inlet sizes for low-pressure turbines are larger than normal (20 cm to 35 cm diameter) and care must be exercised in the piping to avoid pressure drops and places for moisture to become trapped.

Based on a capital cost of about $ 40 000 to $ 50 000 for a 400 kW unit, if a suitable steam source is available, energy savings could amount to almost $ 300 000 in the first year.

4.4.3 *Vapour turbines using low-grade waste heat*

As an alternative to the low-pressure steam turbine described in Section 4.4.2, the turbine utilizing low boiling-point fluids is attractive. Water is by far the most widely used fluid in vapour-cycle turbines, and the search for alternative fluids which could operate in small turbines at different conditions of temperature and pressure has been going on for a number of decades.

As early as 1912 (Ennis, 1912), nine alternative fluids had been tried in turbines largely in binary systems where both mercury and steam cycles are used, the excellent review of the requirements of vapour-cycle working fluids, covering a wide range of operating pressures and temperatures. In the context of energy conservation, however, it is the tubine running on fluorocarbons such as Freon II or Freon 113 which is of interest.

Several factors have helped to generate interest in these fluids for turbines. Much more is known about the thermodynamic properties of these synthetic compounds, also, improvements in technology, particularly in the aerospace industries, have made highly reliable sealed systems possible. In parallel with this, the thermodynamics and fluid mechanics of turbines are sufficiently understood to allow designs for other fluids to be carried out. More recently, the desirability to use even low-grade heat as effectively as possible to conserve energy means that new fluids having lower boiling points than water may be applied.

A 'Fron' turbogenerator system is marketed by Ishikawajima-Harima Heavy Industries (IHI) of Japan, with outputs ranging up to 3·8 MW. Primarily designed to utilize process waste heat, the IHI 'Fron' turbine system, shown in Fig. 4.24, is available as a packaged power unit for generating electricity or driving other machinery (omitting the generator).

Liquid fluorocarbon pressurized by the feedpump passes into the gas generator, where it received heat from the heat source and vaporizes under constant pressure. The high-pressure vapour is then fed into the turbine, undergoing adiabatic expansion. Subsequently the exhaut is condensed and recycled through the feedpump.

A wide variety of heat sources have been suggested for this system, including low-temperature waste steam, hot waste water and other liquid effluent, solar energy and

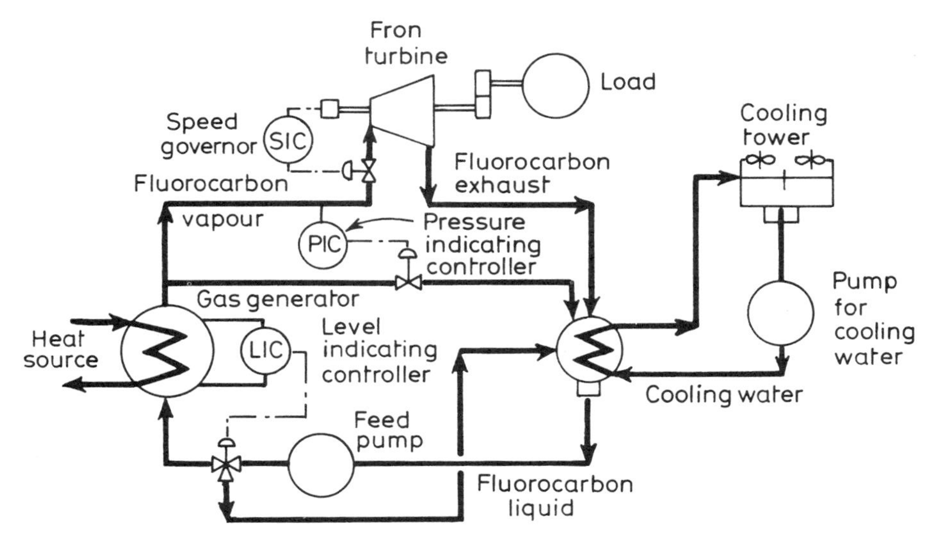

Fig. 4.24 Flow diagram of Fron turbine system.

geothermal energy. The Fron system can also be incorporated in a combined-cycle plant to utilize waste heat from the steam turbine exhaust. The first unit was installed in Okayama at the Mitsubishi Gas Chemicals Company in 1968.

One application of the fluorocarbon turbine is in a sulphuric acid plant, where sulphuric acid at 110° C is the waste heat source (Lewis *et al.*, 1976). Normally in this type of plant, water is used to cool the sulphuric acid, which achieves these high temperatures during concentration processes (involving heat of absorption).

A full technical and economic study was carried out to determine the feasibility of the concept, and it was predicted that a net electricity output of 460 kW to 620 kW would be obtained at an overall thermal efficiency of between 9·7 and 13 per cent. Based on operation for 8400 h/years, it was shown that such a system would be economically attractive if unit costs could be kept below $ 1000 per kilowatt, although the contractors felt that this would not be achieved in the first demonstration. However, if applied universally in the sulphuric acid plants of one company (Allied Chemicals), the systems would save 2·5 million barrels of oil equivalent per year in energy.

References

Boyen, J. L. (1975), *Practical Heat Recovery,* John Wiley, New York.

Bunton, J. F. (1974), 'Recovery of low grade heat', *Proc. Inst. Plant Engrs. Waste Heat Recovery Conf., London, 25–26 September 1974.*

Butler, P. (1975), 'A showpiece to save money by promoting waste heat recovery', *Engineer,* 23 October, 239.

Diamant, R. M. E. (1965), *Total Energy,* Pergamon Press, Oxford.

Ennis, W. D. (1912), *Vapours for Heat Engines,* Van Nostrand, New York.

Frei, D. (1975), 'Gas turbines for the process improvement of industrial thermal power plants', *Sulzer Technical Review,* vol. 56, no. 4, 195–200.

Frei, D. and Holik, H. (1973), 'Total energy supply systems for papermills', *Sulzer Technical Review,* vol. 55, no. 3, 189–194.

Gaggioli, R. A., Wepfer, W. and Chen, H.H. (1977), 'A heat recovery system for process steam industries', *98th ASME Winter Annual Meeting, Georgia, USA, 1977,* ASME Paper 77-WA/ENER-15.

Levers, W. H. (1975), 'The electrical engineer's challenge in energy conservation', *IEEE Trans. Industrial Applications,* 1A-11, 4 *et seq.*

Lewis, G. P. and Smith, R. D. (1976), 'Sulphuric acid plant Rankine cycle waste heat recovery'. *Proc. 11th Intersoc. Energy Conversion Eng. Conf., Nevada, USA, 1976,* vol. 2, 1182–1186.

Mitchell, K. W. S. and Gasparovic, N. (1973), Total energy, *Steam and Heating Engineer,* July, p. 34.

Shearer, A. (1976), 'Selection of prime movers for on-site power generation', *Power Generation Industrial,* no. 3, 26 May.

Stocks, W. J. R. (undated), 'Total energy', Publicity data of Ruston Gas Turbines, Linoln, UK.

Wood, B. D. (1969), *Applications of Thermodynamics,* Addison-Wesley, Reading, Mass. USA.

5 Heat recovery from incineration plant

5.1 Introduction

The incinerator is but one item of plant, like dryers, prime movers and washing machines, which discharges heat energy to the environment, and one may argue that the equipment used to recover heat from incinerators has already been described in previous chapters. To a considerable degree this is true, but the incinerator is too often neglected, unlike much other process plant, as a source of waste heat; and the purpose of this chapter, after briefly describing the three main types of incinerator (in terms of their duty) is to illustrate some cases where waste heat recovery has been successfully applied.

This, together with the manufacturers' data presented in Chapter 12, will hopefully show that existing incinerators are potentially excellent sources of energy, either to reduce the cost of the incineration process or to assist energy savings elsewhere in a plant, and that anyone contemplating investment in incineration plant should seriously consider some form of heat recovery equipment.

5.2 Incinerator types

The incinerator was until recently solely regarded as a piece of equipment to assist disposal of waste products which could not be used in a process or sold for reuse elsewhere. The material to be incinerated may be in the form of a gas, liquid or solid. Incinerators for disposal of gases, to prevent atmospheric pollution with obnoxious or poisonous matter, are commonly known as 'fume incinerators'. Incineration of waste liquid can be used to remove organic pollutants and to recover inorganics. The disposal of solid waste material is the widest used for incinerators, sizes ranging from compact units capable of dealing with a few kilograms per hour, up to very large schemes for disposal of all domestic and industrial waste in a large urban area.

The appreciation of the fact that the waste heat produced by any type of incinerator may be used to preheat the waste or combustion air, or can be fed to other parts of the plant, has led to an interest in heat recovery. Many incinerator manufacturers now offer optional heat recovery equipment. Alternatively the user may select his own recovery system for a special purpose. This section discusses some of the layouts used to date, and reviews some of the incinerators commercially available.

5.2.1 *Fume incinerators*

Fume incinerators are now used, in common with other types of equipment, to remove pollution from gaseous waste before it is discharged to the atmosphere. Most of the air pollutants, identifiable either by their odour, or visually, are combustible and many are in the form of hydrocarbons. The two basic types of fume incinerator used to deal with these gases are the thermal incinerator and the catalytic incinerator.

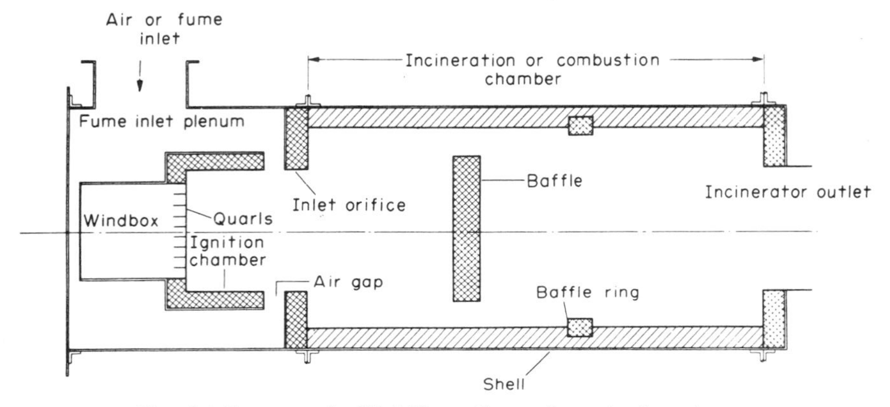

Fig. 5.1 Layout of a Hirt-Hygrotherm fume incinerator.

A typical thermal incinerator, one of which is shown in Fig. 5.1, consists of three major sections. These are the fume-inlet plenum, the burner and the furnace chamber. The unit illustrated, developed in California by Hirt-Hygrotherm, has several attractive features to ensure satisfactory fume incineration and low operating costs. A proportion of the fume input is internally bypassed through the burner for use as combustion air, and the elimination of the need for external-combustion air can save up to 30 per cent of fuel expenditure. Burner design is an important feature of an incinerator, and Hirt use a short, wide flame having a flat profile. By directing the fumes across the flame front as they enter the furnace, rapid heating is achieved, guaranteeing efficient oxidation. Another desirable feature of fume incinerators is the baffle arrangement, which promotes high turbulence in the furnace, leading to effective mixing.

By relating the fuel-gas input to the solvent concentration, efficient use can be made of the calorific value of the gases in the effluent being treated.

Not all fume incinerators burn organic waste. Sulphur dioxide (SO_2) is the most common gaseous atmospheric pollutant, because it is a byproduct of fossil fuel-fired boilers, furnaces, kilns, etc. As one step towards minimizing this form of pollution, fuel oils are treated to remove most of their sulphur content, and some of these

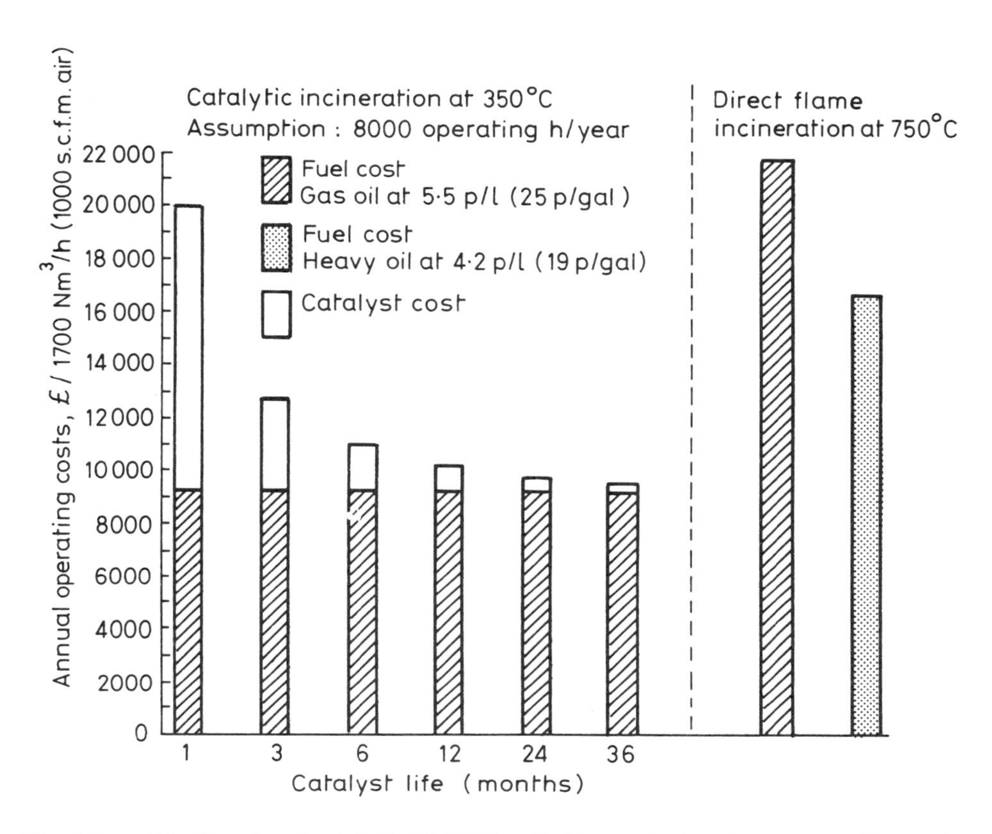

Fig. 5.2 The Engelhard catalytic DEODO pollution-control system – operating costs (by courtesy of Engelhard Industries Ltd.).

desulphurization processes produce hydrogen sulphide (H_2S), itself a toxic and malodorous gas. Treatment of the H_2S by an incineration process, followed by controlled cooling which permits a catalytic reaction to take place, can lead to the recovery of pure sulphur.

Catalysts are becoming increasingly used in fume-incineration processes. A catalyst is a substance which can increase the rate of a chemical reaction without itself being changed during the process. Their use in fume incinerators can accelerate the oxidation reaction necessary to convert organics to carbon dioxide and water. Catalytic incineration can take place at temperatures considerably lower than that needed for conventional thermal incineration, and the quantity of heat required is therefore reduced. Heat recovery systems can be incorporated downstream of such units, however, as will be illustrated later.

A comparison of the operating costs of catalytic and thermal incinerators, prepared by Engelhard to illustrate their DEODO air-pollution control system, is given in Fig. 5.2. The catalyst operates at 350° C, whereas the equivalent thermal incinerator would have to achieve 850° C for satisfactory fume combustion. (Catalyst life has been taken into account, but it is claimed that a life of only 3 months is not typical, even though some economy is still achieved over conventional system running costs.)

Engelhard are finding considerable interest in heat-recovery systems associated with this system, particularly when these units may be used to preheat incoming fumes. A reduction of 50 per cent in fuel costs, with a payback period of 12 months, is considered usual when an efficient heat recovery system is used.

Fume incineration is a major growth area, and it is significant to potential users of heat recovery units on these systems that in some States in the USA the fume-incineration temperature is gradually being raised, by legislation, to meet more stringent pollution-control requirements.

5.2.2 *Liquid waste incinerators*

Liquid waste incinerators may be used to treat spent solvents or liquids containing organic or inorganic matter, and the potential exists for the recovery of organic salts. Typical of systems capable of dealing with a wide variety of liquid wastes is that developed by Trane Thermal Company (Kiang, 1976). This, based on an oxidation system, has been developed and proved in a number of full-scale applications that will efficiently and economically dispose of industrial liquid waste, and at the same time, enable the recovery of energy and valuable chemicals. The heart of this system is a combustion process, which includes a burner, an incinerator and a submerged exhaust quench and scrubbing tank.

Waste liquids in the process industries are numerous in kind and, consequently, defy easy definition. However, based upon chemical content, typical liquid wastes can be grouped into four categories.

(1) Organic wastes: This group covers hydrocarbons that contain only carbon, hydrogen, oxygen (sometimes sulphur) and are self-combustible. They can be used as fuel and, when burned with the proper amount of air, will yield carbon dioxide, oxygen, nitrogen (sulphur dioxide) and water vapour. The combustion products are considered clean (except sulphur dioxide) and can be discharged into atmosphere. Heat in the product gas can be recovered through a boiler or some other waste heat recovery device.
(2) Halogenated waste: This group includes such chemicals as carbon tetrachloride, vinyl chloride, methyl bromide and other halogenated materials. Heating values of these wastes depend upon halogen content and may or may not require auxiliary fuel. When oxidized, the products of combustion will contain either halogens, or hydrogen halides. Hydrogen halides can be either recovered or neutralized, depending on specific needs.
(3) Metallic waste: Included here are such materials as inorganic and organic salts. When oxidized, the combustion products will contain salts. Due to the presence of salts in the flue gas, refractory selection, oxidation temperature and residence time become prominent design considerations. Since this group of wastes usually bypass the burner, auxiliary fuel is required for complete combustion. Also, submicron particulates and mists in the product gas will require secondary gas-treatment equipment. Heat in the system can be recovered by preconcentrating the waste and by using condensing heat

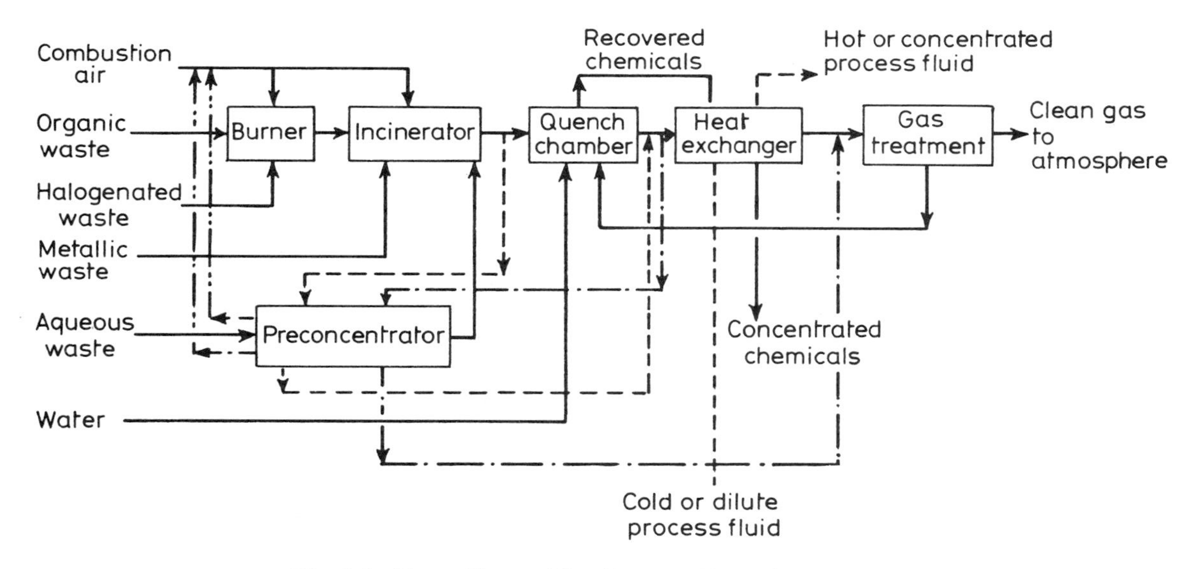

Fig. 5.3 Trane Thermal liquid waste disposal system.

exchange devices. High-purity salts can be recovered by evaporation.
(4) Aqueous waste: This group represents any, or a combination of, the above wastes in a solution of higher than 60 per cent. Due to low heat of combustion, this group of wastes will not support combustion in the burner and must be injected through atomized sprays into the flame of the burner through the incinerator chamber.

An integrated waste-disposal system for the above types of wastes makes use of the heat and material available in each waste to reduce the operational cost for the disposal of other wastes. Waste heat can also be recovered for process use as well as high-purity chemicals. Besides inplant reuse of the recovered chemicals, this system can also be considered as a production plant for commercial chemicals. However, the 'raw material' of this plant is 'waste'.

A typical integrated system for the disposal of these four types of wastes is illustrated in Fig. 5.3. The organic and halogenated wastes are used in the burner as system auxiliary fuel. Metallic waste is introduced into the system through the incinerator downstream of the burner. The aqueous waste is preconcentrated before injection into the combustion flame.

Incinerator product gas is either used as the preconcentration medium or cooled through a quench chamber. Water is used as the primary quench medium. The liquid discharge from the quench chamber is then concentrated through a heat exchange device where flue gas from the quench chamber is used as heating medium.

Depending on specific requirements, the concentrated product of this system can be high-purity crystals, slurry or solution. The cooled gas from the heat exchanger device is then passed into the final gas-cleaning devices for air-pollution control.

The incinerator itself is usually mounted vertically with the burner, firing downwards, located at the top. A feature of the Trane system is that all the salts generated by combustion are in a molten state, and flow down the walls of the incinerator to the quench chamber. This, however, necessitates control of the incineration temperature to prevent premature solidification, or the formation of oxide vapours. Control to within 900°C + or −50°C meets these criteria.

5.2.3 *Solid waste incinerators*

It is the solid waste incinerator which has found very extensive use outside industry, as it is used for the disposal of municipal refuse. Municipal incinerators tend to be very large dealing with, possibly, several thousand tonnes of refuse every day. One such unit, installed by Motherwell Bridge Tacol, in London, is illustrated in Fig. 5.4 (not visible are the five flow lines of incineration equipment, to cater for 24 h/day operation and incineration of up to 1700 tonne/day of refuse).

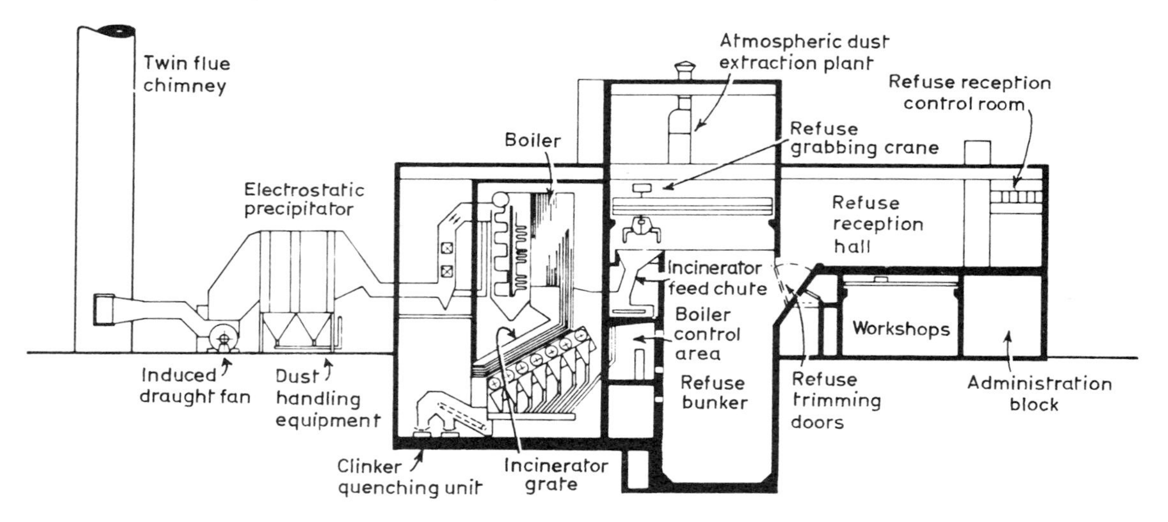

Fig. 5.4 Layout of an incineration plant constructed for the Greater London Council.

After careful consideration of the likely operating economics, it was decided to recover waste heat and to utilize this in the form of steam to drive turbogenerators. The electrical power produced is used for works supplies, the surplus being sold to the electricity authorities. A Yarrow integral three-pass type boiler is used, followed by an economizer, giving steam at 455°C and 44 kg/cm^2. This is sufficient to drive two inhouse 2·5 MW sets and four 12·5 MW sets for 'export' electricity.

In addition to municipal incineration, which may also be linked to a district heating scheme, as shown in Table 5.1, the energy content of much solid waste from industrial processes is high, and if the waste is in a form which cannot be recycled for process

Table 5.1 Heating values of various materials compared with oil fuel

Constituent	Oil-fuel equivalent (l/tonne)
Dust	86
Paper (15% moisture)	286
Wood (20% moisture)	323
PVC	524
Coal (CV 27 900 kJ/kg)	656
Polystyrene	870
Rubber	955

application, an incinerator with waste heat recovery may be the answer to the disposal problem.

A modern incinerator for solid waste processing equipped with a variety of optional ancillary equipment, is illustrated in Fig. 5.5. This machine, called the 'Consumat', and manufactured by Robert Jenkins Systems Ltd., comprises a refractory-lined combustion

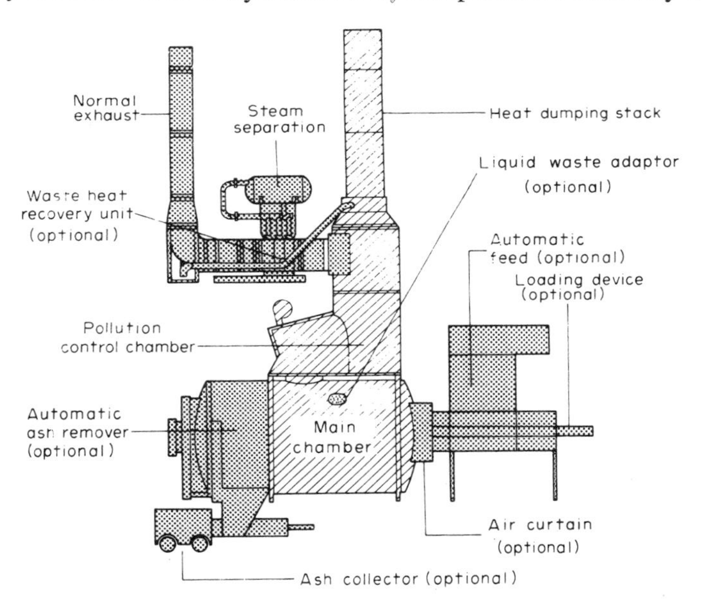

Fig. 5.5 The Consumat solid waste incinerator with heat recovery (by courtesy of Robert Jenkins Systems Ltd.).

chamber into which waste is loaded and air introduced. Initially the waste is heated by small auxiliary burners and undergoes a pyrolysis-type process at a temperature of up to 800°C. Compared to combustion brought about by the introduction of large quantities of air, this procedure allows the waste to decompose under quiescent conditions,

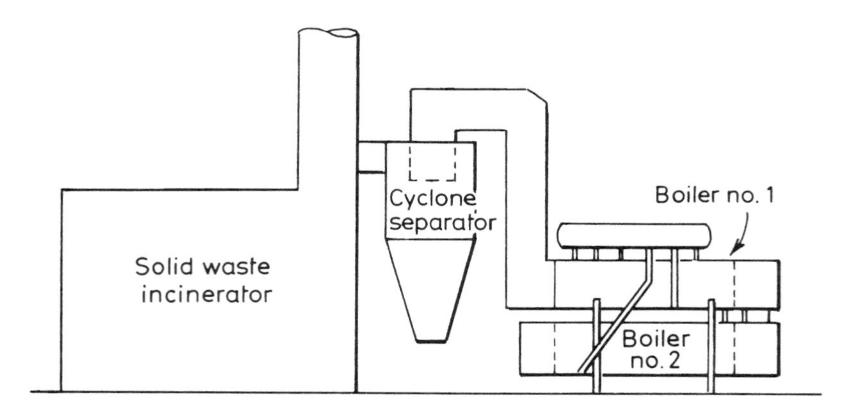

Fig. 5.6 Use of a cyclone separator to protect an incinerator waste heat recovery unit from contamination.

thus minimizing the carryover of particles which would normally be emitted to the atmosphere via the stack.

As a further pollution-control measure, smoke particles are oxidized above the main chamber by being heated to 1000°C to 1200°C using a small burner. The gases are subsequently cooled to 800°C by mixing with cold air before being exhausted to atmosphere.

A waste heat recovery system for providing hot water or steam may be added to the flue, as shown in Fig. 5.5, and the 'Consumat' system is also able to cope with liquid waste with little modification. The waste heat-recovery unit is capable of producing 3·5 tonnes of steam from each tonne of waste processed. The company finds that energy recovery becomes economically viable when disposal rates in excess of 150 kg/h are used. Usually 50 to 60 per cent of the total combustion heat is recoverable.

It is more common for solid waste incinerators to use large quantities of combustion air, and because of the carryover of solid matter, a separator must be installed between the combustion chamber and any heat-recovery equipment. This is illustrated in Fig. 5.6, where a cyclone separator fulfills the cleaning function. In this instance, it was necessary to use a fire tube-type waste heat boiler with high gas velocities to prevent settlement of any solid particles which were not captured by the separator.

A detailed analysis of contamination of a waste heat boiler installed downstream of an incinerator, in this case burning a variety of pumpable liquids, has been published in the USA. Flue gas temperatures at the exit to the secondary combustion chamber varied from 800°C to 1150°C because of the varying nature of the waste, a proportion of which was a chlorinated solvent. This produced hydrogen chloride when burnt, the quantity in the exhaust gas varying from 0·1 to 5 per cent by weight. This, together with fly ash, were the main contaminants (Hung, 1975).

After approximately 600 h of testing, it was found that the carbon-steel boiler tubes were satisfactory, provided that purging of the boiler was regularly carried out and the tube surface was kept clean. It was recommended that the minimum tube diameter be

75 mm, as smaller tubes were more susceptible to fouling. From the fly ash collected in the inlet chamber of the test boiler, it was demonstrated that a cyclone installed upstream of the waste heat boiler, as in the solid incinerator plant illustrated in Fig. 5.6, would greatly reduce the possibility of plugging and the degree of tube fouling. It was also suggested that fairly high gas velocities (20 to 30 m/s) be maintained in the boiler to minimize ash deposit on the surfaces.

5.3 Examples of incinerator heat recovery

Brief descriptions have been given of the three most important areas of waste incineration; fume, liquid and solid. Mention has also been made of heat-recovery devices fitted to these systems, and one or two examples given. One of the best ways of introducing the concept of waste heat recovery on incinerators is by means of examples, and subsequent sections present data on a number of actual and proposed installations.

5.3.1 *Organic fume incineration incorporating recuperators and waste heat boilers*

In a study carried out by Trane Thermal Company (Santoleri, 1975), three basic fume-incineration systems were proposed. Illustrated in Fig. 5.7, and applicable to some other incinerator types also, the two steps following on from a desire to improve the efficiency of a basic incinerator are shown. The examples below show instances of how these may be implemented.

A plant for the production of maleic anhydride, incorporating incineration of organic emissions was constructed, using a Combustion Engineering air preheater and a waste heat boiler, by Amoco in Illinois (Twaddle, 1978). The pollution-control system is based on a thermal oxidizer. (Maleic anhydride, a chemical intermediate and monomer used in polyester synthesis, is prepared in this process by the catalytic oxidation of normal butane.)

The thermal oxidizer converts residual hydrocarbons from the tail gas into carbon dioxide and water vapour. The abatement system oxidizes any carbon monoxide present to carbon dioxide. Particulates are not a problem as long as the plant continues to burn natural gas.

The tail gases out of the maleic anhydride unit are only at about 40°C, because the final step in the process is a wet-scrubbing operation. The tail gases are ducted to the thermal oxidizer where they pass through the inside tubes of the primary heat exchanger. This recuperative heat exchanger preheats the exit gases from the process to 370°C before they enter the combustion chamber, as shown in Fig. 5.8.

In the combustion chamber, the preheated gases are oxidized by direct flame to carbon dioxide and water vapour. The gases remain in the chamber for a minimum residence time of 0·75 s for complete combustion because the tail gases are derived from an oxidation process, and since the oxygen content is high, the burner does not require an outside source of air for fuel combustion. This reduces the fuel requirements

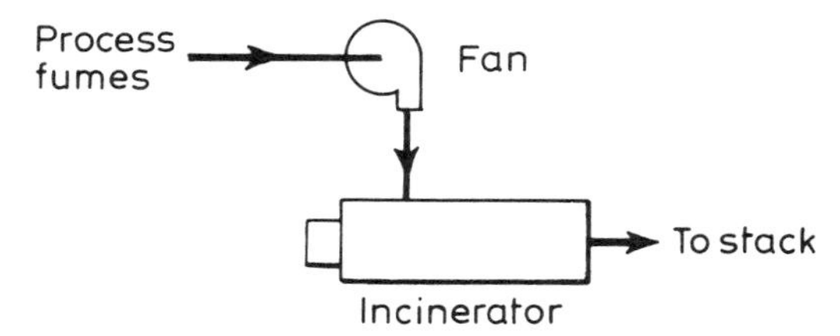

(a) Direct fume incineration without heat recovery. Alternately, incinerator exhaust may be fired into boiler to produce steam or hot water

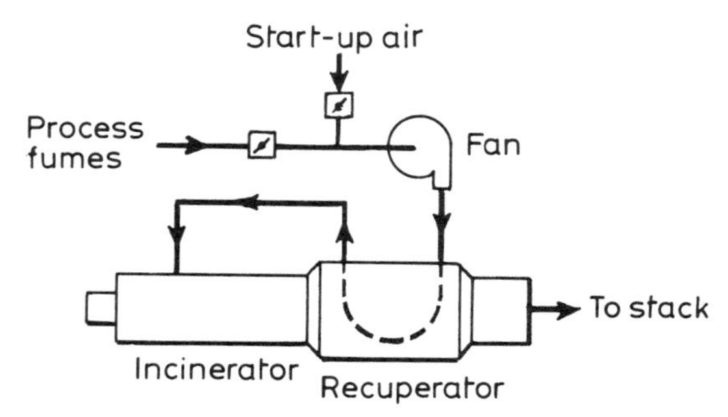

(b) Basic heat recovery scheme with hot exhaust gases from incinerator used to preheat the incoming process fumes. Temperatures up to 650 °C are achieved

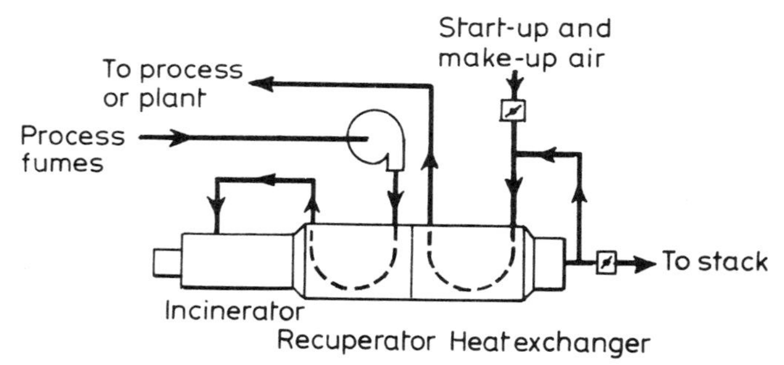

(c) Hot exhaust from incinerator passes through recuperator to preheat process waste fumes and then through second heat exchanger to recover additional heat which may be returned to process or other plant uses

Fig. 5.7 Schematic diagrams of fume systems with heat recovery.

which would have been required to heat the outside air. The hydrocarbons present in the process offgas also reduce fuel needs by a significant amount.

The gases leave the combustion chamber at 760°C and are then passed over the outside tubes of the heat exchanger. Here more than 46 per cent of the heat from the gases is transferred to the fumes inside the tubes. Finally, the gases pass through a

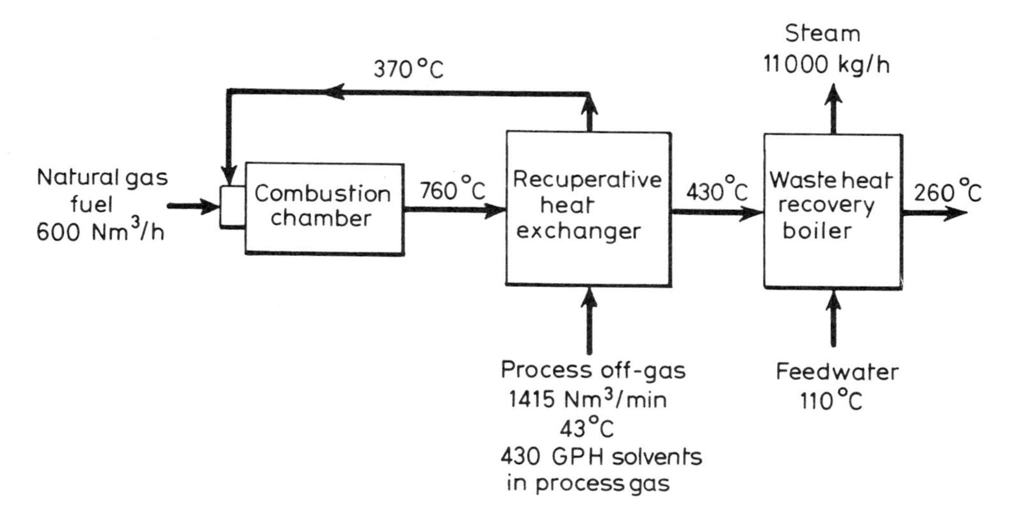

Fig. 5.8 Two-stage heat recovery on a fume incinerator in Illinois.

waste heat recovery boiler and are exhausted through a stack at about 260°C.

The steam from the recovery boiler is employed in the maleic anhydride process. This, along with the heat of oxidation from the process itself, furnishes more than enough heat to run the chemical system. The excess steam is exported to other plant needs as required.

Tail gases from Amoco's maleic anhydride unit are incinerated to meet rigid Illinois air-pollution codes. At the same time, when added to the heat of oxidation from the process itself, more than enough heat is recovered from the incineration process to run the process unit. In fact, there is exportable heat.

Without heat recovery, a total of approximately 15 MW would have been consumed in the process and incinerator. Only 6 MW in fuel gas is now consumed with heat recovery. Exportable 30 bar steam is estimated at 11000 kg/h. Annual savings in fuel costs exceed $970 000.

The Air Preheater Division of Combustion Engineering Inc. cite a number of other applications for the rotating regenerator (see Chapter 1) in incineration plant. Listed in Table 5.2, these include processes in rubber, printing and plastics industries.

An even more complex heat recovery system applied downstream of a thermal oxidizer in printing and curing processes is illustrated in Fig. 5.9. Exhaust gases passing into the thermal oxidizer are preheated to 460°C from 130°C by the hot gases leaving the oxidizer at 760°C. Cooled to 460°C, these gases are then passed across a pressurized water coil, heating water from 120°C to 150°C for process use in the washer. A proportion of the clean exhaust gases at between 290°C and 450°C (the high temperature being achieved when the hot-water boiler is bypassed) is then directly reinjected into the process for heating before being finally rejected, clean, to atmosphere.

The system illustrated in Fig. 5.10, using a Hirt-Hygrotherm incinerator, was added

Table 5.2 Savings achieved using air preheaters on incinerators

Application		Savings		
		Incineration without heat recovery	Cor-Pak system with primary heat recovery	Cor-Pak system with second heat exchanger
Sponge-rubber curing oven				
Fuel: No. 2 fuel oil @ 15 ¢/gal				
Operation: 6000 h/year				
Flow: 4 100 ft^3/min at s.t.p.	Fuel cost	$ 42 450	$ 17 000	
	Fuel savings	(none)	$ 25 450	
Glass-fibre curing oven				
Fuel: Natural gas @ 78.4¢/10^6 Btu				
Operation: 8400 h/year	Incineration-fuel cost	$127 400	$ 29 550	$ 24 650
Flow: 15 000 ft^3/min at s.t.p.	Oven-fuel cost	69 700	69 700	42 550
	Total fuel cost	$197 100	$ 99 250	$ 67 200
	Total fuel savings	(none)	$ 97 850	$129 900
Spray-drying process				
Fuel: No. 4 oil @ 12.3¢/gal	Incineration-fuel cost	$252 200	$ 87 400	* $ 87 400
Operation: 7000 h/year	Oven-fuel cost	94 700	94 700	37 800
Flow: 26 500 ft^3/min at s.t.p.				
	Total fuel cost	$346 900	$182 100	$125 200
	Total fuel savings	(none)	$164 800	$221 700

* Ljungstrom heat exchanger

Table 5.2 Savings achieved using air preheaters on incinerators (*continued*)

Application		Savings		
		Incineration without heat recovery	Cor-Pak system with primary heat recovery	Cor-Pak system with second heat exchanger
Plastic floor-covering process				
Fuel: Natural gas @ 56 ¢/10^6 Btu				
Operation: 5750 h/year	Incineration-fuel cost	$ 116 900	$ 38 900	$ 38 900
Flow: 28 000 ft^3/min at s.t.p.	Oven-fuel cost	126 000	126 000	101 800
	Total fuel cost	$ 242 900	$ 164 900	$ 140 700
	Total fuel savings	(none)	$ 78 000	$ 102 200
Paint-drying ovens				
Fuel: Natural gas @ 56 ¢/10^6 Btu				
Operation: 3200 h/year	Incineration-fuel cost	$ 36 750	$ 15 550	$ 15 550
Flow: 16 000 ft^3/min at s.t.p.	Oven-fuel cost	8 050	8 050	(none)
	Total fuel cost	$ 44 800	$ 23 600	$ 15 550
	Total fuel savings	(none)	$ 21 200	$ 29 250
Lithographic process				
Fuel: Natural gas @ 84 ¢/10^6 Btu				
Operation: 5616 h/year	Incineration-fuel cost	$ 64 400	$ 36 600	$ 35 150
Flow: 10 000 ft^3/min at s.t.p.	Oven-fuel cost	42 150	42 150	29 250
	Total fuel cost	$ 106 550	$ 78 750	$ 64 400
	Total fuel savings	(none)	$ 27 800	$ 42 150

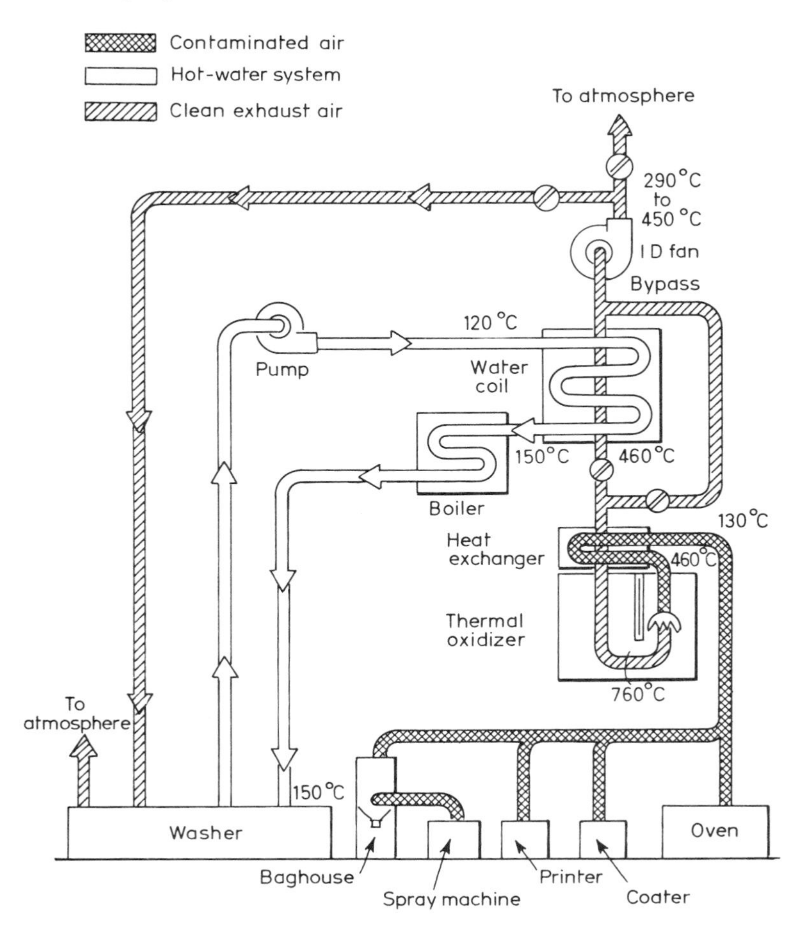

Fig. 5.9 Three stages of heat recovery on a printing line, downstream of a thermal oxidizer.

to aluminium-foil painting ovens where previously direct-gas firing had been applied and the solvent-laden air was being discharged directly to atmosphere. In the new system, the exhaust fume is taken from the fans to a common header duct via pressure controllers which ensure that the system operates satisfactorily under various loadings. The exhaust duct is taken to the incinerator forced-draught fan which blows the fume through the preheater 2, where it is raised in temperature, before entering the incinerator inlet plenum 5. After passing through the incinerator combustion chamber 1, the flue gases re-enter the fume preheater 2 and then the two fresh-air heaters 3 and 4 before entering the chimney. Fresh air to the ovens is blown by forced-draught fans through the air heaters to the oven burner boxes 7 and 8.

Control of the incinerator temperature is achieved by a modulating split-phase system normally operating on the fuel valve, but using the fume preheater bypass if the calorific value of the fume is high. Control of the oven temperature is individually

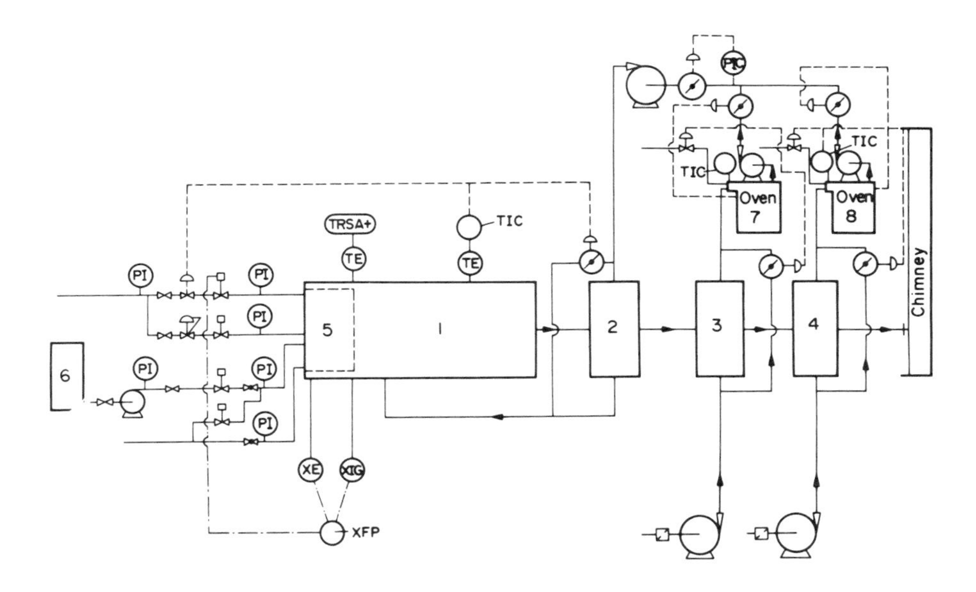

Fig. 5.10 Flow circuit of a Hirt-Hygrotherm fume incinerator used as a heat source for ovens.

attained also by means of modulation of the bypass damper around the heat exchanger and the oven burner fuel valve.

Normally all the heat required is provided by the heat exchangers and the burners are required to operate only during the warmup period.

With total heat demands of 5·2 Gcal/h to raise the temperature of the air entering the ovens, the incinerator was able to supply 4·7 Gcal/h, or 90 per cent of the total heat requirement, while maintaining exhaust to atmosphere at an acceptable contamination level. The incineration temperature in this instance was 700° C.

In even more efficient installations Hygrotherm have used waste heat for fume preheating and heat supply to heat transfer oil, which is in turn used to heat the oven by oil/air heat exchange, thereby eliminating large ducts and high fan heads and giving greater flexibility of operation. As mentioned in Section 5.2, the growth of catalytic incineration has also been accompanied by an increasing use of waste heat-recovery equipment.

The hot-air exhaust from stenters, which are processing synthetic fibres, can contain organics which are a source of atmospheric pollution. Cartwright (1975) reports on an interesting application of catalytic combustion which, when applied to the fumes in stenter exhausts, reduces the pollutants to carbon dioxide and water. In order to sustain the reaction, the exhaust must be *raised* in temperature to 360° C to 380° C, and this is the function of the catalytic burner. In the example quoted, a supplementary oil burner is used to raise the temperature of the exhaust before it reaches the catalyst.

Two possibilities exist where the relative humidity of the exhaust air is at an acceptably

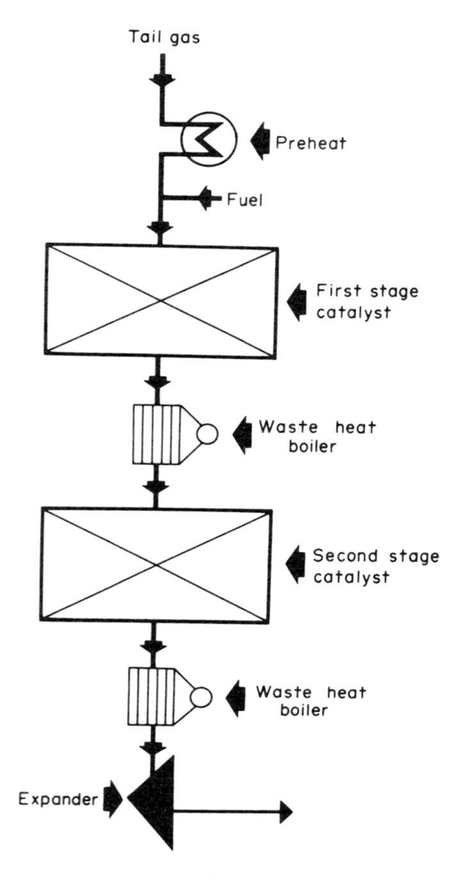

Fig. 5.11 An example of heat recovery downstream of a catalytic reaction.

low level. It could be reintroduced into the stenter once it has been cleaned by the catalyst to transfer heat to the air entering the stenter; in this case, it may be acceptable to use a regenerator if the air is dry. Alternatively recuperators of the type discussed above may be used, and recovery efficiencies of 70 per cent have been achieved.

The purification of exhaust gases using catalysts is carried out also in nitric acid plant. Treated tail gas provides heat for two waste heat boilers in the two-stage catalytic reaction illustrated in Fig. 5.11.

5.3.2 *Liquid waste incinerators as a heat source*

A liquid incinerator with a large heat recovery installation is shown in Fig. 5.12. In this incinerator, the exhaust gas flowing at a rate of 22·9 kg/s is cooled from 1200°C (liquid incinerators often operate at higher temperatures than their fume counterparts) to 168°C, generating 10·4 kg/s of superheated steam at 1930 kN/m^2 and 400°C. The

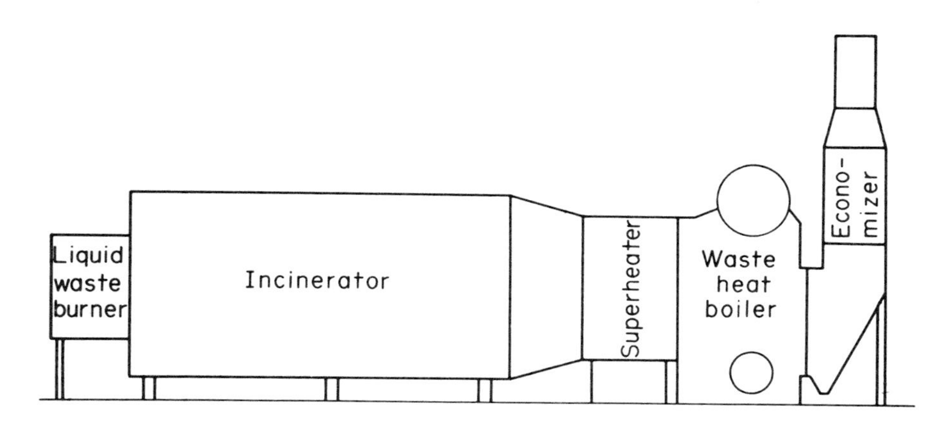

Fig. 5.12 A liquid waste incinerator with a waste heat boiler installation (by courtesy of Deltak Corporation, Minneapolis, Minnesota, USA).

financial savings in terms of reduced oil consumption, amounted to £500 000 per annum (Csathy, 1974).

This is a comparatively straightforward example of heat recovery from a liquid waste incineration plant. However, in his review of waste incineration, Santoleri lists a number of problem areas which should be reviewed when considering the combustion of liquid hydrocarbons for use as a heat source, and these include:

(a) Composition of waste hydrocarbon.
(b) Viscosity of waste hydrocarbon.
(c) Corrosive problems to be considered in pumps, piping, valves and nozzles.
(d) Reaction with other compounds (e.g. steam to waste reaction in the steam-atomization section).
(e) Polymerization (at high temperatures).
(f) Ash formation which would tend to plug valves, orifices, etc. in the piping system.
(g) Ash which would cause reaction with refractory in combustion chamber.
(h) Slag formation which would be detrimental to tube surfaces.
(i) Combustion products that would present problems in the heat exchanger (HCl, H_2SO_4 gases in combustion products).
(j) Nitrogen-bearing wastes that would create problems with regard to NO_x formation.

Having identified these problem areas, however, a competent incinerator manufacturer or heat exchanger manufacturer with incinerator-application experience (those identified in subsequent chapters and the Alphabetical list of manufacturers) should be able to satisfy the needs of most potential users.

In some aqueous and inorganic salt waste treatment plant, a number of alternatives

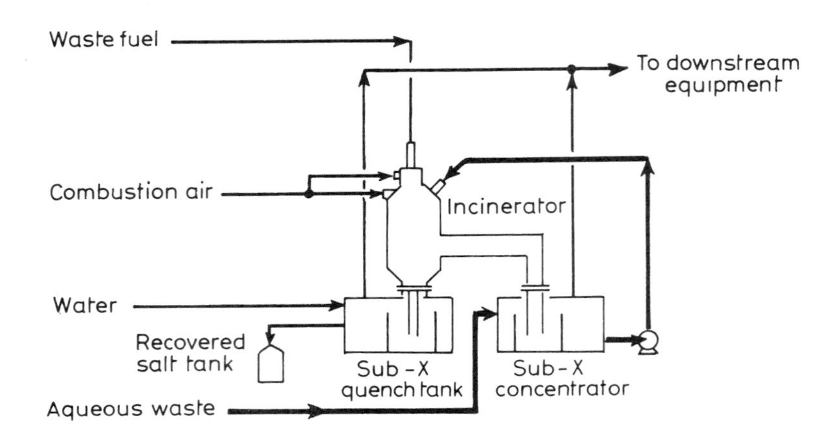

Fig. 5.13 Aqueous waste preconcentration through the Sub-X concentrator.

exist for heat recovery. Kiang (1976) indicates the possibilities of aqueous waste preconcentration using heat produced by the incinerator. This minimizes waste fuel consumption. Three basic preconcentration techniques have been successfully used, which are described below.

The first type makes use of a Sub-X concentrator illustrated in Fig. 5.13. The incinerator exhaust gas will bypass the quench tank and flow into the concentrator. In the concentrator, water in the waste is used to quench the flue gas and the heat contained in the gas is used to evaporate water from the waste.

This system is a direct-contact evaporator and, thus, eliminates fouling of heat transfer surfaces. However, if the waste contains high vapour-pressure chemicals, this approach is not applicable.

The second type is the evaporator (Fig. 5.14). In the heat exchanger section, exhaust gas from the quench tank is cooled and water vapour is condensed. The heat released by the condensed water vapour will be transferred to the incoming aqueous waste. Vacuum is maintained in the separator section to evaporate water from the waste. A double-effect evaporator can also be used.

The third type applies to waste liquids containing large percentage of high vapour-pressure chemicals. The liquid waste is preheated in the same manner as in the second type. A stripping-tower is added to strip out volatile materials which are then oxidized in the incinerator. Combustion air is the stripping medium. These preconcentration techniques may also be used in combination.

Waste heat may also be used, with similar heat exchanger equipment, to concentrate or crystallize salt products. Thus salt solution or crystals of a high concentration and purity can be produced from this system for process-plant reuse or commercial sales. The waste heat can also be used to heat process fluid, and to concentrate or crystallize process chemicals.

The heat exchange device used in this system is different from conventional exchangers, which use steam or hot gas as the heating medium. The heating media are the saturated gases leaving the Sub-X tank, with the saturated temperature range between 80°C and 90°C and with a heat content about 1100 J per gram.

To make use of this heat, the exchangers are designed to lower the temperature of the gas and thus condense the water vapour. The heat available from 1 kg of saturated air by

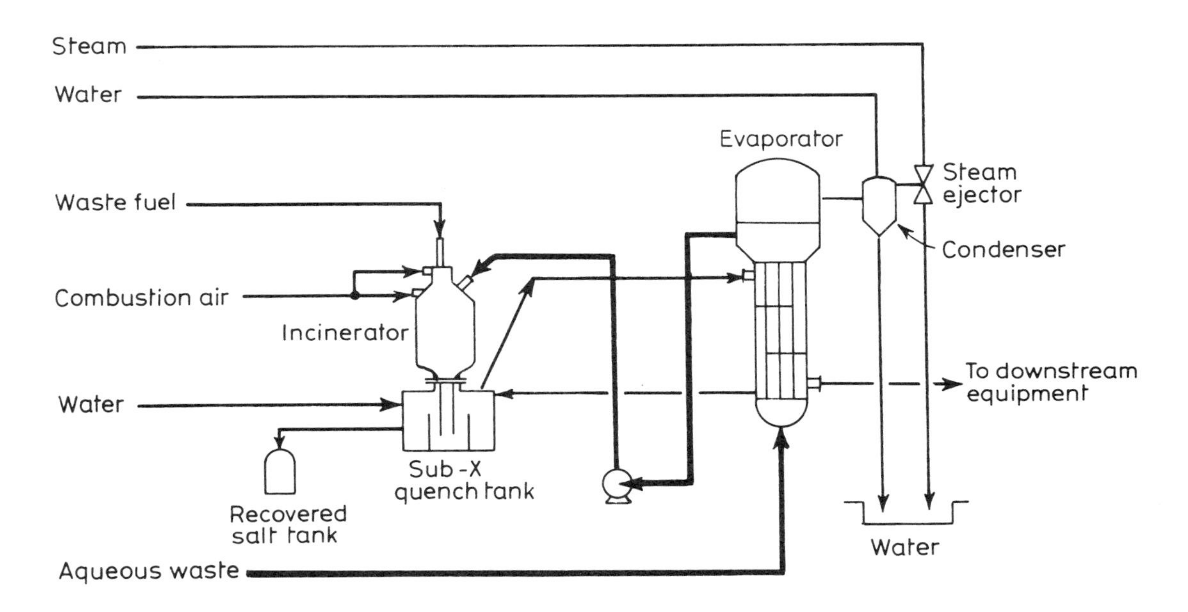

Fig. 5.14 Aqueous waste preconcentration through the evaporator.

decreasing the temperature from 85°C to 80°C is equivalent to the heat released by cooling 1 kg of hot dry air by 450°C. However, since the temperature is relatively low, evaporation must be carried out under reduced pressure, and the heat transfer requirement is relatively large. The latter disadvantages are usually compensated by the high condensing film heat transfer coefficients. Since non-condensible gases are present with the water vapour, this lowers the condensing coefficients available and must be reviewed carefully in the design stage.

Based on the Nittetu process, equipment manufactured by Trane Thermal Company (and Balfour in the UK) can be integrated into a complete liquid waste-disposal system for treating several types of waste, recovering the heat for process and incinerator use, and cleaning the resulting exhaust gases. This is illustrated in Fig. 5.15. This particular unit has a heat recovery capability totalling 3 MW, the total heat content of the waste being 9 MW.

Another type of incinerator which has been successfully applied to the treatment of sludges and slurries is the fluidized bed incinerator (see Chapter 2 for data on fluid beds). The fluid bed has many advantages in the incineration of wastes of these types as

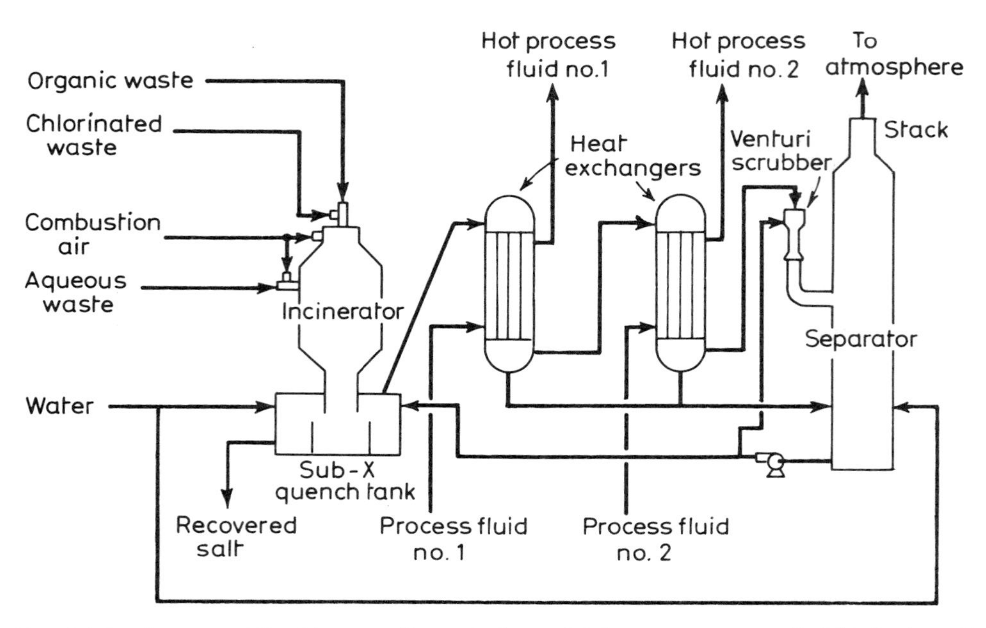

Fig. 5.15 Integrated liquid waste disposal plant with heat recovery applied for incineration and external process uses.

combustion can be maintained at comparatively low temperatures. In addition, very high heat transfer coefficients are obtained, allowing relatively low calorific-values wastes to be burnt without support fuel.

5.3.3 *Solid waste incinerators with heat recovery*

Examples of solid waste incinerators with heat recovery systems have already been given in Section 5.2.3. As has been stated, these may be applied in municipal incinerators and industrial units.

Solid waste incinerators generally cannot make as much use of air preheat as fume incinerators, unless downstream contamination necessitates the use of afterburners. The most popular use of waste heat in solid waste incinerators is the raising of steam, as is done in the Comtro unit illustrated in Fig. 5.16.

However, it is the incineration process itself which, for industrial solid incineration, has received a considerable amount of attention, primarily because of the wide variety of specialized incineration duties to be undertaken, unique to each type of industry.

One area where developments directed at producing economical and environmentally acceptable incineration techniques have proceeded rapidly during the past few years is in the disposal of scrap tyres. Stribling (1977), an international authority on incinerator design, working for Heenan Environmental Systems Ltd., has done much work in this area. He reported tyre-incineration plant such as that owned by Goodyear where steam output is of the order of 13 000 kg/h from the incinerator waste heat

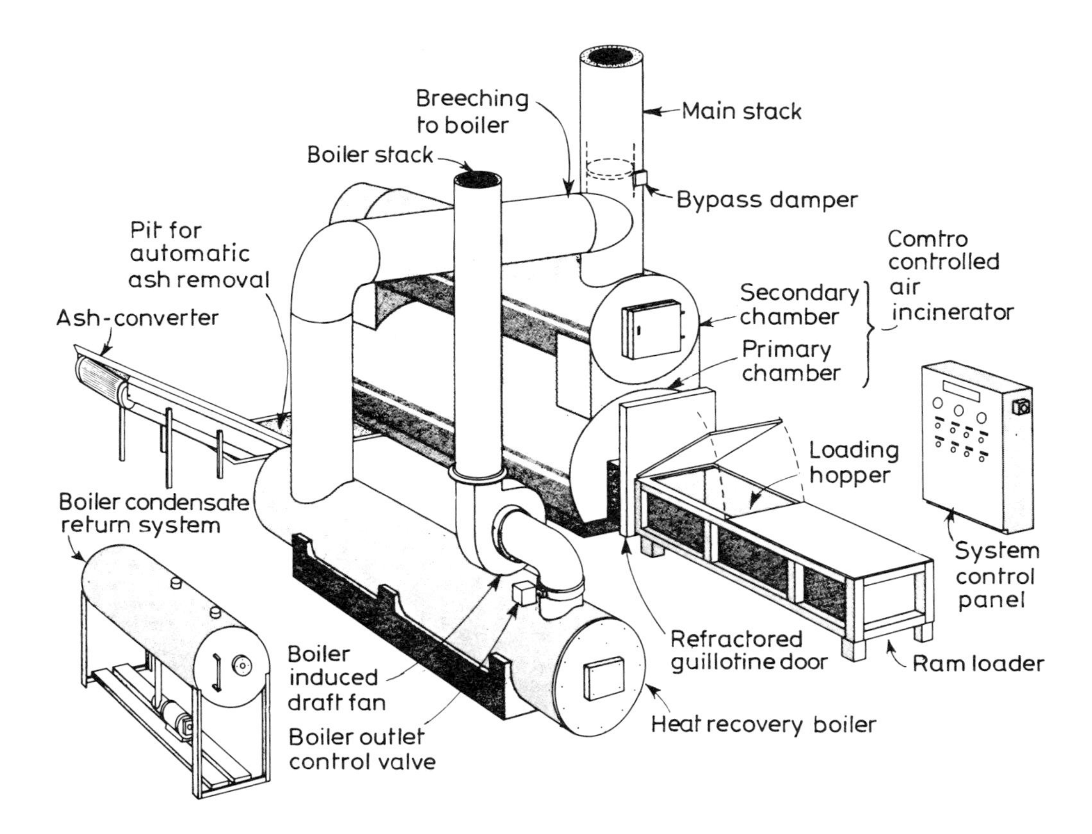

Fig. 5.16 Comtro solid waste incinerator with a waste heat boiler.

boiler. New systems being developed are expected to be available in standard module-form generating 18 140 kg/h of steam with furnace working temperatures of the order of 2000°C.

Fluid bed units are also used for solid waste incineration. A typical fluid bed furnace is illustrated in Fig. 5.17, this being a Babcock-Krauss-Maffei unit.

The fluidized bed furnace comprises a lined, vertical, cylindrical-conical combustion chamber. In the conical part of the chamber, a layer of sand is kept in motion over a flowbase by the combustion air. This air flows, as required, over an air heater in the windbox and is distributed uniformly over the cross section of the furnace by a heat-resistant grid plate. The wastes and any necessary additional fuel are fed directly into the fluidized bed and are burned at a very low level of excess air. The ash is discharged partly with the flue gases and partly with the dust from the fluidized bed packing. The heat content of the waste gases can be reutilized in the successive recovery stages for heating the combustion air or for steam production. Mechanical or electrostatic cleaning plant and wet scrubbers extract dust from the waste gas.

Such a unit forms the combustion system in an incinerator/heat recovery complex in

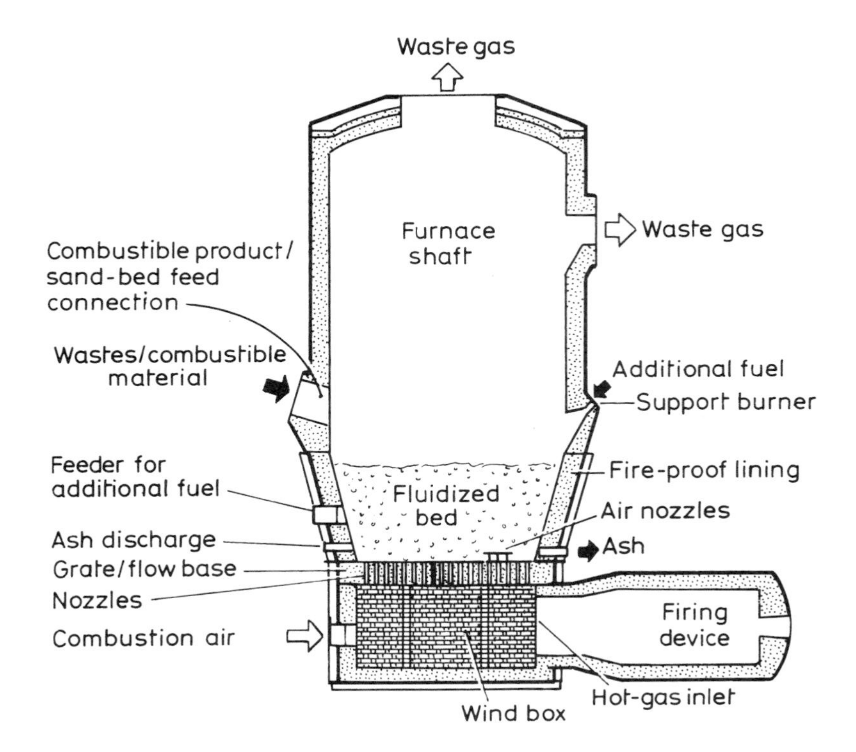

Fig. 5.17 Fluid-bed furnace for incinerator applications.

a pulpmill, and has successfully been used to provide heat to operate a dryer for log fuel, which then was able to replace oil as the boiler heat source. (Although outside the scope of this book, this is approaching the concept of waste as a fuel for use in boilers directly – for example, waste in sawmills is regularly used to fire boiler plant.) The flow diagram for the fluid bed burner and dryer, with heat recovery, is shown in Fig. 5.18, the unit being constructed by the Combustion Power Company (Moody, 1977).

The fluid bed heat source for the drying operation operates on green-log fuel, logyard debris, fly ash from the boiler, and pulpmill sludges. None of the dryer product is required for the fluid bed operation, and the dryer system efficiency is kep high by recycling gases from the dryer exhaust back to the blend zone of the fluid bed burner.

References

Cartwright, K. (1975), Modern air conditioning for the textile industry, *Textile Month,* May.

Csathy, D. (1974), Energy conservation by heat recovery, *9th Intersoc. Energy Conversion Eng. Conf., San Francisco, USA, 1974,* ASME Paper 749086.

Hung, W. (1975), Results of a firetube test boiler in flue gas with hydrogen chloride and fly ash, *Trans. ASME,* Paper 75-WA/HT-39.

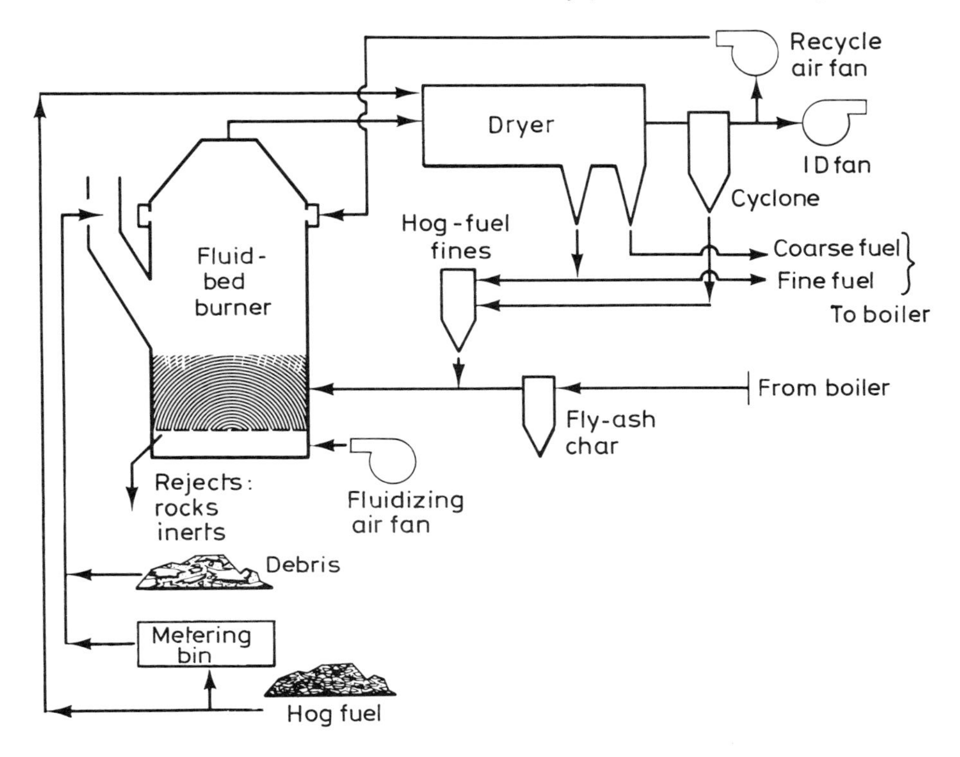

Fig. 5.18 The use of a fluid-bed furnace to incinerate waste and provide heat for a dryer. System specification:

Fuel

Constituents:

Debris
Boiler flyash char
Oversize hog fuel

Typical fuel feed:

Combustible	3 t/h
Moisture	6 t/h
Inert	4 t/h

Burner

Physical size: 5·5 m od.
Maximum thermal output:
20 MW (natural gas equivalent)
Turndown: 3 to 1

Combustion efficiency: 99%
Thermal efficiency: 55·75%

Dryer

Throughput: 15 dry t/h

Moisture:

Input 68% wet basis Output 20% wet basis

Kiang, Y.-H. (1976), Liquid waste disposal system, *CEP*, January, 71–77.
Moody, D. R. (1977), Energy Recovery from log yard waste and fly ash char, *Clemson Univ. Conf. Conservation and Resource Recovery for the Pulp and Paper Industry, South Carolina, USA*, 7–9, December.

Santoleri, J. J. (1975), Energy recovery from low heating value industrial waste, *Trans. ASME, Fuels and Power Divisions,* Paper 75-IPWR-13.

Stribling, J. B. (1977), Scrap tyre incineration – now an economic answer to dumping, *European Rubber J.,* July–August.

Twaddle, W. W. (1978), Heat recovering incineration of organic emissions, *Chemical Processing,* January.

6 Heat pump systems

6.1 Introduction

All of the waste heat recovery techniques described so far in this book are used to recover waste heat which is at a sufficiently high temperature to be reused without the need for upgrading (i.e. raising its temperature still further). This applies equally to heat recovery in air conditioning systems, the use of process heat for space heating and its reuse in the process itself.

There are, however, many areas where heat which could usefully be recovered or even heat which, while of a low grade, exists naturally in the ground, water or air, cannot be used because it is at too low a temperature. (In this context it is worth pointing out that waste heat available in certain processes at temperatures approaching 100°C may be more attractively recovered if it could be upgraded to produce low-pressure steam than solely, for example, for feedwater heating.)

The heat pump is a device which is able to extract energy from a low-temperature heat source and upgrade it to a higher temperature, enabling it to be used more effectively. The principle of the heat pump was put forward by Lord Kelvin in the nineteenth century but, except for its very widespread use as a refrigerator or air conditioning plant, where the primary function is to cool rather than to heat, it has to date found little application in industry world-wide, and is only used in air conditioning to provide both heating and cooling, to any large extent, in the USA. The number of heat pumps in operation in this category in the USA exceeds one million.

Heat pumps are now available for applications in the home, in commercial buildings, swimming-pools, etc., and in industrial processes, with heat outputs from a few kilowatts up to several megawatts. Although most have electric motor-driven compressors, a number of other types of prime movers are available, including reciprocating gas engines (see also Chapter 4), although these are normally custom-built.

Most heat pumps operate on the vapour-compression cycle, described in detail below, but units operating on the absorption cycle, Rankine cycle, Stirling cycle and others are available or under development. The majority of commercially available equipment, however, utilizes the vapour-compression cycle, as will be seen in the data presentation in Chapter 13.

6.2 Heat pump operating cycles

There are several operating cycles for heat pumps, but only the vapour compression cycle will be dealt with in detail here, for reasons listed above. However, the reader who is interested in full data on other cycles will find them fully covered by Macmichael and Reay (1979), and in references cited in the Bibliography.

6.2.1 *The vapour-compression cycle*

The ideal cycle

In a power cycle, heat is received by a working fluid, which then proceeds to do work, rejecting the heat at a lower temperature. In a heat pump or refrigeration cycle, the reverse happens: heat is received at a low temperature, and work is then done on the fluid, which proceeds to reject the heat at a higher temperature. A refrigerator is associated with the removal of heat, or cooling, whereas a heat pump is designed with the principal purpose of supplying heat at a high temperature. Heat pumps are able to provide both heating and cooling, and are widely applied in air conditioning applications for this purpose as mentioned above. In this context, however, we are also concerned with the heat pump operating on a heating-only cycle.

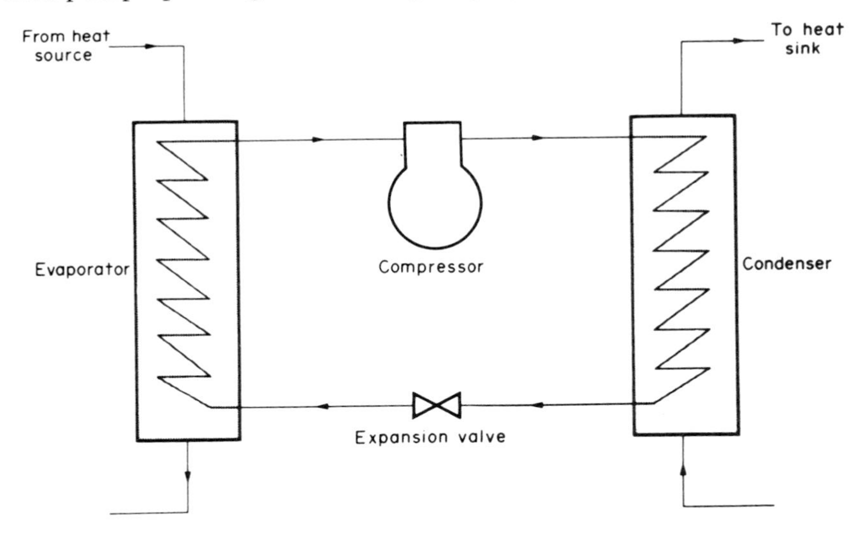

Fig. 6.1 The basic heat pump circuit.

The basic circuit of a heat pump consists of an evaporator, condenser, compressor and expansion valve, connected as shown in Fig. 6.1. The evaporator receives the

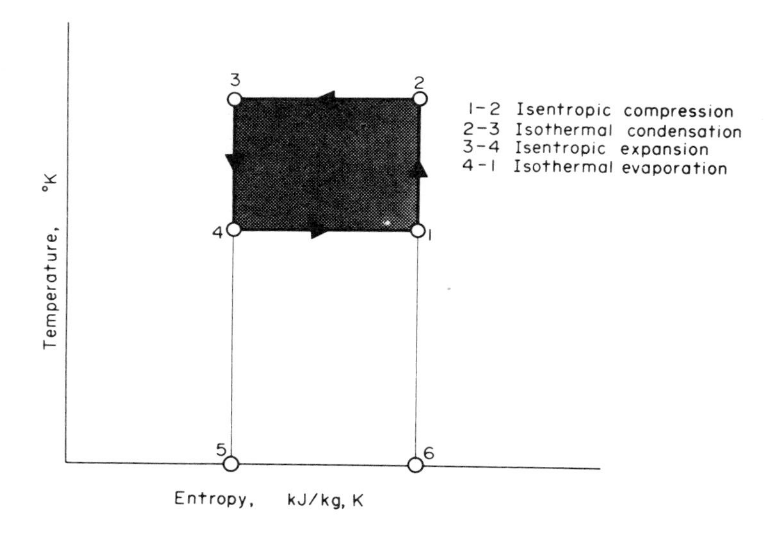

Fig. 6.2 The Carnot cycle – the ideal heat pump operating cycle.

low-grade heat from the waste heat source, the heat being taken up into the heat pump circuit by evaporation of the working fluid being circulated. The resulting vapour is then passed through a compressor, where its pressure and temperature are increased, before being pumped to the condenser, where it gives up the heat collected at the evaporator plus the heat equivalent of the work of compression. As this heat is rejected, the vapour condenses, and hot condensate then passes through an expansion valve, cooling in the process. The working fluid, commonly a fluorinated hydrocarbon, remains in a closed sealed circuit throughout the operation.

The ideal heat pump cycle is represented by a reversed heat engine operating on the Carnot cycle, illustrated in Fig. 6.2. The performance of the heat pump is commonly expressed as the ratio of the heat supplied at the condenser to the heat equivalent of the work put in at the compressor. Alternatively, the 'coefficient of performance' (COP), of the unit may be written thus:

$$COP = \frac{T_1}{(T_1 - T_2)}$$

where T_1 is the condensing temperature
and T_2 is the evaporating temperature, both expressed in absolute degrees.

A typical temperature range over which an industrial heat pump might operate is from 30°C to 80°C. The ideal coefficient of performance is, therefore, 353/(80–30), or 7·06. This means that 7·06 times the quantity of work supplied is delivered to the condenser as heat. In practice, however, the heat pump cycle has a much lower coefficient of performance than that suggested by the reversed Carnot cycle, and is more accurately represented by the Rankine cycle, illustrated in Fig. 6.3. In an actual plant the work done in expanding the fluid would be dissipated as mechanical

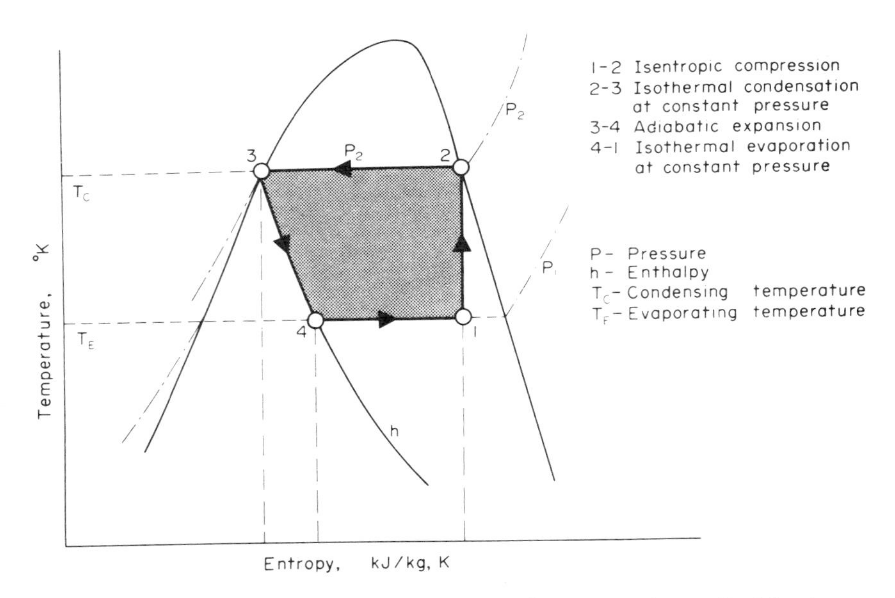

Fig. 6.3 The Rankine cycle, which more accurately represents practical heat pump operation.

friction, and an expansion valve is used, resulting in the stage 3–4 lying on a constant enthalpy line. While no work is now obtained from the expansion process, this does not affect the heat extracted from the condenser, but reduces that taken up in the evaporator. The expansion valve makes the process irreversible. After expansion the working fluid is in the form of a saturated liquid, which then picks up heat in the evaporator, ideally being boiled off to a saturated vapour, absorbing the heat at constant temperature.

A practical heat pump cycle differs from the reversed Carnot cycle in two other respects. The compression is normally carried out in the superheat region, with de-superheating occurring in the condenser. Also, some additional liquid cooling (sub-cooling) occurs between the condenser exit and the expansion valve, prior to the adiabatic expansion between points 3 and 4, whence the cycle is repeated. In actual cycles the pressure of the fluid drops as it flows through the heat exchangers, but this is not shown on the cycle diagram in Fig. 6.4 (Kolbusz, 1975). Subsequent sections in this chapter detail the compressors, prime movers and working fluids used in vapour compression cycles.

6.2.2 *Alternative cycles*

There is a number of other types of heat pump available, or in the process of development. Because they are not in general as fully developed as the vapour compression cycle, they are not described in great detail. However, the reader will find adequate

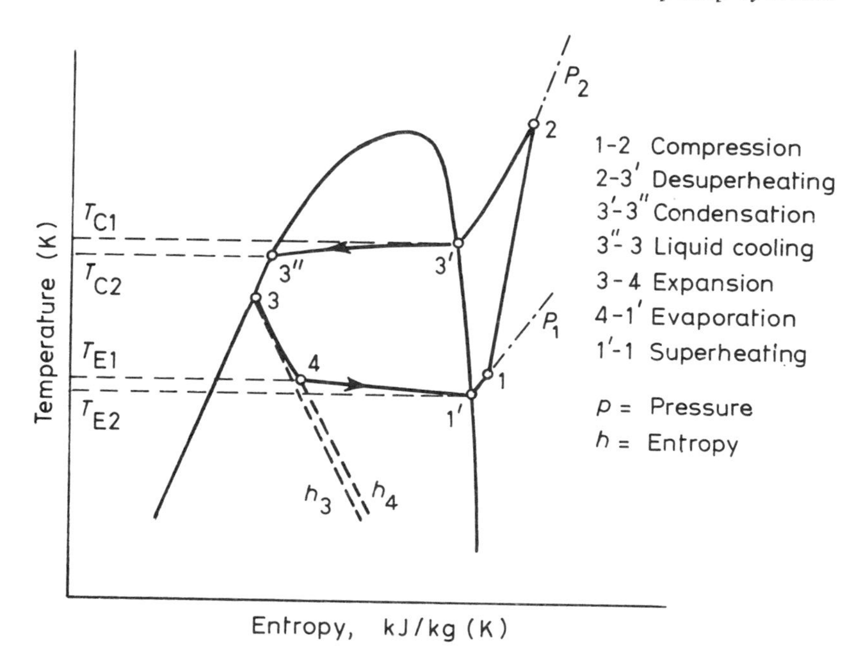

Fig. 6.4 Modified Rankine cycle, taking into account cycle inefficiencies.

information on their performance and status in the references listed.

The thermoelectric heat pump relies on the fact that when an electric current passes through a junction between two different conductors, heat is either absorbed or rejected at the junction, depending on the direction in which the current is flowing. A completed circuit must consist of two of these junctions, and the opposite effect will occur at the second junction. Thus, as a current is passed around the circuit, heat will be taken in at one junction and rejected at the other. This effect can be quite marked when certain combinations of material are used.

At present these devices have very low-power capabilities, and COPs are lower than with vapour compression types. Of potential significance is the fact that they can operate at temperatures well in excess of 100°C (Farrell, 1975).

Absorption cycle machines do not have a compressor. Mechanical processes are replaced by chemical reactions and the working fluid is used in conjunction with a second medium, known as the absorbent, in which it is highly soluble. The solution is heated in a generator, as shown in Fig. 6.5, by an external heat source. High-temperature vapour given off in the generator passes to a condensing heat exchanger and the liquid returns to the absorber. The liquid working fluid is reduced in pressure and evaporates in a low-temperature heat exchanger, which functions in the same way as a vapour compression cycle evaporator. The vapour is then redissolved in the absorbent.

The two most common types of absorption cycle machines, used as air conditioners, operate on lithium bromide and water, or ammonia and water. The lithium bromide system has the highest COP and may be attractive in cases where waste heat such as

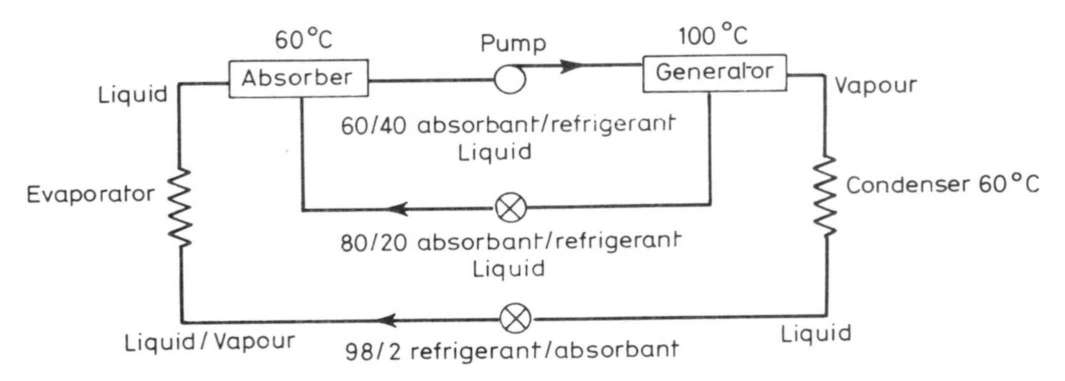

Fig. 6.5 Absorption cycle heat pump.

steam is available to heat the generator (Ellington, 1957).

The United States Institute of Gas Technology has recently undertaken a major study of heat-actuated heat pumps based on Rankine, Stirling and Brayton cycles. The early reports on this work (Wurm, 1974) were particularly comprehensive in their scope, and provide an excellent introduction to the potential of these cycles. Stirling and Rankine cycle heat pumps are now being developed by a number of companies in Europe and North America. Thinking in the UK on alternative cycles is given in a useful Department of Energy publication (1977).

6.3 Heat sources and sinks

The heat pump can operate successfully only if a suitable heat source is available to evaporate the fluid being circulated in the system. With a few notable exceptions, discussed later, the heat pump has to date been applied in situations where both heating and cooling are required, such as year-round air conditioning and space heating using waste heat rejected by refrigeration units.

It is in the USA where most of the major developments in heat pump technology have taken place, and while there have been periods in its evolution when the engineering of the concept has been called into question, particularly from the point of view of reliability, the use of heat pumps in the above applications is now commonplace. A substantial proportion of the market for these units is in domestic air conditioning, and Table 6.1 lists some of the manufacturers of these units, with data on duties together with heat sources used.

We are primarily concerned with the heat pump working on a heating cycle, and the heat sources available are numerous. In air conditioning applications, ambient outside air is the most common external source of heat, being passed over the finned evaporator using a fan. The major drawback of such a heat source is its temperature variation (Macadam, 1974). At the times when most heat is required, the ambient air will

Table 6.1 Heat pumps for domestic air conditioning

	ARI standard rating				COP
	Cooling		Heating		
Manufacturer	Capacity (W)	Power input (W)	Capacity (W)	Power input (W)	Heating
American Air Filter Company Inc. (water source)	2 780*	1 100	2 780	1 150	2·41
	14 600*	6 800	15 000	5 500	2·72
Carlyle (Carrier) Air Conditioning Ltd. (air source)	5 850	3 300	6 450	2 800	2·30
	34 300	14 700	34 300	12 100	2·84
General Electric (air source)	5 300	2 800	5 300	2 500	2·12
	35 000	15 000	35 000	12 600	2·78
Fedders Corp. (air source)	6 450	4 000	6 450	3 800	1·70
	16 400	8 800	17 500	7 200	2·43
York Division of Borg-Warner (air and water source)	5 600*	2 000	6 150	2 000	3·07
	14 600	7 900	15 800	6 600	2·40
Westinghouse Electric Corp. (air source)	5 300	2 900	5 000	2 400	2·08
	17 300	8 900	17 300	6 900	2·50
Temperature Ltd. (UK)	1 460	850	2 280	900	2·53
	3 800	1 860	5 430	1 950	2·78

* Water source heat pump.

obviously be at a low temperature, and this often necessitates the use of a backup heating system to provide a boost to the performance when the heat pump is unable to provide all the heating requirements. In an attempt to overcome this difficulty, a number of natural heat sources which do not succumb so markedly to seasonal and even daily variations have been tried. In the 1950s, it was popular to bury the evaporator coil a few feet below the ground surface, but the extensive amount of piping required to satisfy the heat requirements because of poor heat transfer coefficients between the soil and the evaporator was one of the reasons why interest in this system was not sustained. Solar energy is now increasingly of interest as a heat source for heat pumps in domestic applications, but as this is (like the ambient air) limited by the fact that it is available at times when heat is least needed, energy storage systems are normally also included in the 'package'.

Some heat pumps, as can be seen from Table 6.1, are designed with water as the heat source and/or heat sink. In a heating application, water has proved more attractive than most in this respect as it is rarely subjected to the wide variations in temperature which restrict the use of ambient air heat sources. River water has been used in several well-documented applications of heat pumps in the UK, including the Royal Festival Hall, which used the River Thames as the heat source. Of course, only a small proportion of potential heat pump applications could make use of naturally occurring water heat sources, but such heat sources are commonplace in industry, with important differences: they are normally at a somewhat higher temperature than river water and are even less subjected to seasonal changes in conditions. Also, the duty of any heat pump applied in such a situation will in general remain essentially constant in so far as industrial process requirements are concerned. (This would not of course be true of a heat pump used to provide space heating in an industrial contect, (Montagnon and Ruckley, 1954.)

One way of overcoming the influences of external ambient conditions on the heat source, used by companies such as AAF and Temperature Ltd. (see Chapter 13) is via a decentralized system using a pumped closed-water circuit operating at a constant, or near-constant, temperature. The heat sink in such a system is an air-cooled coil. An example of this system is given in Section 6.6.

6.4 Heat pump working fluids

Normally the working fluid used to transport the heat around the secondary circuit of a heat pump is called the refrigerant. In many cases the fluid is identical to that used in a conventional refrigerator, and in higher-temperature heat pumps derivatives of these fluids, commonly denoted using the prefix 'R' (e.g. R12, R21), and manufactured by companies such as ICI, Du Pont and Imperial Smelting Corporation, are used.

A comparison of the vapour pressures of the various refrigerants, at evaporator and condenser temperatures, for a variety of applications ranging from low-temperature food freezing to high-temperature heat pumps is given in Table 6.2. The only fluid in the listing not belonging to the classification described above is ammonia, which has too high a vapour pressure at heat pump temperatures to be of use in such systems. Pressure is not the only criterion which is used in determining the correct working fluid to be used in a particular application. The amount of heat which can be transported by the refrigerant is a strong function of its latent heat, or enthalpy, values of which are listed in Table 6.3. The specific volume of the working fluid determines the displacement of the compressor needed to deal with the flow of fluid necessary to meet the duty. Critical pressure and temperature influence the limiting conditions beyond which the fluid cannot be successfully used, and the miscibility of the working fluid with the compressor lubricating oil is a significant factor in the selection of fluids (Howden Group, 1976.)

As well as the above properties, such considerations as cost have to be taken into account, although the market for most of the commercially available refrigerants is sufficiently large and competitive to keep prices within reasonable bounds. In addition, as the secondary circuit is in theory a closed system, loss of refrigerant should be minimal and the quantity required in even a large system will not be a significant proportion of the capital or running costs. Safety is an important factor; the refrigerant should be non-toxic, and must be compatible with the components of the circuit into which it is likely to come into contact. It must also not be inflammable. At present one of the factors which is limiting the use of heat pumps at temperatures above 110°C to 120°C

Table 6.2 Saturation pressures with various refrigerants at conditions varying from applications in low-temperature refrigeration to high-temperature heat pumps

Typical application	Temperature		Refri-gerant	Pressures		Remarks
	Evaporator	Condenser		Evaporator (kg/cm^2 abs.)	Condenser (kg/cm^2 abs.)	
Food freezing	−40°C	+ 35°C	R12	0·65	8·6	Usually two stage compression
			R22	1·07	13·8	
			NH3	0·73	13·8	
Food storage	−20°C	+ 35°C	R11	0·16	1·5	Low pressure refrigerants
			R21	0·28	2·5	
			R114	0·38	3·0	
			R12	1·54	8·6	High pressure refrigerants
			R22	2·50	13·8	
			NH3	1·94	13·8	
	−10°C	+ 35°C	R11	0·26	1·5	
			R21	0·45	2·6	
			R114	0·60	3·0	
			R12	2·24	8·6	
			R22	3·60	13·8	
			NH3	2·96	13·8	
Water chilling	+ 1°C	+ 35°C	R11	0·43	1·5	
			R21	0·75	2·6[	
			R114	0·94	3·0	
			R12	3·26	8·6	
			R22	5·25	13·8	
			NH3	4·56	13·8	
	+ 1°C	+ 50°C	R11	0·43	2·4	High condensing temperature from use of air-cooled condenser
			R21	0·75	4·0	
			R114	0·94	4·6	
			R12	3·26	12·4	
			R22	5·25	20·0	
			NH3	4·56	20·7	
Heat pump	+ 25°C	+ 70°C	R11	1·05	4·2	
			R21	1·83	6·7	
			R114	2·18	7·4	
			R12	6·6	19·0	
			R22	10·5	30·5	
			NH3	10·2	35·0	
	+ 25°C	+ 80°C	R11	1·05	5·3	
			R21	1·83	8·4	
			R114	2·18	9·5	
			R12	6·6	23·2	
	+ 25°C	+ 90°C	R11	1·05	6·7	
			R21	1·83	11·3	
			R114	2·18	12·3	
			R12	6·6	29·0	
	+ 25°C	+ 100°C	R11	1·05	8·3	See also Table 6.3
			R21	1·83	14·0	

Table 6.3 Theoretical performance of refrigerants at typical heat pump temperature conditions

Refrigerant	11	21	113	114	12	31/114	12/31
Evaporator pressure (kN/m^2)	220	385	103	425	1190	758	1200
Condenser pressure (kN/m^2)	1000	1610	560	1700	3950	2920	4140
Critical temperature (°C)	198	178	214	146	112	142	118
Compression ratio	4·45	4·2	5·3	3·95	3·35	3·86	3·4
Specific volume (m^3 kg @ 50°C)	0·0801	0·0619	0·1298	0·0320	0·0146	0·0324	0·0170
Weight circulated (kg/s)	0·098	0·126	0·060	0·245	0·530	0·242	0·450
Net refrigerating effect (kJ/kg)	114·5	145·5	86·4	48·5	36·6	118·8	65·0
Heat of compression (kJ/kg)	27·9	35·0	25·6	18·6	18·6	35·0	25·6
Capacity (kW)	13·9	22·8	6·76	13·0	28·3	37·6	41·7
Coefficient of performance	5·1	5·2	4·35	3·5	2·9	4·48	3·55

Evaporator: 50°C
Condenser: 110°C
Assume compressor swept volume of 28·32 m^3/h (1000 ft^3/h)

Note The net refrigerating effect is obtained by subtracting the enthalpy of the refrigerant at the condensing temperature from the enthalpy of the superheated vapour entering the compressor at the evaporator temperature. The heat of compression is the difference between vapour enthalpies upstream and downstream of the compression.

is the thermal stability of the working fluid. Decomposition can occur at temperatures below the critical temperature, and mixing with the compressor lubricant can also be a problem area.

In addition to the more common refrigerants listed in the tables, azeotropic mixtures of 31/114 and 12/21 have been developed and patented by Allied Chemicals. Their capacity is some 60 per cent higher than R21, and both have critical temperatures in excess of 110°C. Working fluids used in absorption cycles are briefly mentioned in Section 6.2.2.

6.5 Heat pump compressors and prime movers

6.5.1 *Compressors*

The main types of compressor available are centrifugal, screw and reciprocating units (which may be 'wet' or 'dry'). In a wet reciprocating compressor there is a significant amount of lubricant oil in the refrigerant, generally any quantity in excess of 15 p.p.m.

Wet reciprocating compressors
Wet reciprocating compressors are single-acting, multicylinder positive displacement machines. The connecting rod is attached directly to the piston by a gudgeon pin, i.e. there is no cross-head and the cylinder walls require a lubricating film of oil. This form of construction is also called 'trunk piston'. The cylinders of the machine are generally positioned radially in banks around the upper half of the crankcase and may have between 2 and 16 cylinders.

The fluid to be compressed enters and leaves the cylinders through suction and discharge valves which are of the ring type and are fitted into the cylinder head of the machine. The valves operate when there is a small pressure differential across them in the appropriate direction (one way valves) and they are returned to the closed position by springs.

Pressurized oil is fed from a centrifugal oil pump (driven from the crankshaft) to the bearings and the crankshaft seal. Oil in the crankcase is splashed on to the cylinder walls where it acts as both a lubricant and a coolant. Capacity reduction of these machines is achieved either by slowing the machine down (the lower limit to this is determined by the prime mover and also by the failure of the oil pump below about half normal speed), or by manually or automatically cutting cylinders. The cutting out can be effected in a number of ways, one of the most common being that of lifting the suction valve such that it remains open throughout the stroke. Typically, a spring-loaded lever holds the suction valve open, when the machine is started. Control oil, from the oil pump is admitted to a servomechanism which compresses the spring and releases the suction valve, allowing it to function normally. The removal of control-oil pressure to the cylinder effects the cutting out of that cylinder for capacity-control purposes. The limitations of this type of machine are as follows:

(i) Oil is present in the cylinder at all times, because the cylinder requires lubrication. Thus, oil and refrigerant are in intimate contact at the discharge temperature. This has several consequences: first, some oil is always carried over into the condenser with the discharged refrigerant. Second, refrigerant breakdown and associated problems will be made worse by the presence of oil. The maximum discharge temperature is limited to around 150°C (depending on the type of oil and refrigerant), since the discharge valves begin to carbonize above this temperature.

(ii) The maximum tolerable suction pressure is limited by the end-thrust on the crankshaft to approximately 0·7 MPa. This is because the crankshaft protudes from one side of the crankcase only. A crankcase pressure which is greatly in excess of atmospheric can, therefore, put an axial loading on to the crank of several hundred kilograms. The load which the thrust bearing is designed to withstand is obviously arbitrary, but generally corresponds to the pressure of 0·7 MPa.

(iii) The maximum discharge pressure is generally about 2 MPa. This is restricted by mechanical strength of the cylinder head and other stressed components. It is standard practice to provide some sort of overload protection, e.g. spring cylinder heads, so that, should the rated pressure be exceeded, then the results are not catastrophic. This condition could arise if, for example, one cylinder was fully filled with liquid refrigerant due to some unforeseen combination of errors.

The wet reciprocating compressor is commonly proposed for heat pumps in process applications and is comparatively cheap and readily available.

Dry reciprocating compressors

Dry reciprocating machines are positive displacement compressors using one or more single or double-acting cylinders. High compression ratios are achieved efficiently by use of two- or three-stage compression with intercooling. The machines are fitted with cross-heads which enable the pistons to be constrained to an accurate vertical oscillation, with no cylinder wall contact. This aspect of the design is most significant because it removes the need for oil to be present in the cylinder as a lubricant and the gas under compression is kept completely free from oil contamination. Furthermore, not only is the machine designed such that oil is not required in the cylinders, but precautions are taken so that oil cannot enter, e.g. the piston rod is sealed with a stuffing-box.

Some means of sealing between the piston and the cylinder wall is required, and in the absence of oil, conventional piston rings are unsuitable. This problem can be solved by either of two methods: polytetrafluoroethylene (PTFE) piston rings provide one solution, the alternative being labyrinth grooves cut into both the side of the piston and the cylinder wall. In both cases the piston and cylinder are a very close fit and the machine is allowed to run in at the correct temperature such that the two surfaces are perfectly matched.

Due to the absence of oil from the cylinder, some additional cylinder cooling is required, usually by circulating water through passages in the cylinder block and cylinder head. Cooling water may also be circulated through passageways adjacent to the cross-head and guide bearing.

These machines usually have a relatively small number of cylinders, and rotate at a comparatively low speed. Significant loads may be transmitted to the foundations and the machine needs to be mounted on a bearing block, typically two-and-a-half times its own mass. They are unlikely to be competitive in most heat pump applications because they are expensive.

Dry screw compressors

The dry screw compressor consists basically of two rotors enclosed by casings. The rotors differ in shape and are distinguished by the terms 'male' and 'female'. The male rotor has a number of lobes on it which are semicircular in profile and formed in a helix along the rotor body. The female rotor has a number of corresponding flutes and channels in it, also semicircular in profile but formed in the opposite helix to the male. The dimensions of the rotors are such that when they are positioned at the appropriate centre distance, they will mesh in a similar manner to a pair of helical gears. As these are rotated, a space is formed, due to the helical angle of the rotor lobes, the space being enclosed by the rotors and the casing. This volume increases as the rotors rotate and gas is drawn in to fill the depression formed. A useful analogy in understanding this process, is to consider the male rotor lobes as pistons and the female flutes as cylinders. Rotation of the rotors causes the piston to slide along the cylinder and so draw in gas. Continued rotation would result in the lobes remeshing and expelling the gas out of the far ends of the casing. However, if a cover is placed over the far end of the machine, then the remeshing of the lobes results in the gas which has been drawn into the machine being compressed between the meshing rotors and the casing.

The basic machine is so arranged that the rotors do not touch and lubrication is not required within the working space. This is achieved by means of timing gears situated at the end of the shaft. The bearings on these machines are, therefore, situated outside the working space, and separated from it by shaft seals.

Due to the absence of the cooling effect of oil within the compressor itself, it is sometimes necessary to provide this externally. Casings are therefore fitted with water jackets, and the rotors themselves may be hollow and have oil continuously passed down them. By this means, discharge temperature of up to 200°C may be accommodated. Expansion problems are overcome via a thrust bearing situated at the discharge-end of the rotors. Thus all expansion is towards the suction-end, where large running clearances are possible without any detrimental effect on performance.

The flow rates which dry screw compressors are capable of delivering broadly cover the range of flow rates under consideration. Capacity control is by means of suction-throttling, i.e. reducing the inlet area.

Dry screw compressors typically have an operating speed range between 500 and 1000 rad/s. A stepup gearbox would, therefore, be required for the machine to be driven by a reciprocating gas engine.

The merit of this type of machine is the fact that there is no contact between the oil and the refrigerant, thus potential problems associated with refrigerant breakdown and oil dilution are avoided. There is also the advantage that this type of compressor can handle a certain amount of liquid without damage to the machine's components.

One of the principle demerits of dry screw compressors is the limit on pressure differential between inlet and outlet conditions. The pressure differential causes the rotors to deflect, and by virtue of the fine clearances involved, it becomes a limiting criterion. This may necessitate the use of several stages of compression.

Wet screw compressors

The principle of operation of wet (oil-injected) screw compressors is identical to that of the 'dry' compressor, already described, with the exception that a fluid is injected into the working space. The main advantage of this injected fluid is that it cools the gas during compression.

In most applications, the fluid adopted is lubricating oil. Other fluids such as water may also be used, but the overall design of an oil-injected compressor is simpler than either dry or other injected sets as there is no necessity to isolate the bearings from the compression space. Due to the relative specific heats of gaseous refrigerant and liquid oil, somewhat less than 1 per cent by volume of injected oil is usually sufficient to give the correct heat balance.

The speed of operation of oil-injected compressors is over a lower range than oil-free machines, because the oil suffers severe churning at high speeds, resulting in loss of efficiency. However, the presence of the oil in the compression space seals off much of the leakage path back to suction and much lower speeds may be used while maintaining the same efficiency. Capacity control of wet screw compressors is efficiently achieved via a sliding valve which alters the effective length of the rotors, a method analogous to shortening the stroke of a reciprocating machine.

Wet screw compressors, by virtue of the cooling oil, can achieve higher pressure ratios than dry screw compressors. Furthermore, because the bearings do not need to be isolated from the compression space and are therefore directly adjacent to the rotor bodies, the machine can withstand a higher-pressure differnetial across the rotors. These pressure differentials can go as high as 2·1 MPa across a single stage. Because of the mixing of oil and refrigerant within the compression space, plus the fact that much of the heat of compression is being absorbed by the oil, these compressors require both an oil separator (to recover the refrigerant from the oil) and an oil cooler.

The upper temperature limit of wet screw compressors may be determined by the life of the white-metal bearings, and the presence of oil in the refrigerant could necessitate the use of separators. This type of compressor is becoming popular for large throughput units.

Centrifugal compressors

Centrifugal compressors are primarily suited to fairly high-volume flow rates and low pressure ratios, although multistage centrifugal machines could be used where higher pressure ratios are required. In a centrifugal compressor the shaft bearings are well isolated from the compression space, so that the refrigerant is dry, with less than 2 p.p.m. of oil leaking inwards. This isolation means that mechanically and thermally, this type of compressor is perfectly suitable.

Because of the high rotational speeds (of the order of 840 rad/s) of these compressors. a stepup gearbox would be necessary when used in conjunction with a gas engine. This will, of course, increase the cost and reduce the efficiency. In the long term, however, it is possible to envisage centrifugal compressors coupled to gas turbine drives, which would ensure suitable speed-matching. That type of installation, however, involves capital expenditure outside the scope of many projects, and would be best suited to installations with upwards of 1 MW drive.

It is interesting to note that the only commercially available purpose-built range of industrial heat pumps, that marketed by Westinghouse (USA) under the name 'Templifier', uses a centrifugal compressor. This compressor was originally developed for refrigeration duties, where it is well proven, and at present is only available in a hermetic unit with electric drive (see Chapter 13). In heat pump duties, COPs of 4·5 to 5·5 are considered readily achievable using this system.

6.5.2 *Prime movers*

There are several options in the selection of the drive unit for a heat pump compressor. These include electric motors, diesel and gas reciprocating engines, and, in some instances, gas turbines.

Electric motors

As will be seen from the data presented in Chapter 13, and the application examples cited in Section 6.6 of this chapter, the electric motor currently dominates the scene as far as heat pump compressor drives are concerned. Having said this, their common

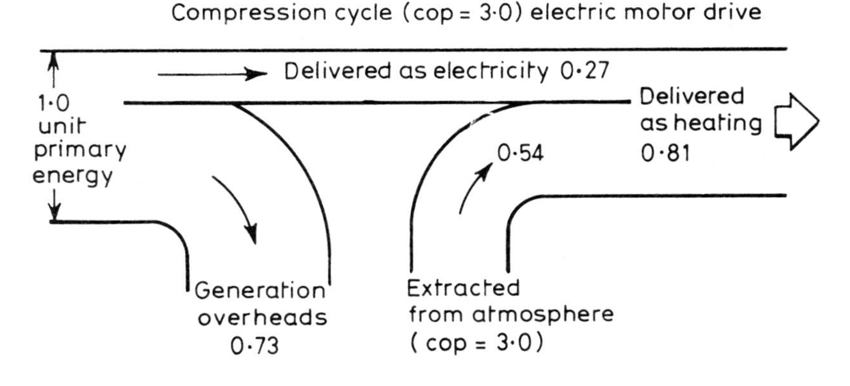

Fig. 6.6 Energy utilization of an electric motor-driven air source heat pump.

use in refrigeration and heat pump systems removes the need to treat their characteristics in detail, as most engineers will be familiar with such equipment.

The electric motor may be used integral with the compressor in a 'hermetic' or 'semihermetic' unit, where the two units are sealed, or it may be mounted in such a way as to drive an open compressore via a shaft connection. Electric motor drives are extremely convenient to use, but, as shown in Fig. 6.6, the electricity generation overhead is only partly recovered on an air-source unit and as users of primary energy they are not the most efficient system.

However, electric motors are used in almost all vapour-compression cycle heat pumps – they are cheap and have low maintenance costs – and do power the only large commercially available industrial-process heat pump system, where the high coefficients of performance do make for an energy-efficient unit.

Other prime movers

Reciprocating diesel and gas engines, gas turbines and Stirling engines (external-combustion units) are all under study for heat pump drives. The reciprocating gas engine is probably the most popular contender at present, with, of course, as much use made of the engine waste heat as possible. A more complete description of reciprocating prime movers is given in Chapter 4, and it is proposed here to comment briefly on the recoverable heat outputs, and to present a circuit diagram.

The economic success of the gas-engine heat pump is completely dependent upon recovery of waste heat from the engine. If a useful amount of heat is not recovered, then the Primary Energy Ratio (PER) or effective fuel utilization of the complete unit will be too low.

The first, and most important, consideration in this design is the thermal efficiency of the engine. This will vary between different types of prime mover (e.g. gas turbine, diesel engine, reciprocating gas engine) and will also vary with speed and load conditions. Typical values of thermal efficiency are tabulated below for a variety of prime movers:

	(%)
Gas turbine	28
Gas engine 4-stroke, non-turbocharged	31
2-stroke, non-turbocharged	34
4-stroke, turbocharged	38
Diesel engine	35

Under off-load operating conditions the efficiency will vary, depending on whether variable speed control is available or not. If it is not available, then power reduction can only be achieved by throttling at constant speed. The efficiency of a typical engine under part-load conditions is such that when there is a reduction in power demand, this can be accommodated in two ways – with or without speed reduction. When the speed is constant, then specific fuel consumption rises as the load is reduced. When the speed is reduced but keeping the engine at maximum power, then fuel consumption actually improves slightly.

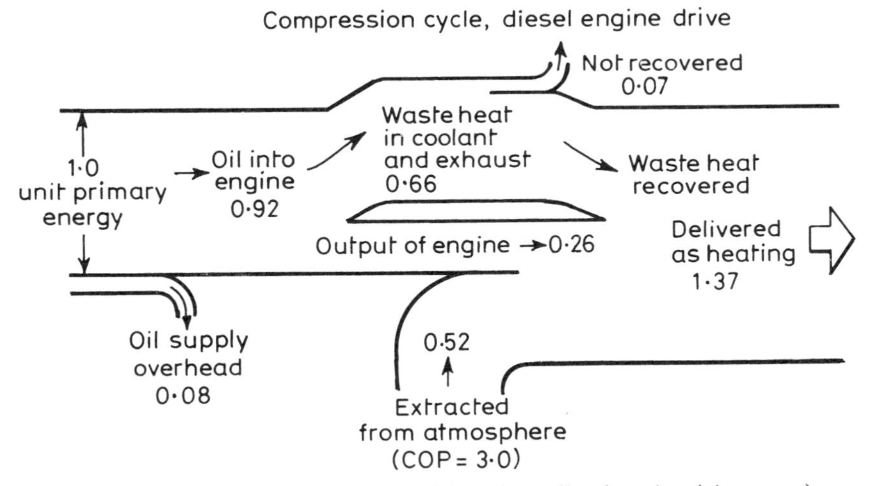

Fig. 6.7 Energy balance of a heat pump driven by a diesel engine (air source).

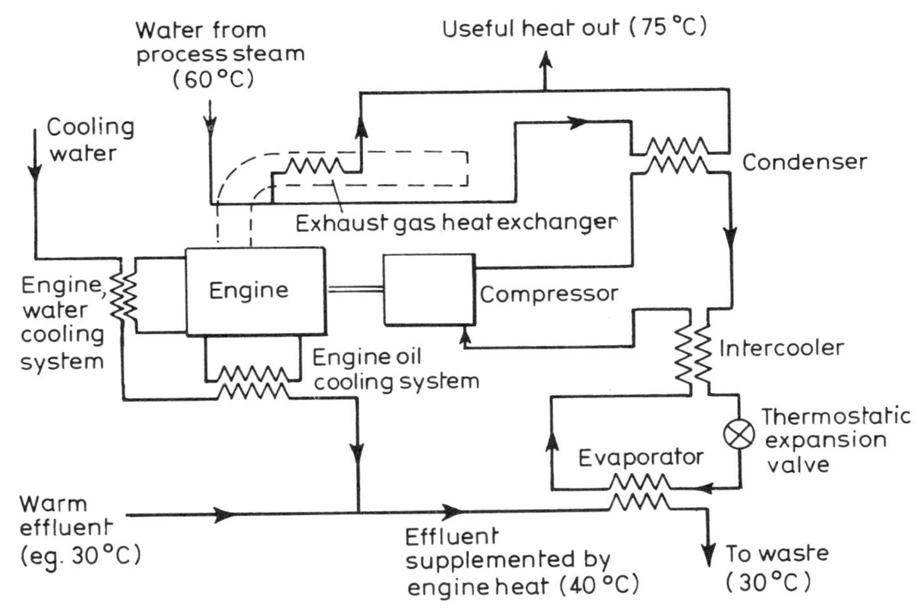

Fig. 6.8 A line diagram of a gas engine-driven heat pump, showing the use of engine waste heat.

For a four-stroke, non-turbocharged engine of 31 per cent efficiency delivering 75 kW, the remaining 69 per cent of input heat emerges from the engine in the form of hot exhaust gas, engine-cooling water, oil-cooling water and stray losses. The relative proportions of these are as follows:

	(%)
Exhaust gas	28
Engine water jacket	30

	(%)
Oil cooler	3
Stray losses	8

Diesel and gas engine-driven heat pumps, particularly with the latter drive, were operated many years ago (e.g. the Festival Hall system), and more recently a considerable number of prototype systems using these prime movers have been designed and constructed – principally in the UK and Europe – where a considerable growth in interest is noticeable. As shown in Fig. 6.6, it is necessary to recover a proportion of the waste heat from the prime mover to make the system viable, but this heat can be at a very high temperature, with steam from the engine water jacket if ebullient cooling is used.

The circuit of a gas engine-driven heat pump, showing how the engine waste heat may be used, is illustrated in Fig. 6.8. This is a water-water heat pump for industrial-process application.

6.6 Applications and economics

There are three major application areas for heat pumps – domestic, commercial and other buildings, and industrial process heating. (It is possible to combine the latter with a refrigeration duty, as will be illustrated.) The domestic heat pump is outside the scope of this book, in so far as it is concerned principally with commercial buildings (larger scale HVAC) and industrial-process plant heat recovery. For more data on these systems, the reader is advised to consult the Bibliography, but it is treated in depth by Macmichael and Reay (1979) and an overview, including economic assessments, is given by Perry (1978).

6.6.1 *Commercial buildings and similar structures*

The use of heat pumps in commercial buildings has particular advantages in cases where both a cooling and heating duty is required. This is because the capital cost of the system is not much different from that of a conventional air conditioning system with electric resistance heaters, but the energy costs are substantially lower.

In the UK, the Electricity Council have produced guidelines for sizing of heat pumps for commercial premises, and while some of the temperature data may be unique to European conditions, their comments are of considerable value (Electricity Council, 1977).

Sizing a heat pump

In sizing a heat pump design, the first consideration is the cooling requirement, which in many instances will be greater or at least equal to the heating load. This is because:

(1) If the cooling capacity is too small, there is no remedy other than to install a

larger heat pump.

(2) If the cooling capacity is too large, humidity control will be poor and the heat pump will shortcycle.

Nevertheless a compromise between cooling and heating requirements should be attempted. Ideally the heat pump should provide as much of the heating requirements as possible since it delivers at least twice the amount of heat per kilowatt input than resistance heaters. Sizing the heat pump to meet the peak heating load at −1°C would be uneconomical, since the number of days when this temperature occurs is low (less than 10 days in an average heating season of 239 days). Moreover, as the outside temperature falls the output of the heat pump drops, while the building's heating requirement increases. So it is recommended that the size of heat pump selected should be such that its output balances the heat requirements of the building at approximately 3°C to 5°C. This point may be termed the Design Balance Point (Fig. 6.9).

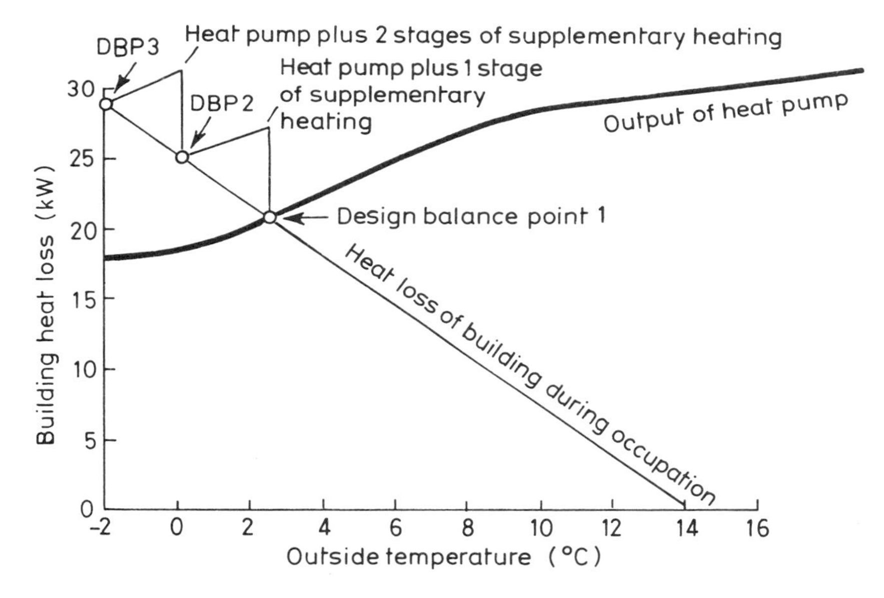

Fig. 6.9 Determination of design balance point of heat pump output and building heat loss.

Where high incidental heat gains occur, from lighting, occupants, etc., they must be taken into account in calculating the heat load of the building. The lower the Design Balance Point the greater the proportion of the heating requirements that can be met by the heat pump. So the sizing of any heat pump design must reconcile:

(a) Cooling load
(b) Air volume and air-change requirements
(c) Heating requirements (low Design Balance Point)
(d) Minimum capital cost

It is current practice that the extra heat required when the outside temperature falls below the Design Balance Point should be provided by direct-acting electric supplementary heating.

How much direct-acting space heating is required?
Having calculated the heating requirements of the building, the output of the heat pump for various temperatures as given by the manufacturers should then be plotted to determine the Design Balance Point (Fig. 6.9).

The additional heating below the Design Balance Point is then provided by incremental stages of supplementary heaters which may be in one or two stages, and are sized to meet the heat load on design day conditions. Manufacturers provide a range of supplementary heater loadings which enable the additional heat requirements to be met – ideally – at at least two stages.

Apart from the extra capital cost, oversizing of the heat pump in its heating mode will also produce unnecessary shortcycling which can be detrimental to the life of the machine. A Design Balance Point of 3°C to 5°C is, therefore, a satisfactory compromise.

Operation of a heat pump
The output of the heat pump is normally controlled by a room thermostat having two-stage heating and one-stage cooling. When the heat pump is heating, the first stage of the thermostat controls the output of the heat pump, but if this is insufficient to meet the heating requirements, then the second stage operates bringing in the supplementary heaters.

These supplementary heaters are also each controlled by an outside thermostat set at Design Balance Point temperatures 1 and 2, shown in Fig. 6.9, so unless the outside temperature is below the Design Balance Point temperature 1, the first stage of the supplementary heating will not come into operation.

Estimated energy consumption
If the heat pump is sized in the manner suggested, an overall COP (heating) of 2 can be expected. In other words, the input energy required for heating only will be reduced to less than 50 per cent of that calculated theoretically, after incidental heat gains have been taken into account. (See Chapter 13 for manufacturers' data on COPs.)

The air-air heat pump system shown in Fig. 6.10, arguments concerning the sizing of which are given above, utilizes low-grade heat recovered from the air for heating in winter, and uses the same air as a heat sink in summer. Figure 6.11 shows an air-air heat pump acting as a heat recovery unit in a double-duct air conditioning system, capable of simultaneously heating and cooling (where, for example, the building may contain a computer room). In this system, the condenser and evaporator are both fulfilling useful functions within the same building. When there is no internal heat in the building to be recovered, e.g. during periods of low or zero occupancy, heat may be taken from the outside air instead.

In specifying a heat pump of this type for building temperature control, it is useful

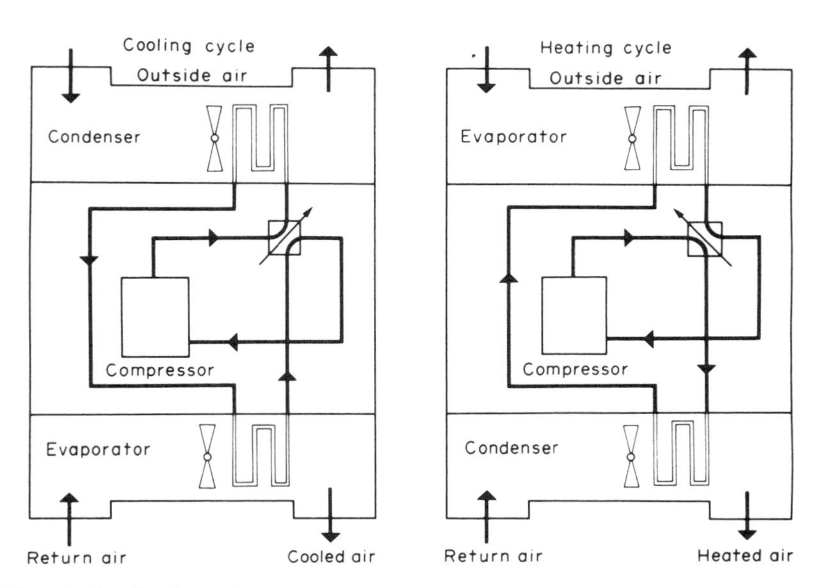

Fig. 6.10 A simple air-to-air pump system, using outside air as a heat source in winter and a heat sink in summer (by courtesy of the US Electric Energy Association).

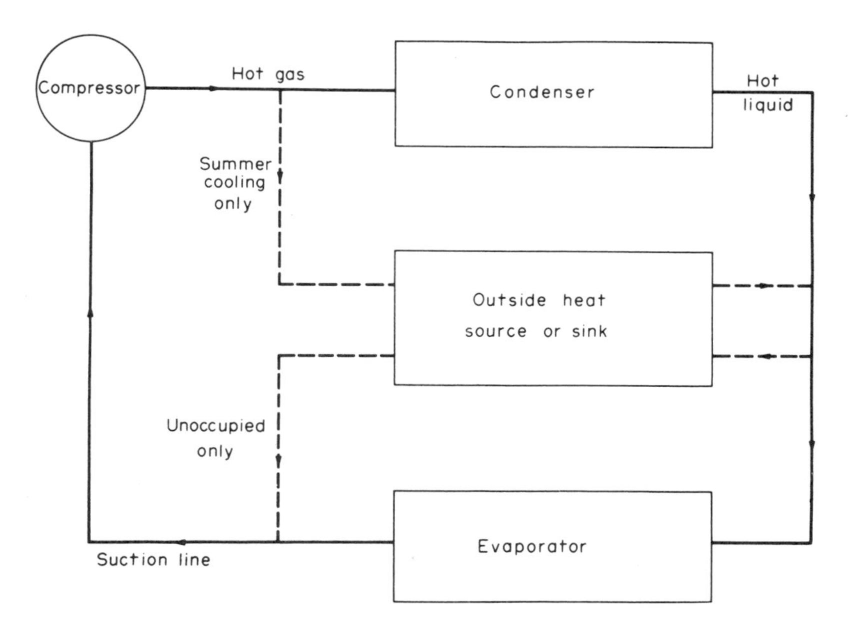

Fig. 6.11 A heat pump applied to simultaneous heating and cooling duties within one building shell.

to know the role of the components and extras available. Discussed at some length in the introduction to Chapter 13, where manufacturers present data on the systems, the principal system components are as follows (excluding compressor and drive):

(1) Indoor-outdoor coils: each coil, indoor and outdoor, must serve as an evaporator or condenser and must satisfy each of these functions – this is known as 'coil-optimization', or in other words, the matching of the area of the coil either as an evaporator, or as a condenser.
(2) Four-way changeover valve: four-way valve has the function of reversing the flow of refrigerant in the circuit and subsequently reverses the role of the indoor and outdoor coils, either to meet a cooling or heating requirement within the building or to defrost the outdoor coil during the heat pump's heating mode.
(3) Check and expansion valves: check valves and expansion valves are fitted to ensure that the refrigerant flow takes place through the proper expansion valve, so that it is always metered into whichever coil is operating as the evaporator.
(4) Accumulator: heat pumps are generally fitted with an accumulator which stores excess liquid refrigerant occurring when the heat pump changes from heating to cooling, and which prevents liquid refrigerant entering the compressor.
(5) Compressor crankcase heater: a heater is required to boil off any liquid refrigerant that might enter the compressor and cause starting difficulties.
(6) Defrost mechanism: as the temperature of the evaporator must always be lower than the outside air temperature, it is clear that on cold days frost or ice will be deposited on the outdoor coil, which seriously reduces its efficiency. Controls fitted to the outdoor coil detect this ice buildup, operate the four-way reversing valve and direct hot gas from the compressor to the outdoor coil for a period (normally 4 min), long enough to melt the accumulated ice. During this operation the fan on the outdoor coil is stopped. Also, most designs incorporate a direct-acting electric heating battery which may be switched on during the defrost cycle. Double evaporators may be used, one being defrosted while the other maintains its heat recovery function.
(7) Supplementary heaters: proprietary heat pumps are fitted with a series of electric heater batteries normally controlled by outside thermostats to supplement the output of the heat pump on cold days. The loading of these heaters is determined when sizing the heat pump to meet a given building heat demand (see section on heat pump sizing).

As an alternative to air, water may be circulated through the evaporator and condenser for heating and cooling respectively, in conjunction with a cooling-tower, as illustrated in Figs. 6.12 and 6.13. Use is made of a double-bundle condenser in the second of these circuits, with two separate water circuits enclosed in the same shell. The heat given up during condensation may be put into either or both of the water circuits, as required.

The size of the heat reclaim 'package' obviously depends upon the duty required, but as shown in Figs. 6.14 and 6.15 (a double-bundle unit and a chiller employing two separate condensers for heat recovery duties) the assembly can be neatly arranged for convenient site location.

A number of manufacturers produce water–air heat pump air conditioning systems.

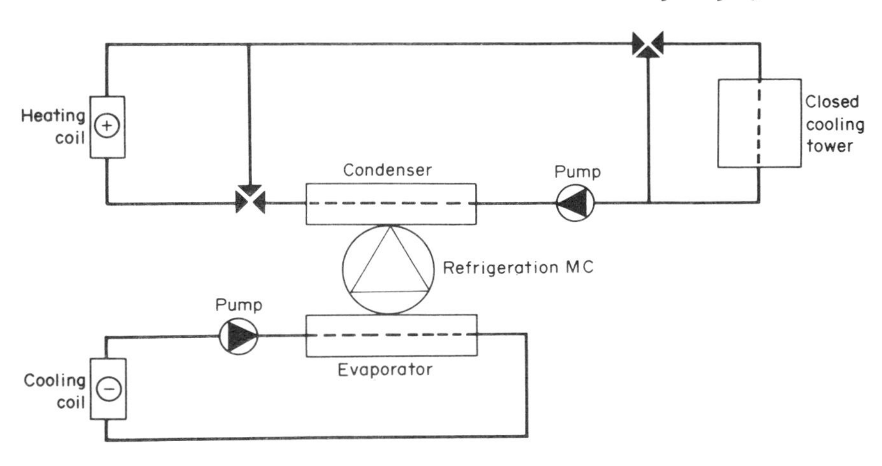

Fig. 6.12 Heat recovery system with a closed circuit cooling tower (by courtesy of The Electricity Council).

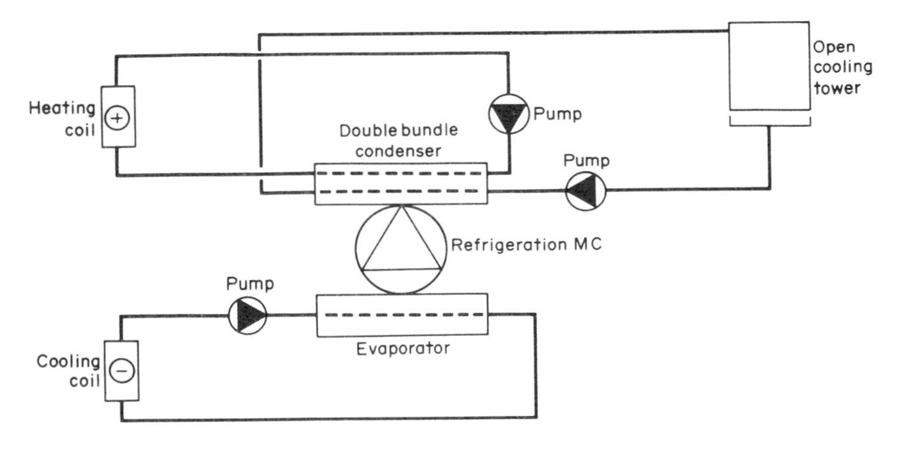

Fig. 6.13 An alternative to the system shown in Fig. 6.12 – use of a double-bundle condenser (by courtesy of The Electricity Council).

Temperature Ltd., first on the market with this decentralized air conditioning systems which is called 'Versa-Temp', in common with companies such as AAF, are able to offer a system which can have any number of individual water-air heat pump air conditioners, situated in individual rooms if required. Each conditioner is connected to a common two-pipe closed-water system by means of hose connections. Each individual unit incorporates a refrigeration circuit with hermetic compressor, a finned refrigerant-to-air heat exchanger, a refrigerant-to-water heat exchanger, a reversing valve, adjustable thermostate for control, fans, filters and protection devices. As shown in Fig. 6.16, the water circuit connects the units with the ancillary cooling-tower and boiler plant to form a closed circuit to operate at a supply temperature between 20°C and 30°C, normally 27°C.

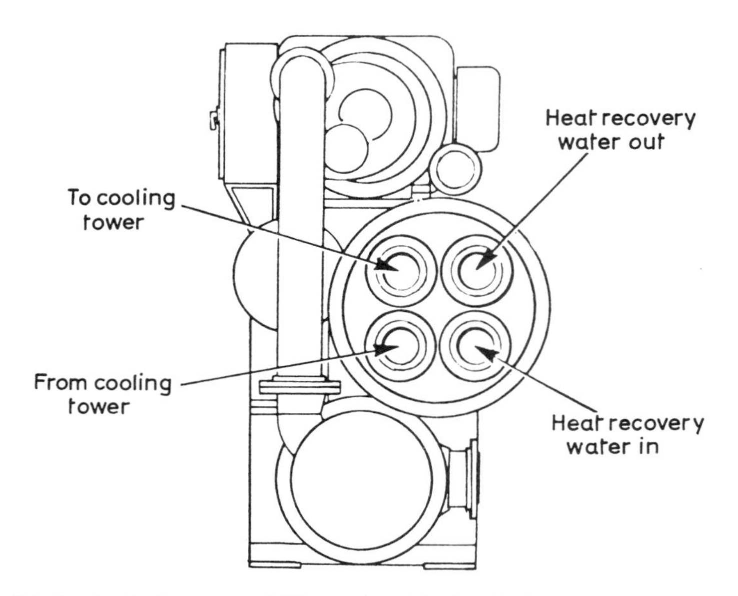

Fig. 6.14 'Packaging' of a water chilling unit with double-bundle condenser (end view).

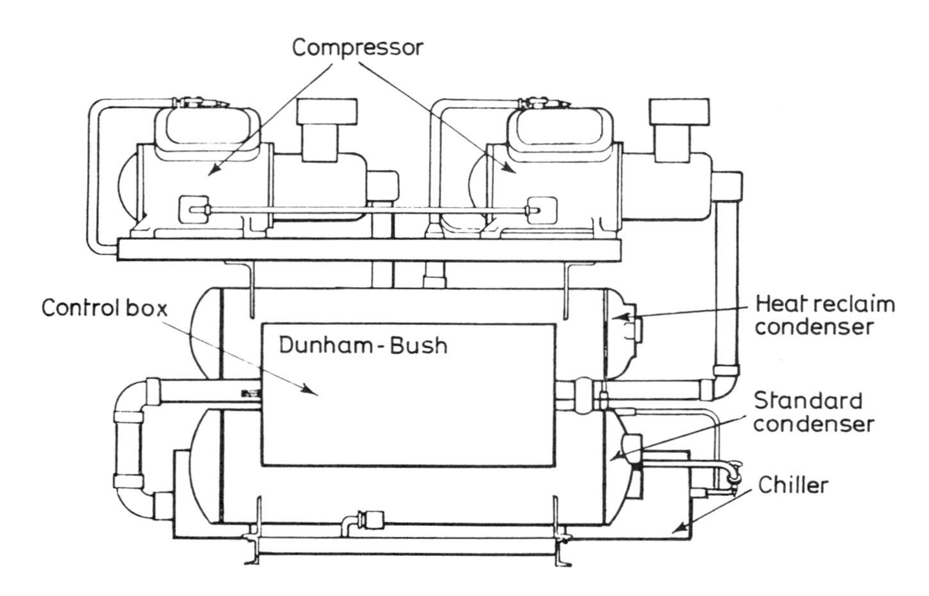

Fig. 6.15 'Packaging' of a chiller using two separate heat recovery condensers (side view).

When the majority of the units are operating on the cooling cycle, the excess heat is transferred to the water circuit and is used either in other areas of the building where a heat demand exists, or rejected to the atmosphere via the cooling-tower. When the heating cycle predominates, any heat deficiency in the water circuit is made good by a small boiler (or alternative heat source). This maintains the water circuit temperature at

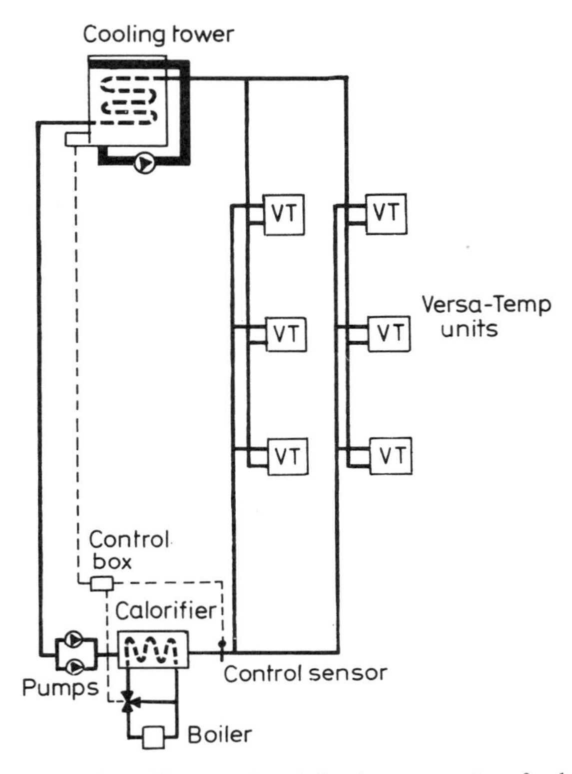

Fig. 6.16 The Versa-Temp water-air heat pump system for buildings.

the set value. However, for a large proportion of the operating time, the boiler and the cooling-tower can remain inactive because the system can be thermally balanced by transferring energy from areas requiring cooling to those needing heating. Operation of these two modes is shown in Fig. 6.17.

Systems of the above type can lead to very significant savings on plant space compared with centralized equipment, and the ability to connect each unit with small-diameter water pipes has benefits when considering the application of heat pumps to older buildings. COPs are given in Table 6.1.

Swimming-pools

As well as serving as heat recovery systems for conventional buildings, the use of heat pumps in swimming-pools is a considerable growth area. Studies in the UK have shown that the use of heat pumps either to recover heat from exhaust air, or for internal dehumidification, can reduce the annual energy requirement from 10 500 kWh/m^2 of pool area to 5000 and 2000 kWh/m^2 respectively. Supplementary-heating requirements are minimal, and can be best provided by conventional boiler plant.

Sulzer have used other heat sources, in conjunction with a heat pump, for heating swimming-pools. The system shown in Fig. 6.18 uses fresh water from a canal as the

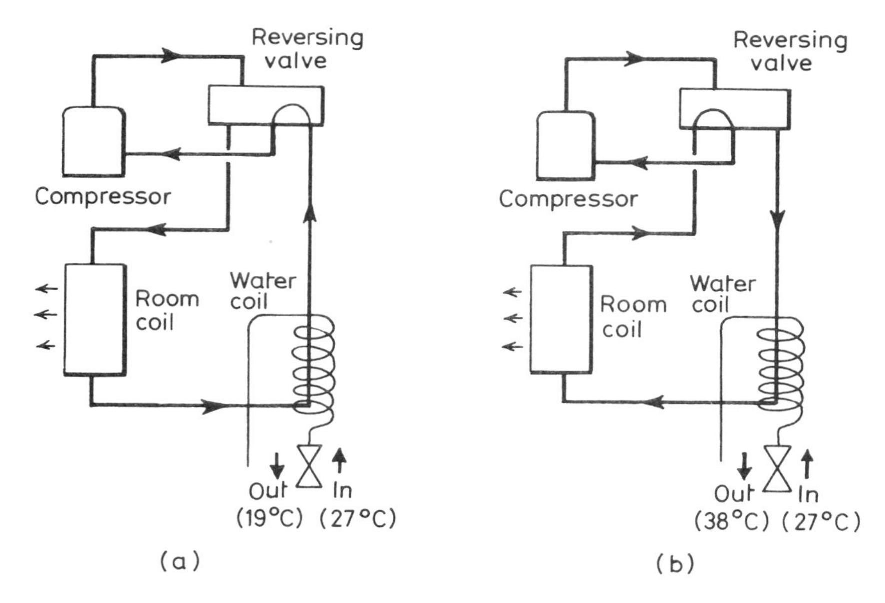

Fig. 6.17 (a) Versa-Temp heating; and (b) cooling operation, switching implemented by a reversing valve.

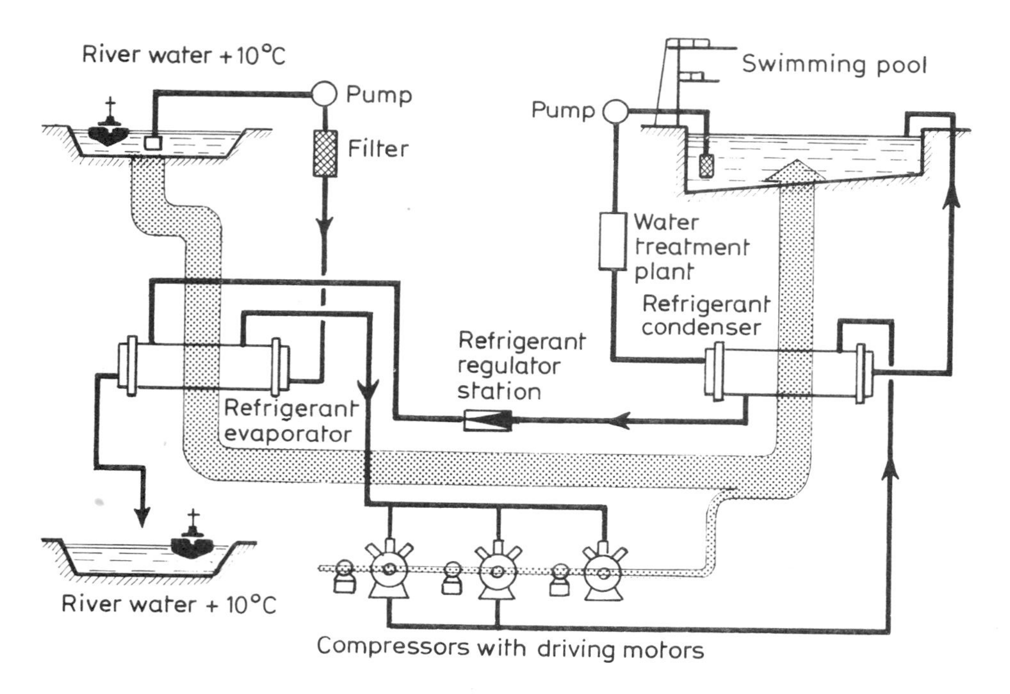

Fig. 6.18 Schematic diagram of a heat pump installation for an outdoor swimming-pool with fresh water as the heat source.

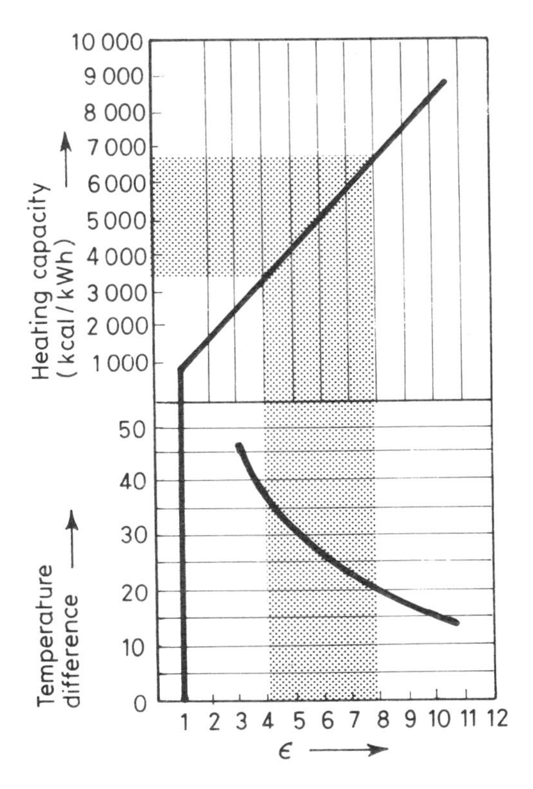

Fig. 6.19 Coefficient of performance as a function of the temperature difference between evaporation and condensation temperatures.

heat source. (Combining ice-rinks and swimming-pools in the same complex has also proved advantageous.) The system shown in Fig. 6.18 cools canal water from 10°C to 7°C, heating the pool water from 23°C to 25°C. Fig. 6.19 shows the expected coefficient of performance of the heat pump as a function of the difference between evaporation and condensation temperatures. The figure also indicates how much heating capacity can be obtained from 1 kWh of electrical energy at the various COPs. The shaded area in the graph represents the range applicable to swimming-pool heating systems, depending on local conditions and the pool temperature required.

The sources of heat available for this and other commercial heat pump applications are several. Fig. 6.20 shows how the temperature of these sources can vary throughout the year.

Ground

Heat stored in the soil is only usable to a limited degree, and only for smaller installations. In the case of artificial skating-rinks, the effect of solar radiation on the concrete of the empty rink can be used as a supplementary heat source.

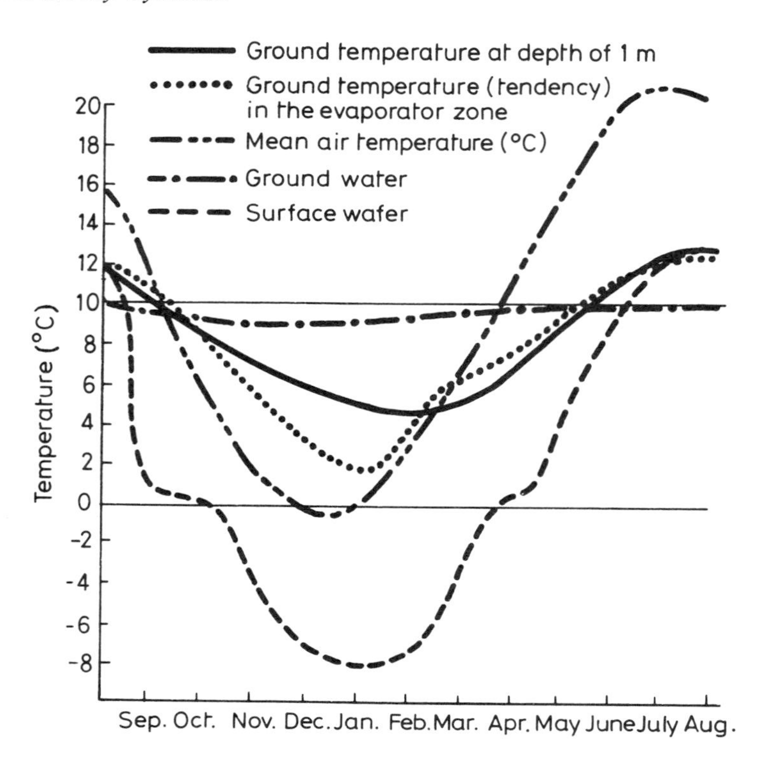

Fig. 6.20 Mean temperature curves for various heat sources.

Table 6.4 Economics of swimming-pool heat pump

Heat pump	Night power rate		0·05	DM/kWh
	Evaporation temperature	+0	+0	°C
	Condensation temperature		+30	°C
	Performance coefficient		5	
	Price per Gcal (1 million kcal)		12	DM
Oil	Price		0·30	DM/kg
	Calorific value		10 000	kcal/kg
	Efficiency	approx.	0·75	
	Price per Gcal		40	DM
Gas	Price		0·30	DM/Nm^3
	Calorific value (natural gas)		8000	kcal/Nm^3
	Efficiency	approx.	0·9	
	Price per Gcal		42	DM

Water

Groundwater is an ideal source of heat, because it is available at a virtually unchanging temperature. Surface water is especially usable in summer time. Another good source is warm waste water, i.e. cooling water from refrigeration systems or industrial processes (water–water heat pump). And the residual heat in swimming-pool drain water can be recovered.

Air
During the summer, outside air can be used to heat swimming-pools (air-water heat pump).

Solar energy
Water heated by solar energy can be raised to a higher temperature level with a heat pump. If an outdoor artificial skating-rink is available, its concrete surface is a suitable heat source because it is subject to solar radiation and represents a good storage mass.

The economics of a heat pump, compared with oil and gas heating, for an outdoor swimming-pool are given in Table 6.4 (German data – Sulzer Bros Ltd.). This illustrates that considerable savings can be made by employing a heat pump in this particular application (having taken into account capital costs).

6.6.2 *Industrial process use*

The use of heat pumps in industrial processes is a major potential growth area, limited at present largely by the number of systems available (see Chapter 13) and the difficulty in achieving high condensing temperatures (in excess of 120°C). Three systems will be considered here, illustrating the variety of process use:

Combined heating and refrigeration
It is possible to make use of both the heat recovered, in the form of useful energy rejected at the condenser, and the cooling effect which results from the heat removed from the evaporator primary circuit. Obviously this is economically the most desirable state of affairs, and can frequently be used to good effect in plants where refrigeration, or even less demanding water-chilling requirements exist.

Typical of such a requirement is an injection-moulding plant, where chilled water is required to cool the moulds, hence solidifying the plastic in the component being moulded. In addition, heating is required in the factory to maintain comfort conditions. A heat pump capable of fulfilling both of these functions has been installed in the factory of the Link 51 Plastic Division at Telford in the UK. Based on a Prestcold compressor, the unit replaces a conventional cooling-tower, and supplies water at 7·2°C at a rate of up to 1140 l/min to the moulding machines. The heat extracted from the water is used for factory space heating. By using the controlled conditions afforded by the heat pump, the rate of production of the injection-moulding machines could be maintained at a high rate, independent of ambient conditions which could affect the performance of the cooling tower (Williams, 1976).

With a condenser heat output of 325 kW, the system, illustrated in Fig. 6.21, is estimated to save about £15 000 per annum, and the heat pump, in conjunction with a number of other minor heat and light conservation measures, has a payback period

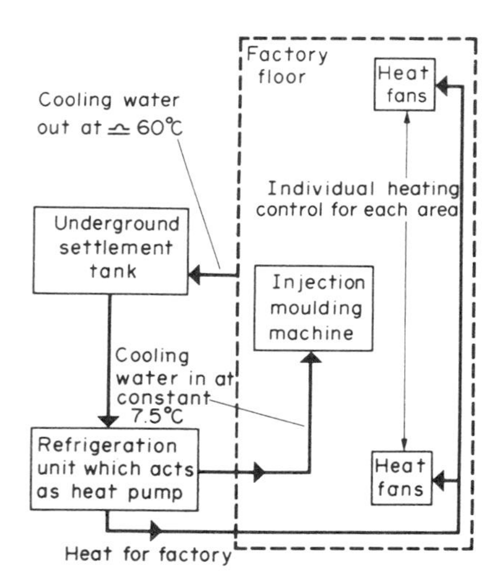

Fig. 6.21 A heat pump functioning as a water chiller and space heater in a factory (by courtesy of *The Engineer*).

of about three years.

Water–water heat pumps for process heat recovery
The use of electric-driven heat pumps to recover heat from waste water or other liquids for reuse in heating process streams, or heating water for space heating, is exemplified by the Westinghouse 'Templifier' heat pump, the compressor/motor assembly having been described above. 'Templifier' units are either single or two-stage heat pumps, based on centrifugal compressors, covering a range of heating capacities up to approximately $1{\cdot}3 \times 10^6$ kcal/h with liquid-leaving temperatures (on the primary side of the condenser) of up to 110°C.

Figure 6.22 shows the 'Templifier' in an application where water is to be delivered at 82·2°C, using heat available from waste water being rejected at 32°C. In this instance two stages of compression are used, and the use of a flash collector tank from which a proportion of the vapour is bled to the inlet of the second-stage compressor, acts to improve the process efficiency (Ross, 1975). The coefficient of performance of the 'Templifier' varies with conditions at the evaporator and condenser, and where the temperature differences between the heat source and heat sink are large, the two-stage compression arrangement is normally recommended. The curves, shown in Fig. 6.22 illustrate the COPs that can be obtained with the 'Templifier' system cooling water acting as the heat source by approximately 6°C and heating process water through a range of approximately 6°C to 20°C to the outlet temperatures shown. If a heat pump is required to provide process hot water at 82°C using returning water at 74°C, using waste water available at 43°C, from Fig. 6.23 a 'Templifier' having a single-stage compressor will meet this duty, and will have a COP of 3·7. The operating cost of the heat pump, assuming electricity is available at 1·5p/kWh, can be compared to that of

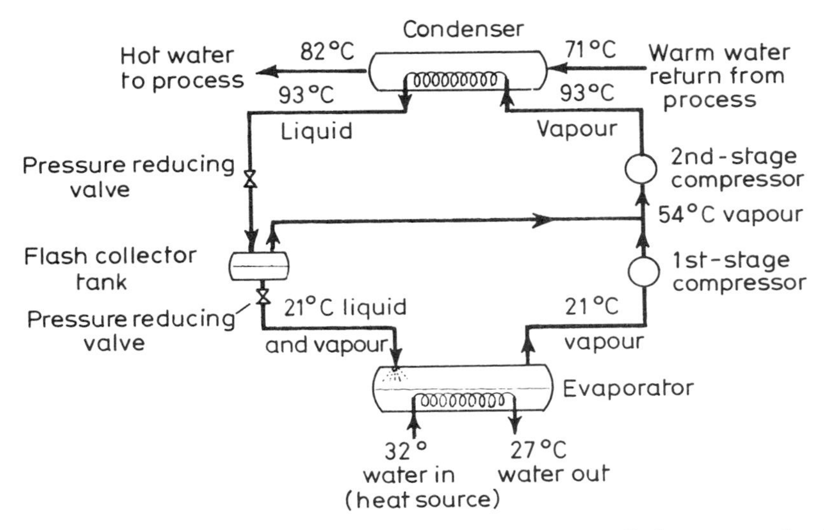

Fig. 6.22 Schematic diagram of the Templifier heat pump applied to recover heat from a waste water stream. Two-stage compression: 32°C hot water delivered. (By courtesy of Westinghouse Electric Corporation).

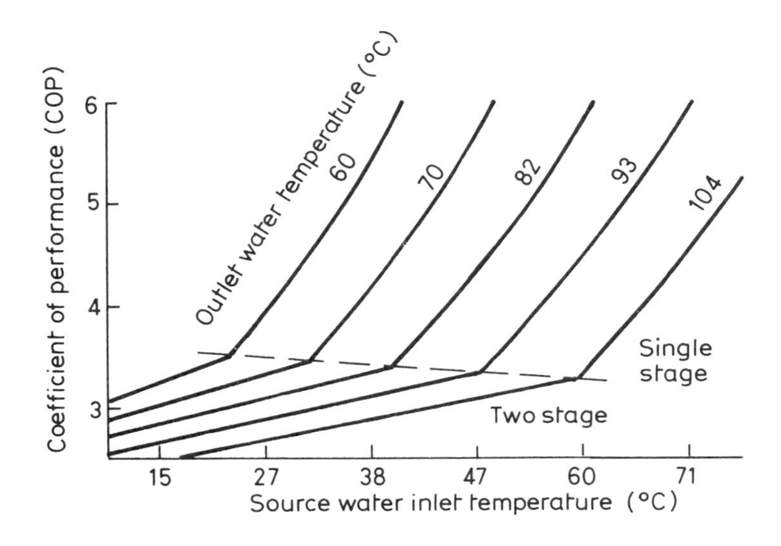

Fig. 6.23 COPs attainable with a Templifier unit in water–water heat recovery (by courtesy of Westinghouse Electric Corporation).

an oil-fired process heater operating as a seasonal efficiency of 70 per cent:

Medium fuel-oil cost	5·2 p/l
Electricity cost	1·5 p/kWh
Conversion efficiency of oil heater	70 per cent

Fuel-oil heat content	$7{\cdot}56 \times 10^3$ kcal/l
Heat pump COP	3·7
Total process heat required	500 kWh

(i) *Direct electric heating*

Cost = 500 x 1·5 = £7·50

(ii) *Oil heating*

$$\text{Cost} = 5{\cdot}2\text{ p} \times \frac{500 \times 10^3 \times 0{\cdot}86\text{ kcal}}{7{\cdot}56 \times 10^3\text{ kcal}} = £2{\cdot}96$$

(iii) *Heat pump*

$$\text{Cost} = \frac{500 \times 1{\cdot}5}{3{\cdot}7\text{ (COP)}} = £2{\cdot}03$$

A further illustration of the 'Templifier' performance may be given by relating heat outputs at the condenser to the power input required by the compressor motor. These are given in Table 6.5 for a single-stage unit heating water from 54°C using source water at 35°C. (In these instances, the source would be cooled to 29·4°C.) Four 'Templifier' models, having motors ranging from 63 kW to 291 kW, are shown. The power required to overcome primary-circuit pressure drops in the evaporator and condenser is not included in any of the above analyses.

Table 6.5 Example of 'Templifier' performance

'Templifier' model	Water flow 35°C source	Water flow 65·5°C delivery	'Templifier' heating capacity	Power input	Approx. operating weight
	m^3/h	m^3/h	Thousands kcal/h	kW	kg
TPO50	33	21	228	63	2300
TPO63	71	44	486	132	5300
TPO79	106	66	728	289	6700
TP100	169	105	1167	291	10 100

Note: Nominal size single-stage 'Templifier' heating delivery water from 54°C to 65·5°C while cooling source water supplied at 35°C to 29·4°C. (Courtesy: Westinghouse Electric Corporation).

Westinghouse have studied a number of applications of the 'Templifier' unit, and Table 6.6 sets out the type of industry and the processes in the industry where the 'Templifier' could be applied. Temperatures given in the table are those required at the condenser outlet.

A second liquid–liquid heat pump system, which can operate with an electric motor or reciprocating engine as a prime mover, is that proposed by Milpro NV, (U.K.)

Table 6.6 Potential applications and temperature requirements of a 'Templifier' heat pump

Major process use (Temperature requirements are indicated in °C) / *Industry*	Meat products (Food products)	Dairy products (Food products)	Canned, cured and frozen foods (Food products)	Grainmill products (Food products)	Bakery products (Food products)	Sugar (Food products)	Confectionary products (Food products)	Beverages (Food products)	Miscellaneous (Food products)	Textile-mill products	Apparel and other textile products	Lumber and wood products	Pulp and papermills	Chemical and allied products	Petroleum refining	Fabricated rubber products	Fabricated metal products	Machinery, except electrical	Electrical equipment and supplies	Transportation equipment
Washing/sanitizing/cleanup (60)	\	\	\	\	\	\	\	\	\											
Cooking (100–115)	\		\				\	\												
Pasteurization (65)		\																		
Blanching (85)			\																	
Dye heating (88)										\										
Pressing (100)											\									
Log soaking (90)												\								
Metal cleaning and plating (60–90)																	\	\	\	\
Paint drying (80–120)																	\	\	\	\
Drying (105)										\										
Mold-release solution (85)																\				
Vessel heating														\	\					
Heat tracing															\					
Evaporators		\											\							
Process heat (hot water)														\						
Paper machine													\							
Pulp heating													\							
Sterilization (110)		\	\																	
Scalding (50–60)	\																			

(Courtesy: Westinghouse Electric Corporation)

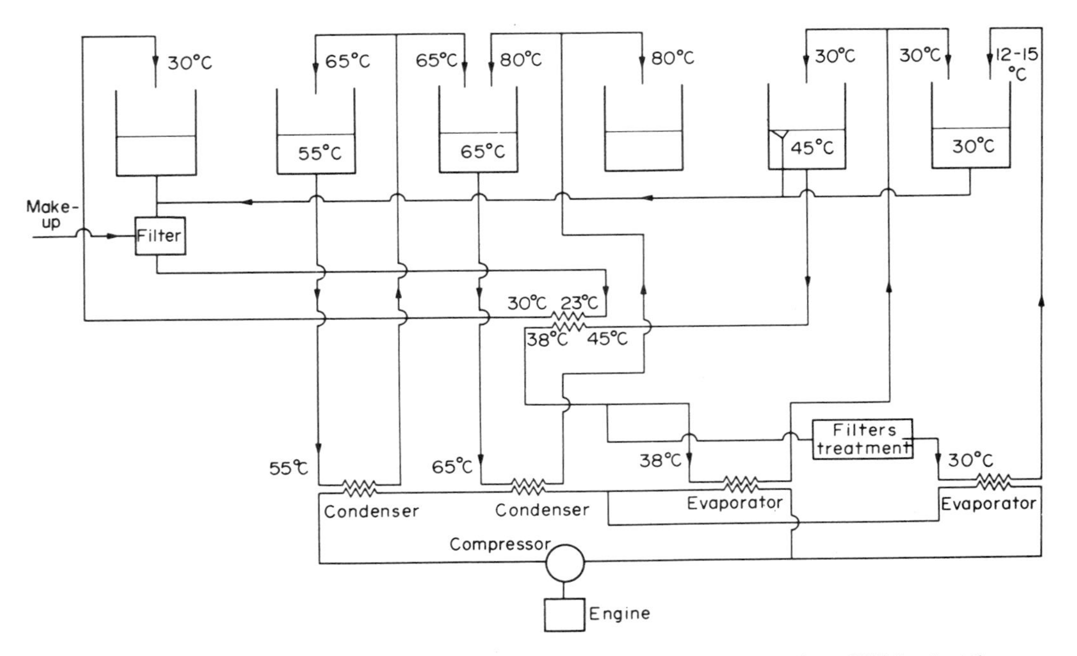

Fig. 6.24 A heat pump and water recovery system developed by Milpro NV for bottle-washing machines.

Patent 1494780). Incorporating water treatment as well as heat recovery, the patent details an application in a bottle-washing machine of the type used in dairies, breweries, etc., although the unit is not restricted to this application. The layout is illustrated in Fig. 6.24. Bottles pass through from left to right, being heated by sprays of water and detergent before being cooled prior to leaving the machine.

A conventional heat exchanger is used to recover heat from the final detergent tank, this heat being used to raise the temperature of the prerinse spray water. The input to the prerinse sprays is supplemented from the final rinse tank (30° C) and also by overflow from the final detergent tank (45° C). A filtration system, which rejects a proportion of the prerinse water, is used to maintain prerinse water in a comparatively clean condition.

Once some of the heat has been removed from the water passing out of the final detergent tank for transfer into the prerinse water, the final detergent-tank water is then passed through an evaporative heat exchanger, serving as one of the heat sources for the heat pump, before being returned at 30° C to the final warm rinse and final detergent tanks. Prior to passing through the evaporator, a proportion of this flow is taken for filter treatment and cooled to 12° C to 15° C in a second evaporator. This clean water is then used as the final cold-rinse water supply.

The heat recovered in these evaporators is taken up in the form of refrigerant vapour, which is then compressed, raising its temperature by the addition of the work of compression. This heat is then rejected at the higher temperature in a condenser to water from the first detergent tank (raised from 55° C to 65° C) and also in a second condenser, where water for the high-temperature section is heated from 65°C to 80°C.

Any makeup water required is added to the prerinse water circuit. Based on a machine washing 30 000 bottles per hour, consuming, with conventional heating and water usage, 13 600 l and 600 kW/h heat energy, Milpro claim that the heat pump and filter system can reduce consumption to 70 kW and 2600 l/h. Payback periods of two to three years are claimed for the unit, which can run either on natural gas or electricity.

The washing cycle of conventional large dishwashing machines is similar to that of bottle-washers, and it is natural that the heat pump is applied here also. In this case the rinse water is heated to 90°C, somewhat higher than that in bottle-washing machines. The waste heat is discharged in part as hot vapour, and it is this which is used as the heat source in the heat pump system. The resulting cooling of this vapour improves comfort conditions in the vicinity of the washer, as well as saving energy.

The heat pump operating with a COP raises the rinse water to 60°C, and affect the power input to the washer as follows:

	Input (kW)		Output (kW)	
Conventional	Motors	5·5	Heat dissipation	5·5
	Calorifier	67	Waste water	26·5
	Tank heating	54	Dishes	41·5
	Total	126	Vapour	53·0
Heat pump design	Input		Output	
	Motors	5·5	Wasted heat	5·5
	Heat for final rinse	25	Drain	26·5
	Heat pump compressor	24	Dishes	22·5
	Total	54·5	Recovered heat	96

The capital cost of this unit is recovered in 31 months.

This is one of the very few cases seen to date where a heat pump is offered as a standard energy-conservation device on a process ‘package’.

Air–air heat pumps in processes

One of the successful heat pump applications arising out of work at the Electricity Council Research Centre at Capenhurst in the UK is in the drying of timber, although the technique is equally applicable to other drying processes (Kolbusz, 1972; Kolbusz, 1975); see also later in this section.

The efficiency of a dryer can be increased with the use of heat recovery systems. These systems all rely on heat recovery from the exhaust (see Chapter 1), which is then vented to atmosphere, the heat being transferred into fresh incoming air. However, in theory the efficiency of a dryer in which the air is completely recirculated could approach 100 per cent, but this state is prevented from being achieved by the increasing humidity of the circulating air. Full recirculation can be implemented in practice if a dehumidifier is located in the drying kiln to remove the moisture from the exhaust air,

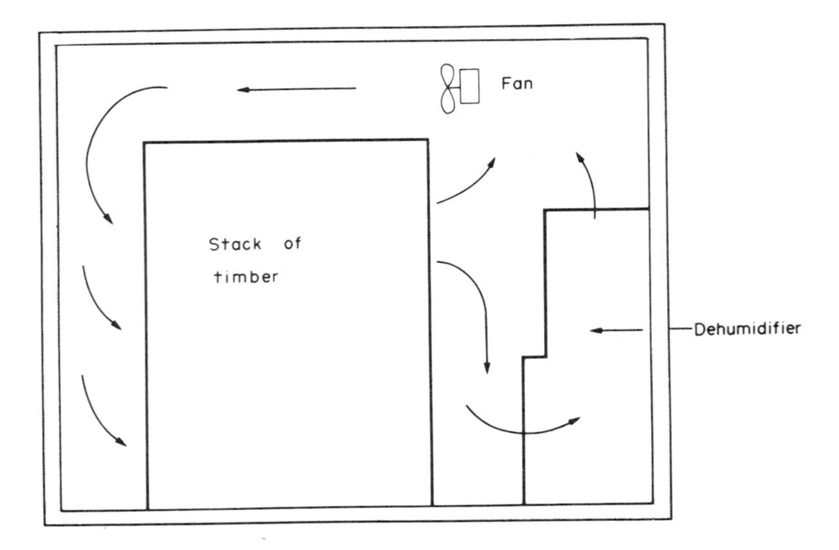

Fig. 6.25 A heat dehumidifier used in a drying kiln.

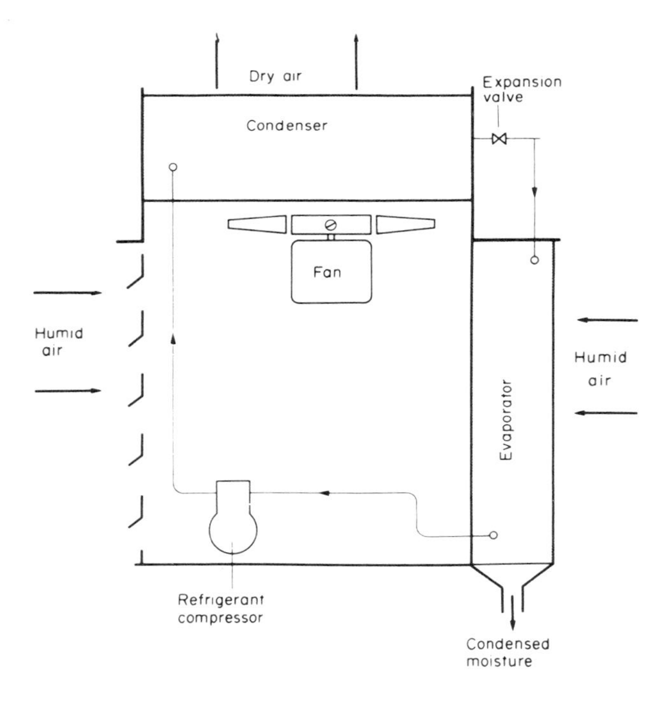

Fig. 6.26 The flow circuit in the dehumidifier (by courtesy of The Electricity Council).

as shown in Fig. 6.25. The dehumidifier used is a heat pump, and the flow circuit is illustrated diagrammatically in Fig. 6.26. The humid air from the timber stack is passed across the heat pump evaporator, which cools this air below the dew point, removing both sensible and latent heat. As a result of this cooling, condensation of a proportion of the water in the air occurs, and this is drained off, the condensate flow rate indicating the rate of drying. The heat removed from the air during dehumidification is then upgraded and given up in the condenser to the resulting dry air. This air is then passed through the timber stack, removing more moisture. In experimental kilns, coefficients of performance of about 3·6 have been obtained, implying fuel savings of approximately 50 per cent. Westair Dynamics (see Chapter 13), who pioneered commercial application of the system in timber drying, have also exploited other areas of industrial drying in the UK (Anon., 1978).

Portacel Ltd. have obtained more uniform moisture contents during the production of ceramic filters. More importantly, drying-cycle times have been reduced by over two-thirds and fuel costs halved. Drying times for slugs (lengths of wet clay rod) and candles (completed filters) have been cut from 72 to 24 h – and for long drying runs a time of only 15 h has been achieved. The slug-drying chamber has been designed for an output of 3000 units per day from each of which 0·5 kg of water has to be evaporated. Three TS5 dehumidifiers, with compressor ratings of 8·5 kW each, can meet this demand. Capital and installation costs of the Portacel installation were around £10 000, but the previous oil and gas fuel costs of £6500 per annum have been reduced to only £3000 – giving a payback period of under three years.

Even more substantial savings have been obtained by paper-yarn manufacturers Somic Ltd. at their Preston works. Using fan-assisted heaters in three drying kilns, the company were faced with an annual energy bill of around £3600 for drying the yarn at a temperature of 32°C to 38°C. Westair Dynamics recommended installation of one of their mobile TS2 units. Costing only around £1800, this cut drying time from 48 to 24 h and reduced energy costs to about £400 per annum, giving a payback period of only a few months. With plans to install a further TS2 unit, Somic anticipate that two dehumidifier kilns – with total energy cost of about £500 per annum – will be equivalent to four operating under the old system, when energy costs would have amounted to nearly £5000 per annum.

Industrial heat pumps driven by reciprocating engines

The use of reciprocating engines (diesel or natural gas-fuelled) for driving heat pump compressors has been dealt with in Section 6.5.2. While at the time of writing no fully developed industrial gas engine-driven heat pumps were available, prototypes are running and systems for use in particular processes are under detail design. These include units capable of producing low-pressure steam at the condenser. This is an area, together with other industrial heat pumps, to watch with interest.

References

Anon, (1976), Heat pumps. A new application for Howden screw compressors. *Howden Journal*, Glasgow, No. 14, 33–39.

Anon (1977), *UK Workshop on heat pumps. 30 June – 2 July 1976. A report prepared by the organising committee.* ETSU Report R1, HL77/683, Harwell.

Anon (1977), *Guidelines to the selection of external source air/air heat pumps in commercial premises.* Electricity Council Data Sheet EC3612/R/4.77.

Anon (1977), *Improvements in or relating to a method and apparatus for washing articles.* UK Patent 1494780 (Milpro N.V. and Cenzad Management A.G.).

Anon (1978), Pumps and pipes put the heat on drying. *Processing,* August, 46, 48.

Ellington, R. T. (1957), *The absorption cooling process.* IGT Research Bulletin.

Kolbusz, P. (1972), *The improvement of drying efficiency.* Electricity Council Research Centre, (UK), Report ECRC/R 476.

Kolbusz, P. (1975), *Industrial applications of heat pumps.* Electricity Council Research Centre (UK), Report ECRC/N845.

Macadam, J. A. (1974), *Heat pumps – the British experience.* Building Research Establishment Note N/117/74.

Macmichael, D. B. A., and Reay, D.A., (1979), *Heat Pumps, Theory, Design, and Applications.* Pergamon Press, Oxford.

Montagnon, P. E. and Ruckley, A. L. (1954), The Festival Hall heat pump. *J. Institute Fuel,* January, 1–17.

Perry, E. J. (1978), Heat pumps – the future. *Heating and Air Cond. J.,* September, 24–30.

Ross, N. N. (1975), The Templifier for process heat recovery, *Proc. EEI Conservation and Energy Management Division Conf.,* Atlanta, 16–18 March.

Williams, E. (1976), Keep the factory fires burning by extracting heat from cooling fluid. *The Engineer,* 1/8 Jan, 18.

Wurm, J. (1974), *An assessment of selected heat pump systems.* Institute of Gas Technology Annual Report, Project HC-4-20, (Period Feb. 1973-Feb 1974), IGT, Chicago.

7 Heat recovery from lighting systems

7.1 Introduction

The growing popularity of heat recovery in air conditioning systems, and the trend towards attempting to use the internal heat gains in a building as effectively as possible in contributing towards space heating (and in many cases, minimizing local heat gains to relieve the refrigeration system) has led to integrated environmental design concepts.

One area upon which this has had an influence is the design of light fittings (or luminaires). Modern fluorescent tubes have a light output which varies as a function of the tube wall temperature, peaking at a wall temperature of about 40°C. This is obviously also affected by the ambient temperatures, and the luminaire itself, as a heat source, may need to be cooled.

This chapter briefly discusses the air-cooled and water-cooled luminaire, which may be regarded as one element of a heat recovery system. Chapter 14 lists manufacturers' data on these systems, and their addresses are in the alphabetical list of manufacturers.

7.2 Air-cooled luminaires

Figure 7.1 shows a typical air-cooled luminaire which is used to minimize heat gains from the lights within the room, and to ensure that the lights themselves are operating at the temperature giving maximum light output. By careful design of the ducting, the air will be drawn through the luminaire, around and over the tubes and control gear, collecting up to 80 per cent of the lighting heat which would otherwise enter the occupied space. The air flow has a cooling effect on the fluorescent tube, thereby increasing its light output. Optimum conditions are achieved when the maximum light output is obtained with maximum removal of heat.

Using this method, increases in luminaire light output of between 12 and 15 per cent can be achieved, and the removal of the heat leads to a number of other benefits:

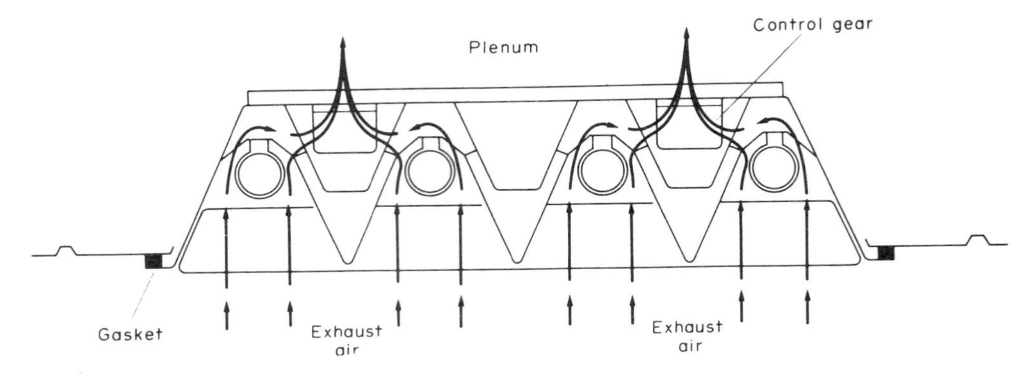

Fig. 7.1 An air-handling luminaire for fluorescent tubes.

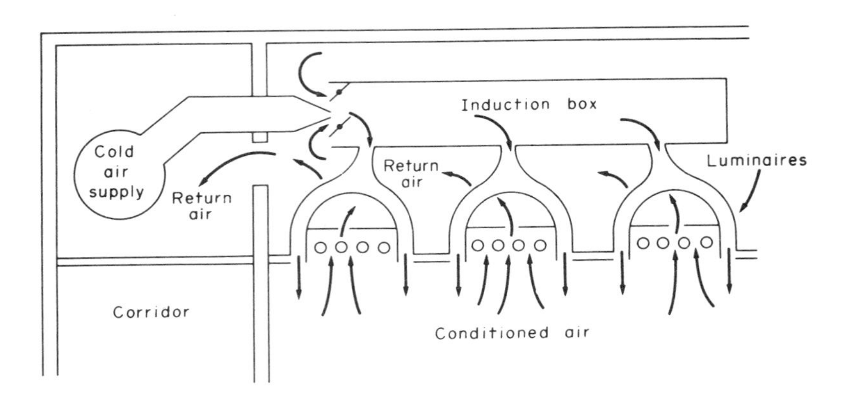

Fig. 7.2 A single-duct heat recovery system applied to luminaires.

(i) Reduction in the number of air changes.
(ii) Reduction in the fan power.
(iii) Reduction in temperature differentials.
(iv) Reduction in luminaire and ceiling temperature, hence lowering radiant heat input.
(v) Reduction in size of refrigeration plant.

Although some of the heat transferred in this manner into the ceiling space is lost, either back into the room below or through the floor into the offices above, at least 50 per cent of the heat will be retained within the ceiling plenum.

Figure 7.2 shows a portion of a single-duct heat recovery system used in the interior zone of a building which may require simultaneous winter heating and cooling. The induction box with thermostatically controlled dampers is the most important element in this system. When full cooling is called for, 100 per cent cold primary air is delivered to the space. When warmer air is needed, warm air from the ceiling cavity is gradually induced into the unit and is mixed with a proportion of cold air. Alternatively warm plenum air can be used to heat areas close to the building perimeter, with supplementary

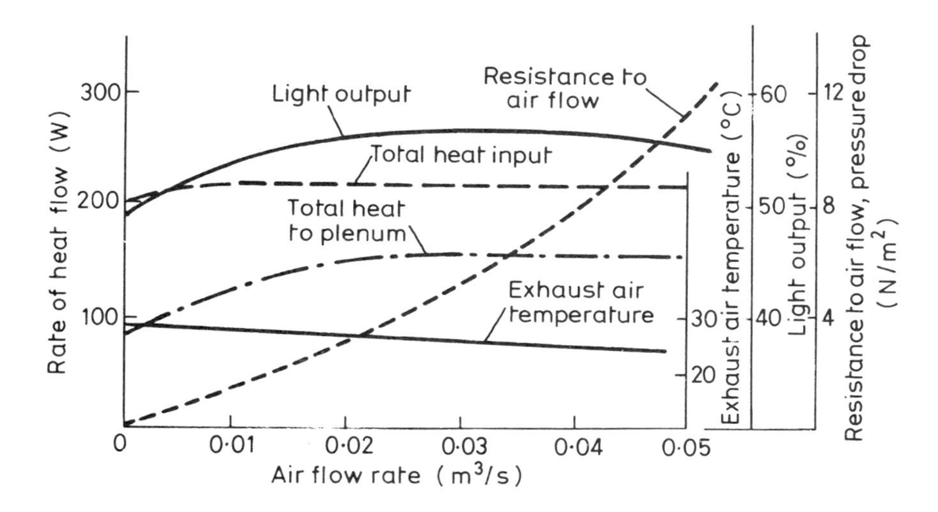

Fig. 7.3 Light and heat performance of a heat transfer luminaire (2 x 1·8 m, 85 W units).

heating supplied as required.

In summer the use of air-cooled luminaires is particularly beneficial, as the lighting heat can be removed and dissipated to the external atmosphere, thus leading to a reduction in the cooling load of the air conditioning system. An advantage resulting from their use, not always appreciated, is the fact that lighting efficiency is improved. By operating at a lower temperature, the lamp can give up to 13 per cent more light output for the same power, compared to an uncooled lamp. (Conversely an overcooled lamp will give a lower output. For a conventional fluorescent tube, the maximum light output is achieved when the tube wall is at 40° C.)

The performance, in terms of heat and light output, of a heat transfer luminaire manufactured by Osram-GEC is detailed in Fig. 7.3. These results were obtained using a purpose-built calorimeter at the GEC Hirst Research Laboratories. Room temperature was 22° C, and the pressure drop is that measured across the luminaire. The effect of overcooling can be seen in this figure, the light output decreasing as the air flow rate rises above 0·035 m^3/s.

7.3 Water-cooled luminaires

Water cooling of luminaires is possible also, but is not nearly so extensive as air cooling. One method for implementing water cooling is associated with a refrigeration-type heat recovery system, as shown in Fig. 7.4. The luminaires used have special aluminium reflector housings which are formed in such a way as to provide integral water passages. Using this method, up to 70 per cent of the heat generated may be transferred to the water.

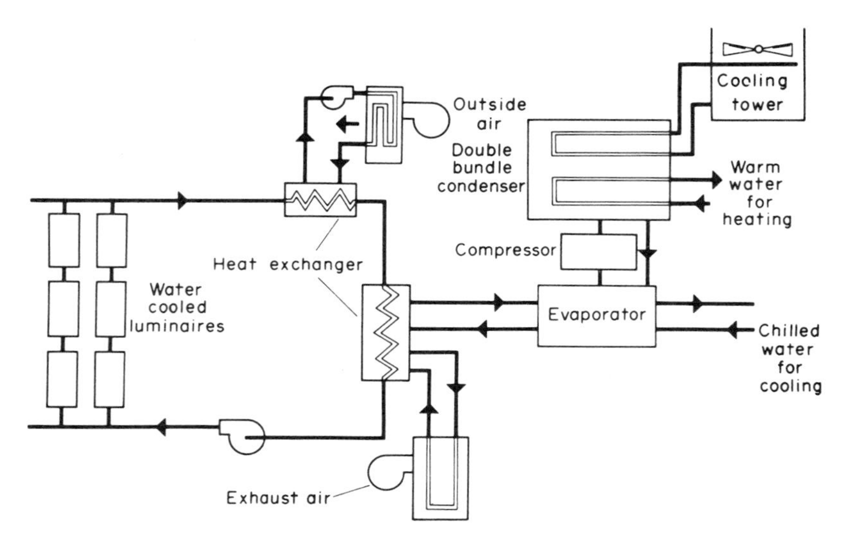

Fig. 7.4 Water-cooled luminaires linked to a comprehensive refrigeration-type heat recovery system.

In one design, the fluid from the luminaires is pumped through an evaporative heat exchanger, minimizing internal heat gain. If lighting accounts for 50 per cent of the heat gain in a building, then absorbing 70 per cent of this heat directly from the fixtures can lead to a 35 per cent reduction in the total air conditioning load, a not inconsiderable quantity. In the winter, this system would be turned off to allow the heat to be used for comfort-conditioning.

8 Gas-to-gas heat recovery equipment

This chapter, the first of seven chapters detailing the equipment available from manufacturers throughout the world, contains the data obtained from questionnaires sent out to firms specializing in the manufacture of gas-to-gas waste heat recovery equipment of the types described in Chapter 1.

The questionnaires were set out in such a way as to enable each manufacturer to present data on his equipment in a concise form, yet in sufficient detail to enable a potential user to identify whether that particular manufacturer has the product base to satisfy his requirements.

Many types of waste heat recovery equipment, both in this category and in those detailed in later chapters, are available as catalogued items. This may consist solely of the heat exchanger itself, or a 'package' containing equipment such as fans, controls and ducting. In some cases firms responding to our questionnaires have presented extremely detailed information on these, enabling the reader to select the particular item he requires. However, in many cases it is not possible to be completely specific concerning the performance of each heat exchanger – indeed, to attempt this would require a book extending to many volumes. In this case, manufacturers have indicated, as best as possible, their capabilities, with, in some cases, selected data on units which they have installed. This will enable the reader to judge for himself the likelihood of any particular company being able to satisfy his requirements.

8.1 Data presentation

The equipment covered by this chapter includes:

Heat pipe heat exchangers;
Rotating regenerators;
Runaround coils;
Plate heat exchangers;

Convection (tubular) recuperators;
Radiation recuperators;
Recuperative burners.

While air–air heat pumps could be categorized under an identical heading, it has been decided, because of their unique nature, to classify all heat pump systems, whatever the heat source or sink medium, under one heading, and these are detailed in Chapter 13.

Difficulties also arise with regard to heat recovery from incinerator plant and, to a lesser degree as far as gas–gas heat recovery equipment is concerned, prime mover heat exchangers. Many heat exchangers used in these applications are of a specialized nature – for example, they may incorporate silencers if used on a reciprocating engine. It was, therefore, felt advisable to present manufacturers' data on these systems in separate chapters. However, several of the gas–gas heat recovery equipment types listed here could be used, for instance, in combustion air preheat for incinerators, and this is indicated by some companies in their data sections and in the list of manufacturers in Appendix 1. However, the reader is also advised to consult Chapters 11 and 12 if his interests lie in these particular fields.

8.1.1 *Format of data*

The data on equipment used for gas–gas heat recovery is arranged as follows:

(1) *Equipment manufacturer*: The prime manufacturer of the equipment is given by name. In some special cases a subsidiary company or agent is given, and in such cases this is indicated. The address of the manufacturer, the country of origin, and where known, the telephone number preceded by the area code, are all given in alphabetical order in Appendix 1, together with the equipment categories in which the manufacturer specializes.

Also included are the manufacturer's agents, covering primarily Europe and North America, subsidiary companies, and manufacturing plant in countries other that the main base. These addresses are followed by a reference to the parent company, so that the reader can identify the source of a particular product in his country. (Only those companies responding to the data questionnaires are listed but most of the major manufacturers are included.)

(2) *Model name and number*: Many manufacturers give names to each range of heat recovery equipment, and these are given. Some companies specialize in 'custom-built' heat exchangers, and either omit a name, or market the product under their company name only. (In such cases the data given in the subsequent sections either represents a typical unit (or units) manufactured by that company, or gives a broad indication of the size range covered.) In order to assist the potential user to narrow down his selection of heat exchanger even further, many heat exchangers are given a model number, or some other form of identification.

(3) *Type*: The type of gas–gas heat exchanger is given. Note that some manufacturers offer three or four different types of heat recovery units. Detailed descriptions of each

type, and their relative advantages and disadvantages, are given in Chapter 1.
(4) *Application*: Data presented under this heading is directed at indicating the broad area of application (i.e. air conditioning, process-to-process or process-to-space heating) rather than detailing all the possible process-application areas for such equipment.
(5) *Operation*: Many gas–gas heat exchangers can only be used beneficially in the counter-flow mode (i.e. the exhaust gases are passing through in the opposite direction to the supply air). Because this is not universally applicable, heat exchangers capable of operating with parallel gas flows or cross-flow are indicated.
(6) *Temperature operating ranges*: Of particular interest to most potential users of heat recovery equipment, is the operating temperature range over which the equipment will function. As well as the maximum operating temperature, gas–gas heat exchangers can have a minimum operating temperature, determined particularly in HVAC applications by outside ambient temperature. Also, in process heat recovery applications, heat exchangers using an intermediate heat transfer medium (such as heat pipe heat exchangers and runaround coils) can have a somewhat higher minimum operating temperature, determined by factors such as the viscosity of the liquid/vapour heat transfer medium.

In order to cater for instances where minimum, as well as maximum, operating temperature will be of significance, provision is made in some cases for giving the temperature range of both the hot (exhaust) and cold (inlet) gases.
(7) *Gas volume flow rate and velocity*: Data is given on the volume or mass flow rate of gas which the heat exchangers are capable of accepting. This enables the potential user to determine whether the units are sufficiently large (or small) enough to meet his process flow requirements.

With regard to gas velocity, this has a bearing on both heat transfer capability and pressure drop. Also, where normal flue gas velocities differ much from the quoted acceptable velocities for the heat exchanger, severe expansion or contraction ducts may be required to lead into the heat recovery unit.
(8) *Pressure drop*: The pressure drop of a heat recovery unit is of considerable importance. While it may be acceptable in many cases to put in extra fans to cater for the increase in pressure drop brought about by the inclusion of a heat exchanger, many plant engineers are concerned about the effect any change in downstream conditions may have on the process operation, and a low pressure drop is an important selling feature of units of this type.
(9) *Regenerator features*: Data unique to rotating regenerators is included here. Rotational speed (which may be variable for control purposes – see Chapter 1), and the size of the drive motor for the wheel are included. Particularly on large heat wheels, an economic evaluation often takes into account the running costs of the drive motor.
(10) *Duty*: The heat transport capability of the heat exchanger is in many cases difficult to define, but data is given in as many cases as possible. As with some subsequent information, the duty is a function of a large number of variables, many of which are unique to the particular application of each heat exchanger. In some cases. therefore, the duty based on specific temperature differences between exhaust and

inlet gas streams is given. One can also, of course, make a good assessment of the duty by using the temperature, volume flow and efficiency figures.

(11) *Efficiency*: Heat exchanger duty and efficiency are inter-related, the efficiency also being a function of numerous variables. However, the data given will enable the reader to check the relative efficiency of different types of gas-gas heat recovery equipment, as well as comparing similar product from different manufacturers. However, as with figures for the duty, the efficiency may vary from one specific application to another, and the manufacturer is the only one who can, in the final analysis, provide a precise figure, once all the process conditions are accurately known.

(12) *Typical payback periods*: No energy manager can make a case for investment in waste heat recovery equipment unless the return on the investment is acceptable to the company management. A function of the cost of the energy saved and the capital and operating cost of the waste heat recovery unit itself, the payback period (discussed in Appendix 2 in more detail) is the most common way for expressing the economic viability of the heat exchanger assembly. Remember that HVAC equipment has a longer payback period than process equipment (due to seasonal use, unless precooling of supply air in summer is required), and because of the long life of HVAC systems, a longer payback period is generally acceptable. The figures quoted are in most cases for installed cost, which may be several times the cost of the basic heat exchanger unit.

(13) *Heat exchanger construction*: Information on materials used in the heat exchanger, the range of weights of the units and the main dimensions (excluding associated connecting ductwork) are given.

(14) *Features*: This section, which is quite detailed, by necessity, covers many of the features of gas–gas heat recovery equipment which a potential user must assess for each application:

(i) Available custom-built: some manufacturers will build heat exchangers specifically to suit a customer's requirements. In general they tend to be more expensive than 'off the shelf' units, but some manufacturers can readily adapt basic heat exchanger coils to meet unique process conditions. A complete engineered factory 'package' containing all necessary control equipment, etc. may also be available.

(ii) Performance guarantees: the type of guarantee varies considerably. Most suppliers quote performance figures, based on data given by the user (which should be as accurate as possible, preferably obtained by carrying out an energy 'audit' on the process or in the building). Guarantees on longevity of the equipment may be given, and standard one-year guarantees are common.

(iii) Installation method: many gas–gas units can be mated directly to the ductwork supplying the two gas flows. Some HVAC systems require rooftop mounting, and units such as some rotating regenerators and most heat pipe heat exchangers, because they are sensitive to orientation, have to be mounted in a specific way. Data given under this heading indicates such factors.

(iv) Installer: installation of the unit may be carried out by the manufacturer, his subcontractor, the user, or the user's subcontractor. Some manufacturers indicate their preferences in this section.

(v) Cleaning: while techniques for cleaning gas–gas heat exchangers are well known, some manufacturers recommend particular methods, depending upon the type of finning or the core form of regenerators. Also, several heat exchangers, particularly when sold as a 'package' have built-in cleaning systems which may operate automatically. Some manufacturers include or recommend filters.

(vi) Corrosion resistance: most engineers will be able to relate the corrosion resistance of the heat exchanger to the known gas temperatures and likely pollution in the exhaust gas stream with a knowledge of the materials of construction of the heat exchanger. Some manufacturers, however, offer coatings of external surfaces which give an added protection against corrosion, and most, of course, rely on cooling the exhaust gases to a temperature insufficiently low to reach the acid dew point.

(vii) Performance control and turndown range: performance control is of prime importance in HVAC applications, as discussed in Chapter 1. Control may be required in some process applications. This section briefs the potential user on the control method used. Turndown, often to 10 per cent of maximum design flow, can be achieved in several ways, the most common being by dampers or dilution of the air stream.

(viii) Cross-contamination: the rotating regenerator is the only unit inherently liable to suffer from cross-contamination (carryover of gas and/or liquid or solid matter from one stream to the other). Most heat wheel manufacturers are able to quantify cross-contamination, and sealing methods are becoming increasingly effective.

(ix) Ability to withstand a differential pressure: effective recuperative heat transfer depends to some extent on the thickness of the solid interface between the two gas streams. While a strong heat exchanger will withstand large pressure differences between the two gases, the temperature gradient through the necessarily thick wall may be unacceptable. For most applications a pressure differential in excess of a few millimetres of water is unlikely to occur, but many manufacturers are able to quantify the maximum pressure which the walls will withstand. Specials may be made to cater for higher pressures. In the case of rotating regenerators, where cross-contamination may be influenced by the differential, as may operation of the purge section, figures are quoted routinely.

(x) Condensate collection: where condensation is anticipated within the heat recovery unit (e.g. HVAC and dryer applications), provision may be made by the manufacturer for collection and removal of the condensate, e.g. by means of a sump or drain pan. The ability to collect condensate can be affected by the geometry of the extended surface used, the orientation of the heat exchanger and the gas velocity. Unless some form of collection is used, carryover into downstream duct sections may prove an embarrassment.

(xi) Availability of fans: manufacturers of the heat recovery equipment may themselves supply fans to cater for the anticipated extra flow resistances. Some include these in the 'package', others may recommend suppliers (and specify the duty required).

(xii) Frost prevention: significant mainly to HVAC heat recovery systems, frost prevention (or deicing) mechanisms are offered by suppliers of heat exchangers. This may be done via a control circuit, supplementary heating or diverting of exhaust gas streams.

(15) *Extras*: Many manufacturers offer a range of 'optional extras', and some are listed in this section. Cleaning equipment, 'packaging' of the system, heavy-duty motors on regenerators and additional corrosion protection are common options available. A section is also devoted to additional comments which manufacturers wished to include to assist potential users to assess their equipment and capabilities.

Manufacturers' data sheets relating to gas-to-gas heat recovery equipment follow on the next page.

MANUFACTURER	*AAF LTD.*
MODEL NAME AND NUMBER	Runaround coil systems
TYPE	Runaround coil
APPLICATIONS	Process-to-space and airconditioning
OPERATION	Exhaust-air coil(s) can be remote from supply-air coil(s). Up to 18 separate coils may be linked together to form a single system
TEMPERATURE (°C)	
Operating range	15–150
AIR	
Volume (m^3/h)	Max. 200 000 per coil
Velocity (m/s)	4
PRESSURE DROP (mm w.g.)	5-30, dependent on requirements
REGENERATOR	
Rotation speed (rev/min)	
Rotation motor size (W)	
DUTY (kW)	To suit
EFFICIENCY (%)	45–65
TYPICAL PAYBACK PERIOD (years)	
Airconditioning	3
Process plant	3
CONSTRUCTION	
Materials	Normally copper tubes and aluminium or copper fins, steel casings
Weight (kg)	To suit
Dimensions (mm)	To suit
FEATURES	
Available custom-built,	Yes
Performance guarantees	Yes
Installation method	Simply bolted into new or existing ductwork
Installed by?	Contractor
Cleaning	Hose, if required
Corrosion resistance	Protective finishes can be applied
Performance control and turndown range	Efficiency virtually constant regardless of load
Cross-contamination	Nil
Withstands differential pressure between ducts	
Condensate collection	Yes, if dehumidification occurs
Fans: included/available separately	
Frost prevention	Yes
EXTRAS	
FURTHER COMMENTS	AAF Ltd. have special computer program available to optimize heat transfer and provide an accurate prediction of the amount of heat recovered.

MANUFACTURER	*ACOUSTICS AND ENVIROMETRICS*			
MODEL NAME AND NUMBER	Econovent (Munters)	Econovent (Munters)	Econovent (Munters	AEL High Temperature
TYPE	Desiccant matrix heat recovery wheel	Metal matrix heat recovery wheel	Static recuperator, plate type	Static recuperator, tube type
APPLICATIONS	Process-to-process; process-to-air airconditioning	Process-to-process; process-to-air airconditioning	Process-to-process; process-to-air airconditioning	Process-to-process process-to-air
OPERATION (°C)	Counter-flow	Counter-flow	Counter-flow	Cross-flow
TEMPERATURE (°C)				
Operating range	Max.200	Max. 200	Max. 100	Max. 450
AIR				
Volume (m^3/h)	Max. 86400	Max. 64800	Max. 4500	Custom-built
Velocity (m/s)	4.5	4.5		Custom-built
PRESSURE DROP (mm w.g.)	Max. 25.4	Max. 25.4	Max. 50	Custom-built
REGENERATOR				
Rotation speed (rev/min)	Max. 101	Max. 101	N/a	N/a
Rotation motor size (W)	Max. 750	Max. 750		
DUTY (kW)	N/a	N/a	N/a	N/a
EFFICIENCY (%)	Approx. 75	Approx. 75	Max. 90	Max. 60
TYPICAL PAYBACK PERIOD (years)				
Airconditioning	Max. 24 months	1-2 years	1-2 years	N/a
Process plant	Max. 1-2 years	1-2 years	1-2 years	1-2 years
CONSTRUCTION				
Materials	Steel/ Munters patent	Steel/ aluminium	Steel/ aluminium	Mild steel
Weight (kg)	Up to 2000	Up to 1550	175	N/a
Dimensions (mm)	4200 (dia.)	3500 (dia.)	1264 x 1180 x 1137	N/a
FEATURES				
Available custom-built?	No	No	No	Yes
Performance guarantees	If required	No	No	Yes
Installation method	Contractor	Contractor	Contractor	Contractor
Installed by?	Contractor	Contractor	Contractor	Contractor
Cleaning	As required	As required	As required	As required
Corrosion resistance	N/a	N/a	N/a	Independently controlled
Performance control and turndown range	Infinite	Infinite	N/a	N/a
Cross-contamination	Should not be a problem	Should not be a problem	Should not be a problem	Should not be a problem
Condensate collection	N/a	N/a	Yes	Yes
Fans: included/available separately	No	No	No	No
Frost prevention	No	No	Available	No
EXTRAS	Speed controls	Speed controls	N/a	N/a
FURTHER COMMENTS	Great care is exercised at the design stage, to ascertain that the equipment is correctly matched to the specific problem.			

MANUFACTURER	*AIR ENERGY RECOVERY CO. INC.*
MODEL NAME AND NUMBER	Type D; Type SB; Futura
TYPE	Runaround: each system complete with automatic start-stop controls, automatic frost-prevention controls and automatic over-recovery prevention controls; Futura system in addition automatically controls supplemental heating and cooling to maintain year-round temperature of supply air; Type D and SB systems have separate prepiped and prewired pump/control cabinets; Futura systems have prepiped and prewired air-handler units; each system requires field-installation (by others) of interconnecting wiring and piping.
APPLICATIONS	Futura: airconditioning; Type D: airconditioning and process-to-process; Type SB: process-to-process
OPERATION	Liquid flow in both supply-air and exhaust-air units must be counter-flow of liquid-to-air; supply- and exhaust-air units separated so that their relationship is not concerned with counter- or parallel-flow
TEMPERATURE ($^{\circ}$C)	-28-425
AIR	
Volume (m^3/h)	Up to 170 000
Velocity (m/s)	2-4.5
PRESSURE DROP (mm w.g.)	10-200
REGENERATOR	
Rotation speed (rev/min)	
Rotation motor size (W)	
DUTY (kW)	Depends on specification
EFFICIENCY (%)	Up to 65
TYPICAL PAYBACK PERIOD (years)	
Airconditioning	3-5
Process plant	6 months-2 years
CONSTRUCTION	
Materials	Coil tubes: copper; cupro-nickel; steel; stainless steel Coil fins: copper; copper solder coated; aluminium; steel
Weight (kg)	Each coil up to 1500; pump assemblies, 400-1200; Futura; 4500-12 000
Dimensions (mm)	Limited only by space and duty requirements
FEATURES	
Available custom-built,	All systems custom-built
Performance guarantees	Yes
Installation method	Coils installed by standard methods; pump control cabinet can be placed any where; Futura installed by standard methods for air-handling units
Installed by?	Contractor
Cleaning	Coils cleaned by standard methods for coils
Corrosion resistance	Coils may be coated with phenolic for most acids and pH values up to 10.5
Performance control and turndown range	Automatically from maximum efficiency to zero
Cross-contamination	None
Withstands differential pressure between ducts	Not applicable
Condensate collection	Drain pans (by others); use standard method for coils
Fans: included/available separately	May be included
Frost prevention	Automatically
EXTRAS	
FURTHER COMMENTS	

MANUFACTURER	*AIR FROEHLICH AG FUR ENERGIERUCKGEWINNUNG*
MODEL NAME AND NUMBER	Glass Tube (GRS); Glass Plate (GPS); Alu Plate (APS) heat exchangers
TYPE	Recuperators
APPLICATIONS	GRS, Process-to-process or process-to-space GPS, APS, airconditioning
OPERATION	Cross-flow
TEMPERATURE (°C)	
Operating range	GRS, Max. 500; GPS, APS, Max. 70
AIR	
Volume (m^3/h)	1000-50 000
Velocity (m/s)	
PRESSURE DROP (mm w.g.)	15-25
REGENERATOR	
Rotation speed (rev/min)	
Rotation motor size (W)	
DUTY (kW)	Depends on specification
EFFICIENCY (%)	60-80
TYPICAL PAYBACK PERIOD (years)	
Airconditioning	2-4
Process plant	1-3
CONSTRUCTION	
Materials	APS, aluminium GPS, GRS, glass
Weight	
Dimensions	
FEATURES	
Available custom-built?	
Performance guarantees	
Installation method	
Installed by?	Air handling
Cleaning	Water (cold or hot)
Corrosion resistance	GRS, GPS, no problem
Performance control and turndown range	
Cross-contamination	0
Withstands differential pressure between ducts	GRS, Max. 1000 APS, GPS, Max. 100
Condensate collection	Yes
Fans: included/available separately	No
Frost prevention	No
EXTRAS	No
FURTHER COMMENTS	

MANUFACTURER	*ASET*
MODEL NAME AND NUMBER	F-type
TYPE	Recuperators; shell and tube exchangers
APPLICATIONS	Process-to-process
OPERATION	Counter-flow
TEMPERATURE (°C)	
Operating range	Up to approx. 300
AIR	
Volume (m^3/h)	Various capacities
Velocity (m/s)	Various capacities
PRESSURE DROP (mm w.g.)	Various capacities
REGENERATOR	
Rotation speed (rev/min)	
Rotation motor size (W)	
DUTY (kW)	
EFFICIENCY(%)	
TYPICAL PAYBACK PERIOD (years)	
Airconditioning	
Process plant	Various capacities
CONSTRUCTION	
Materials	Carbon steel, stainless steel
Weight (kg)	
Dimensions (mm)	
FEATURES	
Available custom-built?	Yes
Performance guarantees	Yes
Installation method	
Installed by?	
Cleaning	
Corrosion resistance	
Performance control and turndown range	
Cross-contamination	
Withstands differential pressure between ducts	
Condensate collection	Yes
Fans: included/available separately	
Frost prevention	
EXTRAS	
FURTHER COMMENTS	

MANUFACTURER	*ATLAS INDUSTRIAL MANUFACTURING CO.*
MODEL NAME AND NUMBER	Engineered units only
TYPE	Shell and tube heat exchanger
APPLICATIONS	Chemical; petrochemical; nuclear power plants
OPERATION	Counter- and parallel-flow
TEMPERATURE (°C)	
Operating range	No limit except as dictated by materials
AIR	
Volume (m^3/h)	As required
Velocity (m/s)	As required
PRESSURE DROP (mm w.g.)	As required
REGENERATOR	
Rotation speed (rev/min)	
Rotation motor size (W)	
DUTY (kW)	
EFFICIENCY (%)	
TYPICAL PAYBACK PERIOD (years)	
Airconditioning	
Process plant	
CONSTRUCTION	
Materials	All metals
Weight (kg)	45 000
Dimensions (mm)	6 x 6 x 100
FEATURES	
Available custom-built?	Always
Performance guarantees	Yes
Installation method	Contractor
Installed by?	Contractor
Cleaning	Maintenance Crews
Corrosion resistance	Part of design features
Performance control and turndown range	
Cross-contamination	
Withstands differential pressure between ducts	
Condensate collection	
Fans: included/available separately	
Frost prevention	
EXTRAS	
FURTHER COMMENTS	

MANUFACTURER	*SA BABCOCK BELGIUM NV*
MODEL NAME AND NUMBER	Ljungstrom
TYPE	Rotating regenerative system
APPLICATIONS	Preheating of combustion air in connection with industrial power and process plants
OPERATION	Counter-flow or parallel-flow
TEMPERATURE (°C)	
Operating range	300-450
AIR	
Volume (m^3/h)	10 000-1 000 000
Velocity (m/s)	10
PRESSURE DROP (mm w.g.)	On air-side as well as on gas-side : 50-150
REGENERATOR	
Rotation speed (rev/min)	1-3
Rotation motor size (W)	2000-20 000
DUTY (kW)	Up to 100 000
EFFICIENCY (%)	98
TYPICAL PAYBACK PERIOD (years)	
Airconditioning	
Process plant	1
CONSTRUCTION	
Materials	Carbon steel, corten steel
Weight (kg)	Up to 500 000
Dimensions (mm)	Up to 12 000 (dia.)
FEATURES	
Available custom-built?	Standard or custom-built
Performance guarantees	Gas outlet temperature
Installation method	Package or site-assembled
Installed by?	Manufacturer
Cleaning	Sootblowers and water lances
Corrosion resistance	Cold parts can be enamelled
Performance control and turndown range	Follows the furnace or boiler
Cross-contamination	0.5
Withstands differential pressure between ducts	Up to 1000 mm w.g.
Condensate collection	
Fans: included/available separately	Available separately
Frost prevention	Not necessary
EXTRAS	
FURTHER COMMENTS	Above ranges are typical and not considered as limits

MANUFACTURER	*FIVES-CAIL BABCOCK*
MODEL NAME AND NUMBER	Custom-built
TYPE	Tubular heat exchanger and regenerator
APPLICATIONS	Process-to-process
OPERATION	Counter- or parallel-flow
TEMPERATURE (°C)	
Operating range	Up to 500
AIR	
Volume (m^3/h)	No limit
Velocity (m/s)	
PRESSURE DROP (mm w.g.)	
REGENERATOR	
Rotation speed (rev/min)	2.5
Rotation motor size (W)	3000
DUTY (kW)	No limit
EFFICIENCY (%)	
TYPICAL PAYBACK PERIOD (years)	
Airconditioning	
Process plant	2
CONSTRUCTION	
Materials	Steel
Weight (kg)	
Dimensions (mm)	
FEATURES	
Available custom-built?	Yes
Performance guarantees	Temperatures
Installation method	Package or erection at site
Installed by?	Manufacturer
Cleaning	Yes
Corrosion resistance	Yes
Performance control and turndown range	
Cross-contamination	
Withstands differential pressure between ducts	
Condensate collection	Yes
Fans: included/available separately	Included, if required
Frost prevention	Not in operation
EXTRAS	
FURTHER COMMENTS	See Fives-Cail Babcock reference book on custom-built exchangers with data on typical units

MANUFACTURER	*BAHCO VENTILATION*
MODEL NAME AND NUMBER	ABC;DCD 1-9 (dependent on air volume)
TYPE	Recuperator/regenerator
APPLICATIONS	Airconditioning
OPERATION	Normally counter-flow on cooling
TEMPERATURE (^{o}C)	
Operating range	4-50
AIR	
Volume (m^3/h)	0.5-23 m^3/s (regenerator); 0.5-23 m^3/s (recuperator)
Velocity (m/s)	4 Max.
PRESSURE DROP (mm w.g.)	33 Max.
REGENERATOR	
Rotation speed (rev/min)	7-10
Rotation motor size (W)	750
DUTY (kW)	Variable
EFFICIENCY (%)	85 Max.
TYPICAL PAYBACK PERIOD (years)	
Airconditioning	Variable
Process plant	
CONSTRUCTION	
Materials	Aluminium
Weight (kg)	1450 Max.
Dimensions (mm)	3.8 m x 3.8 m x 510 mm
FEATURES	
Available custom-built?	Yes
Performance guarantees	Yes
Installation method	Integral with AHU
Installed by?	Customer or contractor
Cleaning	Self-cleaning
Corrosion resistance	Dependent on contaminants
Performance control and turndown range	By thyristor
Cross-contamination	
Withstands differential pressure between ducts	Coils are in excess of 140 p.s.i.; wheel 5 in plus
Condensate collection	Normal for coils; not recommended for wheel
Fans: included/available separately	Included or available separately
Frost prevention	Available
EXTRAS	Filters; manometers
FURTHER COMMENTS	DCD/DCC heat recovery units are designed to form an integral part of an airconditioning system; there is also a new unit reference AEK incorporating plate exchanger or thermal wheel designed for low-volume air flow 500-1800 m^3/h.

MANUFACTURER	*BARRIQUAND*
MODEL NAME AND NUMBER	Welded plate
TYPE	Recuperator
APPLICATIONS	Process-to-process; process-to-space
OPERATION	Counter- and parallel-flow
TEMPERATURE (°C)	60-600
AIR	
Volume (m^3/h)	Without limit
Velocity (m/s)	
PRESSURE DROP (mm w.g.)	Adjustable
REGENERATOR	
Rotation speed (rev/min)	
Rotation motor size (W)	
DUTY (kW)	
EFFICIENCY (%)	Up to 95
TYPICAL PAYBACK PERIOD (years)	
Airconditioning	
Process plant	
CONSTRUCTION	
Materials	Stainless steel
Weight (kg)	
Dimensions (mm)	
FEATURES	
Available custom-built?	
Performance guarantees	
Installation method	
Installed by?	
Cleaning	
Corrosion resistance	
Performance control and turndown range	
Cross-contamination	
Withstands differential pressure between ducts	
Condensate collection	
Fans: included/available separately	
Frost prevention	
EXTRAS	
FURTHER COMMENTS	

MANUFACTURER	*BELTRAN & COOPER LTD.*
MODEL NAME AND NUMBER	BCHE 011; 021; 041; 042; 082; and other multiples of 011 module
TYPE	Modular plate fin block recuperator
APPLICATIONS	All applications to 900°C
OPERATION	Cross-flow
TEMPERATURE (°C)	
Operating range	Aluminium to 200; chrome steel to 900
AIR	011 021 041 042 082
Volume (m^3/h)	2500; 5000; 10 000; 5 000 (2-pass); 10 000 (2-pass)
Velocity (m/s)	Depending on application
PRESSURE DROP (mm w.g.)	Approx. 25-50 on each side
REGENERATOR	
Rotation speed (rev/min)	
Rotation motor size (W)	
DUTY (kW)	100 120 120 200 400 depending on ETD
EFFICIENCY (%)	65 65 65 65 65 on equal mass flow
TYPICAL PAYBACK PERIOD (years)	
Airconditioning	3 2 2 2 2 Assuming T 100°C
Process plant	1½ 1 1 1 1 Aluminium T 200°C
	3 2 2 2 2 Chrome steel T 300°C
CONSTRUCTION	
Materials	Aluminium; Chrome steel (20% Ni; 20 + % Cr)
Weight (kg)	15 100 per 011 block (082 = 8 times this)
Dimensions (mm)	300 x 300 x 600 per 011 block (082 = 600 x 600 x 1200, for example)
FEATURES	
Available custom-built?	Modular design
Performance guarantees	Yes - on heat recovered
Installation method	In - duct or top of oven or outside: very easy
Installed by?	Manufacturer or customer
Cleaning	In-situ cleaning option available
Corrosion resistance	Aluminium: poor; chrome steel: excellent
Performance control and turndown range	Performance excellent at low and high flows
Cross-contamination	Resists blockage, preferably mounted for exhaust flow down
Withstands differential pressure between ducts	Yes (aluminium to 5 p.s.i.; chrome steel to 100 p.s.i.)
Condensate collection	Yes (exhaust flow downward preferred)
Fans: included/available separately	As required
Frost prevention	Not required
EXTRAS	In-situ cleaning Full turnkey installation Pollution-control equipment when required (electrostatic) (mist eliminators)
FURTHER COMMENTS	

MANUFACTURER	*BERNER INTERNATIONAL CORP.*	
MODEL NAME AND NUMBER	Senex SA15 - SA490	Enthalpy Wheel VX15 - VX490
TYPE	Regenerator: sensible heat only	Regenerator: total heat
APPLICATIONS	Process-to-process; process-to-space space-to-space	Airconditioning; space-to-space
OPERATION	Counter-flow →	
TEMPERATURE (°C)		
Operating range	-50-200	-50-70
AIR		
Volume (m^3/h)	1000-83 000 per unit →	
Velocity (m/s)	2-5 →	
PRESSURE DROP (mm w.g.)	6-70	6-33
REGENERATOR		
Rotation speed (rev/min)	10-20 →	
Rotation motor size (W)	100-400 →	
DUTY (kW)		
EFFICIENCY (%)	70-85 →	
TOTAL PAYBACK PERIOD (years)		
Airconditioning	Depends on climate; fuel costs; operating time; and other local factors	
Process plant	Depends on climate; fuel costs; operating time; and other local factors	
CONSTRUCTION		
Materials	Steel housing; aluminium wheel	Steel housing; aluminium wheel coated with a desiccant
Weight (kg)	100-1500 →	
Dimensions (mm)	965-4000 square x 320-480 deep →	
FEATURES		
Available custom-built?		
Performance guarantees	Yes	Yes
Installation method		
Installed by?	Contractor →	
Cleaning	Hot water; steam; compressed air	
Corrosion resistance	Yes, various grades available	
Performance control and turndown range	Yes, zero to nominal effectiveness using SCR control	
Cross-contamination	Without purge sector approx. 2-3%; with purge sector less than 0.1% easily achievable	
Withstands differential pressure between ducts	Standard up to 150 mm Special up to 500 mm	Standard up to 150 mm Special up to 500 mm
Condensate collection	Available, depends on application	Not required
Fans: included/available separately	No	No
Frost prevention	Air preheat, air bypass or reduce effectiveness with SCR control	
EXTRAS		
FURTHER COMMENTS	Nine sizes of heat recovery wheels available for model SA and model VX.	

Size	Air volume (m^3/h)
15	1 000 - 2 600
30	2 300 - 5 800
50	3 400 - 8 500
90	6 000 -15 000
140	9 300 -23 000
180	12 000 -30 000
220	15 000 -38 000
330	23 000 -57 000
490	34 000 -83 000

MANUFACTURER	*BEVERLEY CHEMICAL ENGINEERING LTD.*
MODEL NAME AND NUMBER	Custom-built
TYPE	Recuperator
APPLICATIONS	Process-to-process; process-to-air
OPERATION	Counter- and/or parallel-flow
TEMPERATURE ($^{\circ}$C)	
Operating range	Up to 700
AIR	
Volume (m^3/h)	1000-80000
Velocity (m/s)	
PRESSURE DROP (mm w.g.)	0-500
REGENERATOR	
Rotation speed (rev/min)	
Rotation motor size (W)	
DUTY (kW)	Up to 100000
EFFICIENCY (%)	Up to 60
TYPICAL PAYBACK PERIOD (years)	
Airconditioning	
Process plant	1-2
CONSTRUCTION	
Materials	Mild or stainless steel
Weight (kg)	Depends on duty
Dimensions (mm)	Depends on duty
FEATURES	
Available custom-built?	Yes
Performance guarantees	Available
Installation method	Supplied as package
Installed by?	Manufacturer
Cleaning	Facility built in
Corrosion resistance	
Performance control and turndown range	Incorporate to requirements
Cross-contamination	
Withstands differential pressure between ducts	
Condensate collection	
Fans: included/available separately	As required
Frost prevention	
EXTRAS	
FURTHER COMMENTS	

MANUFACTURER	*BRONSWERK HEAT TRANSFER BV*
MODEL NAME AND NUMBER	Heat exchangers: tubular, finned and plain tubes
TYPE	Recuperators
APPLICATIONS	Process-to-process; process to heating of rooms (airconditioning)
OPERATION	Counter-, cross- and parallel-flow
TEMPERATURE (°C)	
Operating range	-20-2000
AIR	
Volume (m^3/h)	1 000 000
Velocity (m/s)	3.5
PRESSURE DROP (mm w.g.)	50-3000
REGENERATOR	
Rotation speed (rev/min)	
Rotation motor size (W)	
DUTY (kW)	10 000 000
EFFICIENCY (%)	95-99.9
TYPICAL PAYBACK PERIOD (years)	Dependent on design; large variety of large heat recovery heat exchangers
Airconditioning	
Process plant	
CONSTRUCTION	
Materials	Ferrous; non-ferrous
Weight (kg)	Max. 20 tonne = 2000 kg
Dimensions (mm)	15 m x 3 m
FEATURES	
Available custom-built?	Yes
Performance guarantees	Yes
Installation method	Custom-built
Installed by?	Manufacturer or contractor
Cleaning	By hand or automatically
Corrosion resistance	Dependent on materials
Performance control and turndown range	Customer's requirements
Cross contamination	N/a
Withstands differential pressure between ducts	Normally up to 1 bar
Condensate collection	If under dew point
Fans: included/available separately	Included
Frost prevention	Included
EXTRAS	Preheaters for steam; hot water; thermo oil
FURTHER COMMENTS	

MANUFACTURER	*BRUNNSCHWEILER SPA*					
MODEL NAME AND NUMBER	1000;	1500;	2000;	3000;	4500;	6000;
TYPE	Glass tube heat recuperator					
APPLICATIONS	Process-to-process					
OPERATION	Cross-flow					
TEMPERATURE (°C)						
Operating range	Up to 120					
AIR						
Volume (m^3/h)	10 000-14 000	14 000-20 000	20 000-28 000	28 000-44 000	44 000-65 000	65 000-90 000
Velocity (m/s)	5-8					
PRESSURE DROP (mm w.g.)	10-40					
REGENERATOR						
Rotation speed (rev/min)						
Rotation motor size (W)						
DUTY (kW)	Dependent on specification					
EFFICIENCY (%)	60-75					
TYPICAL PAYBACK PERIOD (years)						
Airconditioning						
Process plant	1-2					
CONSTRUCTION						
Materials	Glass tubes in rubber plates on stainless-steel grid in galvanised steel case					
Weight (kg)	650-1000	800-1350	1000-1700	1400-2300	1900-3200	2400-4100
Dimensions (mm)	1000 1000x1500x2000 2000	1000 1500x1500x2000 2000	1000 2000x1500x2000 2000	1000 3000x1500x2000 2000	1000 4500x1500x2000 2000	1000 6000x1500x2000 2000
FEATURES						
Available custom-built?	No					
Performance guarantees	Supplied					
Installation method						
Installed by?	Manufacturer or contractor					
Cleaning	Easy, automatic cleaning device available					
Corrosion resistance	Excellent					
Performance control and turndown range						
Cross-contamination	None					
Withstands differential pressure between ducts	Up to 300					
Condensate collection	Included					
Fans: included/available separately	Available separately					
Frost prevention	None					
EXTRAS	Automatic water-cleaning device					
FURTHER COMMENTS						

MANUFACTURER	*BRY-AIR INC.*
MODEL NAME AND NUMBER	Bry-Therm: 7 standard sizes; HR-2500; HR-5000; HR-10000; HR-15000; HR-20000; HR-25000; HR-30000
TYPE	Heat pipe (Q-Dot coil, or equal)
APPLICATIONS	Process-to-space; exhaust-to-process; exhaust-to-supply (comfort conditioning)
OPERATION	Counter-flow
TEMPERATURE (°C)	
Operating range	Model 120: -20-20; Model 125: -40-20; Model 225 - 1325: 20-600 Max.
AIR	
Volume (m^3/h)	2100-54 600 (7 standard sizes)
Velocity (m/s)	2.5
PRESSURE DROP (mm w.g.)	25 approx., or as designed
REGENERATOR	
Rotation speed (rev/min)	
Rotation motor size (W)	
DUTY (kW)	HR-2500: 4; fans only plus controls: 45 (Max.)
EFFICIENCY (%)	62-72, as designed
TYPICAL PAYBACK PERIOD (years)	
Airconditioning	As designed
Process plant	As designed
CONSTRUCTION	
Materials	Steel frame base with lifting eyescasing and roof of zinc coated 18-gauge steel modular panel construction
Weight (kg)	HR-2500; 1536.8; HR-30000; 8893 (basic package)
Dimensions (mm)	350.5-cm long x 233.8-cm wide x 172.7-cm high; 1036.3-cm long x 355.6-cm wide x 335-cm high
FEATURES	
Available custom-built?	Yes
Performance guarantees	Yes
Installation method	Factory assembled and tested package
Installed by?	Customer
Cleaning	Entire package accessible through access doors, filters all replaceable media
Corrosion resistance	Casing material is corrosion resistant, coil and fans may be
Performance control and turndown range	Q-Dot coil may be tiled for capacity control or frost protection
Cross-contamination	Complete separation of the two air streams
Withstands differential pressure between ducts	88 mm w.g. or more if required
Condensate collection	Drain pans where necessary
Fans: included/available separately	Two included, installed in package; backward inclined, airfoil wheel, non-overloading, V-belt driven
Frost prevention	Coil tilt or bypassed
EXTRAS	Supplementary heating coils (steam or electric) and/or cooling coils (chilled water or refrigerant) available
FURTHER COMMENTS	Bry-Therm units (7 sizes) may be used as a prime mover for either supply or exhaust systems (or both) where heat recovery is required.

MANUFACTURER	*BURKE THERMAL ENGINEERING LTD.*
MODEL NAME AND NUMBER	Q-Pipe TRU 120 to 1325
TYPE	Heat pipe
APPLICATIONS	Process-to-process; Process-to-space; airconditioning
OPERATION	Counter-flow
TEMPERATURE (°C)	
Operating range	20–700
AIR	
Volume (m^3/h)	60 000
Velocity (m/s)	2.6
PRESSURE DROP (mm w.g.)	20
REGENERATOR	
Rotation speed (rev/min)	
Rotation motor size (W)	
DUTY (kW)	Dependent on specifications; units may be mounted in series
EFFICIENCY (%)	70
TYPICAL PAYBACK PERIOD (years)	
Airconditioning	2.5
Process plant	1
CONSTRUCTION	
Materials	Aluminium; copper; steel;
Weight (kg)	Various
Dimensions (mm)	Various
FEATURES	
Available custom built?	Yes
Performance guarantees	Yes
Installation method	
Installed by?	Manufacturer
Cleaning	Various
Corrosion resistance	Coatings, if necessary
Performance control and turndown range	
Cross-contamination	
Withstands differential pressure between ducts	
Condensate collection	Drain pan
Fans: included/available separately	Available
Frost prevention	Yes
EXTRAS	
FURTHER COMMENTS	A range of modular units to meet site requirements

MANUFACTURER	*CARGOCAIRE ENGINEERING CORP.*
MODEL NAME AND NUMBER	Econalloy Rotary Regenerative Sensible Heat Exchanger ECA-7500; 12000; 16000; 20000; 30000; 40000; 60000
TYPE	Regenerator (wheel)
APPLICATIONS	Process-to-process; process-to-space; environmental control packages
OPERATION	Counter-flow
TEMPERATURE (°C)	
Operating range	-40-190
AIR	
Volume (m^3/h)	3300-119 000
Velocity (m/s)	1-5.5
PRESSURE DROP (mm w.g.)	2-40
REGENERATOR	
Rotation speed (rev/min)	0-25
Rotation motor size (W)	250-350
DUTY (kW)	
EFFICIENCY (%)	96 (at 1 m/s)-45 (at 6.45 m/s)
TYPICAL PAYBACK PERIOD (years)	
Airconditioning	On installation (by replacement of heating/cooling capacity)
Process plant	Less than 1
CONSTRUCTION	
Materials	Metal media: aluminium or stainless steel
Weight (kg)	180-3270
Dimensions (mm)	14 860 x 14 860 x 4838-60 140 x 60 140 x 10 700
FEATURES	
Available custom-built?	Yes
Performance guarantees	One year from date of shipment
Installation method	Built up onsite or installed in air-handling package
Installed by?	Manufacturer or contractor
Cleaning	Detergent; steam; or compressed air
Corrosion resistance	Standard considerations for aluminium and stainless steel
Performance control and turndown range	100:1 turndown control by SCR drive motor-speed control
Cross-contamination	Less than 0.04% of rated volume
Withstands differential pressure between ducts	3466
Condensate collection	Built-in drain pan
Fans: included/available separately	Available separately
Frost prevention	Optional variable face and bypass preheat coils
EXTRAS	Available as complete air-handling/conditioning package
FURTHER COMMENTS	

MANUFACTURER	*CARGOCAIRE ENGINEERING CORP.*
MODEL NAME AND NUMBER	Econocaire Rotary Air-Air Total Enthalpy Heat Exchanger EC-3000; 7500; 12000; 16000; 20000; 40000; 60000
TYPE	Regenerator (wheel)
APPLICATIONS	Process-to-process; process-to-space; environmental control packages
OPERATION	Counter-flow
TEMPERATURE (°C)	
Operating range	-40-190
AIR	
Volume (m^3/h)	3300-119000
Velocity (m/s)	1-5
PRESSURE DROP (mm w.g.)	69.3
REGENERATOR	
Rotation speed (rev/min)	0-25
Rotation motor size (W)	250-350
DUTY (kW)	
EFFICIENCY (%)	90 (at 1 m/s)-65 (at 5 m/s); equal latent & sensible transfer at all effectivenesses
TYPICAL PAYBACK PERIOD (years)	
Airconditioning	On installation (by replacement of heating/cooling capacity)
Process plant	Less than 1
CONSTRUCTION	
Materials	Steel frame, proprietary transfer surface
Weight (kg)	130-2725
Dimensions (mm)	14 860 x 14 860 x 435-60 140 x 60 140 x 10 700
FEATURES	
Available custom-built?	Yes
Performance guarantees	One year from date of shipment
Installation method	Built up onsite or installed in air-handling package
Installed by?	Manufacturer or contractor
Cleaning	Compressed air
Corrosion resistance	Available on request
Performance control and turndown range	100:1 turndown control by SCR drive motor-speed control
Cross-contamination	Less than 0.04% of rated volume
Withstands differential pressure between ducts	3466
Condensate collection	Not required, moisture transferred
Fans: included/available separately	Available separately
Frost prevention	Optional variable face and bypass preheat coils
EXTRAS	Available as complete air-handling/conditioning package
FURTHER COMMENTS	

MANUFACTURER	*CARGOCAIRE ENGINEERING CORP.*
MODEL NAME AND NUMBER	Econopipe thermosiphon heat exchanger
TYPE	Heat pipe
APPLICATIONS	Process-to-process; process-to-space; environmental-control packages
OPERATION	Counter-flow
TEMPERATURE (^{o}C)	
Operating range	-45-468
AIR	
Volume (m^3/h)	3300-119 000
Velocity (m/s)	1-5
PRESSURE DROP (mm w.g.)	5-150
REGENERATOR	
Rotation speed (rev/min)	
Rotation motor size (W)	
DUTY (kW)	
EFFICIENCY (%)	85-38
TYPICAL PAYBACK PERIOD (years)	
Airconditioning	1
Process plant	Less than 1
CONSTRUCTION	
Materials	Steel casing: aluminium, copper, steel fins and tubes
Weight (kg)	
Dimensions (mm)	
FEATURES	
Available custom-built?	Yes
Performance guarantees	One year from date of shipment
Installation method	Built into system onsite or in factory package
Installed by?	Manufacturer or contractor
Cleaning	Steam; detergent; or compressed air
Corrosion resistance	Standard considerations, but optional coatings available
Performance control and turndown range	By withdrawal of the working fluid
Cross-contamination	None
Withstands differential pressure between ducts	3466
Condensate collection	Optional drain pan
Fans: included/available separately	Available separately
Frost prevention	Included in capacity-control system
EXTRAS	Capacity control; frost prevention; corrosion resistant coatings; drain pan; available as factory-built air-handling/conditioning package
FURTHER COMMENTS	

MANUFACTURER	*CARGOCAIRE ENGINEERING CORP.*
MODEL NAME AND NUMBER	Econoplate Modular Cross-flow Recuperator EPL-2-M
TYPE	Recuperator
APPLICATIONS	Process-to-process; process-to-space; environmental-control packages
OPERATION	Cross-flow
TEMPERATURE (°C)	
Operating range	-40-190
AIR	
Volume (m^3/h)	3400 (per module)
Velocity (m/s)	2.54
PRESSURE DROP (mm w.g.)	25.4
REGENERATOR	
Rotation speed (rev/min)	
Rotation motor size (W)	
DUTY (kW)	
EFFICIENCY (%)	73
TYPICAL PAYBACK PERIOD (years)	
Airconditioning	1 (varies with volume, temperature and location)
Process plant	Less than 1
CONSTRUCTION	
Materials	Aluminium surface (coatings available)
Weight (kg)	36
Dimensions	14 860 x 16 934 x 16 934 (each module)
FEATURES	
Available custom-built?	Yes
Performance guarantees	One year from date of shipment
Installation method	Built up onsite or included in factory-built air-handling package
Installed by?	Manufacturer or contractor
Cleaning	Steam; detergent; or compressed air
Corrosion resistance	Standard aluminium considerations
Performance control and turndown range	Volume dampers (optional)
Cross-contamination	No measurable gross contamination
Withstands differential pressure between ducts	3466
Condensate collection	Built-in drain pan
Fans: included/available separately	Available separately
Frost prevention	By optional capacity control
EXTRAS	Coating to resist corrosion Available as complete air-handling/conditioning package
FURTHER COMMENTS	

MANUFACTURER	*C-E AIR PREHEATER*
MODEL NAME AND NUMBER	Cor-Pak modules
TYPE	Shell and tube recuperative, single or multiple pass
APPLICATIONS	Process-to-process; process-to-space
OPERATION	Cross counter-flow
TEMPERATURE (°C)	
Operating range	Up to 815
AIR	
Volume (m^3/h)	Max. 69 700; min. 4 000
Velocity (m/s)	9-21
PRESSURE DROP (mm w.g.)	76-365
REGENERATOR	
Rotation speed (rev/min)	
Rotation motor size (W)	
DUTY (kW)	Dependent on specification
EFFICIENCY (%)	Up to 70
TYPICAL PAYBACK PERIOD (years)	
Airconditioning	
Process plant	1-3
CONSTRUCTION	
Materials	Cor-Ten, 304 S.S., and 316 S.S. plus special applications
Weight (kg)	
Dimensions (mm)	
FEATURES	
Available custom-built?	Yes
Performance guarantees	$\pm$5% air-outlet temperatures
Installation method	
Installed by?	
Cleaning	System available
Corrosion resistance	Standard for materials of construction
Performance control and turndown range	
Cross-contamination	Less than 0.1% available
Withstands differential pressure between ducts	22 mm w.g.-250 mm w.g.
Condensate collection	N/a
Fans: included/available separately	Available
Frost prevention	
EXTRAS	
FURTHER COMMENTS	

MANUFACTURER	*C-E AIR PREHEATER*
MODEL NAME AND NUMBER	Hazen Recuperator
TYPE	Sheet and tube; gas-to-gas recuperator
APPLICATIONS	Industrial primary and secondary metals
OPERATION	Counter - cross-flow
TEMPERATURE (°C)	
Operating range	Up to 1287
AIR	
Volume (m^3/h)	Air flows to 280 000 Nm^3/h
Velocity (m/s)	9-12
PRESSURE DROP (mm w.g.)	Air side 254 to 381 mm w.g.; gas side 2.5 to 22.9 mm w.g.
REGENERATOR	
Rotation speed (rev/min)	
Rotation motor size (W)	
DUTY (W)	
EFFICIENCY (%)	40
TYPICAL PAYBACK PERIOD (years)	
Airconditioning	N/a
Process plant	2
CONSTRUCTION	
Materials	All grades of stainless
Weight (kg)	Range from 1814-90 700
Dimensions (mm)	Ranges up to 8000 across individual flues
FEATURES	
Available custom-built?	Yes, 100%
Performance guarantees	Yes
Installation method	Process air header box placed on flue with individual heating elements bolted in place
Installed by?	Normally furnace supplier or client
Cleaning	Yes, when required
Corrosion resistance	High
Performance control and turndown range	Thermocouple: 5 to 1 normal
Cross-contamination	Normal air leakage not measurable; some particulate removal may occur in waste gas flow, due to tube fouling
Withstands differential pressure between ducts	Design as required
Condensate collection	N/a
Fans: included/available separately	Not supplied
Frost prevention	
EXTRAS	
FURTHER COMMENTS	No expansion points required.

MANUFACTURER	*C-E AIR PREHEATER*
MODEL NAME AND NUMBER	Ljungstrom air preheater
TYPE	Regenerator
APPLICATIONS	Process-to-process; process-to-space
OPERATION	Counter- or parallel-flow
TEMPERATURE (^{o}C)	
Operating range	Up to 538
AIR	
Volume (m^3/h)	3516-4 922 817 for each air preheater
Velocity (m/s)	
PRESSURE DROP (mm w.g.)	Up to 703
REGENERATOR	
Rotation speed (rev/min)	1-4
Rotation motor size (W)	1118.5-55927.5
DUTY (kW)	
EFFICIENCY (%)	Up to 93 air-side effectiveness
TYPICAL PAYBACK PERIOD (years)	
Airconditioning	
Process plant	1-3
CONSTRUCTION	
Materials	Cor-Ten and mild steel standard; enamelling iron enamelled and stainless steel also available
Weight (kg)	
Dimensions (mm)	
FEATURES	
Available custom-built?	Yes
Performance guarantees	$\pm$2.78^{o}C. gas leaving uncorrected temperature
Installation method	Small units shop-assembled; all others field-erected
Installed by?	Contractor
Cleaning	System available
Corrosion resistance	Materials designed to minimize corrosion
Performance control and turndown range	
Cross-contamination	Approx. 5% leakage from air to gas
Withstands differential pressure between ducts	Max. 703 mm w.g.
Condensate collection	
Fans: included/available separately	Separate
Frost prevention	Separate
EXTRAS	
FURTHER COMMENTS	

MANUFACTURER	*CORNING GLASS WORKS*
MODEL NAME AND NUMBER	Cercor Ceramic Heat Wheel
TYPE	Regenerator
APPLICATIONS	Process-to-process; preheating combustion air; indirect heating
OPERATION	Counter-flow
TEMPERATURE (°C)	
Operating range	530-1100
AIR	
Volume (m^3/h)	2550-17000
Velocity (m/s)	
PRESSURE DROP (mm w.g.)	Max. 254 each side
REGENERATOR	
Rotation speed (rev/min)	20
Rotation motor size (W)	4000
DUTY (kW)	
EFFICIENCY (%)	70-75
TYPICAL PAYBACK PERIOD (years)	
Airconditioning	N/a
Process plant	Dependent on exhaust temperature, recovered temperature used; number of operational hours/years; price of fuel. Typically 1-3 years
CONSTRUCTION	
Materials	Ceramic wheel; steel frame; insulated hot ducts
Weight (kg)	
Dimensions (mm)	
FEATURES	
Available custom-built?	Basic unit; controls and bypass options extra
Performance guarantees	
Installation method	
Installed by?	
Cleaning	
Corrosion resistance	Materials can be specified for acid resistance
Performance control and turndown range	Good control of recovered temperature over wide turndown range
Cross-contamination	1-3, depending on requirements
Withstands differential pressure between ducts	
Condensate collection	
Fans: included/available separately	Quoted separately
Frost prevention	
EXTRAS	
FURTHER COMMENTS	Corning also make glass tubes for recuperators.

MANUFACTURER	*CURWEN & NEWBERY LTD.*
MODEL NAME AND NUMBER	Twin-coil Runaround Heat Recovery System
TYPE	Runaround
APPLICATIONS	Certain process-to-process applications; process-to-space; heating, ventilating and airconditioning applications
OPERATION	Counter-flow
TEMPERATURE (^{o}C)	
Operating range	-20-200
AIR	
Volume (m^3/h)	No standard range
Velocity (m/s)	2-4
PRESSURE DROP (mm w.g.)	Up to 25
REGENERATOR	
Rotation speed (rev/min)	
Rotation motor size (W)	
DUTY (kW)	Dependent on specification
EFFICIENCY (%)	Typically 60
TYPICAL PAYBACK PERIOD (years)	
Airconditioning	0.5-4
Process plant	0.5-4
CONSTRUCTION	
Materials	Copper tubes; aluminium fins
Weight	
Dimensions	
FEATURES	
Available custom-built?	All units are custom-built
Performance guarantees	Units to perform as specified within measuring tolerances
Installation method	Can be mounted in horizontal or vertical ducts
Installed by?	Schools; hospitals; office blocks; industrial plants; hotels; etc.
Cleaning	Units should have a prefilter
Corrosion resistance	Units may be nickel-chrome coated
Performance control and turndown range	Control effected by motorized valve arrangement located in pipework system
Cross-contamination	Nil
Withstands differential pressure between ducts	Not applicable
Condensate collection	Drip trays provided as extras
Fans: included/available separately	Fans not supplied
Frost prevention	Preheater battery/control system
EXTRAS	Anti-corrosion coatings; drip trays
FURTHER COMMENTS	

MANUFACTURER	*CURWEN & NEWBERY LTD.*
MODEL NAME AND NUMBER	CN Heat Regenerator TH type models: SW1 TH-SW130 TH; DW7 TH-DW130 TH
TYPE	Heat regenerator
APPLICATIONS	Heating and ventilating; airconditioning
OPERATION	Counter-flow
TEMPERATURE (°C)	
Operating range	-20-65
AIR	
Volume (m^3/h)	600-144 000
Velocity (m/s)	2-4.5
PRESSURE DROP (mm w.g.)	8-26
REGENERATOR	
Rotation speed (rev/min)	0-20
Rotation motor size (W)	100-750
DUTY (kW)	Up to 850
EFFICIENCY (%)	Up to 20 total heat efficiency
TYPICAL PAYBACK PERIOD (years)	
Airconditioning	2-3
Process plant	Not applicable
CONSTRUCTION	
Materials	Casing: mild steel; rotor: aluminium; heat transfer media: non-metallic
Weight	165-816 kg
Dimensions	700-4800 mm square
FEATURES	
Available custom-built?	Variations from standard include regenerators outside listed flow ranges or any special requirements, as well as knockdown units
Performance guarantees	Units to perform as specified within measuring tolerances
Installation method	Units may be mounted in 12 different orientations
Installed by?	Industrialists; local authorities; commerce; hospital authorities; DOI, etc.
Cleaning	Vacuum or water spray
Corrosion resistance	Not applicable
Performance control and turndown range	To vary performance, a variable speed drive is fitted as standard equipment, giving almost infinite turndown from optimum speed
Cross-contamination	Less than 1% by volume with a purger
Withstands differential pressure between ducts	250 mm
Condensate collection	BSP tapping located in base to remove condensate
Fans: included/available separately	Packaged units incorporating fans available
Frost prevention	Frost prevention must be considered below supply air temperature of -12°C
EXTRAS	Mounting feet; horizontal units; 3-phase drive; constant-speed drive; Thermostats; kindling switch
FURTHER COMMENTS	TH range of CN heat regenerators is able to transfer total heat (sensible plus latent heat): designed purposely for heating, ventilating and airconditioning applications.

MANUFACTURER	*CURWEN & NEWBERY LTD.*
MODEL NAME AND NUMBER	CN Heat Regenerator SH type models SW1 A,B,C or D-SW130 A,B,C or D; DW7 A,B or C -DW130 A,B or C
TYPE	Heat regenerator
APPLICATIONS	Process-to-process; process-to-space; airconditioning, including all forms of driers, kilns, boiler plant, stenters, etc.
OPERATION	Counter-flow
TEMPERATURE (°C)	Type A: Type B & C: Type D:
Operating range	-20-200 -20-400 20-800
AIR	
Volume (m^3/h)	500-115 000 (double-wheel units: 3000-230 000)
Velocity (m/s)	2-5
PRESSURE DROP (mm w.g.)	8-50
REGENERATOR	
Rotation speed (rev/min)	0-20
Rotation motor size (W)	100-750
DUTY (kW)	Up to 12.6×10^3
EFFICIENCY (%)	Up to 90 sensible heat, but typically 80
TYPICAL PAYBACK PERIOD (years)	
Airconditioning	2-5
Process plant	0.5-3
CONSTRUCTION	
Materials	All casings: mild steel; rotors: aluminium, mild steel, stainless steel; heat transfer media: aluminium, stainless steel,incolloy
Weight	70-3500 Kg, for SW units
Dimensions	700 mm square to 4800 mm square for SW units
FEATURES	
Available custom-built?	Units available custom-built; also knockdown units for site-assembly
Performance guarantees	Units to perform as specified within measuring tolerances
Installation method	Units may be mounted in 12 different orientations
Installed by?	Industrialists; local authorities; commerce; hospital authorities; DOE, etc
Cleaning	Compressed air; vacuum; water and steam sprays
Corrosion resistance	Materials selected to suit application
Performance control and turndown range	To vary performance, a variable speed drive and controller is available to give almost infinite turndown from optimum speed
Cross-contamination	Less than 1% by volume; bacteriological cross-contamination 1 : 15 000; solvent cross-contamination less than 1.25%
Withstands differential pressure between ducts	Up to 400 mm, dependent on type of media
Condensate collection	BSP tappings located in base; slot drains also available on request
Fans: included/available separately	Packaged units incorporating fans available
Frost prevention	Frost protection not normally required
EXTRAS	Variable speed drive and controller; thermostats; special paint finishes; ductwork transition pieces; cleaning sprays; purger; fans-assisted purger
FURTHER COMMENTS	SH range of CN heat regenerators is designed for sensible heat transfer; for total-heat (enthalpy) transfer TH range of units should be used.

MANUFACTURER	*CURWEN & NEWBERY LTD.*	
MODEL NAME AND NUMBER	CN Heat Recuperator, diagonal and cross-flow models	
TYPE	Recuperator	
APPLICATIONS	Heating and ventilation; airconditioning; process-to-space	
OPERATION	Counter-flow	
TEMPERATURE (°C)		
Operating range	-20-70	
AIR		
Volume (m^3/h)	Diagonal-flow range: 500-46 000 (12 models)	Cross-flow range: 500-62 000 (15 models)
Velocity (m/s)	3-10	3-10
PRESSURE DROP (mm w.g.)	8-28	6-360
REGENERATOR		
Rotation speed (rev/min)		
Rotation motor size (W)		
DUTY (kW)	Up to 1429	
EFFICIENCY (%)	Depends on RH of exhaust air stream, but typically 60-75	
TYPICAL PAYBACK PERIOD (years)		
Airconditioning	2-3	
Process plant	0.5-3	
CONSTRUCTION		
Materials	Copper/stainless-steel/glass/aluminium; heat transfer surface/MS casing	
Weight (kg)	20-500	
Dimensions (mm)	600 x 800 x 300-2700 x 2400 x 1350	
FEATURES		
Available custom-built?	Yes	
Performance guarantees	Units to perform as specified within measuring tolerances	
Installation method	Ducts normally horizontal with diagonal-flow unit and horizontal and vertical with cross-flow unit	
Installed by?	Certain process plant: all heating and ventilating; airconditioning installations	
Cleaning	Water spray	
Corrosion resistance	Heat transfer surface may be protected with vinyl finish or CRP	
Performance control and turndown range	Performance adjusted by face and bypass damper arrangement	
Cross-contamination	Nil	
Withstands differential pressure between ducts	Max. pressure differential:100 mm w.g.	
Condensate collection	BSP connections located in base	
Fans: included/available separately	Not supplied	
Frost prevention	Frost protection by preheater battery or bypass arrangement	
EXTRAS	Special anti-corrosion finishes; built-in water sprays; access doors are standard	
FURTHER COMMENTS		

MANUFACTURER	*CURWEN & NEWBERY LTD.*
MODEL NAME AND NUMBER	CN Heat Pipe Recovery Unit
TYPE	Heat Pipe
APPLICATIONS	Process-to-process; process-to-space; heating and ventilation
OPERATION	Counter-flow
TEMPERATURE (°C)	
Operating range	Type: 15 40 50 60 85 224 280 336
AIR	
Volume (m^3/h)	Listed sizes: 650-1000 smallest unit; 1600-2400 largest unit; (other sizes made to order; units ma also be fitted in parallel and seri
Velocity (m/s)	2-3
PRESSURE DROP (mm w.g.)	Min. 2.5; Max. 25
REGENERATOR	
Rotation speed (rev/min)	
Rotation motor size (W)	
DUTY (kW)	Up to 2100 for largest single unit
EFFICIENCY (%)	Dependent on air velocity, typically 60-75
TYPICAL PAYBACK PERIOD (years)	
Airconditioning	2-3
Process plant	0.5-3
CONSTRUCTION	
Materials	Aluminium plate fins/copper tubes; copper plate fins, tubes; spirally wound copper fins, copper tubes; spirally wound MS fins, tubes; special anti-corrosion finishes; mild-steel casings
Weight (kg)	45-700
Dimensions (mm)	610 x 305-3558 x 1219
FEATURES	
Available custom-built?	Any size of unit can be supplied to suit customer's requirements
Performance guarantees	Units to perform as specified within measuring tolerances
Installation method	Units may be mounted with ducts side by side/under and over
Installed by?	Contractor
Cleaning	Low-pressure steam; water; or chemical cleaning
Corrosion resistance	Materials and finishes selected to combat corrosion
Performance control and turndown range	Performance adjusted by face and bypass damper
Cross-contamination	Nil
Withstands differential pressure between ducts	Special centre plate on special tube units is capable of withstanding high-pressure differentials
Condensate collection	BSP drain connections available
Fans: included/available separately	Fans excluded
Frost prevention	Preheater battery to be installed, if severe frosting expected
EXTRAS	Special finishes
FURTHER COMMENTS	

MANUFACTURER	*DESCHAMPS LABORATORIES INC.*
MODEL NAME AND NUMBER	Z-Duct Series PK, PZ, 75 and 74
TYPE	Plate
APPLICATIONS	Process-to-process; process-to-space; airconditioning
OPERATION	Counter-flow
TEMPERATURE (^{o}C)	
Operating range	Max. 750
AIR	
Volume (m^3/h)	Unlimited, 20 000 modules
Velocity (m/s)	
PRESSURE DROP (mm w.g.)	
REGENERATOR	
Rotation speed (rev/min)	
Rotation motor size (W)	
DUTY (kW)	Dependent on number of modules used
EFFICIENCY (%)	70
TYPICAL PAYBACK PERIOD (years)	
Airconditioning	3
Process plant	1
CONSTRUCTION	
Materials	Aluminium or steel
Weight	
Dimensions	
FEATURES	
Available custom-built?	Yes
Performance guarantees	Yes
Installation method	Sheet-metal ducts
Installed by?	Contractor
Cleaning	Easy access; water washing and detergents available
Corrosion resistance	Aluminium; stainless steel; corrosion coating
Performance control and turndown range	Infinite
Cross-contamination	Nil
Withstands differential pressure between ducts	Yes, 380 mm w.g.
Condensate collection	Yes
Fans: included/available separately	Yes
Frost prevention	Yes
EXTRAS	All necessary equipment: e.g. dampers; filters; water washing systems; supplementary heaters and controls
FURTHER COMMENTS	

MANUFACTURER	*DRAVO CORP.*
MODEL NAME AND NUMBER	Ventilation Heat Recovery Units VHR 2-18
TYPE	Rotary regenerative
APPLICATIONS	Process-to-air; air-to-air
OPERATION	Counter-flow
TEMPERATURE (°C)	
Operating range	Up to 66
AIR	
Volume (m^3/h)	3398-30 568
Velocity (m/s)	2.12-5.43
PRESSURE DROP (mm w.g.)	12.7-30.99
REGENERATOR	
Rotation speed (rev/min)	15
Rotation motor size (W)	250 115/60/1
DUTY (kW)	Various
EFFICIENCY (%)	68-82 effectiveness
TOTAL PAYBACK PERIOD (years)	
Airconditioning	
Process plant	Varies
CONSTRUCTION	
Materials	Wheel: aluminium; housing: steel
Weight (kg)	1134-2721
Dimensions (mm)	Smallest: 1956 x 1448 x 1448; Largest: 3810 x 2464 x 2464
FEATURES	
Available custom-built?	Standard package with variations
Performance guarantees	1-5 years
Installation method	Rooftop with curb or indoor
Installed by?	Customer or contractor
Cleaning	Compressed air; steam; hot water
Corrosion resistance	Special coatings available
Performance control and turndown range	SCR variable speed: 0-15 rev/min
Cross-contamination	1%
Withstands differential pressure between ducts	Up to 50% differential
Condensate collection	Drain pan
Fans: included/available separately	Supply and exhaust fans in package
Frost prevention	Yes, pressure differential across wheel
EXTRAS	Reheat; preheat; motorized dampers; purge section; motor option; filter options; control options
FURTHER COMMENTS	

MANUFACTURER	*ECLIPSE LOOKOUT CO.*
MODEL NAME AND NUMBER	Eclipse Heat Recuperators, Type AA (range 3-AA-30-AA)
TYPE	Tubular recuperator
APPLICATIONS	Process-to-process (including fume incineration); process-to-space
OPERATION	Cross-flow (plain tubes); counter- and parallel-flow models available
TEMPERATURE (^{o}C)	
Operating range	Up to 1000 (specials);840 (standard continuous operation) Tube-side gas temperature 426 (max.) (standard units)
AIR	
Volume (m^3/h)	60 000
Velocity (m/s)	1-5
PRESSURE DROP (mm w.g.)	109 on 3-AA model, tube-side, 85 m^3/h, 7.7 on shell-side; 220 on 2-pass 30-AA model, tube-side, 850 m^3/h, 34 on shell side
REGENERATOR	
Rotation speed (rev/min)	
Rotation motor size (W)	
DUTY (kW)	200 up to several MW, depending on number of passes, modules, etc.
EFFICIENCY (%)	60-80
TYPICAL PAYBACK PERIOD (Years)	
Airconditioning	N/a
Process plant	Less than 1
CONSTRUCTION	
Materials	Tubes: 304 stainless steel; ducts, carbon steel; plus block insulation (standard)
Weight (kg)	600 (3-AA)-8000 (30-AA) approx.
Dimensions (mm)	1473 x 1422 (face) x 940 (depth) for 3-AA; 3912 x 2540 x 2616 for 30-AA
FEATURES	
Available custom-built?	Yes
Performance guarantees	Rated duty
Installation method	Installed onsite
Installed by?	Manufacturer or contractor
Cleaning	Off-or online systems
Corrosion resistance	Standard related to materials: special materials available
Performance control and turndown range	Dampers (not supplied), large turndown related to operating requirements
Cross-contamination	N/a
Withstands differential pressure between ducts	Yes
Condensate collection	If required
Fans: included/available separately	Available separately
Frost prevention	N/a
EXTRAS	Multipass systems; special designs and materials
FURTHER COMMENTS	

MANUFACTURER	*ENERGY RECOVERY CO.*
MODEL NAME AND NUMBER	Enreco Wheel Series 100 (models EW106; 109; 114; 120; 125; 135; 137; 145; 154)
TYPE	Regenerator (heat wheel)
APPLICATIONS	Normally airconditioning
OPERATION	Counter-flow
TEMPERATURE (°C)	-4-80
AIR	
Volume (m^3/h)	4000-100 000
Velocity (m/s)	
PRESSURE DROP (mm w.g.)	10-30
REGENERATOR	
Rotation speed (rev/min)	8-12
Rotation motor size (W)	Less than 1000
DUTY (kW)	Dependent on specification
EFFICIENCY (%)	68-80 (equal flows)
TYPICAL PAYBACK PERIOD (years)	
Airconditioning	1-3
Process plant	
CONSTRUCTION	
Materials	Fibre-reinforced phenolic resin core coated with lithium chloride; steel support structure
Weight (kg)	250-1400
Dimensions (mm)	1245-3710 (wheel diameter)
FEATURES	
Available custom-built?	No
Performance guarantees	One year from date of shipment
Installation method	Factory-assembled unit ready for fitting to ductwork
Installed by?	Contractor
Cleaning	
Corrosion resistance	
Performance control and turndown range	
Cross-contamination	Limited to 0.04% when static pressure difference between air flows 25 mm w.g. or greater
Withstands differential pressure between ducts	Yes
Condensate collection	Moisture transfer
Fans: included/available separately	Available separately
Frost prevention	Preheating required, if face temperature less than -4°C
EXTRAS	Packaged rooftop energy recovery system available
FURTHER COMMENTS	Laminar flow-type core permits passage of 100-diameter particles.

MANUFACTURER	*ENERGY RECOVERY CO.*
MODEL NAME AND NUMBER	Enreco Wheel Series 200 (models EW 206; 209; 214; 220; 225; 231; 237; 245; 254)
TYPE	Regenerator (heat wheel)
APPLICATIONS	Process-to-process; process-to-space
OPERATION	Counter-flow
TEMPERATURE (^{o}C)	
Operating range	Up to 200
AIR	
Volume (m^3/h)	4000-100 000
Velocity (m/s)	
PRESSURE DROP (mm w.g.)	10-30
REGENERATOR	
Rotation speed (rev/min)	8-12
Rotation motor size (W)	Less than 1000
DUTY (kW)	
EFFICIENCY (%)	In excess of 70
TYPICAL PAYBACK PERIOD (years)	
Airconditioning	N/a
Process plant	
CONSTRUCTION	
Materials	Aluminium or stainless steel core (corrugated matrice); steel frame
Weight (kg)	245-1390
Dimensions (mm)	1245-3710 (wheel diameter)
FEATURES	
Available custom-built?	Stainless-steel core specials
Performance guarantees	One year from date of shipment
Installation method	Factory-assembled unit ready for fitting to ductwork
Installed by?	Contractor
Cleaning	Steam lance (but upstream filtration preferred)
Corrosion resistance	
Performance control and turndown range	By SCR variable-speed drive
Cross-contamination	Purge section: when supply to exhaust pressure differential 25 mm w.g. or greater, carryover limited to 0.04% of rated wheel volume
Withstands differential pressure between ducts	Yes
Condensate collection	
Fans: included/available separately	Available separately
Frost prevention	N/a
EXTRAS	SCR variable speed drive; rotation detector alarm; packaged rooftop energy-recovery system available
FURTHER COMMENTS	

MANUFACTURER	*ENERGY SYSTEMS INC.*
MODEL NAME AND NUMBER	Component Wheels: SW-64; 80; 90; 100; 110; 120; 130; 142; Package Systems: TR-4; 8; 9; 12; 15; 20; 25; 45; 75
TYPE	Component wheels operate as regenerators; package systems employ all types of energy-recovery devices
APPLICATIONS	Heating; ventilating; airconditioning; process-to-space; process-to-process
OPERATION	Counter-flow operation is norma; special designs may be parallel-flow
TEMPERATURE (°C)	
Operating range	-57-371
AIR	
Volume (m^3/h)	67 088
Velocity (m/s)	4.57
PRESSURE DROP (mm w.g.)	22.35
REGENERATOR	
Rotation speed (rev/min)	0-20
Rotation motor size (W)	745.7
DUTY (kW)	
EFFICIENCY (%)	85
TYPICAL PAYBACK PERIOD (years)	
Airconditioning	2-4
Process plant	6 months-2 years
CONSTRUCTION	
Materials	Aluminium; copper; monel; carbon steel; stainless steel
Weight	570-1965 kg
Dimensions	1.626-3.66 m square x 0.5 m deep
FEATURES	
Available custom-built?	Component wheels and package systems both available custom-built for individual applications
Performance guarantees	1 year, parts and labour; 10 years, package system structure and housing
Installation method	Set in place and connected to mating equipment
Installed by?	Normally contractor
Cleaning	By steam; hot water; detergent solution; compressed air; vacuuming
Corrosion resistance	Superior
Performance control and turndown range	0-100%
Cross-contamination	No cross-contamination from component wheels or package systems; equipment protected from contamination by proper air filtration
Withstands differential pressure between ducts	Yes
Condensate collection	Yes
Fans: included/available separately	Fans included in all package systems
Frost prevention	Yes
EXTRAS	System casings; supply and exhaust fans; supply and exhaust filters; auxiliary coils; controls all extra with component wheels and standard with package systems
FURTHER COMMENTS	Equipment for every project individually designed for application.

MANUFACTURER	*FGF EQUIPMENT*	*AEROPLAST*
MODEL NAME AND NUMBER	RC01-RC24 (24 models)	RX760 (1 model)
TYPE	Heat pipe	Heat pipe
APPLICATIONS	Airconditioning; process-to-process	Airconditioning for houses
OPERATION	Counter-flow →	→
TEMPERATURE ($^{\circ}C$)		
Operating range	-40-250	-40-80
AIR		
Volume (m^3/h)	500-25 000	150-350
Velocity (m/s)	1-5	2-4
PRESSURE DROP (mm w.g.)	5-15	5-10
REGENERATOR		
Rotation speed (rev/min)		
Rotation motor size (W)		
DUTY (kW)		
EFFICIENCY (%)	30-70	45-55
TYPICAL PAYBACK PERIOD (years)		
Airconditioning		
Process plant		
CONSTRUCTION		
Materials	aluminium; copper	aluminium; pipe in Fulton
Weight (kg)		15
Dimensions (mm)	420 x 240 x 180-2060 x 2400 x 450	360 x 300 x 150
FEATURES		
Available custom-built?		
Performance guarantees		
Installation method		
Installed by?		
Cleaning	Water or air →	→
Corrosion resistance	Possibility to have epoxy resin →	→
Performance control and turndown range		
Cross-contamination	0% →	→
Withstands differential pressure between ducts	150 mm c.e.	20 mm c.e.
Condensate collection		
Fans: included/available separately		Fans are included with the heat pipes unit with 3 speeds
Frost prevention	No frost until $-15^{\circ}C$ →	→
EXTRAS		
FURTHER COMMENTS	Very cheap system: e.g. Q-Unit heat pipe with 24 000 m^3/h 50% of efficiency, costs 17 300 French francs.	Very cheap system: the heat pipes with the fans and equipment costs about 1500 French francs in export prices.

MANUFACTURER	*GAYLORD INDUSTIRES INC.*
MODEL NAME AND NUMBER	Heat Reclaim Unit, model series HRU
TYPE	Packaged unit, incorporating heat pipe coil; supply and exhaust fans; controls and automatic wash system
APPLICATIONS	Kitchen exhaust, to temper makeup air; process-to-space
OPERATION	Counter-flow
TEMPERATURE (°C)	
Operating range	-34-107
AIR	
Volume (m^3/h)	Normally up to 20 000, exhaust; 16 000, supply
Velocity (m/s)	2.5-3.0
PRESSURE DROP (mm w.g.)	Normally designed to handle 37 ext. SP on exhaust; 12 on supply
REGENERATION	
Rotation speed (rev/min)	
Rotation motor size (W)	
DUTY (kW)	Dependent on fan motor size, normally approx. 600 W per 2000 m^3/h exhaust; 400 W supply
EFFICIENCY (%)	60-65
TYPICAL PAYBACK PERIOD (years)	
(Kitchen, dining room ventilation)	2-5
CONSTRUCTION	
Materials	
Weight (kg)	Approx. 150 per 2000 m^3/h combined airflow (exhaust; supply)
Dimensions (mm)	3000-4000 long x 2100-2400 wide x 1200-2100 high
FEATURES	
Available custom-built?	Yes
Performance guarantees	1 year
Installation method	Mount unit: connect ductwork; wiring; piping
Installed by?	Mechanical contractor
Cleaning	Automatic integral water wash
Corrosion resistance	
Performance control and turndown range	Fan and bypass damper, modulation: 100%
Cross-contamination	None
Withstands differential pressure between ducts	Yes
Condensate collection	Yes
Fans: included/available separately	Included
Frost prevention	Allowed for in individual project design
EXTRAS	Supplementary heater when needed
FURTHER COMMENTS	Savings in air tempering cost over typical winters vary from 95-99%.

MANUFACTURER	*GRANCO EQUIPMENT INC.*
MODEL NAME AND NUMBER	Heavy Duty Ceramic Wheel
TYPE	Regenerator
APPLICATIONS	Process-to-process (see also Fume incineration)
OPERATION	Counter-flow
TEMPERATURE (°C)	
Operating range	780-1370
AIR	
Volume (m^3/h)	600-17000
Velocity (m/s)	2-4
PRESSURE DROP (mm w.g.)	(Fans) 2-50 kW
REGENERATOR	
Rotation speed (rev/min)	20
Rotation motor size (W)	
DUTY (kW)	Several thousand, dependent on preheat required
EFFICIENCT (%)	70-75
TYPICAL PAYBACK PERIOD (years)	
Airconditioning	N/a
Process plant	1-2
CONSTRUCTION	
Materials	Glass ceramic wheel; steel housing
Weight (kg)	
Dimensions (mm)	355-1778 (wheel dia.)
FEATURES	
Available custom-built?	Yes
Performance guarantees	Yes
Installation method	Supplied as package for connecting to services and mating to ductwork
Installed by?	Contractor
Cleaning	Intermittent reverse flushing every 3 s; organic contamination automatically oxidized
Corrosion resistance	Good
Performance control and turndown range	Normally constant speed, but acceleration control and slow-speed cutout
Cross-contamination	Ceramic seals: not quantified, but zero expansion
Withstands differential pressure between ducts	Some
Condensate collection	N/a
Fans: included/available separately	Included
Frost prevention	N/a
EXTRAS	System flow balancing; air blend to control maximum inlet temperature; indirect heating; numerous temperature-control systems
FURTHER COMMENTS	See also Incineration.

MANUFACTURER	*E. GREEN & SON LTD.*	
MODEL NAME AND NUMBER	Glass Tube Airheater	Cast Iron Airheater
TYPE	Recuperator	Plate
APPLICATIONS	Process-to-process (primarily boilers)	Process-to-process (primarily boilers)
OPERATION	Cross-flow →	
TEMPERATURE (^{o}C)		
Operating range	Up to 200 (approx.)	Up to 760
AIR		
Volume (m^3/h)	Unlimited →	
Velocity (m/s)		
PRESSURE DROP (mm w.g.)		
REGENERATOR		
Rotation speed (rev/min)	N/a →	
Rotation motor size (W)		
DUTY (kW)	Unlimited →	
EFFICIENCY (%)	Depends upon application →	
TYPICAL PAYBACK PERIOD (years)		
Airconditioning		
Process plant	Normally 1-3 →	
CONSTRUCTION		
Materials	Glass tubes; cast-iron or steel tube plates	Cast-iron finned plates
Weight (kg)	Various →	
Dimensions (mm)	Various →	
FEATURES		
Available custom-built?	Yes →	
Performance guarantees		
Installation method	On-site connection to ductwork →	
Installed by?	Manufacturer →	
Cleaning	Conventional techniques	Sootblowing and other techniques
Corrosion resistance	Excellent: glass and PTFE seals	In corrosive streams: operate above dew point
Performance control and turndown range	N/a →	
Cross-contamination	None →	
Withstands differential pressure between ducts	Yes →	
Condensate collection	Can be installed	Operation above dew point
Fans: included/available separately	Separate →	
Frost prevention	N/a →	
EXTRAS	Complete installation of units onsite, including: necessary modification to existing ductwork and supply of bypass; controls and other ancillaries to suit particular application	
OTHER COMMENTS	While not producing complete regenerators, replacement elements are manufactured for rotary air preheaters; materials include: cold-rolled, close-annealed mild steel; Corten (high tensile, corrosion resistant) and enamelling steel (suitable for vitreous enamelling).	

MANUFACTURER	*HARRIS THERMAL TRANSFER PRODUCTS INC.*	
MODEL NAME AND NUMBER		
TYPE	Recuperator	Recuperator
APPLICATIONS	Process-to-process	Solar research project
OPERATION	Counter-flow	Cross-flow
TEMPERATURE (°C)		
Operating range	650-20	1200-20
AIR		
Volume (m^3/h)	Tube-side: 141,800 shell-side: 50,500	Tube-side: 2200 shell-side: 6600
Velocity (m/s)	Tube-side: 22; shell-side: N/a	Tube-side: 30; shell-side: N/a
PRESSURE DROP (mm w.g.)	Tube-side: 76; shell-side:106	Tube-side: 15.5; shell-side: 1.3
REGENERATOR		
Rotation speed (rev/min)	N/a →	
Rotation motor size (W)	N/a →	
DUTY (kW)	5100	250
EFFICIENCY (%)	54	41
TYPICAL PAYBACK PERIOD (years)		
Airconditioning		
Process plant	0.3	Research project
CONSTRUCTION		
Materials	Carbon steel; stainless steel	Stainless steel; hastelloy
Weight (kg)	28500	500
Dimensions (mm)	6200 x 3000 x 2800	1900 x 900 x 690
FEATURES		
Available custom-built?	Yes →	
Performance guarantees	Yes →	
Installation method		
Installed by?	Customer →	
Cleaning	Mechanical with high pressure water	Mechanical
Corrosion resistance	None required →	
Performance control and turndown range	By client →	
Cross-contamination	Fly ash from incineration of wood	None
Withstands differential pressure between ducts	Yes →	
Condensate collection	None required →	
Fans: included/available separately	By client →	
Frost prevention	None required →	
EXTRAS		
FURTHER COMMENTS		

MANUFACTURER	*HIRT COMBUSTION ENGINEERS*
MODEL NAME AND NUMBER	Hirt Rotary Heat Exchanger
TYPE	Regenerator
APPLICATIONS	Process-to-process; incineration plant
OPERATION	Counter-flow
TEMPERATURE (°C)	
Operating range	Up to 870
AIR	
Volume (m^3/h)	Up to 240 000
Velocity (m/s)	1.5-6
PRESSURE DROP (mm w.g.)	Dependent on duty, but very low pressure drop claimed
REGENERATOR	
Rotation speed (rev/min)	
Rotation motor size (W)	
DUTY (kW)	Dependent on duty
EFFICIENCY (%)	Up to 80, but higher achieved (max. 87)
TYPICAL PAYBACK PERIOD (years)	
Airconditioning	N/a
Process plant	1-3
CONSTRUCTION	
Materials	Stainless-steel core; refractory-lined carbon steel housing
Weight (kg)	
Dimensions (mm)	Up to 4700 (diam.)
FEATURES	
Available custom-built?	Yes
Performance guarantees	Yes
Installation method	As package for mating to supply and exhaust ducts
Installed by?	Manufacturer or contractor
Cleaning	Removable segments facilitate remote washing
Corrosion resistance	Normal for stainless steel
Performance control and turndown range	Speed modulation, if required
Cross-contamination	Cross-contamination strong function of wheel size and differential temperature
Withstands differential pressure between ducts	Some
Condensate collection	Generally not required
Fans: included/available separately	
Frost prevention	N/a
EXTRAS	
FURTHER COMMENTS	

MANUFACTURER	*HIRT COMBUSTION ENGINEERS*
MODEL NAME AND NUMBER	Hirt Shell and Tube Heat Exchanger
TYPE	Tubular recuperator
APPLICATIONS	Process-to-process; incinerators
OPERATION	Cross-flow
TEMPERATURE (°C)	
Operating range	Up to 870
AIR	
Volume (m^3/h)	Up to 360 000
Velocity (m/s)	1.5-8
PRESSURE DROP (mm w.g.)	Dependent on duty
REGENERATOR	
Rotation speed (rev/min)	
Rotation motor size (W)	
DUTY (kW)	Dependent on specification
EFFICIENCY (%)	50-65 (depending on use of 1 or 2 passes)
TYPICAL PAYBACK PERIOD (years)	
Airconditioning	N/a
Process plant	1-3
CONSTRUCTION	Stainless-steel tubes, tube sheets and expansion joints; refractory-lined carbon-steel shell
Materials	
Weight (kg)	
Dimensions (mm)	
FEATURES	
Available custom-built?	Yes
Performance guarantees	Yes
Installation method	As package connected to exhaust and supply ducts
Installed by?	
Cleaning	Hinged access doors for conventional cleaning methods
Corrosion resistance	Normal for materials used
Performance control and turndown range	Flow modulation (not included), matches process conditions
Cross-contamination	None
Withstands differential pressure between ducts	Yes
Condensate collection	Normally not required
Fans: included/available separately	Not included
Frost prevention	N/a
EXTRAS	
FURTHER COMMENTS	

MANUFACTURER	*HITACHI ZOSEN (HITACHI SHIPBUILDING & ENGINEERING CO. LTD.)*
MODEL NAME AND NUMBER	Welded Plate Heat Exchanger
TYPE	Welded plate-type heat exchanger for recuperator and regenerator
APPLICATION	Gas-to-gas exchanger such as air preheater by flue gas; air-to-air or process gas-to-process gas
OPERATION	Counter-flow
TEMPERATURE (°C)	
Operating range	Max. 850
AIR	
Volume (m^3/h)	10 000-400 000 Nm^3/h per shell
Velocity (m/s)	5-25 in heating elements
PRESSURE DROP (mm w.g.)	10-300 mm w.g. (due to customer requirement)
REGENERATOR	
Rotation speed (rev/min)	
Rotation motor size (W)	
DUTY (kW)	Max. 40 000, but due to customer requirement
EFFICIENCY (%)	Temperature efficiency 40-90
TYPICAL PAYBACK PERIOD (years)	
Airconditioning	
Process plant	Approx. 0.5-1.5 years, but due to process or other conditions
CONSTRUCTION	
Materials	Mild steel; copper bearing steel; any type of stainless steel
Weight (kg)	200 kg/m^2 of heating surface area
Dimensions	Up to 12 000 m^2 of surface area per shell
FEATURES	
Available custom-built?	Available
Performance guarantees	Performance guarantees by Hitachi, maker
Installation method	On ground or structure, by crane
Installed by?	Maker or customer (as per customer requirement)
Cleaning	Steamblowing or water cleaning (steam sootblow is available under operation)
Corrosion resistance	As customer requirement
Performance control and turndown range	By gas bypass 0-100%
Cross-contamination	Without internal and external leakage
Withstands differential pressure between ducts	About 3000 mm w.g.
Condensate collection	
Fans: included/available separately	
Frost prevention	
EXTRAS	Welded plate-type exchanger for gas-to-gas heat exchanger, developed by Hitachi Zosen in 1977
FURTHER COMMENTS	Without any internal or external leakage.

MANUFACTURER	*HITACHI ZOSEN (HITACHI SHIPBUILDING & ENGINEERING CO. LTD.)*
MODEL NAME AND NUMBER	Hitachi Zosen/Rothemuhle Heat Exchanger
TYPE	Regenerator
APPLICATIONS	Gas-to-gas, such as air preheater by flue gas; air-to-air; process gas-to-process gas
OPERATION	Counter-flow
TEMPERATURE	
Operating range	Max. 500 (normally)
AIR	
Volume (m^3/h)	100 000-2 000 000 Nm^3/h per heat exchanger
Velocity (m/s)	2-30
PRESSURE DROP (mm w.g.)	10-400 w.g. (due to customer requirement)
REGENERATOR	
Rotation speed (rev/min)	0.5-1.0
Rototation motor size (W)	
DUTY (kW)	
EFFICIENCY (%)	Temperature efficiency up to 95
TYPICAL PAYBACK PERIOD (years)	
Airconditioning	
Process plant	0.1-1.0, but due to process or other condition
CONSTRUCTION	
Materials	Ceramic; enamel-coated steel; Corten, mild steel
Weight (kg)	10 000-1 200 000
Dimensions (mm)	2 000-24 000 diam.
FEATURES	
Available custom-built?	Available
Performance guarantees	Hitachi Zosen, maker
Installation method	On ground or structure by crane, etc.
Installed by?	Maker or customer, (as per customer requirement)
Cleaning	Sootblowing by steam or air (during plant operation, and water washing)
Corrosion resistance	As customer requirement; ceramic can also be used
Performance control and turndown range	By gas bypass or recycle 0-150%
Cross-contamination	No gas leakage towards outside
Withstands differential pressure between ducts	About 3000 mm w.g.
Condensate collection	
Fans: included/available separately	
Frost prevention	
EXTRAS	
FURTHER COMMENTS	Huge amount of gas can be cooled or heated within compact heat exchanger with small pressure drop; technique for denox plant application also carried out.

MANUFACTURER	*HITACHI ZOSEN (HITACHI SHIPBUILDING & ENGINEERING CO. LTD.)*
MODEL NAME AND NUMBER	Heat Pipe Exchanger
TYPE	Regenerator by heat pipe
APPLICATIONS	Gas-to-gas exchanger
OPERATION	Counter-flow
TEMPERATURE (°C)	
Operating range	Max. 350
AIR	
Volume (m^3/h)	10 000-400 000 Nm^3/h per shell
Velocity (m/s)	5-15
PRESSURE DROP (mm w.g.)	10-300 mm w.g. (due to customer requirement)
REGENERATOR	
Rotation speed (rev/min)	
Rotation motor size (W)	
DUTY (kW)	Max. 40 000, but due to customer requirement
EFFICIENCY (%)	Temperature efficiency 0-0.8
TYPICAL PAYBACK PERIOD (years)	
Airconditioning	
Process plant	Approx. 0.5-3.0, but due to process or other condition
CONSTRUCTION	
Materials	Mild steel; copper alloy; nickel alloy
Weight (kg)	
Dimensions (mm)	50 mm heat pipe: max. 10 m long
FEATURES	
Available custom-built?	Available
Performance guarantees	Performance guarantee by Hitachi Zosen, manufacturer,
Installation method	On the ground or structure by crane
Installed by?	Customer's manufacturer, due to customer requirement
Cleaning	Steam blow or water cleaning
Corrosion resistance	As client's requirements
Performance control and turndown range	By-pass control, 0-100
Cross-contamination	Without internal or external leakage
Withstands differential pressure between ducts	10 kg/cm^2
Condensate collection	
Fans: included/available separately	
Frost prevention	
EXTRAS	
FURTHER COMMENTS	

MANUFACTURER	*HOLCROFT*	
MODEL NAME AND NUMBER	Holcroft EM Recuperator type EM30-EM3500	
TYPE	Recuperator	
APPLICATIONS	Process-tp-process	
OPERATION	Parallel- and counter-flow	
TEMPERATURE (°C)	Hot gas:	Cold gas:
Operating range	650-1095	Up to 540
AIR		
Volume (m^3/h)	EM30: 85; EM45: 127; EM65: 184; EM150: 425; EM250: 708; EM500: 1416; EM1000: 2832	
Velocity (m/s)	Proprietary	
PRESSURE DROP (mm w.g.)	Typically 500	
REGENERATOR		
Rotation speed (rev/min)		
Rotation motor size (W)		
DUTY (kW)	Dependent on specification	
EFFICIENCY (%)		
TYPICAL PAYBACK PERIOD (years)		
Airconditioning		
Process plant		
CONSTRUCTION		
Materials	EM30; EM45; EM65; EM150; EM250; EM500; EM1000: stainless steel/steel	
Weight (kg)	65; 70; 75; 150; 255; 350; 460	
Dimensions (mm)	1.25 m x 0.32 m; 1.25 m x 0.39 m; 1.25 m x 0.42 m; 1.56 m x 0.57 m; 2.32 m x 0.69 m; 3.08 m x 0.84 m; 3.39 m x 0.91 m	
FEATURES		
Available custom-built?	Yes	
Performance guarantees	Materials and workmanship for 2-year period	
Installation method	Original and/or retrofit	
Installed by?	Manufacturer or customer	
Cleaning	None	
Corrosion resistance	Available in corrosive-resistant alloys	
Performance control and turndown range	Dependent only on burner turndown range	
Cross-contamination	None	
Withstands differential pressure between ducts	Yes	
Condensate collection	None (high-temperature recuperation)	
Fans: included/available separately	Available separately	
Frost prevention	None	
EXTRAS	Various adaptors to mount recuperator to existing furnace	
FURTHER COMMENTS		

MANUFACTURER	*HOTWORK DEVELOPMENT LTD.*
MODEL NAME AND NUMBER	Hotwork Self-recuperative Burners/model nos 1;2;3;5; 5;10;15;25 (model numbers indicate burner rating in thermo/h)
TYPE	Recuperator: furnace flue gas/combustion air; recuperator being an integral part of burner
APPLICATIONS	Process-to-process in high-temperature heating applications, range 700°C-1500°C: e.g. forging furnaces; soaking pits, reheat furnaces, glass tanks, etc.
OPERATION	Counter-flow with metal heat exchanger (highest temperature areas, refractory protected)
TEMPERATURE (°C)	
Operating range	Normal max. 1300; up to 1500 in some cases; minimum temperature dependent on economics (not normally less than 700)
AIR	
Volume (m^3/h)	Burner size designation x 28.5 gives air rate in m^3/h; available in air rates up to 750
Velocity (m/s)	In range 5-8, dependent on burner size
PRESSURE DROP (mm w.g.)	At max. air rate cold: 150-400, dependent on operating temperature
REGENERATOR	
Rotation speed (rev/min)	
Rotation motor size (W)	
DUTY (kW)	N/a
EFFICIENCY (%)	Dependent on operating temperature: heating process at 1200°C is 37% efficient with cold air; 61% efficient using recuperative burner
TYPICAL PAYBACK PERIOD (years)	
Airconditioning	N/a
Process plant	12-18 months
CONSTRUCTION	
Materials	Various grades heat-resisting steel
Weight (kg)	Model no. 5: 150; 10: 250; 15: 300; 25: 500
Dimensions (mm) (approx. overall)	750 long x 380 x 380; 850 long x 460 x 460; 920 long x 510 x 510; 1020 long x 710 x 710
FEATURES	
Available custom-built,	Standard range, but custom-built modifications available
Performance guarantees	Yes, if performance monitoring equipment fitted (i.e. thermocouples, recorders, etc.)
Installation method	Fitted into furnace wall as normal fuel-fired burners
Installed by?	Manufacturer; or furnace builder; pipework contractor, etc.
Cleaning	Not normally necessary, but heat exchanger is removable with burner in position in furnace
Corrosion resistance	Suitable for use with all normal gas and oil fuels; special applications require further investigation
Performance control and turndown range	Burner thermal turndown: 10:1 with conventional air/fuel ratio control; in some applications compensation required for varying air back-pressure
Cross-contamination	Not quantifiable - normally only suitable for clean furnace atmospheres, but possible to blow out loose deposits periodically with built in compressed air jets
Withstands differential pressure between ducts	N/a air and flue gas pressure differentials usually less than 700 mm w.g.
Condensate collection	None
Fans: included/available separately	Available separately or use existing
Frost prevention	Not necessary
EXTRAS	Burner supplied complete, any extras normally incorporated in ancillary valve train, etc., dependent on user preference
FURTHER COMMENTS	1, 2 and 3.5 therm sizes suitable for gaseous fuels only; 5, 10, 15 and 25 therm sizes can be gas only or dual fuel; all include: igniter; gas pilot; sight glass and facility for flame monitoring.

MANUFACTURER	*JAMES HOWDEN & CO. LTD.*
MODEL NAME AND NUMBER	Ljungstrom Air Preheater; Ljungstrom Package Air Preheater
TYPE	Regenerator: transfer surface is alternately heated by flue gases passing through it; cooled by air passing through it
APPLICATIONS	Gas-to-air applications for: power station; process plants; oil refineries; steelworks; brick industry; marine; industrial plants
OPERATION	Counter-flow
TEMPERATURE (°C)	
Operating range	Gas: Max. 750, Min. 100; Air: Min. ambient, Max. 700
AIR	
Volume (m^3/h)	Vary 8000-100 000
Velocity (m/s)	Approx. 8 through heating surface
PRESSURE DROP (mm w.g.)	100 Ave.
REGENERATOR	
Rotation speed (rev/min)	Varies 0.75-4 (dependent on preheater type)
Rotation motor size (W)	Varies 750-15 000 (dependent on preheater type)
DUTY (kW)	Varies 500-75 000 (dependent on preheater type)
EFFICIENCY (%)	Up to 92
TYPICAL PAYBACK PERIOD (years)	
Airconditioning	
Process plant	1.5
CONSTRUCTION	
Materials	Dependent on fuel and temperature conditions: mild steel; corten or stainless steel
Weight	Varies 2-600 tonne
Dimensions	
FEATURES	
Available custom-built?	Yes
Performance guarantees	Performance guarantees given on thermal performance; pressure drop; leakage
Installation method	Package preheaters lifted on to prepared foundations; larger preheaters site-erected
Installed by?	Manufacturer or customer
Cleaning	Steam/air sootblowing and/or water-washing equipment supplied
Corrosion resistance	Cold end heating surface material normally Corten, can be vitreous enamel
Performance control and turndown range	Performance turndown to approx. 10%
Cross-contamination	Air leakage into gas stream: 5-15%; gas entrainment into air stream: approx. 1-2%
Withstands differential pressure between ducts	Yes; fabricated steel sealing strips fitted radially-axially and on circumference of rotor; preheaters operating at 70 mm w.g. air/gas differential
Condensate collection	None
Fans: included/available separately	Available separately
Frost prevention	Yes
EXTRAS	Emergency drive motors; fire detection/protection systems; seal-sensing equipment
FURTHER COMMENTS	

MANUFACTURER	*INTERNATIONAL RESEARCH & DEVELOPMENT CO. LTD.*
MODEL NAME AND NUMBER	Specials only
TYPE	Heat pipe heat exchanger
APPLICATIONS	Process-to-process; process-to-space
OPERATION	Counter-flow
TEMPERATURE (°C)	
Operating range	Up to 300
AIR	
Volume (m^3/h)	Up to 20 000
Velocity (m/s)	2-4
PRESSURE DROP (mm w.g.)	Typically 10-20
REGENERATOR	
Rotation speed (rev/min)	
Rotation motor size (W)	
DUTY (kW)	Up to 50
EFFICIENCY (%)	Up to 65
TYPICAL PAYBACK PERIOD (years)	
Airconditioning	N/a
Process plant	1-3
CONSTRUCTION	
Materials	Normally copper; possibly coated for high-temperature use; aluminium fins on lower-temperature units
Weight (kg)	Various
Dimensions (mm)	Typically 1300 x 500 x 300
FEATURES	
Available custom-built?	Yes
Performance guarantees	Yes
Installation method	Fitted to supply and exhaust ducts
Installed by?	Manufacturer or contractor
Cleaning	Conventional systems
Corrosion resistance	Materials selected to meet user requirements
Performance control and turndown range	As required by customer
Cross-contamination	None
Withstands differential pressure between ducts	Yes
Condensate collection	Can be provided
Fans: included/available separately	Available separately
Frost prevention	Not normally required
EXTRAS	
FURTHER COMMENTS	Specials made to order only.

MANUFACTURER	*IONICS INC.*
MODEL NAME AND NUMBER	
TYPE	Radiation recuperators: Double-Shell Type; Tube Type. Tubular recuperators: Flue gas through tubes; Flue gas around tubes
APPLICATIONS	Process-to-process
OPERATION	Counter-, parallel-flow; or combination of both
TEMPERATURE (°C)	
Operating range	Flue temperatures: up to 1400 Preheat temperatures: up to 800
AIR	
Volume (m^3/h)	Air flow rates: 200-100 000
Velocity (m/s)	1-50
PRESSURE DROP (mm w.g.)	100-800
REGENERATOR	
Rotation speed (rev/min)	
Rotation motor size (W)	
DUTY (kW)	Up to 100×10^6
EFFICIENCY (%)	40-80
TYPICAL PAYBACK PERIOD (years)	
Airconditioning	
Process plant	0.5-3
CONSTRUCTION	
Materials	Austenitic stainless steels; higher alloys
Weight (kg)	Up to 40 000
Dimensions (mm)	
FEATURES	
Available custom-built?	Yes
Performance guarantees	Yes
Installation method	
Installed by?	Contractor
Cleaning	Not required
Corrosion resistance	Yes
Performance control and turndown range	No dilution air required for most radiation recuperator applications Turndown range: 1:10
Cross-contamination	None
Withstands differential pressure between ducts	Yes
Condensate collection	N/a
Fans: included/available separately	No
Frost prevention	N/a
EXTRAS	Stacks; dampers
FURTHER COMMENTS	

MANUFACTURER	*ISOTERIX LTD.*	
MODEL NAME AND NUMBER	All units custom-built	
TYPE	Heat pipe	Thermosyphon
APPLICATIONS	Airconditioning	Process-to-process
OPERATION	Counter-flow ⟶	
TEMPERATURE (°C)		
Operating range	-20-60	10-200
AIR		
Volume (m^3/h)	Approx. 2500	Approx. 13 500
Velocity (m/s)	2.5	2.7
PRESSURE DROP (mm w.g.)	14	16
REGENERATOR		
Rotation speed (rev/min)		
Rotation motor size (W)		
DUTY (kW)	8.5	126
EFFICIENCY (%)	56	66
TYPICAL PAYBACK PERIOD (years)		
Airconditioning	1.5	
Process plant		1.5
CONSTRUCTION		
Materials	Copper tubes; aluminium fins; galvanized steel	
Weight (kg)	75	320
Dimensions (mm)	500 x 1000 overall face	1400 x 1800 overall face
FEATURES		
Available custom-built?	Yes ⟶	
Performance guarantees	Yes ⟶	
Installation method	Flange mounted ⟶	
Installed by?	OEM	Consultant subcontractor
Cleaning	Vacuum cleaner	Steam
Corrosion resistance	N/a ⟶	
Performance control and turndown range	Face and bypass ⟶	
Cross-contamination	None ⟶	
Withstands differential pressure between ducts	Yes ⟶	
Condensate collection	N/a ⟶	
Fans: included/available separately	Available separately ⟶	
Frost prevention	N/a ⟶	
EXTRAS	None ⟶	
FURTHER COMMENTS	Other units suitable for use in corrosive environments available, including: hot tin-dipped all copper units; stainless-steel systems; exchangers may also be manufactured from discrete spirally-gilled heat pipes.	

MANUFACTURER	*ISOTHERMICS INC.*
MODEL NAME AND NUMBER	Iso-Vent models: IV-1; IV-2; IV-3; IV-4; IV-5
TYPE	Heat pipe (packaged ventilation system with heat recovery)
APPLICATIONS	Airconditioning (commercial and industrial) process-to-process, process-to-space
OPERATION	Counter-flow
TEMPERATURE (°C)	
Operating range	Up to 340; covers all airconditioning conditions
AIR	
Volume (m^3/h)	1500-12500
Velocity (m/s)	2-4
PRESSURE DROP (mm w.g.)	Fans included in package to meet requirements, nominally 25
REGENERATOR	
Rotation speed (rev/min)	
Rotation motor size (W)	
DUTY (kW)	11.4-88.4 (based on external ambient: -17°C; exhaust 24°C)
EFFICIENCY (%)	59-76
TYPICAL PAYBACK PERIOD (years)	
Airconditioning	1-3
Process plant	1-3
CONSTRUCTION	
Materials	Internal aluminium finned tubes; galvanized steel casing
Weight (kg)	750-1200
Dimensions (mm)	Coil size: 381 x 914-990 x 1829
FEATURES	
Available custom-built?	Normally standard packages, but special tube bundles available
Performance guarantees	Yes
Installation method	Package connected to supply and exhaust ducts
Installed by?	Contractor
Cleaning	Filters included; provision for internal access
Corrosion resistance	Good for normal airconditioning applications; corrosion protection available
Performance control and turndown range	Sensor monitors temperature/adjusts damper to give constant preset supply-air temperature
Cross-contamination	None
Withstands differential pressure between ducts	Yes
Condensate collection	Yes, drain pan provided
Fans: included/available separately	Included
Frost prevention	Yes, automatic bypassing of part of supply air for defrosting
EXTRAS	Hot water and steam heating coils; optional higher fin spacing for dirty air flows; 2-speed motors; external aluminium construction; 440 v motors; electric reheat coils; larger fan motors for higher pressure drops
FURTHER COMMENTS	Can be used to precool incoming air in summer.

MANUFACTURER	*ISOTHERMICS INC.*
MODEL NAME AND NUMBER	Iso-Vent models IV-6; IV-8; IV-10
TYPE	Heat pipe (packaged ventilation system with heat recovery)
APPLICATIONS	Airconditioning (commercial and industrial)
OPERATION	Counter-flow
TEMPERATURE (°C)	
Operating range	Covers all airconditioning conditions
AIR	
Volume (m^3/h)	10 000-25 000
Velocity (m/s)	2-4
PRESSURE DROP (mm w.g.)	Fans included in package to meet requirements, nominally 25
REGENERATOR	
Rotation speed (rev/min)	
Rotation motor size (W)	
DUTY (kW)	76-176.7 (based on external ambient: 17°C; exhaust 24°C)
EFFICIENCY (%)	59-64
TYPICAL PAYBACK PERIOD (years)	
Airconditioning	1-3
Process plant	N/a
CONSTRUCTION	
Materials	Integral aluminium fins on copper or aluminium tubes; galvanized steel casing
Weight (kg)	1744-2061
Dimensions (mm)	Coil size: 762 x 1524-990 x 1829 (2 coils required)
FEATURES	
Available custom-built?	Normally standard packages, but special tube bundles available
Performance guarantees	Yes
Installation method	Package connected to supply and exhaust ducts
Installed by?	Contractor
Cleaning	Filters included; provision for internal access
Corrosion resistance	Good for normal air conditioning applications; corrosion protection available
Performance control and turndown range	Sensor monitors temperature/adjusts damper to give constant preset supply-air temperature
Cross-contamination	None
Withstands differential pressure between ducts	Yes
Condensate collection	Yes, drain pan provided
Fans: included/available separately	Included
Frost prevention	Yes, automatic bypassing of part of supply air for defrosting
EXTRAS	As for IV-1-IV-5 models
FURTHER COMMENTS	As for IV-1-IV-5 models; The Iso-fin 'Thermo-Coils' - used in the above units are available independently for counter-flow heat recovery in space heating and airconditioning applications.

MANUFACTURER	*ISOTHERMICS INC.*
MODEL NAME AND NUMBER	Energy recovery package units, models SDF-500-CR-RTP to SDF-800-CF-RTP(1)
TYPE	Heat pipe (in a packaged heat recovery unit)
APPLICATIONS	Process-to-process
OPERATION	Counter-flow
TEMPERATURE (°C)	
Operating range	Ambient to 260 (SDF 500) ambient to 427 (SDF 800)
AIR	
Volume (m^3/h)	1700-17 000
Velocity (m/s)	2-4
PRESSURE DROP (mm w.g.)	Up to 25 (fans included in package)
REGENERATOR	
Rotation speed (rev/min)	
Rotation motor size (W)	
DUTY (kW)	(2), (3) SDF-500-CF-1 92 kW (58% efficiency, supply-air in -18°C) SDF-500-CF-10 917 kW (on same basis)
EFFICIENCY (%)	42-58 (depending on fin pitch)
TYPICAL PAYBACK PERIOD (Years)	
Airconditioning	N/a
Process plant	1-2
CONSTRUCTION	
Materials	Steel tubes, fins, frames and casing
Weight (kg)	
Dimensions (mm)	1727 x 1118 x 1168 (SDF-500-800-CF-1); 3785 x 3010 x 1473 (SDF-500-700-CF-10)
FEATURES	
Available custom-built?	Normally standard package, but available custom-built
Performance guarantees	Yes
Installation method	Package connected to supply and exhaust ducts and services
Installed by?	Contractor
Cleaning	Spray-wash system incorporated in supply and exhaust ducts
Corrosion resistance	Maintains flue-gas temperature above sulphur dew point
Performance control and turndown range	Face and bypass dampers fitted
Cross-contamination	None
Withstands differential pressure between ducts	Yes
Condensate collection	Normally operates above dew point
Fans: included/available separately	Included
Frost prevention	N/a
EXTRAS	Inlet supply filters, louvres, insulation. Optional fin spacing and efficiency
FURTHER COMMENTS	
NOTES	(1) SDF denotes: 'steel-disc fin'; 500-800 denotes: operating exhaust temperature in °F ;CF denotes: counter-flow; and RTP denotes: 'return-to-process'; a similar package, denoted MA, provides large volumes of make-up warm air using process waste heat.
	(2) Performance of SDF-550; 600; 650; 700; 750; 800 similar.
	(3) The final figure in the model designation (e.g. 1 or 10) indicates SCFM handled by unit: 1 = 1000 SCFM; 10 = 10 000 SCFM.
	The Iso-Fin 'Thermo-Coils' used in the industrial heat recovery packages are available independently for use in industry on process-reheat and process-to-space heating applications.

MANUFACTURER	*ITT REZNOR*
MODEL NAME AND NUMBER	HR: HRA: HRB (3 sizes: 7; 12; 16)
TYPE	Recuperator
APPLICATIONS	Process-to-space
OPERATION	Cross-flow
TEMPERATURE (°C)	
Operating range	37.78-787.77
AIR	
Volume (m^3/h)	117.6-1881.6
Velocity (m/s)	1-3
PRESSURE DROP (mm w.g.)	0.25-38.1
REGENERATOR	
Rotation speed (rev/min)	
Rotation motor size (W)	
DUTY (kW)	Up to 60
EFFICIENCY (%)	30-55
TYPICAL PAYBACK PERIOD (years)	
Airconditioning	
Process plant	3-5
CONSTRUCTION	
Materials	Aluminized: opt. 409 stainless steel; opt. 321 stainless steel
Weight (kg)	59.9-176.6
Dimensions (mm)	Each section: 457.2 x 457.2 x 69.85; HR7 = 7 sections; HR12 = 12 sections; HR16 = 16 sections
FEATURES	
Available custom-built?	N/a
Performance guarantees	1 year warranty
Installation method	Air and exhaust ducts
Installed by?	Contractor
Cleaning	By contractor
Corrosion resistance	Excellent (see Materials)
Performance control and turndown range	N/a
Cross-contamination	None
Withstands differential pressure between ducts	Up to 50 mm w.g.
Condensate collection	By contractor
Fans: included/available separately	On HRA; HRB
Frost prevention	N/a
EXTRAS	Drip tray; safety guards; starters; motors (varying duties and sizes); heavy-duty transformers
FURTHER COMMENTS	

MANUFACTURER	*JAEGGI LTD.*
MODEL NAME AND NUMBER	
TYPE	Recuperator: with tubes (stainless steel); with plates (aluminium)
APPLICATIONS	Process-to-process (tubes); airconditioning (plates)
OPERATION	Counter-, cross-flow
TEMPERATURE (°C)	
Operating range	Plates 100 Tubes 300
AIR	
Volume (m^3/h)	40 000
Velocity (m/s)	
PRESSURE DROP (mm w.g.)	
REGENERATOR	
Rotation speed (rev/min)	
Rotation motor size (W)	
DUTY (kW)	
EFFICIENCY (%)	
TYPICAL PAYBACK PERIOD (years)	
Airconditioning	
Process plant	
CONSTRUCTION	
Materials	Stainless steel; aluminium
Weight	
Dimensions	
FEATURES	
Available custom-built?	Yes
Performance guarantees	
Installation method	
Installed by?	
Cleaning	
Corrosion resistance	
Performance control and turndown range	
Cross-contamination	
Withstands differential pressure between ducts	
Condensate collection	
Fans: included/available separately	
Frost prevention	
EXTRAS	
FURTHER COMMENTS	

MANUFACTURER	*JOHNSON CONSTRUCTION CO. AB*
MODEL NAME AND NUMBER	INKA Radiation Recuperator; INKA Metal Recuperator
TYPE	Metal tube recuperator
APPLICATIONS	90% as recuperator for glass-melting furnaces; 10% for other purposes such as hot working furnaces in the steel industry, direct-fired air and gas preheater for different processes, direct-fired pressed air preheater
OPERATION	Both counter-, parallel-flow
TEMPERATURE (°C)	Flue gas: / Air (gas):
Operating range	1700 / 0 (in); 600 / 800 (out)
AIR	
Volume (m^3/h)	100-17000 Nm^3/h; at bigger quantities 2 parallel working recuperators must be used
Velocity (m/s)	Velocity is depending on requirement for permitted pressure drop and maximum tube temperature
PRESSURE DROP (mm w.g.)	50-250; normally 130
REGENERATOR	
Rotation speed (rev/min)	
Rotation motor size (W)	
DUTY (kW)	The only additional duty is needed for overcoming a pressure drop of 50-250 mm w.g. at the actual air quantity
EFFICIENCY (%)	Radiation recuperators have limited efficiency as they cannot cool flue gases lower than 600°C (approx. 50%). High efficiency can be achieved by combination with a convection recuperator or a waste gas boiler
TYPICAL PAYBACK PERIOD (years)	
Airconditioning	At glass furnaces, which earlier has worked without recuperator (cold-air operation): 8-15 months, never more than 2 years
Process plant	
CONSTRUCTION	
Materials	
Weight	Approx. 45 tonne
Dimensions	3.6 m diam. 11.0 m high: valid for recuperator with 750°C air preheat at a 100 tonne/day glass furnace with oil consumption 16 tonne/day
FEATURES	
Available custom-built?	Yes
Performance guarantees	1 year, regarding materials and manufacturing
Installation method	
Installed by?	If desired, installed by us, otherwise by the customer on basis of a detailed erection inscription
Cleaning	A program for melting down the dust, sometimes by press airblowing
Corrosion resistance	High corrosion resistance
Performance control and turndown range	Manufacturing control and pressure test made by Swedish authorities, customer may be present
Cross-contamination	None
Withstands differential pressure between ducts	Yes
Condensate collection	After cleaning by melting, liquid slag can be easily taken out at the recuperator bottom; otherwise no problems with other kinds liquid condensate
Fans: included/available separately	Fans can be included in delivery; otherwise figures (data) given for required fan
Frost prevention	No problems with frost
EXTRAS	
FURTHER COMMENTS	

MANUFACTURER	*KLEINEWEFERS GMBH.*
MODEL NAME AND NUMBER	Cast-iron and Steel Recuperators
TYPE	Radiation and convection recuperators
APPLICATIONS	Process-to-process (furnaces, gas coolers)
OPERATION	Counter-, parallel-flow
TEMPERATURE (°C)	
Operating range	Up to 1500
AIR	
Volume (m^3/h)	Up to 250 000 Nm^3/h
Velocity (m/s)	
PRESSURE DROP (mm w.g.)	5-100 (on exhaust); 150-500 (on inlet)
REGENERATOR	
Rotation speed (rev/min)	
Rotation motor size (W)	
DUTY (kW)	Up to 30 000
EFFICIENCY (%)	80 approx.
TYPICAL PAYBACK PERIOD (years)	
Airconditioning	
Process plant	1-2
CONSTRUCTION	
Materials	Cast-iron or steel plain (or finned) tubes; steel plates
Weight (kg)	Dependent on duty
Dimensions (mm)	Dependent on duty
FEATURES	
Available custom-built?	Yes
Performance guarantees	1 year
Installation method	Assembled onsite
Installed by?	Manufacture or customer
Cleaning	Shotcleaning or steamblower
Corrosion resistance	Uses alloyed steel: good
Performance control and turndown range	Control by diluting air control system
Cross-contamination	None
Withstands differential pressure between ducts	Yes
Condensate collection	If needed
Fans: included/available separately	Available
Frost prevention	N/a
EXTRAS	
FURTHER COMMENTS	

MANUFACTURER	*LAMINAIRE PRODUCTS*
MODEL NAME AND NUMBER	Laminaire types S2, S1H; S1V
TYPE	Plate
APPLICATIONS	Process-to-process; process-to-space; airconditioning
OPERATION	Counter-flow
TEMPERATURE (°C)	
Operating range	100 max. (standard) 150°C max. (special high temperature)
AIR	
Volume (m^3/h)	Non-standard, all units custom-built
Velocity (m/s)	Max. 14.25; min. 4.8
PRESSURE DROP (mm w.g.)	Max. 55; min. 6
REGENERATOR	
Rotation speed (Rev/min)	
Rotation motor size (W)	
DUTY (kW)	Non-standard
EFFICIENCY (%)	Min. 50; Max. 75 (sensible heat only), but up to 90 with latent recovery
TYPICAL PAYBACK PERIOD (years)	
Airconditioning	2-5 years (not UK); can often be paid for by reduction in initial capital cost of cooling plant
Process plant	Process-to-process: often under 1; Process-to-space: 1-3 years; space-to-space 2-5 years
CONSTRUCTION	
Materials	All aluminium with resin bonding
Weight (kg)	Non-standard
Dimensions (mm)	Type S1:915 x 915 x variable; types S2H, S2V: 915 x 458 x variable
FEATURES	
Available custom-built?	All units custom-built from standard plate modules to custom's specific requirements
Performance guarantees	Yes (sensible heat)
Installation method	Direct duct connections
Installed by?	Customer's own ductwork installation contractor; or Associated Installation Co. if required
Cleaning	By compressed air blowthrough
Corrosion resistance	Good for most lightly contaminated atmospheres, including: Swimming-pools, not for heavy acid or alkali contamination
Performance control and turndown range	No mechanical controls provided, turndown can be achieved via dampered bypass arrangements incorporated into ductwork by installer
Cross-contamination	No cross-contamination
Withstands differential pressure between ducts	80 mm w.g.
Condensate collection	Incorporated by installer into duct connections; exhaust air flow must not be upwards through unit
Fans: included/available separately	Not included, to be selected and supplied by installer
Frost prevention	Not included, but if required, can be incorporated by installer using bypass arrangement and thermostat control
EXTRAS	Special high-temperature bonding for operation up to 150°C; flange connections fitted to standard M/F duct spigots
FURTHER COMMENTS	Airflow through unit is laminar flow non-turbulent, minimizing the precipitation of dust within the unit; see custom-built example
CUSTOM-BUILT UNITS	
Size and performance details on typical examples	<u>Unit</u>: S2-120 <u>Size</u>: 915 x 915 x 508 (nominal) <u>Duty</u> 1: Air flow, 1.0 m^3/s; inlet and exhaust; pressure drop, 10mm w.g.; recovery, 70% sensible heat <u>Duty</u> 2: Air flow, 1.575 m^3/s; pressure drop, 25mm w.g.; recovery, 60% sensible heat <u>Unit</u>: S1H-S1V-152 <u>Size</u>: 915 x 458 x 645 <u>Duty</u> 1: Air flow, 0.633 m^3/s; inlet and exhaust; pressure drop, 10mm w.g.; recovery, 70% sensible heat <u>Duty</u> 2: Air flow, 1.0 m^3/s; pressure drop, 25mm w.g.; recovery 60% sensible heat
NOTES	Laminaire unit type is indicated by the prefix S2, S1H or S1V; the unit size is then indicated following number indicating the number of plates in unit; thus, a unit designated S2-100 has 100 heat exchanger plates of the S2 type.

MANUFACTURER	*METALLURGICAL ENGINEERS LTD.*
MODEL NAME AND NUMBER	
TYPE	Recuperator
APPLICATIONS	Process-to-process; process-to-space
OPERATION	Counter-, parallel-flow, and combination
TEMPERATURE (°C)	
Operating range	
Inlet	Up to 1450
Exhaust	Down to dew point
AIR	
Volume (m^3/h)	500 Nm^3 upwards
Velocity (m/s)	Variable
PRESSURE DROP (mm w.g.)	Variable
REGENERATOR	
Rotation speed (rev/min)	
Rotation motor size (W)	
DUTY (kW)	Variable
EFFICIENCY (%)	50-85
TYPICAL PAYBACK PERIOD (years)	
Airconditioning	
Process plant	6 months-2 years
CONSTRUCTION	
Materials	Metallic
Weight (kg)	
Dimensions (mm)	
FEATURES	
Available custom-built?	Always custom-built
Performance guarantees	Yes
Installation method	Usually complete shop-assembly for installation in flue system
Installed by?	Main contractor; user; supplier
Cleaning	Provision for cleaning can be built in
Corrosion resistance	Designs for
Performance control and turndown range	As required by process
Cross-contamination	Designs for dusty gases are special feature of range of designs available
Withstands differential pressure between ducts	Yes
Condensate collection	If required
Fans: included/available separately	Available, if required
Frost prevention	Not required
EXTRAS	
FURTHER COMMENTS	

MANUFACTURER	*MUNTERS ECONOVENT AB*		
MODEL NAME AND NUMBER	RT 950-RT 5000 (1) normal aluminium foil	ET 950-ET 5000 (1) aluminium foil treated for hygroscopicity	EX 10 aluminium foil
TYPE	Regenerator	Regenerator	Recuperator
APPLICATIONS	Process-to-process process-to-space airconditioning		
OPERATION	Counter-flow	Counter-flow	Cross-flow
TEMPERATURE (°C)			
Operating range	-40-130 (S-version) (2) -40-80 (W-version)	-40-130 (S-version) (2) -40-80 (W-version)	-40-80
AIR			
Volume (m^3/h)	150-6500 →	→	1800-5400/module (5)
Velocity (m/s)	1-5 →	→	1-4
PRESSURE DROP (mm w.g.)	28-80 →	→	10-50
REGENERATOR			
Rotation speed (rev/min)	20	20	N/a
Rotation motor size (W)	150-750	150-750	N/a
DUTY (kW)	Dependent on specification →	→	→
EFFICIENCY (%)	70-85 →	→	50-75
TYPICAL PAYBACK PERIOD (years)			
Airconditioning	0-5 (3)	0-5 (3)	0-5 (3)
Process plant	0-2	0-2	0-2
CONSTRUCTION			
Materials	Normal aluminium foil	Aluminium foil treated to give hygroscopicity	Aluminium foil
Weight (kg)	160-2100 →	→	125/module
Dimensions (mm)	950-5000 dia.	950-5000 dia.	1130 x 1130 x 1100
FEATURES			
Available custom-built?	No, (standard units) →	→	→
Performance guarantees	Yes, (2 years) →	→	→
Installation method	Normal procedure - fit to existing ducts as package →	→	→
Installed by?	Contractor →	→	→
Cleaning	Each year according to instructions. (vacuum cleaning) →	→	→
Corrosion resistance	High →	→	→
Performance control and turndown range	Thyrovent control (4) →	→	Bypass
Cross-contamination	0.05%	0.05%	None
Withstands differential pressure between ducts	1000 Pa →	→	
Condensate collection	Not necessary →	→	Drain pan in unit
Fans: included/available separately	Not included →	→	→
Frost prevention	Available →	→	→
EXTRAS			
FURTHER COMMENTS			

NOTES:

(1) The actual numbers refer to rotor diameters in millimetres

(2) The W-version has a segmented rotor

(3) Payback periods of zero are mentioned to show cases where the investment in Econovent could be substantially less expensive than the extension of a heating plant already existing, in addition to which one adds the yearly energy saving.

(4) The temperature-control equipment of Econovent has to be rather sophisticated, since it is important to provide accurate control year-assured with maximal utilization of the Econovent and in full coordination with the supplementary conditioning equipment and its control. Therefore, refer to the Thyrovent manual and the chapter on temperature control in the manual, available from the manufacturer.

(5) Several modules may be mounted in series.

MANUFACTURER	*NOREN PRODUCTS, THERMAL DIVISION*
MODEL NAME AND NUMBER	Noren Heat Pipe Heat Exchangers
TYPE	Heat pipe Quik-watt type
APPLICATIONS	Process-to-process; process-to-space; airconditioning
OPERATION	Mostly counter-, but some parallel-flow
TEMPERATURE (°C)	
Operating range	Models from 100-530
AIR	
Volume (m^3/h)	Varies dependent on size
Velocity (m/s)	
PRESSURE DROP (mm w.g.)	10 stand fin spacing and row spring to control to meet desired pressure drop
REGENERATOR	
Rotation speed (rev/min)	
Rotation motor size (W)	
DUTY (kW)	Smallest unit, 0.137; Largest unit to date, 23,529 kW
EFFICIENCY (%)	47%-83% depending on application
TYPICAL PAYBACK PERIOD (years)	
Airconditioning	8 months-4.7 years; ave. (60% of jobs) 1.7
Process plant	
CONSTRUCTION	Copper; aluminium; steel; stainless steel; cupon
Materials	Copper; aluminium; steel; stainless steel; cupon
Weight	
Dimensions	36 standard sizes to choose from
FEATURES	
Available custom-built?	Yes
Performance guarantees	Yes
Installation method	Fitted as unit to ductwork
Installed by?	Customer or contractor
Cleaning	Easy access available; fin spacing can be custom-designed to needs
Corrosion resistance	Excellent material and custom-designed to reduce corrosion
Performance control and turndown range	3 methods available
Cross-contamination	None
Withstands differential pressure between ducts	100% leak proof design; 12 bar pressure differential available
Condensate collection	Yes
Fans: included/available separately	Fans not available
Frost prevention	Units with automatic frost prevention available
EXTRAS	Specific coating for corrosion resistance available to reduce cost over more expensive materials
FURTHER COMMENTS	

MANUFACTURER	*NOVENCO LTD.*
MODEL NAME AND NUMBER	LFA; LFB
TYPE	Runaround; water with 20% glycol
APPLICATIONS	Process-to-space; airconditioning
OPERATION	Counter-, parallel-flow
TEMPERATURE (°C)	
Operating range	-10-90
AIR	
Volume (m^3/h)	Up to 25 000
Velocity (m/s)	Up to 3.0
PRESSURE DROP (mm w.g.)	Up to 30
REGENERATOR	
Rotation speed (rev/min)	
Rotation motor size (W)	
DUTY (kW)	Up to 100
EFFICIENCY (%)	Up to 60
TYPICAL PAYBACK PERIOD (years)	
Airconditioning	2-2.5
Process plant	Variable
CONSTRUCTION	
Materials	Copper tubes; aluminium or steel fins
Weight (kg)	Variable
Dimensions (mm)	Up to 2000 mm wide x 1000 mm high
FEATURES	
Available custom-built?	Yes
Performance guarantees	Yes
Installation method	Runaround system with pumped circulation
Installad by?	Normal HHV, airconditioning contractor
Cleaning	Vacuum or compressed air
Corrosion resistance	High
Performance control and turndown range	Fully modulating
Cross-contamination	None
Withstands differential pressure between ducts	Yes
Condensate collection	Yes
Fans: included/available separately	Separate
Frost prevention	Thermostatic
EXTRAS	
FURTHER COMMENTS	

MANUFACTURER	*ENGINEERING CONTROLS DIVISION POTT INDUSTRIES*
MODEL NAME AND NUMBER	ECX, EPV series
TYPE	ECX: tube-type high fin EPV: tube-type, plate fin
APPLICATIONS	Dust-laden or fouled flue-gas streams to preheat combustion-air exhaust; or flue-gas streams to heat circulating air
OPERATION	Counter-flow
TEMPERATURE (^{o}C)	
Operating range	157-466
AIR	
Volume (m^3/h)	40 639-812 800
Velocity (m/s)	
PRESSURE DROP (mm w.g.)	94-285
REGENERATOR	
Rotation speed (rev/min)	
Rotation motor size (W)	
DUTY (kW)	Dependent on specification
EFFICIENCY (%)	
TYPICAL PAYBACK PERIOD (years)	
Airconditioning	
Process plant	
CONSTRUCTION	
Materials	Carbon; stainless steel
Weight (kg)	907-27 215
Dimensions (mm)	2438-12198 high; 4876-7620 long
FEATURES	
Available custom-built?	All custom-built
Performance guarantees	Yes
Installation method	
Installed by?	Contractor
Cleaning	Flue brush or sootblowers
Corrosion resistance	Standard corrosion allowances
Performance control and turndown range	Flue-gas stream bypass valve
Cross-contamination	
Withstands differential pressure between ducts	55.9 mm w.g.
Condensate collection	No
Fans: included/available separately	No
Frost prevention	No
EXTRAS	
FURTHER COMMENTS	

MANUFACTURER	*Q-DOT INTERNATIONAL CORP.*
MODEL NAME AND NUMBER	Q-Pipe Thermal Recovery Unit
TYPE	Heat pipe
APPLICATIONS	Process-to-process; process-to-comfort; comfort-to-comfort
OPERATION	Designed for counter-flow; will operate less efficiently in parallel-flow
TEMPERATURE (^{o}C)	
Operating range	-51-440
AIR	
Volume (m^3/h)	135 915 at velocity of 1.52 m/s
Velocity (m/s)	1.52
PRESSURE DROP (mm w.g.)	25.4 or less
REGENERATOR	
Rotation speed (rev/min)	
Rotation motor size (W)	
DUTY (kW)	Dependent on specification; up to MW range
EFFICIENCY (%)	60-80
TYPICAL PAYBACK PERIOD (years)	
Airconditioning	2
Process plant	1
CONSTRUCTION	
Materials	Aluminium; copper; steel
Weight (kg)	Variable between 220-11022
Dimensions (mm)	330 x 609-1371 x 4877
FEATURES	
Available custom-built?	Yes
Performance guarantees	Yes
Installation method	Vertical or horizontal air flow
Installed by?	Contractor
Cleaning	As required
Corrosion resistance	Copper and Q-Site coating available
Performance control and turndown range	Full range
Cross-contamination	None
Withstands differential pressure between ducts	680 mm w.g.
Condensate collection	Drain/pan (contractor furnished)
Fans: included/available separately	None available from factory
Frost prevention	Tilt control
EXTRAS	Flow-through water-wash system available as an option
FURTHER COMMENTS	

MANUFACTURER	*RECUPERATOR SPA..*									
MODEL NAME AND NUMBER HRX; HDRX; HVRX	021	027	033	042	047	060	078	095	116	146
TYPE	Plate									
APPLICATIONS	Process-to-process; process-to-space; airconditioning									
OPERATION	Counter-flow									
TEMPERATURE (°C)										
Operating range	HRX: -30-60 HDRX: -30-150 HVRX: -30-200									
AIR										
Volume (m^3/h)	2100;	2700;	3300;	4200;	4700;	6000;	7800;	9500;	11600;	14600
Velocity (m/s)	4 →									
PRESSURE DROP (mm w.g.)	25 →									
REGENERATOR										
Rotation speed (rev/min)										
Rotation motor size (W)										
DUTY (kW)	Dependent on specification									
EFFICIENCY (%)	53 →									
TYPICAL PAYBACK PERIOD (years)										
Airconditioning	2 →									
Process plant	0.5 →									
CONSTRUCTION										
Materials	Heat exchanger: aluminium; casing: galvanized steel									
Weight (kg)	75	85	95	115	125	145	175	200	240	285
Dimensions (mm)	450 x 946 x 1020 →	550 x 946	650 x 946	800 x 946	650 x 1306	800 x 1306	1000 x 1306	1200 x 1306	1450 x 1306	1800 x 1306 x 1020
FEATURES										
Available custom-built?	No									
Performance guarantees	Yes									
Installation method										
Installed by?	Contractor									
Cleaning	Possible									
Corrosion resistance	Epoxy or PVF paint protection (optional at extra cost)									
Performance control and turndown range	Bypass									
Cross-contamination	None									
Withstands differential pressure between ducts	70 mm w.g.									
Condensate collection	Yes									
Fans: included/available separately	No									
Frost prevention	Bypass									
EXTRAS										
FURTHER COMMENTS										

MANUFACTURER	*RECUPERATOR SPA.*
MODEL NAME AND NUMBER	HR Recuperators (models HR021-2100 m^3/h HRB684-68400 m^3/h)
TYPE	Recuperator: plate-type; air-to-air heat exchanger
APPLICATIONS	Ventilation; aircondition; all kinds of industrial air-drying processes
OPERATION	Counter-flow
TEMPERATURE (°C)	
Operating range	-30-200
AIR	
Volume (m^3/h)	Nominal: 2100-68400
Velocity (m/s)	4
PRESSURE DROP (mm w.g.)	35
REGENERATOR	
Rotation speed (rev/min)	
Rotation motor size (W)	
DUTY (kW)	Dependent on air volume; temperatures
EFFICIENCY (%)	90
TYPICAL PAYBACK PERIOD (years)	
Airconditioning	1-3
Process plant	
CONSTRUCTION	
Materials	High-purity aluminium heat exchanger plates; galvanized steel
Weight (kg)	HR021: 100; HRB684: 2650
Dimensions (mm)	HR021: 450 x 946 x 1020; HRB684: 3400 x 2986 x 267C
FEATURES	
Available custom-built?	Standard range of models
Performance guarantees	Guarantee given that in case the efficiency we state in literature should not be true we will reinburse not only the price paid per unit but also the installation expenses
Installation method	Horizontal; vertical installation within ductwork layout
Installed by?	Supplied by Climate Equipment Ltd.; installed by customer or their contractor
Cleaning	Removable casing panels
Corrosion resistance	High-purity aluminium heat exchanger plates; also special corrosive paint treatments
Performance control and turndown range	Optional face/bypass damper arrangement
Cross-contamination	None
Withstands differential pressure between ducts	70 mm w.g.
Condensate collection	2 off 1 in. BSP drain connection fitted to base panel of casing
Fans: included/available separately	Fans can be provided
Frost prevention	Yes, face damper arrangement
EXTRAS	Heat exchanger plate spacing: 4 mm x 8 mm; special corrosive protection treatments; special stainless-steel housings
FURTHER COMMENTS	For more detailed information, see model catalogues.

MANUFACTURER	*RECUPERATOR SPA.*									
MODEL NAME AND NUMBER HRH; HDRH; HVRH	021	027	033	042	047	060	078	095	116	146
TYPE	Plate									
APPLICATIONS	Process-to-process; process-to-space; airconditioning									
OPERATION	Counter-flow									
TEMPERATURE (°C)										
Operating range	HRH: -30-60; HDRH: -30-150; HVRH: -30-200									
AIR										
Volume (m^3/h)	2100;	2700;	3300;	4200;	4700;	6000;	7800;	9500;	11600;	14600
Velocity (m/s)	4 →									
PRESSURE DROP (mm w.g.)	34 →									
REGENERATOR										
Rotation speed (rev/min)										
Rotation motor size (W)										
DUTY (kW)	Dependent on specification									
EFFICIENCY (%)	72 →									
TYPICAL PAYBACK PERIOD (years)										
Airconditioning	2 →									
Process plant	0.5 →									
CONSTRUCTION										
Materials	Heat Exchanger: aluminium; casing: galvanized steel									
Weight (kg)	100	120	135	165	180	215	265	315	380	460
Dimensions (mm)	450 x 946 x 1020 →	550 x 946	650 x 946	800 x 946	650 x 1306	800 x 1306	1000 x 1306	1200 x 1306	1450 x 1306	1800 x 1306 x 1020
FEATURES										
Available custom-built?	No									
Performance guarantees	Yes									
Installation method										
Installed by?	Contractor									
Cleaning	Possible									
Corrosion resistance	Epoxy or PVF paint protection (optional at extra cost)									
Performance control and turndown range	Bypass									
Cross-contamination	None									
Withstands differential pressure between ducts	70 mm w.g.									
Condensate collection	Yes									
Fans: included/available separately	No									
Frost prevention	Bypass									
EXTRAS										
FURTHER COMMENTS										

MANUFACTURER	*RECUPERATOR SPA.*						
MODEL NAME AND NUMBER HRC; HDRC; HVRC	176	235	277	346	456	570	684
TYPE	Plate						
APPLICATIONS	Process-to-process; process-to-space; airconditioning						
OPERATION	Counter-flow						
TEMPERATURE (°C)							
Operating range	HRC: -30-60 HDRC: -30-150 HVRC: -30-200						
AIR							
Volume (m^3/h)	17 600	23 500	27 700	34 600	45 600	57 000	68 400
Velocity (m/s)	4 →						
PRESSURE DROP (mm w.g.)	25 →						
REGENERATOR							
Rotation speed (rev/min)							
Rotation motor size (W)							
DUTY (kW)	Dependent on specification						
EFFICIENCY (%)	53 →						
TYPICAL PAYBACK PERIOD (years)							
Airconditioning	2 →						
Process plant	0.5 →						
CONSTRUCTION							
Materials	Heat exchanger: aluminium; casing: galvanized steel						
Weight (kg)	475	600	730	860	1270	1520	1770
Dimensions (mm)	1800 x 1546 x 1960	1800 x 2026 x 1960	2100 x 2026 x 2130	2100 x 2506 x 2130	3400 x 2026 x 2850	3400 x 2506 x 2850	3400 x 2986 x 2850
FEATURES							
Available custom-built?	No						
Performance guarantees	Yes						
Installation method							
Installed by?	Contractor						
Cleaning	Possible						
Corrosion resistance	Epoxy or PVF paint protection (optional at extra cost)						
Performance control and turndown range	Bypass						
Cross-contamination	None						
Withstands differential pressure between ducts	70 mm w.g.						
Condensate collection	Yes						
Fans: included/available separately	No						
Frost prevention	Bypass						
EXTRAS							
FURTHER COMMENTS							

MANUFACTURER	*RECUPERATOR SPA.*						
MODEL NAME AND NUMBER HRB; HDRB; HVRB	176	235	277	346	456	570	684
TYPE	Plate						
APPLICATIONS	Process-to-process; process-to-space; airconditioning						
OPERATION	Counter-flow						
TEMPERATURE (°C)							
Operating range	HRB: -30-60; HDRB: -30-150; HVRB: -30-200						
AIR							
Volume (m^3/h)	17 600;	23 500;	27 700;	34 600;	45 600;	57 000;	68 400
Velocity (m/s)	4 →						
PRESSURE DROP (mm w.g.)	34 →						
REGENERATOR							
Rotation speed (rev/min)							
Rotation motor size (W)							
DUTY (kW)	Dependent on specification						
EFFICIENCY (%)	72 →						
TYPICAL PAYBACK PERIOD (years)							
Airconditioning	2 →						
Process plant	0.5 →						
CONSTRUCTION							
Materials	Heat exchanger: aluminium; casing: galvanized steel						
Weight (kg)	700;	900;	1080;	1300;	1850;	2250;	2650
Dimensions (mm)	1800 x 1546 x 1960;	1800 x 2026 x 1960;	2100 x 2026 x 2130;	2100 x 2506 x 2130;	3400 x 2026 x 2850;	3400 x 2506 x 2850;	3400 x 2986 x 2850
FEATURES							
Available custom-built?	No						
Performance guarantees	Yes						
Installation method							
Installed by?	Contractor						
Cleaning	Possible						
Corrosion resistance	Epoxy or PVF paint protection (optional at extra cost)						
Performance control and turndown range	Bypass						
Cross-contamination	None						
Withstands differential pressure between ducts	70 mm w.g.						
Condensate collection	Yes						
Fans: included/available separately	No						
Frost prevention	Bypass						
EXTRAS							
FURTHER COMMENTS							

MANUFACTURER	*SMITH ENGINEERING CO.*
MODEL NAME AND NUMBER	Custom-built
TYPE	Recuperator
APPLICATIONS	Air-to-air; processes
OPERATION	Counter-flow
TEMPERATURE (°C)	
Operating range	Up to 760
AIR	
Volume (m^3/h)	56 000
Velocity (m/s)	N/a
PRESSURE DROP (mm w.g.)	Approx. 175
REGENERATOR	
Rotation speed (rev/min)	
Rotation motor size (W)	
DUTY (kW)	112
EFFICIENCY (%)	78
TYPICAL PAYBACK PERIOD (years)	
Airconditioning	
Process plant	
CONSTRUCTION	
Materials	304 stainless steel
Weight (kg)	
Dimensions (mm)	
FEATURES	
Available custom-built?	Yes
Performance guarantees	$\pm$ 3%
Installation method	
Installed by?	
Cleaning	Yes
Corrosion resistance	
Performance control and turndown range	
Cross-contamination	
Withstands differential pressure between ducts	Yes
Condensate collection	
Fans: included/available separately	Available separately
Frost prevention	
EXTRAS	
FURTHER COMMENTS	

MANUFACTURER	*STANDARD & POCHIN LTD.*
MODEL NAME AND NUMBER	Heat Recovery Coils
TYPE	Runaround
APPLICATIONS	Airconditioning; warm-air heating; low-temperature process
OPERATION	Counter-flow
TEMPERATURE (°C)	
Operating range	Sub-zero-180
AIR	
Volume (m^3/h)	850-32 000 for minimum and maximum blocks; multiples available
Velocity (m/s)	Typically 2.5
PRESSURE DROP (mm w.g.)	Typically 15-25
REGENERATOR	
Rotation speed (rev/min)	
Rotation motor size (W)	
DUTY (kW)	3.6-4 kW per 1000 m^3/h
EFFICIENCY (%)	50-60
TYPICAL PAYBACK PERIOD (years)	
Airconditioning Process plant	As building usage; fuel cost; building and plant-life, etc. are all variables in determining this period, there is no clear answer. however a high fuel cost, long life, full usage building could have payback period of 1 year or less, or up to 3 years for same with low fuel cost. either of these are likely to be viable.
CONSTRUCTION	
Materials	Copper tubes and headers; aluminium or copper fins; mild steel casing
Weight (kg)	Depends on size and liquid content
Dimensions (mm)	Smallest, 300 x 300; largest, 935 x 3660 duct size; multiples available
FEATURES	
Available custom-built?	Yes, within parameters of tube and fin parameters
Performance guarantees	Coil performance based on well-established data
Installation method	Duct-mounted coils; flanged as necessary with standard pipe connections
Installed by?	Contractors usually or as part of OEM plant
Cleaning	Closed circuit on liquid-side; brush; wash or vacuum/air jet on air-side
Corrosion resistance	On AC plant, above materials are conventional and acceptable
Performance control and turndown range	Self-balancing
Cross-contamination	Coils are/can be completely remote from each other
Withstands differential pressure between ducts	Not applicable
Condensate collection	Conventional drain features can be incorporated
Fans: included/available separately	Available separately; or as part of package plant including normal features
Frost prevention	Ethylene glycol; electric element; or conventional system-linked frost coils
EXTRAS	Such coils can be supplied as part of an air-handling unit package, incorporating the usual cooling; heating filtration; fan; etc. or as duct-mounted coils with other services separate
FURTHER COMMENTS	General description of Standard & Pochin coils in Publication no. 305; runaround coils described in Data Sheet DS306.

MANUFACTURER	*STEIN ATKINSON STORDY LTD.*		
MODEL NAME AND NUMBER	Escher Austeel Recuperators		
TYPE	Radiation and convection type stack recuperators		
APPLICATIONS	Process-to-process		
OPERATION	Counter-, parallel-flow		
TEMPERATURE (°C)	Waste gas	Air reheat	
Operating range	To 1370	0	(inlet)
	To 780	700	(exhaust)
AIR			
Volume (m^3/h)	To suit customer's requirements; largest to date = 77000 Nm^3/h		
Velocity (m/s)			
PRESSURE DROP (mm w.g.)	Normally 380 (max.)		
REGENERATOR			
Rotation speed (rev/min)			
Rotation motor size (W)			
DUTY (kW)			
EFFICIENCY (%)	Approx. 70		
TYPICAL PAYBACK PERIOD (years)			
Airconditioning			
Process plant	18 months		
CONSTRUCTION			
Materials	Steel		
Weight (kg)			
Dimensions (mm)			
FEATURES			
Available custom-built?	Yes		
Performance guarantees	Yes		
Installation method	Package or site assembled		
Installed by?	Contractor or manufacturer		
Cleaning			
Corrosion resistance	Good - designed to suit environment of exhaust		
Performance control and turndown range			
Cross-contamination	None		
Withstands differential pressure between ducts	Yes		
Condensate collection	Generally not required		
Fans: included/available separately	Separate		
Frost prevention	N/a		
EXTRAS			
FURTHER COMMENTS			

MANUFACTURER	*STRUTHERS WELLS CORP.*
MODEL NAME AND NUMBER	Equipment Custom-Designed; standard sizes and model numbers not available
TYPE	Fire tube and water tube designs are provided; the latter are provided with both bare and extended-surface tubes; in addition, standard shell and tube heat exchanger designs available
APPLICATIONS	Process-to-process; process-to-flue gas; process-to-combustion air; flue gas-to-combustion air
OPERATION	Both counter-, parallel-flow designs available
TEMPERATURE (°C)	
Operating range	200-1200
Exhaust	200-1200
Inlet	100-200
AIR	
Volume (m^3/h)	18 000-600 000
Velocity (m/s)	15-30
PRESSURE DROP (mm w.g.)	100-800
REGENERATOR	
Rotation speed (rev/min)	
Rotation motor size (W)	
DUTY (kW)	60 000-1 200 00
EFFICIENCY (%)	85-92
TYPICAL PAYBACK PERIOD (years)	
Airconditioning	
Process plant	0.5-3.5
CONSTRUCTION	
Materials	Full range
Weight	50 000-2 500 000
Dimensions	Shop-assembled to large units requiring field-assembly
FEATURES	
Available custom-built?	Yes
Performance guarantees	Yes
Installation method	
Installed by?	Manufacturer or customer
Cleaning	Normally through use of sootblowers
Corrosion resistance	Materials can be selected for any corrosive conditions
Performance control and turndown range	Variable, depending on application
Cross-contamination	N/a
Withstands differential pressure between ducts	Yes, on certain designs differential pressure may be several pounds
Condensate collection	Yes
Fans: included/available separately	Normally provided with heat recovery equipment
Frost prevention	Yes
EXTRAS	Full range of controls; bypass arrangements; dump stacks; auxiliary firing on gas turbine applications
FURTHER COMMENTS	Struthers Wells provides a wide range of industrial heat recovery equipment.

MANUFACTURER	*AB SVENSKA FLAKTFABRIKEN*
MODEL NAME AND NUMBER	Duoterm
TYPE	Heat pipe
APPLICATIONS	Process-to-process; airconditioning
OPERATION	Counter-flow
TEMPERATURE (°C)	
Operating range	Up to 50
AIR	
Volume (m^3/hr)	2 800-25 000
Velocity (m/s)	1.5-3
PRESSURE DROP (mm w.g.)	2-20
REGENERATOR	
Rotation speed (rev/min)	
Rotation motor size (W)	
DUTY (kW)	Dependent on specification
EFFICIENCY (%)	40-60
TYPICAL PAYBACK PERIOD (years)	
Airconditioning	3-5
Process plant	1-5
CONSTRUCTION	
Materials	Copper/aluminium; copper/copper
Weight (kg)	80-945
Dimensions (mm)	800 x 750-2000 x 1050
FEATURES	
Available custom-built?	Yes
Performance guarantees	Yes
Installation method	
Installed by?	Manufacturer
Cleaning	Yes, possible if necessary by compressed air
Corrosion resistance	Good for most applications
Performance control and turndown range	
Cross-contamination	None
Withstands differential pressure between ducts	Yes (determined by duct itself)
Condensate collection	
Fans: included/available separately	Available separately
Frost prevention	Bypass
EXTRAS	
FURTHER COMMENTS	

MANUFACTURER	*AB SVENSKA FLAKTFABRIKEN*
MODEL NAME AND NUMBER	Ecoterm
TYPE	Runaround
APPLICATIONS	Airconditioning; process-to-process
OPERATION	Counter-flow
TEMPERATURE (°C)	
Operating range	Up to 300
AIR	
Volume (m^3/hr)	1000-58 000
Velocity (m/s)	2-3
PRESSURE DROP (mm w.g.)	6-13
REGENERATOR	
Rotation speed (rev/min)	
Rotation motor size (W)	
DUTY (kW)	
EFFICIENCY (%)	Temperature efficiency: 45-70
TYPICAL PAYBACK PERIOD (years)	
Airconditioning	2-5
Process plant	1-3
CONSTRUCTION	
Materials	Copper/aluminium; copper/copper;
Weight (kg)	43-310
Dimensions (mm)	550 x 400-3200 x 1800
FEATURES	
Available custom-built?	Yes
Performance guarantees	Yes
Installation method	
Installed by?	Manufacturer
Cleaning	Compressed air
Corrosion resistance	Good, coil materials chosen according to the air quality
Performance control and turndown range	Full control (0 Max.)
Cross-contamination	Zero
Withstands differential pressure between ducts	Yes (determined by duct itself)
Condensate collection	Yes
Fans: included/available separately	Available separately
Frost prevention	Yes (3-way valve)
EXTRAS	
FURTHER COMMENTS	

MANUFACTURER	*AB SVENSKA FLAKTFABRIKEN*
MODEL NAME AND NUMBER	Heat Recovery Unit type RDAA-a-b-c
TYPE	Complete unit with two filters; exhaust fan; supply-air fan; plate heat exchanger; automatic defrosting device
APPLICATIONS	Ventilation in combination with heat recovery in single houses; or part of bigger buildings
OPERATION	Cross-flow
TEMPERATURE (°C)	
Operating range	Normal outdoor and indoor temperature
AIR	
Volume (m^3/h)	120-360
Velocity (m/s)	
PRESSURE DROP (mm w.g.)	
REGENERATOR	
Rotation speed (rev/min)	
Rotation motor size (W)	
DUTY (kW)	Rated power: 0.15 or 0.25; consumption about 60% of rated power
EFFICIENCY (%)	55-80, dependent on air volume
TYPICAL PAYBACK PERIOD (years)	
Airconditioning	5-7 (dependent on energy prices and many other factors)
Process plant	
CONSTRUCTION	
Materials	Casing: galvanized sheet steel; Heat exchanger: aluminium
Weight (kg)	37, for complete unit
Dimensions (mm)	1020 x 380 x 480 (whole unit)
FEATURES	
Available custom-built?	No, complete unit
Performance guarantees	Normally 2 years
Installation method	Prefabricated unit; direct-connected to ducts ø 160
Installed by?	BHV contractor
Cleaning	2filters 2-3 times per year
Corrosion resistance	Galvanized sheet steel and aluminium
Performance control and turndown range	Automatic defrosting by means of thermostat and electric preheater
Cross-contamination	N/a
Withstands differential pressure between ducts	Normal applications: 200 Pa; tested at factory against leakage at 1000 Pa
Condensate collection	Yes, connection to 12 mm pipe
Fans: included/available separately	Exhaust and supply air fans are included in the unit
Frost prevention	Automatic defrosting by electric heating
EXTRAS	Fire-insulated casing against 15-min fire (class A15)
FURTHER COMMENTS	

MANUFACTURER	*AB SVENSKA FLAKTFABRIKEN*
MODEL NAME AND NUMBER	Regoterm PAB (A,B) (Sizes: 095; 130; 170; 200; 240; 290; 350)
TYPE	Regenerator
APPLICATIONS	Airconditioning
OPERATION	Counter-flow
TEMPERATURE (°C)	
Operating range	-30-80
AIR	
Volume (m^3/h)	1 000-83 000 (7 sizes)
Velocity (m/s)	1-5
PRESSURE DROP (mm w.g.)	4-26
REGENERATOR	
Rotation speed (rev/min)	10
Rotation motor size (W)	45, 180, 250
DUTY (kW)	
EFFICIENCY (%)	65-85
TYPICAL PAYBACK PERIOD (years)	
Airconditioning	1-3
Process plant	
CONSTRUCTION	
Materials	Aluminium rotor, sheet-steel casing
Weight (kg)	Size 095: 145; Size 350: 1470
Dimensions (mm)	Size 095: 1100 x 1100 x 355; Size 350: 3650 x 3650 x 460
FEATURES	
Available custom-built?	No
Performance guarantees	According to catalogue
Installation method	See catalogue
Installed by?	Contractor
Cleaning	Vacuum cleaning; cleaning by air or water
Corrosion resistance	See catalogue
Performance control and turndown range	Yes
Cross-contamination	0.1%
Withstands differential pressure between ducts	100 mm w.g.
Condensate collection	Yes
Fans: included/available separately	Available separately
Frost prevention	Rotor speed control or preheater
EXTRAS	Speed detector, smoke damper
FURTHER COMMENTS	Available with hygroscopic or non-hygroscopic rotor, with constant or variable rotor speed; PABA is for horizontal air flow; PABB is for vertical air flow.

MANUFACTURER	*AB SVENSKA FLAKTFABRIKEN*
MODEL NAME AND NUMBER	VLF
TYPE	Recuperator (plate)
APPLICATIONS	Process-to-process; process-to-air; airconditioning
OPERATION	Cross-flow for the unit, plant in cross- , co- or counter-flow
TEMPERATURE (°C)	
Operating range	To 150 (aluminium); 300 (stainless steel)
AIR	
Volume (m^3/h)	No upper limit for plant
Velocity (m/s)	1.5-6
PRESSURE DROP (mm w.g.)	Dependent on application
REGENERATOR	
Rotation speed (rev/min)	
Rotation motor size (W)	
DUTY (kW)	Dependent on specification
EFFICIENCY (%)	70-90
TYPICAL PAYBACK PERIOD (years)	
Airconditioning	
Process plant	0.5-1 (aluminium) 1.5-3 (stainless steel)
CONSTRUCTION	
Materials	
Weight (kg)	
Dimensions (mm)	970 long x 1700 wide x 1970-3000 high per unit
FEATURES	
Available custom-built?	Yes
Performance guarantees	Yes
Installation method	
Installed by?	Flakt
Cleaning	Water or air
Corrosion resistance	Dependent on material used
Performance control and turndown range	
Cross-contamination	Negligible
Withstands differential pressure between ducts	Up to 50 mm w.g. (aluminium); up to 70 mm w.g. (stainless steel)
Condensate collection	No
Fans: included/available separately	Separate
Frost prevention	No
EXTRAS	
FURTHER COMMENTS	

MANUFACTURER	*AB SVENSKA FLAKTFABRIKEN*
MODEL NAME AND NUMBER	QGA
TYPE	Recuperator (tubular)
APPLICATIONS	Process-to-process; process-to-air; airconditioning
OPERATION	Cross-flow for the unit, plant in cross- , co- or counter-flow
TEMPERATURE (°C)	
Operating range	To 80 (aluminium); 400 (stainless steel)
AIR	
Volume (m^3/h)	No upper limit for plant
Velocity (m/s)	3, outside pipes; 16, in pipes (dry conditions)
PRESSURE DROP (mm w.g.)	Dependent on application
REGENERATOR	
Rotation speed (rev/min)	
Rotation motor size (W)	
DUTY (kW)	Dependent on specification
EFFICIENCY (%)	70-90
TYPICAL PAYBACK PERIOD (years)	
Airconditioning	
Process plant	0.5-1 (aluminium) 1.5-3 (stainless steel
CONSTRUCTION	
Materials	
Weight (kg)	
Dimensions (mm)	1000 long x 2000 wide x 3300 high
FEATURES	
Available custom-built?	Yes
Performance guarantees	Yes
Installation method	
Installed by?	Flakt
Cleaning	Water or compressed air
Corrosion resistance	
Performance control and turndown range	
Cross-contamination	Negligible
Withstands differential pressure between ducts	100 mm w.g. (aluminium); 200 mm w.g. (stainless steel)
Condensate collection	No
Fans: included/available separately	Separate
Frost prevention	No
EXTRAS	
FURTHER COMMENTS	

MANUFACTURER	*AB SVENSKA FLAKTFABRIKEN*
MODEL NAME AND NUMBER	QGB
TYPE	Recuperator (pipes)
APPLICATIONS	Process-to-process; process-to-air; airconditioning
OPERATION	Cross-flow for the unit, plant in cross- , co- or counter-flow
TEMPERATURE (°C)	
Operating range	150 (aluminium)
AIR	
Volume (m^3/h)	No upper limit for plant
Velocity (m/s)	To 3, outside pipes; 16, in pipes (dry conditions)
PRESSURE DROP (mm w.g.)	Dependent on application
REGENERATOR	
Rotation speed (rev/min)	
Rotation motor size (W)	
DUTY (kW)	Dependent on specification
EFFICIENCY (%)	70-90
TYPICAL PAYBACK PERIOD (years)	
Airconditioning	
Process plant	0.5-1
CONSTRUCTION	
Materials	
Weight (kg)	
Dimensions (mm)	970 long x 1700 wide x 1970-3000 high, per unit
FEATURES	
Available custom-built?	Yes
Performance guarantees	Yes
Installation method	
Installed by?	Flakt
Cleaning	Water or compressed air
Corrosion resistance	Material is aluminium
Performance control and turndown range	
Cross-contamination	Negligible
Withstands differential pressure between ducts	200 mm w.g.
Condensate collection	No
Fans: included/available separately	Separate
Frost prevention	No
EXTRAS	
FURTHER COMMENTS	

MANUFACTURER	*AB SVENSKA FLAKTFABRIKEN*
MODEL NAME AND NUMBER	QGH
TYPE	Recuperator (tubular)
APPLICATIONS	Process-to-process; process-to-air
OPERATION	Cross-flow for the unit, plant on cross- , co- or counter-flow
TEMPERATURE (^{O}C)	
Operating range	800
AIR	
Volume (m^3/h)	No upper limit for plant
Velocity (m/s)	3m outside pipes; 16, in pipes (dry conditions)
PRESSURE DROP (mm w.g.)	Dependent on application
REGENERATOR	
Rotation speed (rev/min)	
Rotation motor size (W)	
DUTY (kW)	Dependent on specification
EFFICIENCY (%)	50-80
TYPICAL PAYBACK PERIOD (years)	
Airconditioning	
Process plant	1.5-3
CONSTRUCTION	
Materials	
Weight (kg)	
Dimensions (mm)	1000 long x 2000 wide x 3300 high
FEATURES	
Available custom-built?	Yes
Performance guarantees	Yes
Installation method	
Installed by?	Flakt
Cleaning	Water or compressed air
Corrosion resistance	
Performance control and turndown range	
Cross-contamination	Negligible
Withstands differential pressure between ducts	200 mm w.g.
Condensate collection	No
Fans: included/available separately	Separately
Frost prevention	No
EXTRAS	
FURTHER COMMENTS	

MANUFACTURER	*THERMAL EFFICIENCY LTD.*
MODEL NAME AND NUMBER	Plain Steel Tube; Cast Tube
TYPE	Recuperator
APPLICATIONS	Process-to-process
OPERATION	Counter- , parallel-flow
TEMPERATURE (°C)	
Operating range	Air: 20 (inlet), 700 (exhaust); Gas: 1350 (inlet), 350 (exhaust)
AIR	
Volume (m^3/h)	No reasonable limit
Velocity (m/s)	
PRESSURE DROP (mm w.g.)	Up to 500
REGENERATOR	
Rotation speed (rev/min)	
Rotation motor size (W)	
DUTY (kW)	No reasonable limit
EFFICIENCY (%)	95
TYPICAL PAYBACK PERIOD (years)	
Airconditioning	
Process plant	1.5
CONSTRUCTION	
Materials	Heat-resisting alloy steels
Weight	No reasonable limit
Dimensions	No reasonable limit
FEATURES	
Available custom-built?	Yes
Performance guarantees	Yes
Installation method	Package units
Installed by?	Manufacturer or customer
Cleaning	Usually manual
Corrosion resistance	Within specified limits
Performance control and turndown range	5 : 1
Cross-contamination	5%
Withstands differential pressure between ducts	Yes
Condensate collection	Where required
Fans: included/available separately	Not applicable
Frost prevention	Not applicable
EXTRAS	
FURTHER COMMENTS	

MANUFACTURER	*TJERNLUND MANUFACTURING CO.*
MODEL NAME AND NUMBER	Custom 5000
TYPE	Recuperator; plate exchanger; heat pipe; heat wheel
APPLICATIONS	All customized to suit application
OPERATION	As required
TEMPERATURE (^{o}C)	
Operating range	As required
AIR	
Volume (m^3/h)	As required
Velocity (m/s)	As required
PRESSURE DROP (mm w.g.)	As required
REGENERATOR	
Rotation speed (rev/min)	As required
Rotation motor size (W)	As required
DUTY (kW)	As required
EFFICIENCY (%)	As required
TYPICAL PAYBACK PERIOD (years)	
Airconditioning	
Process plant	
CONSTRUCTION	
Materials	As required
Weight	
Dimensions	
FEATURES	
Available custom-built?	Exclusive
Performance guarantees	100%
Installation method	
Installed by?	Customer or contractor
Cleaning	
Corrosion resistance	
Performance control and turndown range	
Cross-contamination	
Withstands differential pressure between ducts	
Condensate collection	
Fans: included/available separately	
Frost prevention	
EXTRAS	
FURTHER COMMENTS	Tjernlund specializes in the design and manufacture of custom heating; ventilating; airconditioning; and energy recovery equipment.

MANUFACTURER	*TORIN CORP.*
MODEL NAME AND NUMBER	HeatBank
TYPE	Heat pipe
APPLICATIONS	Process-to-process; process-to-space; airconditioning
OPERATION	Counter-flow
TEMPERATURE (°C)	
Operating range	Up to 315
AIR	
Volume (m^3/h)	No limit
Velocity (m/s)	1.5-6
PRESSURE DROP (mm w.g.)	Dependent on duty
REGENERATOR	
Rotation speed (rev/min)	
Rotation motor size (W)	
DUTY (kW)	Up to MW size
EFFICIENCY (%)	70
TYPICAL PAYBACK PERIOD (years)	
Airconditioning	3-4
Process plant	1-2
CONSTRUCTION	
Materials	Aluminium; copper; steel; stainless steel
Weight (kg)	
Dimensions (mm)	Up to 3500 (largest dimensions)
FEATURES	
Available custom-built?	Yes
Performance guarantees	Yes
Installation method	Fitted to exhaust and supply ducts
Installed by?	Manufacturer/contractor/customer
Cleaning	Any method
Corrosion resistance	Available
Performance control and turndown rangw	Available
Cross-contamination	None
Withstands differential pressure between ducts	Yes
Condensate collection	Available
Fans: included/available separately	Available
Frost prevention	Available
EXTRAS	
FURTHER COMMENTS	

MANUFACTURER	*TRANE*
MODEL NAME AND NUMBER	Trane Coil Loop; Coil Loop Economics
TYPE	Runaround
APPLICATIONS	Airconditioning-to-air conditioning; process-to-airconditioning
OPERATION	Fluid flow counter to air flow
TEMPERATURE (°C)	-34-43 (for process maximum may be higher)
AIR	
Volume (m^3/h)	Variable 3000-86000
Velocity (m/s)	Variable nominal
PRESSURE DROP (mm w.g.)	Variable
REGENERATOR	
Rotation speed (rev/min)	
Rotation motor size (W)	
DUTY (kW)	Variable, dependent on avaiable energy and economic considerations
EFFICIENCY (%)	60-65, but up to 85
TYPICAL PAYBACK PERIOD (years)	
Airconditioning	From 1, but variable to suit economic requirements
Process plant	From 1, but variable to suit economic requirements
CONSTRUCTION	
Materials	Galvanized steel; copper; aluminium
Weight (kg)	Depends on sizes
Dimensions (mm)	Depends on sizes
FEATURES	
Available custom-built?	Yes
Performance guarnatees	Yes
Installation method	By mechanical services contractor, ductwork
Installed by?	
Cleaning	As for central-station plant
Corrosion resistance	Standard fins are aluminium but copper can be provided
Performance control and turndown range	Preset for maximum efficiency
Cross-contamination	Zero, separate ducts
Withstands differential pressure between ducts	Yes, separate ducts
Condensate collection	Yes
Fans: included/available separately	Available - full range of fans and central-station plant
Frost prevention	Yes
EXTRAS	
FURTHER COMMENTS	Available throughout USA and Europe on computer selection program.

MANUFACTURER	*TRANE THERMAL CO.*
MODEL NAME AND NUMBER	UT Heat Exchanger
TYPE	Recuperator
APPLICATIONS	Process-to-process (incinerator gases used to preheat fumes)
OPERATION	Counter-flow
TEMPERATURE (°C)	
Operating range	500-1000
AIR	
Volume (m^3/h)	15 000-90 000
Velocity (m/s)	
PRESSURE DROP (mm w.g.)	50-100
REGENERATOR	
Rotation speed (rev/min)	
Rotation motor size (W)	
DUTY (kW)	Dependent on specification
EFFICIENCY (%)	70-80
TYPICAL PAYBACK PERIOD (years)	
Airconditioning	
Process plant	1
CONSTRUCTION	
Materials	Stainless steel: 304; 304L; 321; 309; 310
Weight (kg)	
Dimensions (mm)	Custom-designed
FEATURES	
Available custom-built?	Yes
Performance guarantees	Yes, duty and pressure drop
Installation method	Factory packaged in sub-assemblies
Installed by?	Customer or field contractor
Cleaning	N/a
Corrosion resistance	Good
Performance control and turndown range	Control systems supplied; normal turndown: 4 : 1
Cross-contamination	None
Withstands differential pressure between ducts	Yes
Condensate collection	N/a
Fans: included/available separately	Included
Frost prevention	Yes
EXTRAS	Fuel firing may be waste fuel rather than natural-gas or fuel oil
FURTHER COMMENTS	

MANUFACTURER	*UNITED AIR SPECIALISTS INC.*
MODEL NAME AND NUMBER	HD Industrial - HD-12 to HD-96 (17 models)
TYPE	Shell and tube
APPLICATIONS	Process-to-process; process-to-space
OPERATION	Counter/cross-flow
TEMPERATURE (^{o}C)	
Operating range	Up to 425 (standard); up to 650 (high temperature unit)
AIR	
Volume (m^3/h)	600 (HD-12)-100 000 (HD-96)
Velocity (m/s)	3-12
PRESSURE DROP (mm w.g.)	10-100
REGENERATOR	
Rotation speed (rev/min)	
Rotation motor size (W)	
DUTY (kW)	Up to MW capabilities
EFFICIENCY (%)	56-70
TYPICAL PAYBACK PERIOD (years)	
Airconditioning	N/a
Process plant	
CONSTRUCTION	
Materials	Black steel (standard); stainless steel (high temperature)
Weight)kg)	500 approx. (HD-12)-15 000 approx. (HD-96)
Dimensions (mm)	6000 long x 350 dia. approx. (HD-12)-6000 long x 2600 dia. approx. (HD-96)
FEATURES	
Available custom-built?	Yes, including units for operation in excess of 650°C
Performance guarantees	Yes, 1 year
Installation method	Fitted to supply and exhaust ducts - need to accommodate expansion and contraction at high temperatures
Installed by?	Contractor
Cleaning	Conventional methods. If dirty gas steam used, normally passed through tubes
Corrosion resistance	Appropriate to materials used
Performance control and turndown range	Performance function of exhaust and inlet flows
Cross-contamination	None
Withstands differential pressure between ducts	Yes
Condensate collection	No
Fans: included/available separately	Not included
Frost prevention	N/a
EXTRAS	Shorter unit 3000 long available for applications where lower efficiency can be tolerated
FURTHER COMMENTS	

MANUFACTURER	*UNITED AIR SPECIALISTS, INC.*
MODEL NAME AND NUMBER	Temp-X-Changer (Standard-A, High Temperature - VHT)
TYPE	Platè heat exchanger
APPLICATIONS	Process-to-process; process-to-space; airconditioning
OPERATION	Counterflow
TEMPERATURE (°C)	
Operating range	Up to 204 (standard); 815 (high temperature)
AIR	
Volume (m^3/h)	Typically up to 60 000 per module
Velocity (m/s)	2-6.5
PRESSURE DROP (mm w.g.)	5-45
REGENERATOR	
Rotation speed (rev/min)	
Rotation motor size (W)	
DUTY (kW)	
EFFICIENCY (%)	60-80
TYPICAL PAYBACK PERIOD (years)	
Airconditioning	2-5
Process plant	Less than 1 year-2.5 years
CONSTRUCTION	
Materials	Aluminium, mild steel or aluminized steel, (standard); Stainless steel (high temperature); steel casings
Weight (kg)	
Dimensions (mm)	
FEATURES	
Available custom-built?	Generally built up from modules, but 'packaged' systems available
Performance guarantees	Yes
Installation method	Connected by flanges to supply and exhaust ducts - operates horizontally or vertically
Installed by?	Contractor
Cleaning	By water/detergent mix (see extras)
Corrosion resistance	Appropriate to materials used
Performance control and turndown range	Via optional dampers to suit process requirements
Cross-contamination	None
Withstands differential pressure between ducts	Yes
Condensate collection	If required
Fans: included/available separately	Available in 'packaged' heat recovery units
Frost prevention	
EXTRAS	Auxiliary heating, automatic cleaning, filters, face and bypass dampers, various mounting systems (e.g. roof, suspension)
FURTHER COMMENTS	Also available are an air-air counterflow shell and tube heat exchanger (Heavy Duty TXC-HD), operating at up to 871°C and having an efficiency of up to 70%; and a gas-liquid unit (Fluid Air TXC-FA).

MANUFACTURER	*VALMET OY*
MODEL NAME AND NUMBER	LVR
TYPE	Plate
APPLICATIONS	Process-to-process; process-to-space
OPERATION	Cross-flow
TEMPERATURE (°C)	
Operating range	-30-200
AIR	
Volume (m^3/h)	10 000-200 000
Velocity (m/s)	4-10
PRESSURE DROP (mm w.g.)	4-30
REGENERATOR	
Rotation speed (rev/min)	
Rotation motor size (W)	
DUTY (kW)	According to the size and application
EFFICIENCY (%)	Dependent on the case
TYPICAL PAYBACK PERIOD (years)	
Airconditioning	
Process plant	0.3-2
CONSTRUCTION	
Materials	Al; AISI 304; AISI 316
Weight (kg)	According to the application
Dimensions (mm)	According to the application
FEATURES	
Available custom-built?	Yes
Performance guarantees	Yes
Installation method	
Installed by?	Customer or supplier
Cleaning	Brushing; washing
Corrosion resistance	Dependent on the materials
Performance control and turndown range	
Cross-contamination	Minimal on the basis of construction
Withstands differential pressure between ducts	1200 Pa
Condensate collection	Separate collection plates
Fans: included/available separately	Available separately
Forst prevention	Normally not needed
EXTRAS	
FURTHER COMMENTS	

MANUFACTURER	*VAN SWAAY INSTALLATIES BV*
MODEL NAME AND NUMBER	Twin-Cel Kathabar systems
TYPE	Recuperator and regenerator
APPLICATIONS	Airconditioning
OPERATION	Counter- , parallelflow
TEMPERATURE	
Operating range	-30-50
AIR	
Volume (m^3/h)	Approx. from 10 000-115 000
Velocity (m/s)	1.5-2.2
PRESSURE DROP (mm w.g.)	16-34
REGENERATOR	
Rotation speed (rev/min)	
Rotation motor size (W)	
DUTY (kW)	Heat reclaim depending on: outside/return air condition and air quantity ratio
EFFICIENCY (%)	Approx. 65 on enthalpy base, dependent on seasonal heat factor
TYPICAL PAYBACK PERIOD (years)	
Airconditioning	Dependent on outside climate, from 0-4
Process plant	
CONSTRUCTION	
Materials	Special galvanized steel construction/contact surface impregnated cell deck
Weight (kg)	Depending on unit size, 1600-7000
Dimensions (mm)	Depending on unit size, 2000 x 2000 x 2500-10 000 x 20 000 x 2700
FEATURES	
Available custom-built?	Standardized sizes
Performance guarantees	Yes
Installation method	1 unit in outside air/1 unit in return air connected by piping
Installed by?	Airconditioning contractor
Cleaning	Self-cleaning
Corrosion resistance	Longest time in service by now 4 years estimated life-time over 15 years
Performance control and turndown range	Yes, simple thermostats
Cross-contamination	Not possible, used fluid in bacteriostatic; moreover bacteria filtered with effluent over 95% (killed in our liquid)
Withstands differential pressure between ducts	Yes, because of separate units
Condensate collection	No, not required: Twin-Cel is total-enthalpy exchanger
Fans: included/available separately	Can be furnished by any factory or contractor
Frost prevention	Not required, liquid is stable up to -30°C
EXTRAS	Controls will normally be supplied, because this is part of performance guarantee
FURTHER COMMENTS	Both Kathabar and Twin-Cel are using the same bacteriostatic solution, consequently extremely suitable for use in hospitals, pharmaceutical, food industries.

MANUFACTURER	*VENTILATION EQUIPMENT & CONDITIONING LTD.*
MODEL NAME AND NUMBER	Vequip Heat Transfer Coils
TYPE	Suitable for runaround and also gas-to-liquid WHR duties
APPLICATIONS	Process-to-process; process-to-space; airconditioning
OPERATION	N/a
TEMPERATURE (°C)	
Operating range	Up to about 200 gas (copper tubes)
AIR	
Volume (m^3/h)	To customers's requirements
Velocity (m/s)	1-5
PRESSURE DROP (mm w.g.)	Dependent on fin pitch (100-300/m)
REGENERATOR	
Rotation speed (rev/min)	
Rotation motor size (W)	
DUTY (kW)	To customer's requirements
EFFICIENCY (%)	Dependent on duty
TYPICAL PAYBACK PERIOD (years)	
Airconditioning	2
Process plant	Up to 1 possible
CONSTRUCTION	
Materials	Copper tubes; copper or aluminium fins
Weight (kg)	11.9 kg/m^2 per row (aluminium fins); 16.4 kg/m^2 per row (copper fins)
Dimensions (mm)	To customer's requirements; tubes 15.9 mm o.d.
FEATURES	
Available custom-built?	Yes
Performance guarantees	Yes
Installation method	Bolting to adjacent ductwork
Installed by?	Contractor
Cleaning	Compressed air/steam offline
Corrosion resistance	Coatings supplied
Performance control and turndown range	Dependent on fluid circulating rate and air flows
Cross-contamination	N/a
Withstands differential pressure between ducts	N/a
Condensate collection	Yes, moisture eliminators may be fitted
Fans: included/available separately	Available separately
Frost prevention	No
EXTRAS	Turbulators in tubes
FURTHER COMMENTS	

MANUFACTURER	*LEE WILSON ENGINEERING CO. INC.*
MODEL NAME AND NUMBER	Lee Wilson Recuperation Systems
TYPE	Tubular recuperators
APPLICATIONS	Process-to-process (high temperature combustion air preheaters)
OPERATION	Counter-flow
TEMPERATURE (°C)	
Operating range	Example: 11/12 recuperator, 760
AIR	
Volume (m^3/h)	(Burner size: 110-205 kW)
Velocity (m/s)	
PRESSURE DROP (mm w.g.)	75
REGENERATOR	
Rotation speed (rev/min)	
Rotation motor size (W)	
DUTY (kW)	Air preheated to 415-454°C, using exhaust at 760°C
EFFICIENCY (%)	60
TYPICAL PAYBACK PERIOD (years)	
Airconditioning	N/a
Process plant	Less than 2
CONSTRUCTION	
Materials	HF; HT; HX; or CC-50 alloys
Weight (kg)	Up to 70 per recuperator module
Dimensions (mm)	N/a
FEATURES	
Available custom-built?	Yes
Performance guarantees	
Installation method	In package supplied by manufacturer, or as tubes
Installed by?	Contractor
Cleaning	
Corrosion resistance	Normal considerations; pin-fins used on external surface
Performance control and turndown range	N/a
Cross-contamination	Dependent on fuel and furnace duty
Withstands differential pressure between ducts	N/a
Condensate collection	N/a
Fans: included/available separately	N/a
Frost prevention	N/a
EXTRAS	Package - conversion systems for radiant tube and direct-fired equipment; conversions from radiant tube to direct fire with recuperation can also be supplied
FURTHER COMMENTS	

MANUFACTURER	*WING CO.*	
MODEL NAME AND NUMBER	Cortemp WCT 875	Cortemp WCT 1400
TYPE	Regenerator →	
APPLICATIONS	Process-to-process →	
OPERATION	Counter-flow ——	
TEMPERATURE (°C)		
Operating range	Max. 820 →	
AIR		
Volume (m^3/h)	4077-19026 (at NTP)	6599-30197 (at NTP)
Velocity (m/s)	1.5-7.0 →	
PRESSURE DROP (mm w.g.)	3-14.2 →	
REGENERATOR		
Rotation speed (rev/min)	8 →	
Rotation motor size (W)	250 →	
DUTY (kW)	Dependent on customer requirements →	
EFFICIENCY (%)	55-90 →	
TYPICAL PAYBACK PERIOD (years)		
Airconditioning	N/a →	
Process plant	0-1 →	
CONSTRUCTION		
Materials	Stainless steel →	
Weight (kg)	1000	1400
Dimensions (mm)	2100 x 2100 x 675	2515 x 2515 x 675
FEATURES		
Available custom-built?	Yes, special materials →	
Performance guarantees	1 year →	
Installation method	Unit in package for location to adjacent ductwork	
Installed by?	Contractor →	
Cleaning	Provision, if required →	
Corrosion resistance	General stainless-steel considerations →	
Performance control and turndown range	Variable speed with SCR controller and temperature control	
Cross-contamination	Cross-contamination of exhaust air less than 1% by volume (4 kW purge fan included)	
Withstands differential pressure between ducts	305 mm w.g. differential →	
Condensate collection	Generally not required →	
Fans: included/available separately	Not included or available →	
Frost prevention	Not required →	
EXTRAS	Rotation detector; special materials; cleaning systems	
FURTHER COMMENTS	Sensible heat recovery only; core of wheel can transfer solids of up to 2000 μm without special-filter requirements.	

MANUFACTURER	*WING CO.*		
MODEL NAME AND NUMBER	Enthalex WE350-WE6250	Correx WC350-WC4770	Cormed WCM185-WCM3160
TYPE	Regenerator →		
APPLICATIONS	Airconditioning	Process-to-space process-to-process	Process-to-process
OPERATION	Counter-flow →		
TEMPERATURE (°C)			
Operating range	Up to 66	Up to 150	Up to 400
AIR			
Volume (m^3/h)	1105-84 704	1105-107 806	1658-73 526
Velocity (m/s)	1.0-5.5	1.0-7.0	1.5-7.0
PRESSURE DROP (mm w.g.)	5.1-39.1	4.1-31.0	3.0-14.2
REGENERATOR			
Rotation speed (rev/min)	8 →		
Rotation motor size (W)	93-559	93-249	249-373
DUTY (kW)	To customer's requirements		
EFFICIENCY (%)	90-68.5	85-63	82-57
TYPICAL PAYBACK PERIOD (years)			
Airconditioning	0-4		
Process plant		0.10	0-2
CONSTRUCTION			
Materials	Non metallic	Aluminium	Stainless steel
Weight (kg)	190-2041	272-1970	199-1903
Dimensions (mm)	1067 x 1067 x 305- 4420 x 4420 x 500	1067 x 1067 x 305- 3658 x 3658 x 500	914 x 914 x 660- 3359 x 3359 x 813
FEATURES			
Available custom-built?	Modifications available	Yes, special materials →	
Performance guarantees	1 year →		
Installation method	Fitted as package between ducts →		
Installed by?	Contractor →		
Cleaning	Yes, →		
Corrosion resistance	Variable rotation speed →		
Performance control and turndown range	20 or 100:1 →		
Cross-contamination	0.04% by volume .2% by particulate	.1% by volume →	
Withstands differential pressure between ducts	305 mm w.g. →		
Condensate collection	Not required	Yes	Not required
Fans: included/available separately	Not included or available (see below) →		
Frost prevention	Available	Available	Not required
EXTRAS	Rotation detector; summer/winter changeover	Rotation detector; special materials; cleaning systems	Rotation detector
FURTHER COMMENTS	Total-energy recovery: Both sensible; latent; available as 'rooftop' energy recovery package, complete with fans and other fittings	Sensible recovery only →	

9 Gas-to-liquid heat recovery equipment

While the use of heat contained in exhaust gas streams to preheat air or gas is probably the most prolific heat recovery arrangement, certainly judging by the number of manufacturers in the field, the size of such units and their heat transport capability is often exceeded by installations used to recover such heat for transfer to a liquid.

Gas-to-liquid heat recovery data, as that presented in other sections, was collected via questionnaires circulated to manufacturers. For a more detailed description of the general features of the equipment listed in this chapter, the reader is referred to Chapter 2.

9.1 Data presentation

The equipment covered by this chapter includes:

Waste heat boilers (steam and hot water);
Superheaters;
Economizers;
Thermal fluid heaters;
Gas–liquid HVAC systems;
Gas–liquid heat pipe heat exchangers;
Fluid bed gas–liquid heat exchangers.

As with the gas–gas equipment listing, heat pumps transferring heat from gas to a liquid heat sink are reserved for Chapter 13, where all heat pump systems are detailed.

The structure of the boiler industries in many countries, and the manufacturers of associated plant such as economizers and superheaters, does not make identification of the suppliers of waste heat recovery units easy. In order to produce a fully comprehensive listing of potential waste heat recovery boiler suppliers, it would be necessary to list each fired-boiler manufacturer, and in this exercise one would be duplicating much data available in other publications.

Many companies offer waste heat boilers of one form or another, based on the structure of their conventional boilers, and regularly their product literature concentrates on the latter type, giving somewhat sparse information on the waste heat units. Similar criticism, if that is the correct word, can be levelled at some manufacturers of thermal fluid heaters. In obtaining data in this section, an attempt has been made to concentrate on manufacturers who offer economizers for use on process plant other than boilers, and boiler manufacturers who specialize in waste heat recovery systems.

Also in common with gas–gas heat recovery equipment, it can be difficult to draw the line between manufacturers offering equipment for general process use, and those who specialize in products for incineration plant and prime mover heat recovery. In this chapter a number of manufacturers cite applications, for waste heat boilers in particular, in these two areas, and in Appendix 1, these manufacturers are identified in terms of the product classification. In addition, typical applications are listed under the name of each manufacturer listed below.

9.1.1 *Format of data*

The data on equipment used for gas–liquid waste heat recovery is arranged in the following manner:

(1) *Equipment manufacturer*: The prime manufacturer or subsidiary company is given by name. Addresses, country of origin, telephone number and product categories of each manufacturer are listed alphabetically in Appendix 1. Where known, agents for companies in countries other than the home base are also given in this Appendix supplemented by the name of the prime manufacturer for cross-reference.
(2) *Model name and number*: The name of the equipment, and the model number (generally not used as regularly as in gas–gas equipment listings, because a greater number of custom-built units exist in the gas-liquid category) are given.
(3) *Type*: The type of gas-liquid heat exchanger, often including the specific form of the waste heat boiler (e.g. water tube, fire tube, hot water and/or steam-raising) is listed. Many manufacturers produce boilers and economizers, together with superheaters, and full product-range data is given for these.
(4) *Applications*: The application areas for the heat recovery equipment are listed next. These include power-generation plant, commercial boilers, industrial-process plant and furnaces, prime movers and incineration plant (see also Chapter 11, 12 and Appendix 1).
(5) *Gas conditions*: The operating temperature range, generally presented in terms of the maximum exhaust gas temperature which the unit will accept, is given, together with the gas flow rate. In cases where custom-built units may be offered, some manufacturers put no upper limit on the volume flow rate which the unit will accommodate. The temperature of the gas after the heat exchanger is sometimes specified.
(6) *Number of passes*: The number of fluid passes, normally on the gas-side, is given.
(7) *Working pressure*: Working pressure range of the equipment, normally on the

liquid/vapour-side, is given. Most manufacturers quote a maximum operating pressure.
(8) *Pressure drop*: As discussed in Chapter 8, the pressure drop on the gas-side is of particular interest in process-plant applications, especially in 'retrofit' cases. This, together with the liquid/vapour-side pressure drop, is given, where available.
(9) *Liquid-side flow rates*: The flow rate on the liquid-side, or the steam-raising rate, is given where possible.
(10) *Duty*: The heat recovery capability, or the range of duties covered by the manufacturer is listed. In many cases, as with other parameters, units designed to meet a particular customer's requirements are difficult to define, and in some of these instances a typical example of the performance of an installed plant may be given.
(11) *Efficiency*: The thermal efficiency of the heat recovery system is given.
(12) *Typical payback periods*: As emphasized in Chapter 8, the payback period on waste heat recovery equipment of any type is particularly important. The figures quoted are for installed cost, but in most cases neglect any savings which may accrue to the ability to install smaller primary steam-raising plant, a feature of major importance when considering waste heat recovery at the plant-design stage. Examples of the influence this may have on payback periods are given in the descriptive chapters at the beginning of this Directory.
(13) *Heat exchanger construction*: Materials used in the construction of the heat recovery units, weight ranges and dimensions are given. In addition, the types of tubes used are specified, where appropriate. The choice between finned tubes and plain tubes is often determined by fouling and corrosion considerations, and some manufacturers offer both, depending on gas conditions. Finned and plain tubes may also be found in the same heat exchanger – in common with some gas–gas recuperative heat exchangers (see Chapter 1).
(14) *Features*: This section, set out in a similar manner to that described in Chapter 8, although covering a number of features unique to gas–liquid systems, gives data on a number of parameters which will be of interest to the plant designer when selecting his unit, and planning its installation:

- (i) Available custom-built: manufacturers offering custom-built equipment in the gas–liquid category predominate, and standard ranges are the exception rather than the rule.
- (ii) Performance guarantees: all reputable manufacturers offer a guarantee of one form or another. Once a contract is placed by the customer, he is likely to receive a statement giving the performance of the system, based on the figures he is able to provide relating to the operating conditions, including gas flow rate, contamination, quality of steam required (for waste heat boiler), etc. It is in the interest of both parties to know the operating conditions accurately as early as possible.
- (iii) Installer: the manufacturer may accept responsibility for the installation of the unit itself, arrange for a contractor to do it or leave the arrangements to the user. In the latter case, with very large items of plant, the user may

arrange for the plant engineering contractor to include waste heat recovery equipment in his 'turnkey' remit.

(iv) Cleaning methods: techniques recommended by the manufacturers for cleaning the heat exchangers are given. Onstream cleaning equipment is sometimes available as an optional extra.

(v) Corrosion resistance: as with gas–gas heat recovery equipment, plant engineers will be able to relate the corrosion resistance of the units to the gas type and contaminants, knowing the materials used in the exchanger. In cases of doubt, the manufacturer should be consulted. If dual-fuel burners are used in the process, remember that oil and natural gas can give exhaust gases having different corrosion properties, so make the manufacturer aware of your burner system. In economizer units, as in boilers, tube replacement may be recommended every few years, particularly at the back-end where condensation may occur.

(vi) Control range: in some cases the turndown ratio of the equipment is given. On waste heat boilers automatic-control equipment to maintain constant steam conditions may be provided, possibly as an optional extra (see later).

(vii) Design codes applicable: most advanced industrial countries have their own design codes for pressure vessels, including waste heat boilers. Manufacturers marketing outside their own country are able to offer designs which meet the codes required by the customer, and these are listed as appropriate.

(viii) Access for inspection: regular inspection of waste heat recovery equipment, as with all other types of plant, is a necessary part of any preventative maintenance programme. Inspection of the gas-side surfaces is particularly important, from the point of view of fouling, which can seriously reduce the heat transfer effectiveness, and back-end corrosion. A variety of means of access are available, and these are listed.

(ix) Handling and lifting: many items of plant described in the subsequent pages are large and very heavy. In order to facilitate assembly, provision is regularly made by the manufacturer, normally in the form of lifting lugs, to ease assembly. Some manufacturers supply units in modular form for site erection, enabling smaller lifting gear to be used. (Modular systems also have benefits from the point of view of later plant expansion, tube replacement and off-line cleaning.)

(x) Turbulators in tubes: some manufacturers use turbulators inside tubes to enhance heat transfer on the liquid-side.

(xi) Safety cutoffs: safety devices of several types are available on most systems, operating on pressure or low liquid level. On some low-pressure gas–liquid systems, safety cutoffs are not needed. Some offer safety systems as options.

(xii) Liquid/steam gauges: some manufacturers fit gauges giving conditions off the tube-side as standard, others offer them as optional extras.

(xiii) Acceptance of supplementary firing: supplementary firing can be a particularly useful feature of many of these heat recovery systems, making operation largely independent of upstream process conditions. This is very important in situations where the waste heat boiler, for example, is supplying process steam for other services within the factory, as is commonly the case. Manufacturers have indicated whether their systems accept supplementary firing. In some cases this is an optional extra, and should be specified when the equipment specification is made, together with the fuel used (which may affect the choice of heat exchanger material).

(15) *Extras*: Most manufacturers offer a number of extras which may be added to increase control, widen the application of the equipment or assist with maintenance. These include feedwater controls, supplementary firing equipment, superheaters and economizers (on waste heat boilers), type of boiler (natural or forced-circulation options), blowdown systems, level controllers, feedpumps and circulating pumps and cleaning systems; these are listed by some manufacturers.

(16) *Further comments*: Some manufacturers have qualified their data with comments on other equipment available, and their ability to design 'specials'.

Manufacturers' data sheets relating to gas-to-liquid heat recovery equipment follow on the next page.

MANUFACTURER	*ABCO INDUSTRIES INC.*
MODEL NAME AND NUMBER	Single Drum 1-50; Double Drum 1-50
TYPE	Fire tube waste heat recovery boiler
APPLICATIONS	Industrial process plant; all types of incinerators; furnaces
INLET GAS	
Temperature ($^{\circ}$C)	1200 and below; cooling to 250
Flow rate (m^3/s)	100 and below
NUMBER OF PASSES	1 pass is standard; up to 4 passes on special application
WORKING PRESSURE (bar)	Up to 14
PRESSURE DROP (mm w.g.)	
Gas-side	Ave. 100
Liquid-side	Negligible
LIQUID-SIDE FLOW RATES (m^3/s)	0.00656
DUTY (kW)	14 500
EFFICIENCY (%)	up to 85
TYPICAL PAYBACK PERIODS (years)	1 or less
CONSTRUCTION	
Materials	Carbon steel
Tube type (finned or plain)	Bare
Weight (kg)	25 000
Dimensions (mm)	10 000 x 3000 x 4000
FEATURES	
Available custom-built?	All units custom-built
Performance guarantees	Yes
Installed by?	Contractor
Cleaning	Manual
Corrosion resistance	Per specification/application
Control range	
Design codes applicable	ASME Code, section I or VIII
Access for inspection	Yes, manways provided
Handling and lifting	Yes, lugs provided
Turbulators in tubes	None required
Safety cut-offs	LWCO, high limit all standard
Liquid/steam gauges	Standard
Accepts supplementary firing	Yes
EXTRAS	Chevron separators; feedwater controls; auxiliary LWCO; superheaters; auxiliary firing
FURTHER COMMENTS	

MANUFACTURER	*AGA-CTC VAVMEVAXLAVE AB*
MODEL NAME AND NUMBER	CTC-OSBY
TYPE	Waste heat boilers
APPLICATIONS	Industrial process plants
INLET GAS	
Temperature (°C)	800-1200
Flowrate (m^3/s)	10-20
NUMBER OF PASSES	1 or 2
WORKING PRESSURE (bar)	Max. 32
PRESSURE DROP (mm w.g.)	
Gas side	100
Liquid side	No limit
LIQUID SIDE FLOW RATES (m^3/s)	No real limit
DUTY (kW)	2000-10 000
EFFICIENCY (%)	Only thermal limits, if connected with economizer
TYPICAL PAYBACK PERIODS (years)	1-5
CONSTRUCTION	
Materials	Carbon steel
Tube type (finned or plain)	Plain
Weight (kg)	Custom-built
Dimensions (mm)	Custom-built
FEATURES	
Available custom-built?	Yes
Performance guarantees	Yes, capacity and pressure drops
Installed by?	AGA-CTC
Cleaning	Water washing
Corrosion resistance	Good, if rightly handled
Control range	With dumping cond., 0-100
Design codes applicable	All European
Access for inspection	Mandoors on both sides
Handling and lifting	Lifting ears
Turbulators in tubes	Depends
Safety cut-offs	Yes
Liquid/steam gauges	Yes
Accepts supplementary firing	Yes
EXTRAS	
FURTHER COMMENTS	

MANUFACTURER	*AGA-CTC*
MODEL NAME AND NUMBER	CTC Waste Heat Units
TYPE	Waste heat unit, with or without economizer, evaporator and superheater
APPLICATIONS	Industrial process plants, gas turbines, diesel engines
INLET GAS	
Temperature (°C)	200-500 or 600
Flow rate (m^3/s)	Up to 100-200
NUMBER OF PASSES	1
WORKING PRESSURE (bar)	Up to 40-50
PRESSURE DROP (mm w.g.)	
Gas side	No limit
Liquid side	No limit
LIQUID SIDE FLOW RATES (m^3/s)	No limit
DUTY (kW)	Approx. 50 000-100 000
EFFICIENCY (%)	Only thermal limits
TYPICAL PAYBACK PERIOD (years)	1-3
CONSTRUCTION	
Materials	Carbon steel
Tube type (finned or plain)	Finned
Weight (kg)	Custom-built
Dimensions (mm)	Custom-built
FEATURES	
Available custom-built?	Yes
Performance guarantees	Capacity and pressure drops
Installed by?	AGA-CTC
Cleaning	Water washing in fixed nozzles
Corrosion resistance	Good, if service instructions are followed
Control range	With dumping cond., 1-100
Design codes applicable	All European Codes
Access for inspection	Outside heating surface
Handling and lifting	Dependent on size
Turbulators in tubes	No
Safety cut-offs	Yes
Liquid/steam gauges	Yes
Accepts supplementary firing	No, not without extra arrangement
EXTRAS	
FURTHER COMMENTS	

MANUFACTURER	*AIR FROEHLICH AG FUR ENERGIERUCKGEWINNUNG*
MODEL NAME AND NUMBER	GWL-030
TYPE	Glass tube heat exchanger
APPLICATIONS	Industrial process plants
INLET GAS	
Temperature (°C)	Up to 250
Flow rate (m^3/s)	0.6-1.5
NUMBER OF PASSES	1-10
WORKING PRESSURE (bar)	Max. 1
PRESSURE DROP (mm w.g.)	
Gas-side	10-30
Liquid-side	1-50 mm bar
LIQUID SIDE FLOW RATES (m^3/s)	0.1-5
DUTY (kW)	
EFFICIENCY (%)	0-90
TYPICAL PAYBACK PERIOD (years)	1-3
CONSTRUCTION	
Materials	Glass
Tube type (finned or plain)	Plain
Weight (kg)	40
Dimensions (mm)	340 x 300 x 1600
FEATURES	
Available custom-built?	
Performance guarantees	
Installed by?	Air-handling installations
Cleaning	Water, cleaning medium
Corrosion resistance	Optimal
Control range	
Design codes applicable	5333096
Access for inspection	
Handling and lifting	No problem
Turbulators in tubes	No
Safety cut-offs	No
Liquid/steam gauges	No
Accepts supplementary firing	No
EXTRAS	
FURTHER COMMENTS	

MANUFACTURER	*ALFA-LAVAL*	
MODEL NAME AND NUMBER	Spiral types II, III and G	Lamella
TYPE	Condenser or gas cooler	Condenser or gas cooler
APPLICATIONS	Industrial process plant	
INLET GAS		
Temperature (^{o}C)	Up to 400	Up to 500
Flow rate (m^3/s)	50	14
NUMBER OF PASSES	1	1
WORKING PRESSURE (bar)	Max. 18	Max. 35
PRESSURE DROP (mm w.g.)		
Gas-side	Depends on thermal duty →	Depends on thermal duty
Liquid-side	Depends on thermal duty →	Depends on thermal duty
LIQUID-SIDE FLOW RATES (m^3/s)	0.11	0.56
DUTY (kW)	To customer's requirements →	To customer's requirements
EFFICIENCY (%)	Depends on thermal duty →	Depends on thermal duty
TYPICAL PAYBACK PERIODS (years)	Depends on thermal duty →	Depends on thermal duty
CONSTRUCTION		
Materials	Carbon steel, stainless steel, titanium →	Carbon steel, stainless steel, titanium
Tube type (finned or plain)	Not applicable →	Not applicable
Weight (kg)	Varies, custom-built →	Varies, custom-built
Dimensions (mm)	Varies, custom-built →	Varies, custom-built
FEATURES		
Available custom-built?	Yes →	Yes
Performance guarantees		
Installed by?		
Cleaning	High-pressure jet or CIP	CIP (tube-side) Manual (shell-side)
Corrosion resistance		
Control range		
Design codes applicable	ASME, BS	TUV
Access for inspection	Yes →	Yes
Handling and lifting		
Turbulators in tubes	No (not applicable) →	No (not applicable)
Safety cut-offs	No (not applicable) →	No (not applicable)
Liquid/steam gauges	No (not applicable) →	No (not applicable)
Accepts supplementary firing	No (not applicable) →	No (not applicable)
EXTRAS		
FURTHER COMMENTS		

MANUFACTURER	*ARMSTRONG/CHEMTEC GROUP R. ARMSTRONG CO. (USA); CHEMTEC (UK, EUROPE, FAR EAST)*
MODEL NAME AND NUMBER	Many
TYPE	Flue-gas recovery heat exchangers
APPLICATIONS	Waste heat recovered from any gaseous source
INLET GAS	
Temperature (^{o}C)	Up to 1100; designs available for any requirement, varying metals, etc
Flow rate (m^3/s)	Unlimited
NUMBER OF PASSES	Usually single
WORKING PRESSURE (bar)	To 700
PRESSURE DROP (mm w.g.)	
Gas-side	1.13 upwards
Liquid-side	Usually about 0.5
LIQUID-SIDE FLOW RATES (m^3/s)	Unlimited
DUTY (kW)	Unlimited
EFFICIENCY (%)	Varies widely with application; very high efficiency attainable, if needed
TYPICAL PAYBACK PERIOD (years)	2
CONSTRUCTION	
Materials	Steel; stainless steel; nickel; monel
Tube type (finned or plain)	Finned
Weight (kg)	100 kg-20 tonne
Dimensions (mm)	
FEATURES	
Available custom-built?	Yes
Performance guarantees	Yes
Installed by?	Customer or contractor
Cleaning	Blowers
Corrosion resistance	Dependent on metal used and in what service
Control range	
Design codes applicable	ASME TUV Service des Mines; Stoomwezen; Japanese Code; DLI Australia; British Standard; etc.
Access for inspection	As required
Handling and lifting	Reasonable
Turbulators in tubes	Yes
Safety cut-offs	Sometimes
Liquid/steam gauges	As required
Accepts supplementary firing	Yes
EXTRAS	
FURTHER COMMENTS	

MANUFACTURER	*ASET*
MODEL NAME AND NUMBER	
TYPE	Waste heat boiler
APPLICATIONS	Industrial process plants
INLET GAS	
Temperature (^{o}C)	800-1000
Flow rate (m^3/s)	Very much varied
NUMBER OF PASSES	Very much varied
WORKING PRESSURE (bar)	Very much varied
PRESSURE DROP (mm w.g.)	
Gas-side	Very much varied
Liquid-side	Very much varied
LIQUID-SIDE FLOW RATES (m^3/s)	Very much varied
DUTY (kW)	Very much varied
EFFICIENCY (%)	Very much varied
TYPICAL PAYBACK PERIOD (years)	Very much varied
CONSTRUCTION	
Materials	Stainless steel; carbon steel
Tube type (finned or plain)	Finned tubes; bare tubes
Weight (kg)	
Dimensions (mm)	
FEATURES	
Available custom-built?	Yes
Performance guarantees	
Installed by?	
Cleaning	
Corrosion resistance	
Control range	
Design codes applicable	ASME; SNCT; ANCC; TU; TUV
Access for inspection	Yes
Handling and lifting	Yes
Turbulators in tubes	No
Safety cut-offs	Yes
Liquid/steam gauges	Yes
Accepts supplementary firing	
EXTRAS	
FURTHER COMMENTS	

MANUFACTURER	*BABCOCK PRODUCT ENGINEERING LTD.*
MODEL NAME AND NUMBER	Generally custom-built
TYPE	Waste-heat boiler (natural and forced circulation, water tube and shell and tube designs)
APPLICATIONS	Industrial process plant (petrochemicals; smelting; gas manufacture)
INLET GAS	
Temperature (^{o}C)	Up to combustion temperatures
Flow rate (m^3/s)	No limits
NUMBER OF PASSES	Varies
WORKING PRESSURE (bar)	Specials have operated at 350 (N_2H_2 synthesis plant)
PRESSURE DROP (mm w.g.)	
Gas-side	
Liquid-side	
LIQUID-SIDE FLOW RATES (m^3/s)	
DUTY (kW)	E.G. on a 760 tonne/day acid plant, boiler raises 880 tonne/day steam at 44 bar
EFFICIENCT (%)	
TYPICAL PAYBACK PERIODS (years)	
CONSTRUCTION	
Materials	Selected to suit application, normally carbon and/or stainless steel
Tube type (finned or plain)	Plain or finned, according to gases
Weight (kg)	
Dimensions (mm)	
FEATURES	
Available custom-built?	Yes
Performance guarantees	Yes
Installed by?	Manufacturer or contractor
Cleaning	Sootblowers and other more specialized techniques
Corrosion resistance	By selection of correct steels
Control range	
Design codes applicable	BS, ASME and others
Access for inspection	Yes
Handling and lifting	
Turbulators in tubes	
Safety cut-offs	
Liquid/steam gauges	
Accepts supplementary firing	
EXTRAS	
FURTHER COMMENTS	

MANUFACTURER	*SA BABCOCK BELGIUM NV*	
MODEL NAME AND NUMBER	Glass Furnace	Gas Reforming Plant
TYPE	Waste heat boiler with superheat and economizer	Waste heat boilers; superheaters; economizers air preheaters
INLET GAS		
Temperature (°C)	500, cooling to 240-300	1000, cooling to 400 on process gas; 200 on flue gas
Flow rate (m^3/s)	15-50	10-60 on each stream
NUMBER OF PASSES	1	2
WORKING PRESSURE (bar)	10-60	20-40 on process gas-side 20-130 on water steam-side
PRESSURE DROP (mm w.g.)		
Gas side	20-60	2000 (on process gas) 150 (on flue gas)
Liquid side	2 bar in the superheater →	
LIQUID SIDE FLOW RATES (m^3/s)		
DUTY (kW)	4000-2000	10 000-160 000 (on process gas stream) 15 000-130 000 (on flue gas stream)
EFFICIENCY (%)	97	98
TYPICAL PAYBACK PERIOD (years)	1 →	
CONSTRUCTION		
Materials	Mild steel; low alloy	Mild steel; low-alloy steel; austenitic steel
Tube type (finned or plain)	Plain	Plain for flue-gas stream- plain and finned for process-gas stream
Weight (kg)	50 000-150 000	100 000-500 000
Dimensions (mm Max.)	12 000 long x 3000 wide x 10 000 high	Itemized plant
FEATURES		
Available custom-built?	Custom-built →	
Performance guarantees	Steam production; temperature; pressure; gas-outlet temperature →	
Installed by?	Manufacturer →	
Cleaning	Sootblowers	No
Corrosion resistance	No risk, due to choice of metal temperature	No risk, if gas temperature above 200°C
Control range	10-100% →	
Design codes applicable	NBN, TRD, BS, APAVE, ASME, ANCC,LLOYD →	
Access for inspection	Bolted doors →	
Handling and lifting	Site-erected	Package units
Turbulators in tubes	No →	
Safety cut-offs	Yes	Alarms
Liquid/steel gauges	Yes →	
Accepts supplementary firing	Not usual →	
EXTRAS		
FURTHER COMMENTS	Usual equipment proposed: natural-circulation boiler; forced-circulation boilers can be proposed above ranges are typical and not considered as limits.	

MANUFACTURER	*SA BABCOCK BELGIUM NV*	
MODEL NAME AND NUMBER	Sulfuric Acid Plant (pure sulphur)	Nitric Acid Plant
TYPE	Waste heat boiler with superheaters and economizers →	→
APPLICATIONS	Power generation; heating; or process plant →	→
INLET GAS		
Temperature (°C)	1100; cooling to 200	1000; cooling to 300
Flow rate (m^3/s)	6-60 Nm^3/s	10-20 Nm^3/s
NUMBER OF PASSES	1 →	→
WORKING PRESSURE (bar)	40-60 →	→
PRESSURE DROP (mm w.g.)		
Gas side	1000-1500	20
Liquid side	2 bar in the superheater →	→
LIQUID SIDE FLOW RATES (m^3/s)		
DUTY (kW)	6000-60 000	10 000-25 000
EFFICIENCY (%)	97 →	→
TYPICAL PAYBACK PERIOD (years)	1	2
CONSTRUCTION		
Materials	Mild steel; low alloy	Mild steel; low alloy; stainless steel
Tube type (finned or plain)	Plain; helically finned	Plain
Weight (kg)	50 000-200 000 →	→
Dimensions (mm)	Itemized plant	4000 long x 4000 wide x 10 000 high
FEATURES		
Available custom-built?	Standard or custom-built	Custom-built
Performance guarantees	Steam production; temperature; pressure; gas-outlet temperature →	→
Installed by?	Manufacturer →	→
Cleaning	No →	→
Corrosion resistance	No risk (dried gases)	Stainless steel, where needed
Control range	10-100 →	→
Design codes applicable	NBN, TRD, BS, APAVE, ASME, ANCC, LLOYD →	→
Access for inspection	Bolted doors →	→
Handling and lifting	Package units of 100 000 kg max.	Packaged units for small units or site erected
Turbulators in tubes	No →	→
Safety cut-offs	Yes →	→
Liquid/steam gauges	Yes →	→
Accepts supplementary firing	No →	→
EXTRAS		
FURTHER COMMENTS	Natural-circulation or forced-circulation boilers can be proposed. Above ranges are typical and not considered as limits.	

MANUFACTURER	*SA BABCOCK BELGIUM NV*		
MODEL NAME AND NUMBER	Zinc Blende Roasting Plant; Pyrite Roasting Plant	Zinc Fuming Plant; Zinc Cupola	Copper Smelter; Copper Reverberatory Furnace
TYPE	La Mont waste heat boiler with superheater	La Mont waste heat boiler with superheater and economizer	Waste heat boiler with superheater and economizer or air heater
APPLICATIONS	Power generation; heating; industrial process	Power generation; heating; industrial process	Power generation; heating; industrial process
INLET GAS			
Temperature (°C)	900-1000, cooling to 300	1300, cooling to 250 →	
Flowrate (m^3/s)	3-25 Nm^3/s	10-25 Nm^3/s →	
NUMBER OF PASSES	1 →		
WORKING PRESSURE (bar)	26-90	20-90	20-60
PRESSURE DROP (mm w.g.)			
Gas side	8-20	20	10-20
Liquid side	2 bar in the superheater →		
LIQUID SIDE FLOW RATES (m^3/s)			
DUTY (kW)	500-40 000	20 000-35 000	10 000-20 000
EFFICIENCY (%)	97 →		
TYPICAL PAYBACK PERIOD (years)	2 →		
CONSTRUCTION			
Materials	Mild steel; low alloy →		
Tube type (finned or plain)	Plain		
Weight (kg)	150 000-500 000	250 000-500 000	100 000-250 000
Dimensions (mm) Max.	16 000 long x 600 wide x 10 000 high	20 000 long x 5000wide x 10 000 high	15 000 long x 5000 wide x 10 000 high
FEATURES			
Available custom-built?	Standard; custom-built	Custom-built →	
Performance guarantees	Steam production; temperature; pressure; gas outlet temperature →		
Installed by?	Manufacturer →		
Cleaning	Patented mechanical device	Sootblowers →	
Corrosion resistance	Self-protected (min. tube and casing temperature 250°C)	No risk	Self-protected when necessary by the choice of tube and casing temperature
Control range	50-100%	30-100% →	
Design codes applicable	NBN, TRD, BS, APAVE, ASME, ANCC, LLOYD →		
Access for inspection	Hinged doors	Hinged doors	Doors
Handling and lifting	Heating surface in removable bundles	Can be manufactured with removable bundles →	
Turbulators in tubes	No →		
Safety cut-offs	Alarms	Alarms	Yes
Liquid/steam gauges	Yes →		
Accepts supplementary firing	No	Yes →	
EXTRAS			
FURTHER COMMENTS	Above ranges are typical and not considered as limits →		

MANUFACTURER	*FIVES-CAIL BABCOCK*
MODEL NAME AND NUMBER	
TYPE	Waste heat boilers; economizers; superheaters
APPLICATIONS	Industrial process plants; town or industrial refuse incineration plants
INLET GAS	
Temperature (^{o}C)	Up to combustion temperatures, cooling to 150 according to gas type
Flow rate (m^3/s)	No limits
NUMBER OF PASSES	Generally 1
WORKING PRESSURE (bar)	Gas-side: up to 350
PRESSURE DROP (mm w.g.)	
Gas side	According to specification
Liquid side	According to specification
LIQUID SIDE FLOW RATES (m^3/s)	According to specification
DUTY (kW)	According to specification
EFFICIENCY (%)	97
TYPICAL PAYBACK PERIODS (years)	3
CONSTRUCTION	
Materials	Carbon and/or stainless steels
Tube type (finned or plain)	Plain; finned tubes, according to gases
Weight (kg)	According to specification
Dimensions (mm)	According to specification
FEATURES	
Available custom-built?	Yes
Performance guarantees	Temperatures, vaporisation
Installed by?	Manufacturer or contractor
Cleaning	By sootblowers, if necessary
Corrosion resistance	By using specific steel
Control range	
Design codes applicable	ASME, French or others
Access for inspection	Yes
Handling and lifting	
Turbulators in tubes	No
Safety cut-offs	
Liquid/steam gauges	Yes
Accepts supplementary firing	Yes
EXTRAS	
FURTHER COMMENTS	See our Reference Book on custom-built WHB with data on typical units.

MANUFACTURER	*BELTRAN & COOPER LTD.*		
MODEL NAME AND NUMBER	Purpose-Designed	BHR Model	KHR Model
TYPE	Waste heat boiler (Lamont type)	Boiler; economizer	Kiln waste heat unit
APPLICATIONS	Industrial incinerators; HT kilns	Industrial steam boilers	Ceramics and refractory -material kilns (continuous or intermittent)
INLET GAS			
Temperature (^{o}C)	500-1500	200-500	300-1200
Flow rate (m^3/s)	1-10	1-10	5-50
NUMBER OF PASSES	Purpose designed →	→	→
WORKING PRESSURE (bar)	Up to 40 →	→	→
PRESSURE DROP (mm w.g.)			
Gas side	25-50	10-25	10-50
Liquid side	1500-3000 →	→	→
LIQUID SIDE FLOW RATES (m^3/s) (x 10^{-3})	0.63-125	5-500	10-150
DUTY (kW)	1500-15 000	300-30 000	1000-15 000
EFFICIENCY (%) (of heat available over $0^{o}C$)	Approx. 30	Approx. 30	Approx. 46
TYPICAL PAYBACK PERIODS (years)	2-3	0.8-1.0	1.5
CONSTRUCTION			
Materials	Stainless steel (sometimes 316 usually 304)	Carbon steel	Carbon or stainless steel
Tube type (finned or plain)	Plain	Finned	Finned (welded fin)
Weight (kg)	10 000-100 000	2500-25 000	7000-70 000
Dimensions (mm)	To suit application	To suit (usually 300 mm long)	To suit (usually 300, 600 mm long)
FEATURES			
Available custom-built?	Yes →	→	→
Performance guarantees	Yes →	→	→
Installed by?	Local subcontractor →	→	→
Cleaning	Sootblowers	Slide withdrawal; no in-situ cleaning	Sootblowers; side withdrawal
Corrosion resistance	Yes (inherent)	Keep metal temperature above dew point	Keep metal temperature above SO_2 dew point
Control range	Not very good	Self-modulating →	→
Design codes applicable	ASME 8	BS 1113; ASME 8 →	→
Access for inspection	Yes →	→	→
Handling and lifting	Slide rails and lift eyes →	→	→
Turbulators in tubes	No →	→	→
Safety cut-offs	Yes →	→	→
Liquid/steam gauges	Yes →	→	→
Accepts supplementary firing	Yes →	→	→
EXTRAS	Recirculating system when firing high sulphur fuel		
FURTHER COMMENTS			

MANUFACTURER	*BEVERLEY CHEMICAL ENGINEERING LTD.*
MODEL NAME AND NUMBER	
TYPE	Thermal fluid heater
APPLICATIONS	Industrial processes
INLET GAS	
Temperature (°C)	Up to 1200
Flowrate	Up to 80 000 Nm^3/h
NUMBER OF PASSES	As required
WORKING PRESSURE (bar)	Up to 35
PRESSURE DROP (mm w.g.)	
Gas side	0-500
Liquid side	0-40 000
LIQUID SIDE FLOW RATES (m^3/s)	Up to 0.1
DUTY (kW)	Up to 30 000
EFFICIENCY (%)	Up to 75
TYPICAL PAYBACK PERIOD (years)	1-2
CONSTRUCTION	
Materials	Mild or stainless steel
Tube type (finned or plain)	Finned and/or plain
Weight (kg)	Dependent on size
Dimensions (mm)	Dependent on size
FEATURES	
Available custom-built?	Yes
Performance guarantees	Available
Installed by?	Manufacturer
Cleaning	In design
Corrosion resistance	In design
Control range	
Design codes applicable	As required
Access for inspection	Incorporated
Handling and lifting	
Turbulators in tubes	No
Safety cut-offs	Yes
Liquid/steam gauges	Yes
Accepts supplementary firing	Possible
EXTRAS	
FURTHER COMMENTS	

MANUFACTURER	*BONO*
MODEL NAME AND NUMBER	RT Range 50; 75; 100; 120; 150; 200; 250; 300 m^2 heating surface
TYPE	Waste heat boiler for steam or high temperature water
APPLICATIONS	Industrial process plant
INLET GAS	
Temperature (^{o}C)	500-1000 cooling to 200-300°C (inlet temperature)
Flow rate	10 000-40 000 (Nm^3/h)
NUMBER OF PASSES	1
WORKING PRESSURE (bar)	10-20
PRESSURE DROP (mm w.g.)	
Gas side	50-150
Liquid side	
LIQUID SIDE FLOW RATES (m^3/s)	
DUTY (kW)	
EFFICIENCT (%)	50-80
TYPICAL PAYBACK PERIOD (years)	
CONSTRUCTION	
Materials	
Tube type (finned or plain)	Plain
Weight (kg)	7000-30.000
Dimensions (mm)	4500 long x 2000 wide x 2300 high-8000 long x 3000 wide x 3500 high
FEATURES	
Available custom-built?	Custom-built
Performance guarantees	
Installed by?	Contractor
Cleaning	
Corrosion resistance	
Control range	
Design codes applicable	ANCC; ASME
Access for inspection	To gas inlet and outlet box; to inside of boiler
Handling and lifting	
Turbulators in tubes	No
Safety cut-offs	Yes
Liquid/steam gauges	Yes
Accepts supplementary firing	Yes
EXTRAS	
FURTHER COMMENTS	

MANUFACTURER	*BRONSWERK HEAT TRANSFER BV.*
MODEL NAME AND NUMBER	
TYPE	Economizers
APPLICATIONS	Commercial and industrial
INLET GAS	
Temperature (°C)	1000-300, cooling to 300-90
Flow rate (m^3/s)	100 000 (not limited)
NUMBER OF PASSES	Single pass
WORKING PRESSURE (bar)	Exhaust-side: 2 coils (max.); fluid-side: 100 max.
PRESSURE DROP (mm w.g.)	
Gas side	20-300
Liquid side	500-10 000
LIQUID SIDE FLOW RATES (m^3/s)	30 000-1 000 000
DUTY (kW)	300 000
EFFICIENCY (%)	20-50
TYPICAL PAYBACK PERIOD (years)	1
CONSTRUCTION	
Materials	Metals
Tube type (finned or plain)	Plain and finned
Weight (kg)	Max. unit weight: 20 tonne
Dimensions (mm)	Max. 15 x 3 m
FEATURES	
Available custom-built?	Only custom-built
Performance guarantees	By Bronswerk Heat Transfer
Installed by?	Bronswerk Heat Transfer or contractor
Cleaning	Steam cleaners
Corrosion resistance	Material selection; problems below dew point
Control range	Custom-built
Design codes applicable	ASME; BS; AD Merkblatter; Lloyd's; Swedish
Access for inspection	Coils between tuberows
Handling and lifting	Max. 20 tonnes
Turbulators in tubes	Yes
Safety cut-offs	Yes
Liquid/steam gauges	Yes
Accepts supplementary firing	No
EXTRAS	Control; cleaning equipment
FURTHER COMMENTS	

MANUFACTURER	*PETER BROTHERHOOD LTD.*
MODEL NAME AND NUMBER	Waste Heat Recovery Turbogenerators (custom-engineered using steam from waste heat boilers)
TYPE	Waste heat boiler feeding steam turbine, generating electricity
APPLICATIONS	Diesel or gas turbine exhausts, incinerators, process gas streams, furnaces, marine systems
INLET GAS	Peter Brotherhood use WHBs listed elsewhere to raise steam for their WHR turbogenerators
Temperature (^{o}C)	
Flow rate (m^3/s)	
NUMBER OF PASSES	
WORKING PRESSURE (bar)	
PRESSURE DROP (mm w.g.)	
Gas side	
Liquid side	
LIQUID SIDE FLOW RATES (m^3/s)	
DUTY (kW)	Up to 15 000 electrical output
EFFICIENCY (%)	Depends on WHB and turbine efficiency
TYPICAL PAYBACK PERIOD (years)	
CONSTRUCTION	
Materials	N/a
Tube type (finned or plain)	N/a
Weight (kg)	N/a
Dimensions (mm)	N/a
FEATURES	
Available custom-built?	Yes
Performance guarantees	Yes
Installed by?	Supplied as package or installed by manufacturer(s) as turnkey project
Cleaning	N/a
Corrosion resistance	Standard considerations
Control range	Standard considerations
Design codes applicable	BS; ASME; etc.
Access for inspection	N/a
Handling and lifting	N/a
Turbulators in tubes	N/a
Safety cut-offs	N/a
Liquid/steam gauges	N/a
Accepts supplementary firing	N/a
EXTRAS	
FURTHER COMMENTS	

MANUFACTURER	*F. CASINGHINI & FIGLIO*
MODEL NAME AND NUMBER	
TYPE	Waste heat boilers; economizers; heat recovery equipment
APPLICATIONS	Power station plant; commercial boilers; industrial process plant or other
INLET GAS	
Temperature ($^{\circ}$C)	1000-500 for waste heat boilers and from 500-300 for economizers, cooling to 200-150
Flow rate (m^3/s)	Dependent on specific case
NUMBER OF PASSES	Normally 1 for gas-side
WORKING PRESSURE (bar)	7-170
PRESSURE DROP (mm w.g.)	
Gas side	10-200
Liquid side	0.2-2 bar
LIQUID SIDE FLOW RATES (m^3/s)	Dependent on specific case
DUTY (kW)	
EFFICIENCY (%)	
TYPICAL PAYBACK PERIOD (years)	1-3
CONSTRUCTION	
Materials	Steel and steel; steel and cast iron
Tube type (finned or plain)	Finned tubes
Weight (kg)	Dependent on specific case
Dimensions (mm)	Dependent on specific case
FEATURES	
Available custom-built?	Yes
Performance guarantees	As required
Installed by?	Our own firm
Cleaning	By steamblowers; occasionally water washing
Corrosion resistance	Cast-iron protected tubes used when possibility of corrosion
Control range	As required
Design codes applicable	ANCC; ASME; etc.
Access for inspection	Yes
Handling and lifting	As required
Turbulators in tubes	No
Safety cutoffs	No
Liquid/steam gauges	Yes
Accepts supplementary firing	Yes
EXTRAS	
FURTHER COMMENTS	

MANUFACTURER	*CURWEN & NEWBERY LTD.*
MODEL NAME AND NUMBER	Econ-o-Mate; Econ-o-Pack; Econ-o-Pass; Econ-o-Heat
TYPE	Packaged coaxial water-cooled refrigeration heat exchanger
APPLICATIONS	Restaurants; supermarkets; dairy farms; office blocks; factories; any refrigeration users
INLET GAS	
Temperature (^{o}C)	Up to 50 (exhaust) depending on refrigerant
Flow rate (m^3/s)	
NUMBER OF PASSES	Single pass
WORKING PRESSURE (bar)	Depends on refrigerant: approx. 8
PRESSURE DROP (mm w.g.)	
Gas side	3500-9100
Liquid side	
LIQUID SIDE FLOW RATES (m^3/s)	3 x 10^{-3}
DUTY (kW)	3.5-141
EFFICIENCY (%)	Max. 60
TYPICAL PAYBACK PERIOD (years)	1-1.5
CONSTRUCTION	
Materials	Copper heat exchanger; mild steel casing
Tube type (finned or plain)	Convoluted tubing
Weight (kg)	12-26
Dimensions (mm)	533 x 305 x 152 (smallest unit)-660 x 381 x 229 (largest unit)
FEATURES	
Available custom-built?	Econ-o-Pack are custom-made packaged units for special applications
Performance guarantees	
Installed by?	Recognized refrigeration engineers
Cleaning	Not applicable
Corrosion resistance	Not applicable
Control range	
Design codes applicable	ARI standards
Access for inspection	Front casing removable
Handling and lifting	
Turbulators in tubes	Yes
Safety cut-offs	Yes
Liquid/steam gauges	N/a
Accepts supplementary firing	N/a
EXTRAS	
FURTHER COMMENTS	

MANUFACTURER	*CURWEN & NEWBERY LTD.*
MODEL NAME AND NUMBER	Type RO Gas-to-Liquid
TYPE	Waste heat boiler
APPLICATIONS	For use in recovering heat from exhaust flues from industrial process plant and boilers
INLET GAS	
Temperature (^{o}C)	50-800
Flow rate (m^3/s)	To suit application
NUMBER OF PASSES	1
WORKING PRESSURE (bar)	To suit application
PRESSURE DROP (mm w.g.)	
Gas side	25-40
Liquid side	1500
LIQUID SIDE FLOW RATES (m^3/s)	To suit application
DUTY (kW)	To suit application
EFFICIENCY (%)	Typically 70
TYPICAL PAYBACK PERIOD (years)	2-2.5
CONSTRUCTION	
Materials	To suit application but typically: casing and base, cor-ten; tubes, black mild steel/stainless steel
Tube type (finned or plain)	Plain
Weight (kg)	To suit application
Dimensions (mm)	To suit application
FEATURES	
Available custom-built?	Heat exchanger made to measure on basis of technical information supplied by customer
Performance guarantees	Units to perform as specified within measuring tolerances
Installed by?	Heating and ventilating, and process-plant contractors
Cleaning	Cleaning of tubes is through a rear hinged access door
Corrosion resistance	Materials selected to suit application
Control range	
Design codes applicable	ASPCV regulations
Access for inspection	Inspection via rear hinged access door
Handling and lifting	Lifting by slings, using fitted lifting eyes
Turbulators in tubes	No
Safety cut-offs	Yes, as extra
Liquid/steam gauges	Yes, as extra
Accepts supplementary firing	No
EXTRAS	Safety valves; stop valves; blowdown valve/drain cock; check valve; feedpump/circulating pumps; pressure gauges and cocks; level controllers; pressure switches
FURTHER COMMENTS	

MANUFACTURER	*DANKS OF NETHERTON LTD.*
MODEL NAME AND NUMBER	Custom-built
TYPE	Waste-heat shell and water tube boilers
APPLICATIONS	Power station plant; commercial boilers; industrial plant; agricultural; food and pharmaceutical plant
INLET GAS	
Temperature (^{o}C)	Up to 1900; cooling to 200 min.
Flow rate (m^3/s)	11
NUMBER OF PASSES	1; 2; 3
WORKING PRESSURE (bar)	Max. 62
PRESSURE DROP (mm w.g.)	
Gas side	Up to 200
Liquid side	To design
LIQUID SIDE FLOW RATES (m^3/s)	To design
DUTY (kW)	Max. 1800
EFFICIENCY (%)	55-65 (typical)
TYPICAL PAYBACK PERIOD (years)	1 (typical)
CONSTRUCTION	
Materials	To boiler codes
Tube type (finned or plain)	Plain
Weight (kg)	120 000
Dimensions (mm)	
FEATURES	
Available custom-built?	Yes
Performance guarantees	Yes
Installed by?	Danks Technical Services Ltd.
Cleaning	Yes
Corrosion resistance	
Control range	Custom-built
Design codes applicable	BSI, ASME, DIN, etc.
Access for inspection	Yes
Handling and lifting	
Turbulators in tubes	Yes and no
Safety cut-offs	Yes
Liquid/steam gauges	Yes
Accepts supplementary firing	Yes
EXTRAS	
FURTHER COMMENTS	

MANUFACTURER	*DELTAK CORP.*
MODEL NAME AND NUMBER	Delta Waste Heat Boiler
TYPE	Water tube waste heat boiler (natural circulation)
APPLICATIONS	Oil-refinery fired heaters; gas turbines; incinerators
INLET GAS	
Temperature (°C)	Up to 1315
Flow rate	Up to 38 kg/s (mass)
NUMBER OF PASSES	2
WORKING PRESSURE (bar)	1-37
PRESSURE DROP (mm w.g.)	
Gas side	Depends on application
Liquid side	Depends on application
LIQUID SIDE FLOW RATES (m^3/s)	To suit customer's requirements
DUTY (kW)	To suit customer's requirements
EFFICIENCY (%)	
TYPICAL PAYBACK PERIOD (years)	
CONSTRUCTION	
Materials	Steel; refractory-lined casing
Tube type (finned or plain)	Finned (serrated type)
Weight (kg)	
Dimensions (mm)	
FEATURES	
Available custom-built?	Available with 8 gas inlet/outlet configurations and 60 sizes
Performance guarantees	Yes
Installed by?	Manufacturer/contractor
Cleaning	Normally offline
Corrosion resistance	Standard material considerations
Control range	
Design codes applicable	ASME
Access for inspection	Access panels
Handling and lifting	Lifting lugs, assembled as package
Turbulators in tubes	No
Safety cut-offs	Yes
Liquid/steam gauges	Optional
Accepts supplementary firing	Yes
EXTRAS	Sootblowers flue-gas bypass damper control
FURTHER COMMENTS	Other waste heat boilers include Dino (water tube); Dragon (for petrochemical plant); Dart (fire tube); Dwarf (engine silencer plus WHB); Duplex (combined fired and WHB); Dagger (vertical WHB); Diamond (water or thermal fluid heating)

MANUFACTURER	*ECLIPSE LOOKOUT CO.*
MODEL NAME AND NUMBER	Heat Recovery Boilers Type HR, Models 1HR-25HR (Steam) Type HR-W, Models 1HR-W-25HR-W (Water)
TYPE	Waste heat boiler (fire tube) (for steam or hot water)
APPLICATIONS	Recover heat from furnaces, incinerators, turbines, etc.
EXHAUST GAS	
Operating temperature range (°C)	Standard: up to 930, above on special applications
NUMBER OF PASSES	
WORKING PRESSURE	21 kg/cm^2
PRESSURE DROP (mm w.g.)	
Gas side	Up to 151
Liquid side	
LIQUID SIDE FLOW RATES	Production of steam at up to 22 000 kg/h
DUTY (kW)	9945
EFFICIENCY (%)	98
TYPICAL PAYBACK PERIOD (years)	Within 2
CONSTRUCTION	
Materials	SA285C boiler plate and 12 BWG 50 mm tubes
Tube type	
Weight	
Dimensions	
FEATURES	
Available custom-built?	Each unit custom-built to customer requirements; normally 12-14 weeks
Performance guarantees	Rated duty
Installed by?	Manufacturer/contractor
Cleaning	Sootblowing arrangements available
Corrosion resistance	Good - as required
Control range	
Design codes applicable	ASME Section I
Access for inspection	Available
Handling and lifting	Lugs provided
Turbulators in tubes	
Safety cut-offs	Yes
Liquid/steam gauges	Yes
Accepts supplementary firing	Yes
EXTRAS	Any components on boiler and/or hot water equipment
FURTHER COMMENTS	

MANUFACTURER	*ECLIPSE LOOKOUT CO.*
MODEL NAME AND NUMBER	TL Unit (model range 2TL-30TL)
TYPE	Finned tube convection for thermal fluid; gas; water-and-water/glycol solutions
APPLICATIONS	Heat recovery from incinerators; turbines; reciprocating engines; furnaces; etc.
EXHAUST GAS	
Operating temperature range (°C)	Max. 930, unless specified
NUMBER OF PASSES	
WORKING PRESSURE	84 kg/cm^2
PRESSURE DROP	
Gas side	Measured in mm w.g. on shell-side
Liquid side	Measured in kg/m^2 on tube-side
LIQUID SIDE FLOW RATES	200 000 kg/h, determined by pipe size and number of circuits
DUTY (kW)	Unlimited, designed units to 42/157
EFFICIENCY (%)	90 plus
TYPICAL PAYBACK PERIOD (years)	Within 2
CONSTRUCTION	
Materials	SA53 Grade B tube seamless, CS fins (tube-side); SA285C, carbon steel (shell-side)
Tube type	
Weight	
Dimensions	
FEATURES	
Available custom-built?	All units custom-built to requirements, normally 12-14 weeks
Performance guarantees	Rated duty
Installed by?	Manufacturer/contractor
Cleaning	If gas is dirty on shell-side, sootblowers available
Corrosion resistance	Good - as required
Control range	
Design codes applicable	ASME Code, section I
Access for inspection	Yes
Handling and lifting	Lugs provided
Turbulators in tubes	
Safety cut-offs	Yes (see below)
Liquid/steam gauges	Yes
Accepts supplementary firing	Yes
EXTRAS	Transition pieces; relief valves; isolation valves; pumps; skid mounting; diverting valves; etc.
FURTHER COMMENTS	

MANUFACTURER	*FUEL FURNACES LTD.*
MODEL NAME AND NUMBER	Modular Heat Recovery Systems, models HR5, HR9
TYPE	Waste heat boiler
APPLICATIONS	Industrial process plant; engines; boilers
INLET GAS	
Temperature (°C)	200-900
Flow rate	Up to 3500 kg/h
NUMBER OF PASSES	Gas-side: 1; liquid-side: multipass
WORKING PRESSURE (bar)	Normally for hot water or low-pressure steam
PRESSURE DROP (mm w.g.)	
Gas side	2.5-500
Liquid side	
LIQUID SIDE FLOW RATES (m^3/s)	
DUTY (kW)	Up to 180 (HR5); 275 (HR9)
EFFICIENCY (%)	Dependent on specification
TYPICAL PAYBACK PERIOD (years)	Less than 1
CONSTRUCTION	
Materials	Cast iron
Tube type (finned or plain)	Finned
Weight (kg)	HR5: 390; HR9: 539
Dimensions (mm)	HR5: 1193 x 533 x 550; HR9: 3100 x 533 x 550
FEATURES	
Available custom-built?	No, but can be built up using standard modules
Performance guarantees	Yes
Installed by?	Contractor
Cleaning	Conventional techniques
Corrosion resistance	Normal for cast iron
Control range	
Design codes applicable	
Access for inspection	Yes
Handling and lifting	
Turbulators in tubes	No
Safety cut-offs	Limit thermostat
Liquid/steam gauges	
Accepts supplementary firing	Normally not included
EXTRAS	ID fan and motor fittings
FURTHER COMMENTS	Based on Hamworthy Modular Design Acts as silencer on engines.

MANUFACTURER	*GRAHAM MANUFACTURING LTD.*
MODEL NAME AND NUMBER	Flue Gas Unit
TYPE	Economizer using waste heat from hot gas stream to heat single-phase liquid heat transfer medium
APPLICATIONS	Commercial boilers; process plant; total energy systems
INLET GAS	
Temperature (^{o}C)	300-750
Flow rate (m^3/s)	0.2-36
NUMBER OF PASSES	Single-pass gas and liquid
WORKING PRESSURE (bar)	Gas: 1.3 bar abs. ; Liquid: 10 bar
PRESSURE DROP (mm w.g.)	
Gas side	Typically 25-50
Liquid side	Max. 0.5 bar
LIQUID SIDE FLOW RATES (m^3/s)	Min. 0.0005 Max. 0.126
DUTY (kW)	4000
EFFICIENCY (%)	Dependent on operating conditions
TYPICAL PAYBACK PERIODS (years)	1.5-3
CONSTRUCTION	
Materials	Carbon steel; stainless steel
Tube type (finned or plain)	Plain or finned
Weight (kg)	Max. size: 6000 (approx.)
Dimensions (mm)	Min. diam.: 150; Max. diam.: 1000
FEATURES	
Available custom-built?	Yes
Performance guarantees	Yes
Installed by?	Customer or contractor
Cleaning	Chemical clean inside tubes; manual clean outside tubes
Corrosion resistance	Depends on material selected and operating temperature; designed to suit application
Control range	N/a
Design codes applicable	Specials can be built to most European Codes
Access for inspection	N/a
Handling and lifting	Lifting lugs supplied
Turbulators in tubes	No
Safety cut-offs	No
Liquid/steam gauges	No
Accepts supplementary firing	N/a
EXTRAS	
FURTHER COMMENTS	Basic heat exchange unit supplied for incorporation into customer's heat recovery system.

MANUFACTURER	*E. GREEN & SON LTD.*	
MODEL NAME AND NUMBER	ST Boiler	Diesecon
TYPE	Waste heat boiler (natural circulation)	Waste heat boiler (forced circulation)
APPLICATIONS	Industrial process plant; gas turbine; exhausts; incineration	Diesel engine exhaust; industrial process plant; incineration
INLET GAS		
Temperature (°C)	Unlimited	Up to 500-600 (gilled tubes) up to 1100 (plain tubes)
Flow rate (m^3/s)		
NUMBER OF PASSES	Designed to suit requirements	
WORKING PRESSURE (bar)	Normally up to 65, but can be more for special applications	Normally 150 psi
PRESSURE DROP (mm w.g.)		
Gas side Liquid side	Designed to suit requirements	Designed to suit requirements
LIQUID SIDE FLOW RATES (m^3/s)	Steam rates: 3000-30 000 kg/h	See below
DUTY (kW)	See above	Example: on 34200 thp diesel - generates 6500 kg/h superheated and 2000 kg/h saturated steam at 1 kg/h
EFFICIENCY (%)	Dependent on application	Dependent on application
TYPICAL PAYBACK PERIOD (years)	Normally 1-3	Normally 1-3
CONSTRUCTION		
Materials	Mild steel	Mild steel; alloy steels
Tube type (finned or plain)	Combinations of plain and finned	Finned (welded-steel gills)
Weight (kg)		
Dimensions (mm)		
FEATURES		
Available custom-built?	Yes →	
Performance guarantees	Yes →	
Installed by?	Manufacturer →	
Cleaning	Sootblowers; compressed air	Steam or compressed air; rotary sootblowers recommended
Corrosion resistance	Tube type selected to suit	Yes
Control range	Generally kept above dewpoint - can control this	Maintains steam o/p under all engine conditions using gas bypass and damper-control system
Design codes applicable	BSI; ASTM; DIN →	
Access for inspection	Yes →	
Handling and lifting	Provision made →	
Turbulators in tubes	No →	
Safety cut-offs	Yes →	
Liquid/steam gauges	Yes →	
Accepts supplementary firing	Yes	No
EXTRAS	Economizers where justified	Integral economizer and superheater; steam-dumping control system; steam/water cleaning equipment
FURTHER COMMENTS	Complete plant design, supply and installation available for boiler projects.	

MANUFACTURER	*E. GREEN & SON LTD.*	
MODEL NAME AND NUMBER	Welded Steel Gill Economizer	
TYPE	Economizer	
APPLICATIONS	Power generation industry; industrial and marine boiler plant; waste heat steam generators; furnaces; gas turbines; diesel exhaust heat recovery	
	Examples:	
	Economizer on 43700 kg/h boiler giving steam at 115 bar	Economizer on oil-fired boiler generating 500 000 kg/h at 100 bar, 529°C
INLET GAS		
Temperature (°C)	387	582
Flow rate (m^3/s)		
NUMBER OF PASSES		
WORKING PRESSURE (bar)		
PRESSURE DROP (mm w.g.)		
Gas side		
Liquid side		
LIQUID SIDE FLOW RATES (m^3/s)		
DUTY	Water raised, 127°C-191°C; gas reduced, 387°C-181°C	Water raised, 177°C-263°C; gas cooled, 582°C-260°C
EFFICIENCY (%)		
TYPICAL PAYBACK PERIOD (years)	Few weeks to 2 years	
CONSTRUCTION		
Materials	Steel and thermally insulated casing	
Tube type (finned or plain)	Finned (welded-steel gills)	
Weight (kg)	Various	
Dimensions (mm)	Various	
FEATURES		
Available custom-built?	Yes, computerized design service available	
Performance guarantees	Yes	
Installed by?	Manufacturer or contractor; skilled erectors available	
Cleaning	On- or offline	
Corrosion resistance	Welded steel gills ideal for high feedwater temperatures in SO_3-contaminated exhausts (1)	
Control range		
Design codes applicable	BSI: DIN: ASTM	
Access for inspection	Yes	
Handling and lifting	Systems may be modular for easy erection	
Turbulators in tubes	No	
Safety cut-offs		
Liquid/steam gauges	N/a	
Accepts supplementary firing		
EXTRAS	Type 55 economizer tubes (having steel tubes with cast-iron sleeves) are available for use with high sulphur fuels and lower feed-inlet temperatures Welded helical-fin steel tubing (compact) is available for cleaner exhausts	
FURTHER COMMENTS		
NOTE	(1) If satisfactory metal temperatures cannot be maintained, a cast-iron protected heating surface can be installed at the cold end of the unit.	

MANUFACTURER	*HAMWORTHY ENGINEERING CO. LTD.*
MODEL NAME AND NUMBER	
TYPE	Economizer or waste heat boiler
APPLICATIONS	All applications
INLET GAS	
Temperature (°C)	Up to 650
Flow rate	Up to 200 000 m^3/h
NUMBER OF PASSES	1
WORKING PRESSURE (bar)	35
PRESSURE DROP (mm w.g.)	
Gas side	As required
Liquid side	As required
LIQUID SIDE FLOW RATES (m^3/s)	As required
DUTY (kW)	
EFFICIENCY (%)	
TYPICAL PAYBACK PERIODS (years)	
CONSTRUCTION	
Materials	Steel
Tube type (finned or plain)	Finned
Weight (kg)	
Dimensions (mm)	
FEATURES	
Available custom-built?	Yes
Performance guarantees	Yes
Installed by?	Customer
Cleaning	Steam sootblower or water wash
Corrosion resistance	
Control range	
Design codes applicable	Most
Access for inspection	Yes
Handling and lifting	
Turbulators in tubes	No
Safety cut-offs	As required
Liquid/steam gauges	As required
Accepts supplementary firing	No
EXTRAS	
FURTHER COMMENTS	

MANUFACTURER	*HAMWORTHY ENGINEERING CO. LTD.*
MODEL NAME AND NUMBER	FA Type
TYPE	Economizer
APPLICATIONS	Shell boilers
INLET GAS	
Temperature (^{o}C)	150-400
Flow rate (m^3/s)	Up to 25 000 m^3/h
NUMBER OF PASSES	1
WORKING PRESSURE (bar)	35
PRESSURE DROP (mm w.g.)	
Gas side	As required
Liquid side	As required
LIQUID SIDE FLOW RATES (m^3/s)	As required
DUTY (kW)	
EFFICIENCY (%)	
TYPICAL PAYBACK PERIOD (years)	
CONSTRUCTION	
Materials	Steel
Tube type (finned or plain)	Finned
Weight (kg)	
Dimensions (mm)	
FEATURES	
Available custom-built?	No
Performance guarantees	Yes
Installed by?	Customer
Cleaning	Water wash
Corrosion resistance	
Control range	
Design codes applicable	Most
Access for inspection	Yes
Handling and lifting	
Turbulators in tubes	No
Safety cut-offs	Yes
Liquid/steam gauges	Yes
Accepts supplementary firing	No
EXTRAS	
FURTHER COMMENTS	

MANUFACTURER	*HAMWORTHY ENGINEERING CO. LTD.*
MODEL NAME AND NUMBER	HR Type
TYPE	Economizer
APPLICATIONS	All applications
INLET GAS	
Temperature (°C)	Up to 1200
Flow rate (m^3/s)	Up to 6000 m^3/h
NUMBER OF PASSES	1
WORKING PRESSURE (bar)	6
PRESSURE DROP (mm w.g.)	
Gas side	As required
Liquid side	As required
LIQUID SIDE FLOW RATES (m^3/s)	As required
DUTY (kW)	
EFFICIENCY (%)	
TYPICAL PAYBACK PERIOD (years)	
CONSTRUCTION	
Materials	Cast iron
Tube type (finned or plain)	Cast iron
Weight (kg)	
Dimensions (mm)	
FEATURES	
Available custom-built?	No
Performance guarantees	Yes
Installed by?	Customer
Cleaning	Brush
Corrosion resistance	
Control range	
Design codes applicable	Most
Access for inspection	Yes
Handling and lifting	
Turbulators in tubes	No
Safety cut-offs	Yes
Liquid/steam gauges	As required
Accepts supplementary firing	No
EXTRAS	
FURTHER COMMENTS	

MANUFACTURER	*HARRIS THERMAL TRANSFER PRODUCTS INC.*	
MODEL NAME AND NUMBER		
TYPE	Economizer	Waste-heat boiler
APPLICATIONS	Industrial process plant	Industrial process plant
INLET GAS		
Temperature (°C)	Up to 450	Up to 450
Flow rate (m^3/s)	30.3	1.0
NUMBER OF PASSES	Flue gas-side = 1; Tube-side = 10	1
WORKING PRESSURE (bar)	Flue gas-side = 1; tube-side = 35.5	30.3
PRESSURE DROP (mm w.g.)		
Gas side	72.0	460
Liquid side	2300	None
LIQUID SIDE FLOW RATES (m^3/s)	0.0115	0.7×10^{-3}
DUTY (kW)	3000	1720
EFFICIENCY (%)	67	
TYPICAL PAYBACK PERIOD (years)	0.2	1.0
CONSTRUCTION		
Materials	All carbon steel →	
Tube type (finned or plain)	Finned	Plain
Weight (kg)	2900	2500
Dimensions (mm)	3900 x 1400 x 1300	3400 x 610 o.d.
FEATURES		
Available custom-built?	Yes →	
Performance guarantees	Yes →	
Installed by?	Customer →	
Cleaning	Sootblowers, if required (not required on this unit)	Chemical
Corrosion resistance	None required on this unit (stainless steel if required)	No
Control range	By customer	By customer: level devices
Design codes applicable	ASME →	
Access for inspection	Yes →	
Handling and lifting	Yes →	
Turbulators in tubes	No →	
Safety cut-offs	No →	
Liquid/steam gauges	Yes →	
Accepts supplementary firing	No →	
EXTRAS		
FURTHER COMMENTS		

MANUFACTURER	*HAWTHORN LESLIE (ENGINEERS) LTD.*		
MODEL NAME AND NUMBER	Seajoule 500	Seajoule 1000	Seajoule 1500
TYPE	Integrated package plant incorporating waste heat boiler and condensing turbo-alternator		
APPLICATIONS	Diesel or gas turbine exhaust gas, marine or land applications		
INLET GAS			
Temperature (°C)	280-450 →		
Flow rate (m^3/s)	20	37	54
NUMBER OF PASSES			
WORKING PRESSURE (bar)	5.5 bar gauge →		
PRESSURE DROP (mm w.g.)			
Gas side	140 →		
Liquid side	1.2 bar →		
LIQUID SIDE FLOW RATES (Kg/sec)	3.4 kg/s	6.5 kg/s	9.5 kg/s
DUTY (kW) (electric)	500	1000	1500
EFFICIENCY (%)			
Overall thermal	Full load 18.6	18.6	18.6
TYPICAL PAYBACK PERIOD (years)	7	3.5	2.5
CONSTRUCTION			
Materials	Heat exchanger carbon steel		
Tube type (finned or plain)	Finned		
Weight (kg)	Heat exchangers SJ500, 30 000; SJ1000, 55 000; SJ 1500, 77 000		
Dimensions (mm)	SJ500, 3000 long x 2000 wide x 3000 high, SJ1000, 3500 long x 2400 wide x 3500 high SJ1500, 4200 long x 3000 wide x 4200 high		
FEATURES			
Available custom-built?	37		
Performance guarantees	Yes		
Installed by?	Manufacturer		
Cleaning	Automatic water washing		
Corrosion resistance	As for normal marine steam turbine plant		
Control range	Fully automated		
Design codes applicable	Lloyd's class 2		
Access for inspection	Yes		
Handling and lifting	Packaged in manageable units		
Turbulators in tubes	No		
Safety cut-offs	Yes		
Liquid/steam gauges	Yes		
Accepts supplementary firing	Yes		
EXTRAS			
FURTHER COMMENTS			

MANUFACTURER	*HIRAKAWA IRON WORKS LTD.*
MODEL NAME AND NUMBER	Custom-built
TYPE	Superheater
APPLICATIONS	Boilers and process plant
INLET GAS	
Temperature (^{o}C)	Up to 530
Flow rate (m^3/s)	Up to 530 000
NUMBER OF PASSES	
WORKING PRESSURE (bar)	
PRESSURE DROP (mm w.g.)	
Gas side	22
Liquid side	0.54 kg/cm^2
LIQUID SIDE FLOW RATES (m^3/s)	
DUTY (kW)	1272
EFFICIENCY (%)	
TYPICAL PAYBACK PERIOD (years)	Approx. 1
CONSTRUCTION	
Materials	STBA (alloy steel boiler and heat exchanger tubes)
Tube type (finned or plain)	
Weight (kg)	9000
Dimensions (mm)	3000 x 3500 x 1000
FEATURES	
Available custom-built?	Yes
Performance guarantees	1 year
Installed by?	Manufacturer
Cleaning	
Corrosion resistance	
Control range	Temperature control: turndown 1:3
Design codes applicable	
Access for inspection	
Handling and lifting	
Turbulators in tubes	
Safety cut-offs	
Liquid/steam gauges	
Accepts supplementary firing	
EXTRAS	
FURTHER COMMENTS	Fans available separately.

MANUFACTURER	*HIRAKAWA IRON WORKS LTD.*			
MODEL NAME AND NUMBER	MP			
TYPE	Waste heat boiler		Economizer	Heat pipe
APPLICATIONS	Industrial process plant	Industrial process plant	Commercial boilers; industrial process plant	Commercial boilers
INLET GAS				
Temperature (°C)	302-915	200-800	150-265	230
Flow rate (m^3/s)	4.8-144	2.9-87.3	1.6-10.8	0.6
NUMBER OF PASSES	1	1-2		
WORKING PRESSURE (bar)	9.8	9.8	9.8-15.7	0.98
PRESSURE DROP (mm w.g.)				
Gas side	10-145	30-200	20-40	10-20
Liquid side			10-30	10-20
LIQUID SIDE FLOW RATES (m^3/s)	0.07-6.5	0.07-6.1	0.8-6.6	0.3
DITY (kW)	168-14000	185-16360	103-575	22
EFFICIENCY (%)				
TYPICAL PAYBACK PERIOD (years)	Approx. 2-3	Approx. 2-3	Approx. 1	Approx. 1
CONSTRUCTION				
Materials	STB	SB; STB		SUS
Tube type (finned or plain)	Water panel tube	Smoke tube	Finned	Finned
Weight (kg)				
Dimensions (mm)	3000 x 2200 x 1800-10300 x 9000 x 3300	2800 x 1900 x 1500-9700 x 5000 x 3400	3000 x 1000 x 800-3400 x 3000 x 2200	960 x 600 x 270
FEATURES				
Available custom-built?				
Performance guarantees	1 year →	→	→	→
Installed by?	Manufacturer →	→	→	→
Cleaning				
Corrosion resistance				SUS
Control range				
Design codes applicable	Boiler Construction code →	→	→	
Access for inspection	2	2	4	4
Handling and lifting				
Turbulators in tubes	Yes →	→	→	→
Safety cut-offs	Yes →	→	→	→
Liquid/steam gauges	Yes →	→	→	→
Accepts supplementary firing	Yes	Yes	No	No
EXTRAS				
FURTHER COMMENTS				

MANUFACTURER	*HIRT COMBUSTION ENGINEERS*
MODEL NAME AND NUMBER	Hirt Waste Heat Boiler
TYPE	Waste heat boiler/thermal fluid heater
APPLICATIONS	Industrial process plant; incinerators
INLET GAS	
Temperature (°C)	Up to 800
Flow rate (m^3/s)	Designed to meet requirements of plant supplying heat, typically up to 360 000
NUMBER OF PASSES	
WORKING PRESSURE (bar)	Various, typical unit 10 bar
PRESSURE DROP (mm w.g.)	
Gas side	
Liquid side	
LIQUID SIDE FLOW RATES (m^3/s)	Depends on duty
DUTY (kW)	Up to MW size
EFFICIENCY (%)	Depends on duty
TYPICAL PAYBACK PERIOD (years)	1-3
CONSTRUCTION	
Materials	Steel, including stainless
Tube type (finned or plain)	
Weight (kg)	
Dimensions (mm)	
FEATURES	
Available custom-built?	Yes
Performance guarantees	Yes
Installed by?	Manufacturer or contractor
Cleaning	Conventional systems
Corrosion resistance	Appropriate to materials used
Control range	Full control system available
Design codes applicable	ASME
Access for inspection	Yes
Handling and lifting	Lifting lugs fitted
Turbulators in tubes	
Safety cut-offs	Yes, plus safety bypass damper
Liquid/steam gauges	Yes
Accepts supplementary firing	Yes
EXTRAS	
FURTHER COMMENTS	

MANUFACTURER	*HITACHI ZOSEN (HITACHI SHIPBUILDING & ENGINEERING CO. LTD.)*
MODEL NAME AND NUMBER	HCW
TYPE	Natural-circulation type waste heat boiler
APPLICATIONS	Industrial refuse incineration plant
INLET GAS	
Temperature (^{o}C)	1100
Flow rate (m^3/s)	23.3
NUMBER OF PASSES	1
WORKING PRESSURE (bar)	16
PRESSURE DROP (mm w.g.)	
Gas side	70
Liquid side	
LIQUID SIDE FLOW RATES (EVAPORATION) (m^3/s)	0.00239 (2.39 kg/s)
DUTY (kW)	
EFFICIENCY (%)	
TYPICAL PAYBACK PERIOD (years)	
CONSTRUCTION	
Materials	Carbon steel
Tube type (finned or plain)	Plain
Weight (kg)	32 000
Dimensions (mm)	4480 long x 2650 wide x 3950 high
FEATURES	
Available custom-built?	Custom-built
Performance guarantees	Evaporation and outlet-gas temperature
Installed by?	Manufacturer
Cleaning	
Corrosion resistance	
Control range	
Design codes applicable	MOL rules in Japan
Access for inspection	
Handling and lifting	
Turbulators in tubes	No
Safety cut-offs	Yes
Liquid/steam gauges	
Accepts supplementary firing	No
EXTRAS	
FURTHER COMMENTS	

MANUFACTURER	*HITACHI ZOSEN (HITACHI SHIPBUILDING & ENGINEERING CO. LTD.)*	
MODEL NAME AND NUMBER	HWH	HWL
TYPE	Forced-circulation type waste heat boiler	Forced-circulation type waste heat boiler
APPLICATIONS	Dry coke quenching plant	Industrial process plant
INLET GAS		
Temperature (°C)	700	450
Flow rate (m^3/s)	175.2	322.2
NUMBER OF PASSES	1	1
WORKING PRESSURE (bar)	40	14
PRESSURE DROP (mm w.g.)		
Gas side	115	100
Liquid side		
LIQUID SIDE FLOW RATES (EVAPORATION) (m^3/s)	0.01356 (13.56 kg/s)	0.01847 (18.47 kg/s)
DUTY (kW)		
EFFICIENCY (%)		
TYPICAL PAYBACK PERIODS (years)		
CONSTRUCTION		
Materials	Carbon steel; 1% Cr ½ Mo steel	Carbon steel
Tube type (finned or plain)	Finned and plain	Finned
Weight (kg)	410 000	440 000
Dimensions (mm)	8250 wide x 6000 deep x 16 000 high	8500 wide x 11000 deep x 11000 high
FEATURES		
Available custom-built?	Custom-built ⟶	
Performance guarantees	Evaporation; final steam temperature	Evaporation
Installed by?	Manufacturer ⟶	
Cleaning		
Corrosion resistance		
Control range		
Design codes applicable	MITI rules in Japan ⟶	
Access for inspection		
Handling and lifting		
Turbulators in tubes	No ⟶	
Safety cut-offs	Yes ⟶	
Liquid/steam gauges		
Accepts supplementary firing	No ⟶	
EXTRAS		
FURTHER COMMENTS		

MANUFACTURER	*INTERNATIONAL RESEARCH & DEVELOPMENT CO. LTD.*
MODEL NAME AND NUMBER	Specials only
TYPE	Heat pipe heat exchanger
APPLICATIONS	Industrial process plant
INLET GAS	
Temperature (°C)	Up to 300
Flow rate (m^3/s)	As required by user
NUMBER OF PASSES	Gas side: 1; liquid side: several
WORKING PRESSURE (bar)	Can heat high-pressure hot water (e.g. 150°C)
PRESSURE DROP (mm w.g.)	
Gas side	Dependent on duty
Liquid side	
LIQUID SIDE FLOW RATES (m^3/s)	As required by user
DUTY (kW)	As required by user
EFFICIENCY (%)	60-85
TYPICAL PAYBACK PERIOD (years)	1-4
CONSTRUCTION	
Materials	Copper; coated copper; steel
Tube type (finned or plain)	Finned in gas stream; plain in liquid stream
Weight (kg)	Various
Dimensions (mm)	Various
FEATURES	
Available custom-built?	Yes
Performance guarantees	Yes
Installed by?	Manufacturer or contractor
Cleaning	Conventional methods
Corrosion resistance	As determined by materials used
Control range	Determined by flow modulation
Design codes applicable	To user's requirements
Access for inspection	Access doors
Handling and lifting	Lugs provided
Turbulators in tubes	N/a
Safety cut-offs	If required
Liquid/steam gauges	If required
Accepts supplementary firing	Normally not
EXTRAS	
FURTHER COMMENTS	

MANUFACTURER	*JAEGGI LTD.*
MODEL NAME AND NUMBER	
TYPE	Economizer
APPLICATIONS	Industrial process plants; boiler plants
INLET GAS	
Temperature (^{o}C)	Up to 900
Flow rate (m^3/s)	Custom-built, dependent on specifications
NUMBER OF PASSES	Custom-built, dependent on specifications
WORKING PRESSURE (bar)	Custom-built, dependent on specifications
PRESSURE DROP (mm w.g.)	
Gas side	Custom-built, dependent on specifications
Liquid side	Custom-built, dependent on specifications
LIQUID SIDE FLOW RATES (m^3/s)	Custom-built, dependent on specifications
DUTY (kW)	
EFFICIENCY (%)	
TYPICAL PAYBACK PERIOD (years)	
CONSTRUCTION	
Materials	Steel; stainless steel; high-temperature steel
Tube type (finned or plain)	
Weight (kg)	20 tons (20 000 kg)
Dimensions (mm)	
FEATURES	
Available custom-built?	Yes
Performance guarantees	
Installed by?	
Cleaning	
Corrosion resistance	
Control range	
Design codes applicable	
Access for inspection	
Handling and lifting	
Turbulators in tubes	No
Safety cut-offs	No
Liquid/steam gauges	No
Accepts supplementary firing	No
EXTRAS	
FURTHER COMMENTS	

MANUFACTURER	*KLEINEWEFERS*
MODEL NAME AND NUMBER	
TYPE	Economizer
APPLICATIONS	Power station plant; industrial process plant
INLET GAS	
Temperature (°C)	1000, cooling to 150
Flow rate (m^3/s)	100 000 Nm^3/h
NUMBER OF PASSES	
WORKING PRESSURE (bar)	80
PRESSURE DROP (mm w.g.)	
Gas side	10-100
Liquid side	Up to 4 bar
LIQUID SIDE FLOW RATES (m^3/s)	100 000 Nm^3/h
DUTY (kW)	10 000
EFFICIENCY (%)	80
TYPICAL PAYBACK PERIOD (years)	1-2
CONSTRUCTION	
Materials	Cast iron; steel tubes
Tube type (finned or plain)	Finned and plain tubes
Weight (kg)	Dependent on size
Dimensions (mm)	Dependent on size
FEATURES	
Available custom-built?	
Performance guarantees	1 year
Installed by?	Manufacturer or customer
Cleaning	Shot cleaning; steamblower
Corrosion resistance	Alloyed materials
Control range	Thermocouples, values
Design codes applicable	
Access for inspection	Yes
Handling and lifting	
Turbulators in tubes	No
Safety cut-offs	Yes
Liquid/steam gauges	Yes
Accepts supplementary firing	No
EXTRAS	
FURTHER COMMENTS	

MANUFACTURER	*MASKINAFFAREN GENERATOR AB*
MODEL NAME AND NUMBER	Eckrohr Boiler
TYPE	Waste heat boiler
APPLICATIONS	Steam or hot-water production from industrial process plant gaseous exhausts
INLET GAS	
Temperature (^{o}C)	1200, cooled to 320
Flow rate (m^3/s)	8000
NUMBER OF PASSES	
WORKING PRESSURE (bar)	40
PRESSURE DROP (mm w.g.)	
Gas side	Depends on duty
Liquid side	Depends on duty
LIQUID SIDE FLOW RATES (m^3/s)	As required by user
DUTY (kW)	1.25 MW
EFFICIENCY (%)	Dependent on duty
TYPICAL PAYBACK PERIOD (years)	Dependent on duty
CONSTRUCTION	
Materials	Steel
Tube type	Plain
Weight (kg)	Dependent on duty
Dimensions (mm)	Dependent on duty
FEATURES	
Available custom-built?	All custom-built
Performance guarantees	Yes
Installed by?	Manufacturer or contractor
Cleaning	Conventional sootblowers
Corrosion resistance	Appropriate to materials used
Control range	
Design codes applicable	European
Access for inspection	Access ports
Handling and lifting	Designed for ease of erection onsite (only necessary on large units)
Turbulators in tubes	
Safety cut-offs	Yes
Liquid/steam gauges	Yes
Accepts supplementary firing	Yes
EXTRAS	Dust collection; economizers
FURTHER COMMENTS	Derived from range of fired boilers
NOTES	(1) figures quoted are for typical installation.

MANUFACTURER	*METALLURGICAL ENGINEERS*
MODEL NAME AND NUMBER	
TYPE	Waste heat boilers
APPLICATIONS	Industrial process plant
INLET GAS	
Temperature ($^{\circ}$C)	No limit
Flow rate (m^3/s)	No limit
NUMBER OF PASSES	Variable
WORKING PRESSURE (bar)	Up to 150 bar
PRESSURE DROP (mm w.g.)	
Gas side	Variable
Liquid side	Variable
LIQUID SIDE FLOW RATES (m^3/s)	No limit
DUTY (kW)	No limit
EFFICIENCY (%)	To suit process
TYPICAL PAYBACK PERIOD (years)	Down to 1, depending on application
CONSTRUCTION	
Materials	Metallic
Tube type (finned or plain)	Finned; plain
Weight (kg)	
Dimensions (mm)	
FEATURES	
Available custom-built?	Always custom-built
Performance guarantees	Yes
Installed by?	Main contractor, customer, supplier
Cleaning	Sootblowers; water washing; shot cleaning
Corrosion resistance	By design or special materials
Control range	
Design codes applicable	American and European (others, if requested)
Access for inspection	Man doors, etc.
Handling and lifting	Facilities provided
Turbulators in tubes	Sometimes
Safety cut-offs	Yes
Liquid/steam gauges	Yes
Accepts supplementary firing	Sometimes
EXTRAS	
FURTHER COMMENTS	Special designs for high-pressure/temperature and dusty gases.

MANUFACTURER	*NEI THOMPSON COCHRAN LTD.*
MODEL NAME AND NUMBER	Designed to suit specific conditions
TYPE	Waste heat boiler (fire tube type)
APPLICATIONS	Industrial process plant
INLET GAS	
Temperature (^{o}C)	Max. 1250-Min. 300
Flow rate (m^3/s)	To suit application
NUMBER OF PASSES	To suit application
WORKING PRESSURE (bar)	50
PRESSURE DROP (mm w.g.)	
Gas side	To suit application
Liquid side	To suit application
LIQUID SIDE FLOW RATES (m^3/s)	N/a
DUTY (kW)	300-20 000
EFFICIENCY (%)	85
TYPICAL PAYBACK PERIOD (years)	2-3
CONSTRUCTION	
Materials	MS boiler quality, to suit code applicable
Tube type (finned or plain)	Plain HFS
Weight (kg)	80 000
Dimensions (mm)	7620 long x 4270 diam.
FEATURES	
Available custom-built?	Yes
Performance guarantees	Dependent on application
Installed by?	Manufacturer
Cleaning	Tube brush
Corrosion resistance	To suit application
Control range	Approx. 6:1 turndown
Design codes applicable	All American and European codes
Access for inspection	Manholes; handholes
Handling and lifting	Jacking posts; eye lugs
Turbulators in tubes	No
Safety cut-offs	Yes
Liquid/steam gauges	Yes
Accepts supplementary firing	Yes
EXTRAS	
FURTHER COMMENTS	

MANUFACTURER	*NORTHERN ENGINEERING INDUSTRIES, POWER PLANT DIVISION*
MODEL NAME AND NUMBER	Heat Recovery Train
TYPE	Assisted-circulation water tube waste heat boiler of modular construction
APPLICATIONS	Chemical plant
INLET GAS	
Temperature (^{o}C)	Up to 1000
Flow rate (m^3/s)	Typically 77.5
NUMBER OF PASSES	Gas-side: 1
WORKING PRESSURE (bar)	Steam-side: Up to 110
PRESSURE DROP (mm w.g.)	
Gas side	Up to 250
Liquid side	If natural circulation, up to 200 kN/m^2 (evaporation duty)
LIQUID SIDE FLOW RATES (m^3/s)	To customer's requirements
DUTY (kW)	To customer's requirements
EFFICIENCY (%)	Dependent on performance requirements
TYPICAL PAYBACK PERIOD (years)	Dependent on application
CONSTRUCTION	
Materials	Low chrome steel; mild steel
Tube type (finned or plain)	Finned; plain
Weight (kg)	
Dimensions (mm)	Various
FEATURES	
Available custom-built?	Yes (modules assembled at site)
Performance guarantees	Yes
Installed by?	Manufacturer or contractor
Cleaning	Offline techniques
Corrosion resistance	Yes
Control range	
Design codes applicable	ASME; BS; local
Access for inspection	Yes
Handling and lifting	Lifting lugs provided
Turbulators in tubes	No
Safety cut-offs	Safety valves
Liquid/steam gauges	Yes
Accepts supplementary firing	Yes
EXTRAS	
FURTHER COMMENTS	

MANUFACTURER	*NORTHERN ENGINEERING INDUSTRIES, POWER PLANT DIVISION*
MODEL NAME AND NUMBER	Reformed Gas Boiler
TYPE	Assisted-circulation water tube boiler of modular construction
APPLICATIONS	Chemical plant
INLET GAS	
Temperature (°C)	Up to 1000
Flow rate (m^3/s)	Up to 22 kg/s (dry)
NUMBER OF PASSES	1
WORKING PRESSURE (bar)	Steam-side: up to 110
PRESSURE DROP (mm w.g.)	
Gas side	Up to 20
Liquid side	Not critical unless natural-circulation boiler used
LIQUID SIDE FLOW RATES (m^3/s)	Up to 20 kg/s (steam)
DUTY (kW)	Up to 50 000
EFFICIENCY (%)	Dependent on performance requirement
TYPICAL PAYBACK PERIOD (years)	Dependent on application
CONSTRUCTION	
Materials	Low chrome steel; mild steel
Tube type (finned or plain)	Plain
Weight (kg)	60 000-80 000
Dimensions (mm)	Various
FEATURES	
Available custom-built?	Yes
Performance guarantees	Yes
Installed by?	Manufacturer or contractor
Cleaning	Offline
Corrosion resistance	Yes
Control range	
Design codes applicable	ASME; BS; local
Access for inspection	Yes
Handling and lifting	Lifting lugs provided
Turbulators in tubes	No
Safety cut-offs	Yes
Liquid/steam gauges	Yes
Accepts supplementary firing	Yes
EXTRAS	
FURTHER COMMENTS	

MANUFACTURER	*Q-DOT INTERNATIONAL CORP.*
MODEL NAME AND NUMBER	Q-Dot WHRB
TYPE	Waste heat boiler
APPLICATIONS	Commercial boilers; industrial process plant
INLET GAS	
Temperature ($^{\circ}$C)	260-650
Flow rate (m^3/s)	
NUMBER OF PASSES	1
WORKING PRESSURE (bar)	10.3 bar
PRESSURE DROP (mm w.g.)	
Gas side	25.4-76.2
Liquid side	
LIQUID SIDE FLOW RATES (m^3/s)	Steam: 0.423 up to 2.15
DUTY (kW)	1260-6430
EFFICIENCY (%)	82, with exhaust passing through air-side of unit at 650° at 1.52 m/s velocity
TYPICAL PAYBACK PERIOD (years)	Approx. 1
CONSTRUCTION	
Materials	Steel
Tube type (finned or plain)	Finned
Weight (kg)	2177-8845
Dimensions (mm)	3048 long x 4775 wide x 2209 high
FEATURES	
Available custom-built?	Yes
Performance guarantees	Yes
Installed by?	Contractor
Cleaning	As required
Corrosion resistance	Carbon steel
Control range	Full range
Design codes applicable	ASME Boiler code section VIII
Access for inspection	Yes
Handling and lifting	Yes
Turbulators in tubes	No
Safety cut-offs	Yes
Liquid/steam gauges	Yes
Accepts supplementary firing	Yes
EXTRAS	
FURTHER COMMENTS	

MANUFACTURER	*SENIOR ECONOMIZERS LTD.*
MODEL NAME AND NUMBER	Steel Finned Tube; Steel Tube, cast iron clad
TYPE	Economizer
APPLICATIONS	Power station plant; diesel waste heat recovery; industrial boilers
INLET GAS	
Temperature (°C)	600, cooling to approx. 175, dependent on type of fuel and load conditions
Flow rate (m^3/s)	
NUMBER OF PASSES	1; 2; 3
WORKING PRESSURE (bar)	No limit
PRESSURE DROP (mm w.g.)	
Gas side	To suit conditions up to 150
Liquid side	
LIQUID SIDE FLOW RATES (m^3/s)	No limit
DUTY (kW)	Unlimited
EFFICIENCY (%)	
TYPICAL PAYBACK PERIOD (years)	2-3
CONSTRUCTION	
Materials	Mild steel; cast iron
Tube type (finned or plain)	Finned
Weight (kg)	No limit
Dimensions (mm)	No limit
FEATURES	
Available custom-built?	Yes
Performance guarantees	With small tolerances
Installed by?	Manufacturer
Cleaning	Steam sootblowing; water washing; air lancing
Corrosion resistance	Within limits
Control range	Down to approx. 25% of full output
Design codes applicable	BSS 3059
Access for inspection	Yes
Handling and lifting	Yes
Turbulators in tubes	No
Safety cut-offs	
Liquid/steam gauges	
Accepts supplementary firing	
EXTRAS	
FURTHER COMMENTS	

MANUFACTURER	*STIERLE HOCHDRUCK ECONOMIZER KG*
MODEL NAME AND NUMBER	Custom-built
TYPE	Waste heat boilers and economizers
APPLICATIONS	Power stations, commercial boilers, and process plant heat recovery (1)
INLET GAS	
Temperature (°C)	380 (2)
Flow rate (m^3/s)	28
NUMBER OF PASSES	Gas-side: 1
WORKING PRESSURE (bar)	12 (2)
PRESSURE DROP (mm w.g.)	
Gas side	
Liquid side	
LIQUID SIDE FLOW RATES (m^3/s)	Steam output: 10 000 kg/h (2)
DUTY (kW)	
EFFICIENCY (%)	
TYPICAL PAYBACK PERIOD (years)	2-3
CONSTRUCTION	
Materials	Cast iron; cast iron with steel core
Tube type (finned or plain)	Finned and plain, dependent on conditions
Weight (kg)	35 000 (2)
Dimensions (mm)	Approx. 3500 x 3500 x 2750
FEATURES	
Available custom-built?	Yes
Performance guarantees	Yes
Installed by?	Manufacturer or contractor
Cleaning	Rake blowers; rotary tube blowers; and conventional sootblowers
Corrosion resistance	Within limits
Control range	Automatic control system possible
Design codes applicable	European
Access for inspection	Yes
Handling and lifting	Provision made
Turbulators in tubes	No
Safety cut-offs	Yes
Liquid/steam gauges	Yes
Accepts supplementary firing	Yes
EXTRAS	
FURTHER COMMENTS	
NOTES	(1) Stierle products may also be used on diesels and gas turbines. (2) Figures are for refinery installation at Mannheim.

MANUFACTURER	*STONE-PLATT CRAWLEY LTD.*
MODEL NAME AND NUMBER	Stone-Platt
TYPE	Fluidised bed waste heat recovery
APPLICATIONS	Furnaces; kilns; ovens; boilers; incinerators; fired air heaters; gas generators; diesel engines; gas turbines
INLET GAS	
Temperature (°C)	Max. 1000
Flow rate (m^3/s)	As required
NUMBER OF PASSES	1 per fluidized bed
WORKING PRESSURE (bar)	Up to 17.5
PRESSURE DROP (mm w.g.)	
Gas side	None (induced fan draught assisted)
Liquid side	Normally less than 10 p.s.i., dependent on application
LIQUID SIDE FLOW RATES (m^3/s)	As required
DUTY (kW)	75-1500 for single units
EFFICIENCY (%)	Normally 55-85, dependent on application
TYPICAL PAYBACK PERIOD (years)	Usually less than 2 (can be considerably shorter, dependent upon duty cycle of supply source)
CONSTRUCTION	
Materials	Mild steel refractory-lined casing; certain components in stainless steel
Tube type (finned or plain)	Finned
Weight (kg)	Typically 1000-10 000
Dimensions (mm)	750 x 750-2500 x 4000; 3500 high
FEATURES	
Available custom-built?	Yes
Performance guarantees	Yes
Installed by?	SPC or others
Cleaning	Not normally required, can be fitted with automatic cleaners
Corrosion resistance	Yes
Control range	2:1 turndown without fan fitted; 1000% MCR to zero with fan fitted
Design codes applicable	Usually BS codes applicable to boilers, but capable of designing to any requirement
Access for inspection	Yes, large inspection doors
Handling and lifting	Yes
Turbulators in tubes	No
Safety cut-offs	Yes
Liquid/steam gauges	Yes
Accepts supplementary firing	Yes
EXTRAS	Automatic cleaning gear; steam drums; circulating pumps
FURTHER COMMENTS	

MANUFACTURER	*STRUTHERS WELLS CORP.*
MODEL NAME AND NUMBER	Custom-built
TYPE	Waste heat boiler and economizer
APPLICATIONS	Power plant, process plants such as for ammonia production, combined cycle with gas turbine for prime movers
INLET GAS	
Temperature (^{o}C)	400-1200, cooled to 200
Flow rate (m^3/s)	18 000-600 000
NUMBER OF PASSES	Single and multiple
WORKING PRESSURE (bar)	Gas-side: 1-40 Liquid-side: 50-150
PRESSURE DROP (mm w.g.)	
Gas side	200-800
Liquid side	150-8000
LIQUID SIDE FLOW RATES (m^3/s)	20-150
DUTY (kW)	30 000-1 200 000
EFFICIENCY (%)	85-92
TYPICAL PAYBACK PERIOD (years)	2-5
CONSTRUCTION	
Materials	Full range
Tube type (finned or plain)	Finned; plain
Weight (kg)	25 000-200 000
Dimensions (mm)	Variable
FEATURES	
Available custom-built?	Yes
Performance guarantees	Yes
Installed by?	Customer
Cleaning	Gas-side: sootblowers Liquid-side: chemical or mechanical cleaning
Corrosion resistance	Yes
Control range	Full range
Design codes applicable	ASME section I or VIII
Access for inspection	Yes
Handling and lifting	Yes
Turbulators in tubes	No
Safety cut-offs	Yes
Liquid/steam gauges	Yes
Accepts supplementary firing	Yes
EXTRAS	Full range of controls
FURTHER COMMENTS	

MANUFACTURER	*STUDSVIK ENERGITEKNIK AB*
MODEL NAME AND NUMBER	Retherm AS; Retherm AM (close-tube pitch)
TYPE	Heat exchanger for heating and cooling
APPLICATIONS	Corrosive environments
INLET GAS	
Temperature (^{o}C)	-30-40 (1)
Flow rate (m^3/s)	1.7-4.2
NUMBER OF PASSES	20
WORKING PRESSURE (bar)	4
PRESSURE DROP (mm w.g.)	
Gas side	3-50
Liquid side	1200-5300
LIQUID SIDE FLOW RATES (m^3/s)	0.0005-0.007
DUTY (kW)	Depends on flow
EFFICIENCY (%)	50-67
TYPICAL PAYBACK PERIOD (years)	1
CONSTRUCTION	
Materials	High-density polyethylene
Tube type (finned or plain)	Plain
Weight (kg)	150
Dimensions (mm)	1500 x 700
FEATURES	
Available custom-built?	No
Performance guarantees	Yes
Installed by?	Contractor or agent
Cleaning	By mechanical means; liquid flushing
Corrosion resistance	Yes, to acids, salts, and bases
Control range	Dictated by flows
Design codes applicable	
Access for inspection	Yes, also provison for draining
Handling and lifting	Yes, lifting equipment available
Turbulators in tubes	No
Safety cut-offs	N/a
Liquid/steam gauges	N/a
Accepts supplementary firing	N/a
EXTRAS	Plastic ductwork, fans, etc. for aggressive environments
FURTHER COMMENTS	
NOTES	(1) Higher temperature materials becoming available. Retherma may be used as the recovery unit on a gas-gas run-around coil system, where its plastic construction may be of advantage.

MANUFACTURER	*AB SVENSKA MASKINVERKEN*
MODEL NAME AND NUMBER	Sunrod
TYPE	Waste heat recovery; steam; hot water; or other liquid
APPLICATIONS	Diesel engines; gas turbines; exhaust from process plant; boiler
INLET GAS	
Temperature (^{o}C)	Range up to 650°C (higher temperatures available)
Flow rate (m^{3}/s)	Up to approx. 150
NUMBER OF PASSES	Dependent on requirements
WORKING PRESSURE (bar)	Up to 200
PRESSURE DROP (mm w.g.)	
Gas side	According to specification
Liquid side	According to specification
LIQUID SIDE FLOW RATES (m^{3}/s)	Dependent on technical data
DUTY (kW)	
EFFICIENCY (%)	According to customer's specification
TYPICAL PAYBACK PERIOD (years)	Dependent on total installation, for heat exchanger, normally approx. 1
CONSTRUCTION	
Materials	Steel tubing according to DIN or other standard
Tube type (finned or plain)	Sunrod extended surface (pins)
Weight (kg)	
Dimensions (mm)	
FEATURES	
Available custom-built?	Yes
Performance guarantees	Yes
Installed by?	According to customer's specification
Cleaning	Sootblowers (steam or air) or water nozzles
Corrosion resistance	
Control range	0-100%
Design codes applicable	Most national codes, including ASME; all marine classification societies
Access for inspection	To meet requirements
Handling and lifting	To meet requirements
Turbulators in tubes	No
Safety cut-offs	Yes
Liquid/steam gauges	Yes
Accepts supplementary firing	Yes, special design
EXTRAS	Special design to meet specified noise reduction guarantees
FURTHER COMMENTS	

MANUFACTURER	*VALMET OY*
MODEL NAME AND NUMBER	AHR
TYPE	Economizer
APPLICATIONS	Industrial process plant
INLET GAS	
Temperature (^{o}C)	
Flow rate (m^3/s)	Dependent on application
NUMBER OF PASSES	
WORKING PRESSURE (bar)	2.0
PRESSURE DROP (mm w.g.)	
Gas side	5-30
Liquid side	200-3000
LIQUID SIDE FLOW RATES (m^3/s)	Dependent on application
DUTY (kW)	Dependent on application
EFFICIENCY (%)	Dependent on application
TYPICAL PAYBACK PERIOD (years)	0.2-2
CONSTRUCTION	
Materials	AISI 316
Tube type (finned or plain)	Designed plate
Weight (kg)	Dependent on application
Dimensions (mm)	Dependent on application
FEATURES	
Available custom-built?	Yes
Performance guarantees	Yes
Installed by?	Customer or supplier
Cleaning	Automatic washer equipment available
Corrosion resistance	See Material
Control range	
Design codes applicable	
Access for inspection	
Handling and lifting	By truck or crane
Turbulators in tubes	None
Safety cut-offs	None
Liquid/steam gauges	None
Accepts supplementary firing	
EXTRAS	
FURTHER COMMENTS	

MANUFACTURER	*WANSON CO. LTD.*
MODEL NAME AND NUMBER	Wanson Waste Heat Recovery Unit
TYPE	Waste heat boiler (including water and thermal fluid heating duties)
APPLICATIONS	Industrial process plant and marine use
INLET GAS	
Temperature (^{o}C)	
Flow rate (m^3/s)	
NUMBER OF PASSES	1
WORKING PRESSURE (bar)	
PRESSURE DROP (mm w.g.)	
Gas side	
Liquid side	
LIQUID SIDE FLOW RATES (m^3/s)	
DUTY (kW)	500-5000
EFFICIENCY (%)	
TYPICAL PAYBACK PERIODS (years)	
CONSTRUCTION	
Materials	
Tube type (finned or plain)	
Weight (kg)	
Dimensions (mm)	
FEATURES	
Available custom-built?	
Performance guarantees	
Installed by?	
Cleaning	
Corrosion resistance	
Control range	
Design codes applicable	
Access for inspection	
Handling and lifting	
Turbulators in tubes	
Safety cut-offs	
Liquid/steam gauges	
Accepts supplementary firing	
EXTRAS	
FURTHER COMMENTS	

10 Liquid-to-liquid heat recovery equipment

Most processes involve liquid-liquid heat exchangers, the two most common types being the shell and tube heat exchanger and the plate heat exchanger (see Chapter 3), and both of these types are predominant in the equipment available for heat recovery from liquids listed in this chapter.

It is in the process industries, rather than HVAC, where liquid effluents are at a sufficiently high temperature to warrant heat recovery, unless upgrading of the heat is carried out by means of a heat pump (see Chapters 6 and 13), but boiler plant used in commercial buildings can benefit from blowdown heat recovery and may use other liquid sources as heat for feedwater preheating.

10.1 Data presentation

The equipment listed in this chapter includes the following:

Shell and tube heat exchangers[1];
Plate heat exchangers;
Concentric heat exchangers;
Lamella heat exchangers;
Condensate control/recovery units;
Storage heat exchangers;
Finned heat exchangers.

As pointed out in Chapter 9, certain industrial sectors are served predominantly by large manufacturers of equipment such as boilers, who may see waste heat units as an incidental part of their main range. A similar situation exists in liquid–liquid heat recovery equipment, in that shell and tube heat exchanger manufacturers serve the petrochemical (and some other process) industry with large numbers of heat exchangers,

[1] Some shell and tube heat exchangers are known as 'spiral tube' heat exchangers.

only a proportion of which may be categorized as waste heat recovery units. However, it is somewhat easier to adapt a shell and tube heat exchange for a waste heat duty than it is to design a waste heat boiler, and most of the heat exchanger manufacturers, even those not listed in this Directory, will be prepared to quote for liquid–liquid units for heat recovery duties.

Liquid–liquid waste heat recovery is a major growth area for heat pumps, and manufacturers offering systems in this area are listed in Chapter 13. If liquid effluent streams are below 40° C, it is worth exploring the potential of these devices for producing useful heat.

An important application area for liquid–liquid heat exchangers is the recovery of heat from the cooling water and oil on reciprocating diesel and natural-gas engines. The user will find that most manufacturers of prime movers of this type offer such heat exchangers as optional extras, and it is generally the engine exhaust-gas heat exchanger which has to be procured from a separate source.

10.1 *Format of data*

In common with the data in other chapters, the following format is adopted for liquid–liquid heat recovery equipment:

(1) *Equipment manufacturer*: The prime manufacturer or subsidiary company is given by name, further details, including address and telephone number, are listed in Appendix 1. Agents are also given in this Appendix, and all companies are identified by product classifications.
(2) *Model name and number*: The name or identification letters/numbers of the heat exchanger range are given. Where manufacturers wish to identify models within each range, the model numbers are given, or the number of models within the range identified. Many shell and tube heat exchangers are custom-built, and the size of plate heat exchangers, may be of almost infinite variability, depending upon the number of plates added.
(3) *Type*: The type of liquid–liquid heat exchanger, as listed at the beginning of this chapter, is given.
(4) *Typical applications*: Broad application areas, not necessarily exhaustive, are given for the equipment. A number of manufacturers have cited domestic, power station and process uses, and a fair proportion have indicated district heating as an application area of significance. Food processing and textile finishing are other major industrial users of liquid-liquid waste heat recovery.
(5) *Operating temperature range*: The operating temperature range of the heat exchangers is given, normally being stated in terms of the maximum operating temperature. Where operation involves approaching both maximum operating temperature and pressure, the manufacturers should be consulted, particularly in the case of plate heat exchangers.
(6) *Liquid flow rates*: The liquid flow rates through the heat exchanger are given,

Don't waste your time on heat recovery

Contact APV and save yourself the energy!

As recovery experts, APV can usefully re-direct surplus heat from industrial liquids back into the process while your thermal equilibrium is restored and the fuel bills reduced.

Manufactured and assembled in Great Britain, APV liquid-to-liquid heat transfer equipment is available in various materials to suit applications in the chemical, pharmaceutical, brewing, dairy, food, offshore and power generation industries.

Paraflow plate heat exchangers, developed originally by APV over 50 years ago, comprise a range of compact, versatile units which complement an equally efficient family of evaporators.

APV Heat Recovery – a plan for all seasons

The A.P.V. Company Limited
P.O. Box 4, Crawley, West Sussex. Tel: (0293) 27777. Telex: 87237.

C259

E&F N Spon

The technical division of Associated Book Publishers Ltd,
11 New Fetter Lane, London EC4P 4EE.

ENERGY BOOKS FROM SPON

Heat Pumps
R. D. HEAP

. . . a clear explanation of the principles behind various types of heat pumps describing their main components and design features. Although written primarily with the heat pump specifier in mind the book will be invaluable to all designers, installers and users of heat pumps and will be of interest to all those concerned with the economical use of energy for heating, whether in buildings or for industrial purposes.

August 1979 240 x 159mm 168 pages
Hardback illustrated 0 419 11330 4 £7.50

The Generation of Electricity by Windpower
E. W. GOLDING

" . . . It is a thorough and unimpeachable volume covering in rigorous detail yet accessible style every major aspect of windpower theory . . ." **Undercurrents**

1976 224 x 145mm 352 pages
Hardback illustrated 0 419 11070 4 £7.50

Geothermal Energy
Its past, present and future contributions to the energy need of man.
H. C. H. ARMSTEAD

" . . . has everything in its 350 pages including an authority, an unmistakable integrity and comprehensive coverage that leaves no energy expert – or amateur – curiosity unsatisfied . . . Mr. Armstead has produced an astonishingly readable and almost exciting recital of the opening chapters of a new era in the exploration of natural, ready made, already-converted, door step energy." **Heating and Ventilating News**

1978 240 x 160mm 350 pages
Hardback illustrated 0 419 11240 5 £10.50

Biological Energy Resources
M. SLESSER and C. LEWIS

. . . provides a concise account of how plants capture energy from the sun through photosynthesis, and the technologies available for their subsequent conversion into fuels for use by man.

October 1979 234 x 156mm 196 pages
Hardback illustrated 0 419 11340 1 £8.50

ALSO AVAILABLE

The Control of Noise in Ventilation Systems
a designers' guide
Atkins Research and Development
1977 252 x 189mm 116 pages
Hardback 0 419 11050 X £10.00

Basic Vibration Control
Sound Research Laboratories
1978 294 x 210mm 138 pages
Paperback illustrated 0 419 11440 8 £6.00

Noise Control in Industry
Sound Research Laboratories
1976 222 x 147mm 430 pages
Hardback 0 419 11220 0 £8.00

Practical Building Acoustics
Sound Research Laboratories
1976 252 x 189mm 250 pages
Paperback illustrated 0 419 11200 6 £6.50

normally in terms of a maximum throughput.
(7) *Pressure drops*: Typical values are given for the pressure drops through the heat exchanger (for shell-side and tube-side where appropriate). In some cases the heat exchanger design is governed by allowable pressure drops in the process stream, and in these cases no detailed data can be given.
(8) *Maximum working pressure*: The maximum pressures which the heat exchanger components shell, tube, etc.) will stand are listed. Where these are insufficient for the potential user's requirements, materials selection may assist if the manufacturer offers options in this area.
(9) *Duty*: The heat transfer capability of liquid–liquid heat exchangers is, in most cases, difficult to specify accurately, depending upon so many variables in each application. Where it has not been possible to quote the duty, the user will find the liquid throughout range of relevance to his size requirements.
(10) *Efficiency*: Similar arguments apply to figures for liquid–liquid heat exchanger efficiency. As illustrated in Chapter 3, the efficiency of a plate heat exchanger can be increased to values approaching 100 per cent, if one is prepared to tolerate an exponential rise in capital cost of the equipment. The optimum efficiency of the heat exchanger is, therefore, a compromise between size (and capital cost) and the return on the investment due to energy savings.
(11) *Typical payback period*: The time taken for the installation to pay for itself, in terms of the value of the energy saved related to the capital cost, is given. This is normally between 0·5 and 4 years for this type of equipment, some units operating with comparatively small approach temperature differences, leading to longer paybacks than are normally found on gas–gas heat recovery units, for example.
(12) *Materials of construction*: Because of the frequent arduous duties of liquid–liquid heat exchangers of all types, particularly in the petrochemical field, most are available in a range of materials suitable for most corrosive liquids, Stainless steel is a standard plate heat exchanger material, with many others being available, while carbon steel may be found in many shell and tube heat exchangers, where stainless may not be necessary.
(13) *Features*: As in previous chapters, data was requested from manufacturers of liquid–liquid heat recovery equipment on a number of features which will assist the potential user in selecting his equipment:

(i) Available custom-built: many manufacturers are able to offer custom-built equipment in the liquid–liquid field. For plate heat exchangers, these are normally built up from standard elements.
(ii) Performance guarantees: these vary considerably, but are generally given. Manufacturers normally provide a specification, based on the user's data, covering performance and pressure drops.
(iii) Cleaning methods: automatic air offline cleaning techniques are available. Chemical or mechanical cleaning may also be used, and some manufacturers have developed their own particular cleaning techniques for heat exchangers.

(iv) Design codes applicable: many manufacturers construct to a particular range of design codes, predominantly European and ASME, but will make units to other codes, in some cases, if required.

(14) *Extras*: Provision is made for manufacturers to list extras available. These include control and cleaning systems, alternative metals and gasket materials, packaged heat recovery systems, storage tanks, thermal insulation, relief valves, etc.

(15) *Further comments*: A space is provided for manufacturers to add comments, possibly pointing out some interesting features of their equipment or indicating appropriate brochures for the user to request.

Manufacturers' data sheets relating to liquid-to-liquid heat recovery equipment follow on the next page.

MANUFACTURER	*AGA-CTC*
MODEL NAME AND NUMBER	CTC-Spiraflow
TYPE OF HEAT EXCHANGER	Concentric
TYPICAL APPLICATIONS	Pharmaceutical; food; dairy; distillation
OPERATION TEMPERATURE RANGE (°C)	Up to 200
LIQUID FLOW RATE (l/h)	Up to 50 000
PRESSURE DROPS (mm w.g.)	Only practical economical limit
MAXIMUM WORKING PRESSURE	32 kg/m^2
DUTY (kW)	4-7 000
EFFICIENCY (%)	Only thermal limits
TYPICAL PAYBACK PERIOD (years)	0.5-1
CONSTRUCTION	
Materials	Stainless steel
FEATURES	
Available custom-built?	Yes
Performance guarantees	Yes, capacity ; pressure drops
Cleaning	Chemical; mechanical
Design codes applicable	All European codes
EXTRAS	
FURTHER COMMENTS	

MANUFACTURER	*AGA-CTC*
MODEL NAME AND NUMBER	CTC-HISAKA
TYPE OF HEAT EXCHANGER	Plate heat exchanger
TYPICAL APPLICATIONS	Marine; dairy; food chemical industries
OPERATING TEMPERATURE RANGE (^{o}C)	Up to 170
LIQUID FLOW RATE (l/h)	Plate: 1 500 000
PRESSURE DROPS (mm w.g.)	No limit
MAXIMUM WORKING PRESSURE	16-25 kg/cm^2
DUTY (kW)	2-300 000
EFFICIENCY (%)	Only thermal limits
TYPICAL PAYBACK PERIOD (years)	0.5-2
CONSTRUCTION	
Materials	Stainless steel; titanium; copper; nickel alloys; etc.
FEATURES	
Available custom-built?	Yes
Performance guarantees	Yes, capacity; pressure drops
Cleaning	Chemical and mechanical (both sides)
Design codes applicable	All European, ZIS and ASME codes through the licensee group
EXTRAS	
FURTHER COMMENTS	

MANUFACTURER	*AGA-CTC*
MODEL NAME AND NUMBER	AGA-CTC
TYPE OF HEAT EXCHANGER	Shell and tube
TYPICAL APPLICATIONS	Domestic heating; power stations; pulp and paper; petrochemical; etc.
OPERATING TEMPERATURE RANGE (°C)	Up to 500
LIQUID FLOW RATE (l/h)	Shell-side: Up to 1000- 1 500 000 (S)
PRESSURE DROPS (mm w.g.)	Only practical economical limits
MAXIMUM WORKING PRESSURE	Up to 300 kg/cm^2
DUTY (kW)	2-300 000
EFFICIENCY (%)	Only thermal limits
TYPICAL PAYBACK PERIOD (years)	0.5-3
CONSTRUCTION	
Materials	Carbon steel, stainless steel, copper, special alloys, etc.
FEATURES	
Available custom-built?	Yes
Performance guarantees	Yes, capacity and pressure drops
Cleaning	Chemical and mechanical (for some models limited)
Design codes applicable	All European and ASME (No stamp)
EXTRAS	
FURTHER COMMENTS	

MANUFACTURER	*EDUARD AHLBORN*		
MODEL NAME AND NUMBER	Ahlborn Free-Flow Shell-type Heat Exchangers (models 157; 159; 161)		
TYPE OF HEAT EXCHANGER	Plate: model 157	Plate: model 159	Plate: model 161
TYPICAL APPLICATIONS	Low-grade waste heat recovery from liquid effluent		
OPERATING TEMPERATURE RANGE (°C)	Up to 135		
LIQUID FLOW RATE (l/h)	10 000	25 000	150 000
PRESSURE DROPS (mm w.g.)	To customer's requirements		
MAXIMUM WORKING PRESSURE (kN/m^2)			
DUTY (kW)	To customer's requirements		
EFFICIENCY (%)	Depends on duty		
TYPICAL PAYBACK PERIOD (years)	Depends on duty		
CONSTRUCTION			
Materials	Plate: stainless steel		
FEATURES			
Available custom-built?	Yes, from standard units		
Performance guarantees	Yes		
Cleaning	CIP; or by simple plate removal		
Design codes applicable			
EXTRAS			
FURTHER COMMENTS			

MANUFACTURER	*ALFA-LAVAL*		
MODEL NAME AND NUMBER	A35,A30,AX30,A20,AM20,A15 A10,AM10,A3-P25,P45,P3,P13,	Types I	Ramen
TYPE OF HEAT EXCHANGER	Plate	Spiral	Lamella
TYPICAL APPLICATIONS	Low-grade heat recovery		
OPERATING TEMPERATURE RANGE (°C)	Up to 250	Up to 400	Up to 500
LIQUID FLOW RATE (l/h)	Up to 2 500 000 (max.) in one exchanger	400 000 (max.)	2 000 000 (max.)
PRESSURE DROPS (mm w.g.)	Designed to customer's requirements →		
MAXIMUM WORKING PRESSURE (kN/m^2)	2500	1800	3500
DUTY (kW)	Designed to customer's requirements →		
EFFICIENCY (%)	Dependent on thermal duty required →		
TYPICAL PAYBACK PERIOD (years)	Dependent on thermal duty		
CONSTRUCTION			
Materials	Stainless steel, hastelloy, titanium, incoloy, aluminium brass, inconnel	Stainless steel, carbon steel, titanium	Carbon steel, stainless steel
FEATURES			
Available custom built?		Yes	
Performance guarantees			
Cleaning	Easily disassembled for manual cleaning; CIP also	CIP; high-pressure hose	CIP, tube-side; manual, shell-side
Design codes applicable	ASME: TUV	ASME: BS	TUV
EXTRAS			
FURTHER COMMENTS	Features: Compact: i.e. one-fifth volume of shell and tube High Heat Transfer Coefficients: Up to 4 times that of shell and tube. Very easy to clean and maintain. Low fouling: Highly turbulent flow keeps solids in suspension.	Features: Compact: Resistance to clogging by solids.	Features: Compact:

MANUFACTURER	*APV COMPANY LTD.*	
MODEL NAME AND NUMBER	Junior Paraflow	HX series Paraflow
TYPE OF HEAT EXCHANGER	Plate	Plate
TYPICAL APPLICATIONS	Low-grade heat recovery	Low-grade heat recovery, some with high-viscosity liquids
OPERATING TEMPERATURE RANGE (°C)	Up to 260 →	
LIQUID FLOW RATE (l/hr)	Up to 3200	Up to 41 000
PRESSURE DROPS (mm w.g.)	To customer's requirements →	
MAXIMUM WORKING PRESSURE (kN/m^2)	1400	400-1240
DUTY (kW)	To customer's requirements →	
EFFICIENCY (%)	Dependent on duty →	
TYPICAL PAYBACK PERIOD (years)	Dependent on duty →	
CONSTRUCTION		
Materials	Stainless steel 316 S 16 (standard); also titanium, incoloy, monel, hastelloy; nickel; aluminium brass; kunifer alloys	
FEATURES		
Available custom built?	Yes, using standard elements	
Performance guarantees		
Cleaning	Cleaning in place; or simple plate removal	
Design codes applicable	BS	
EXTRAS		
FURTHER COMMENTS		

MANUFACTURER	*APV CO. LTD.*				
MODEL NAME AND NUMBER	Paraflow R6 (4 sizes)	R8 (4 sizes)	R10 (4 sizes)	R14 (4 sizes)	R23 (5 sizes)
TYPE OF HEAT EXCHANGER	Plate	Plate	Plate	Plate	Plate
TYPICAL APPLICATIONS	Low-grade heat recovery ⟶				
OPERATING TEMPERATURE RANGE (°C)	Up to 260 ⟶				
LIQUID FLOW RATE (l/h)	E.g. R66: 550 000			e.g. R145: 1 000 000	Up to 2500 000
PRESSURE DROPS (mm w.g.)	To customer's requirements ⟶				
MAXIMUM WORKING PRESSURE (kN/m^2)	690	1030	690	1030	690
DUTY (kW)	To customer's requirements ⟶				
EFFICIENCY(%)	Dependent on duty ⟶				
TYPICAL PAYBACK PERIOD (years)	Dependent on duty ⟶				
CONSTRUCTION					
Materials	Plate: stainless steel (standard) ⟶				
FEATURES					
Available custom-built?	Yes, using standard elements ⟶				
Performance guarantees					
Cleaning	CIP; or simple plate removal ⟶				
Design codes applicable	BS ⟶				
EXTRAS	Other plate and gasket materials available ⟶				
FURTHER COMMENTS					

MANUFACTURER	*APV CO. LTD.*	
MODEL NAME AND NUMBER	Paraflow HMBL (3 sizes)	R40 Paraflow series (3 sizes)
TYPE OF HEAT EXCHANGER	Plate	Plate
TYPICAL APPLICATIONS	Low-grade heat recovery from process effluent →	
OPERATING TEMPERATURE RANGE (°C)	Up to 260 →	
LIQUID FLOW RATE (l/h)	Typically 12 000 →	
PRESSURE DROPS (mm w.g.)	To customer's requirements →	
MAXIMUM WORKING PRESSURE (kN/m²)	690	1370
DUTY (kW)	To customer's requirements →	
EFFICIENCY (%)	Dependent on duty →	
TYPICAL PAYBACK PERIOD (years)	Dependent on duty →	
CONSTRUCTION		
Materials	Plate: stainless steel 316 S 16 (standard) →	
FEATURES		
Available custom-built?	Yes, using standard elements →	
Performance guarantees		
Cleaning	CIP; or simple plate removal →	
Design codes applicable	BS →	
EXTRAS	Alternative plate and gasket materials →	
FURTHER COMMENTS		

MANUFACTURER	*APV CO. LTD.*	
MODEL NAME AND NUMBER	Paraflow type CHF (5 sizes)	Paraflow R5 series (9 sizes)
TYPE OF HEAT EXCHANGER	Plate	Plate
TYPICAL APPLICATIONS	Low-grade heat recovery from process effluent	Low-grade heat recovery from process plant
OPERATING TEMPERATURE RANGE (°C)	Up to 260 →	→
LIQUID FLOW RATE (l/h)	Up to 12 000	e.g. R5, 6: 16 000
PRESSURE DROPS (mm w.g.)	To customer's requirements →	→
MAXIMUM WORKING PRESSURE (kN/m^2)	1400	930-2060
DUTY (kW)	To customer's requirements →	→
EFFICIENCY (%)	Dependent on duty →	→
TYPICAL PAYBACK PERIOD (years)	Dependent on duty →	→
CONSTRUCTION		
Materials	Stainless steel (standard) →	→
FEATURES		
Available custom-built?	Yes, using standard elements →	→
Performance guarantees		
Cleaning	CIP; or simple plate removal →	→
Design codes applicable	BS →	→
EXTRAS	Alternative plate and gasket materials →	→
FURTHER COMMENTS		

MANUFACTURER	*ASET*
MODEL NAME AND NUMBER	Shell and Tube Heat Exchangers
TYPE OF HEAT EXCHANGER	Shell and tube heat exchangers
TYPICAL APPLICATIONS	
OPERATING TEMPERATURE RANGE ($^{\circ}$C)	400
LIQUID FLOW RATE (l/h)	Very varied
PRESSURE DROPS (mm w.g.)	Very varied
MAXIMUM WORKING PRESSURE (kN/m^2)	Very varied
DUTY (kW)	Very varied
EFFICIENCY (%)	Very varied
TYPICAL PAYBACK PERIOD (years)	Very varied
CONSTRUCTION	
Materials	Very varied
FEATURES	
Available custom-built?	Yes
Performance guarantees	Yes
Cleaning	
Design codes applicable	ASME: TUV; TU; SNCT; ANCC; TEMA
EXTRAS	
FURTHER COMMENTS	

MANUFACTURER	*SA BABCOCK BELGIUM NV*
MODEL NAME AND NUMBER	Reboiler
TYPE OF HEAT EXCHANGER	Shell and tube
TYPICAL APPLICATIONS	Production of low-grade steam used in open cycle from turbine exhaust steam at medium pressure (steam is condensed in reboiler and recycled without treatment)
OPERATING TEMPERATURE RANGE (^{o}C)	From 130-250
LIQUID FLOW RATE (l/h)	Tube-side, shell-side: 40 000
PRESSURE DROPS (mm w.g.)	Plate, shell-side: 0 tube-side: 4000
MAXIMUM WORKING PRESSURE (kN/m^2)	Plate, shell-side: 20 000 tube-side: 60 000
DUTY (kW)	40 000
EFFICIENCY (%)	98
TYPICAL PAYBACK PERIOD (years)	1
CONSTRUCTION	
Materials	Plate, shell-side, tube-side: mild steel, low alloy
FEATURES	
Available custom-built?	Standard or custom-built
Performance guarantees	Condensated steam flow; low-grade steam pressure
Cleaning	No
Design codes applicable	NBN; TRD; BS; APAVE; ASME; ANCC; LLOYD
EXTRAS	
FURTHER COMMENTS	Above ranges are typical and not considered as limits.

MANUFACTURER	*BARRIQUAND*	
MODEL NAME AND NUMBER		
TYPE OF HEAT EXCHANGER	Plate with gaskets	Welded plate
TYPICAL APPLICATIONS	Liquid-to-liquid	Steam-to-liquid; Liquid-to-liquid
OPERATING TEMPERATURE RANGE (°C)	140	600
LIQUID FLOW RATE (l/h)	400 000 (400 m^3/h)	Without limit
PRESSURE DROPS (mm w.g.)	Adjustable →	
MAXIMUM WORKING PRESSURE (kN/m^2)	2100 kN/m^2 (21 bar)	6000 kN/m^2 (60 bar)
DUTY (kW)		
EFFICIENCY (%)		
TYPICAL PAYBACK PERIOD (years)		
CONSTRUCTION		
Materials	Stainless steel; titanium; nickel; hastelloy; incolloy	Stainless steel; nickel; hastelloy; incolloy; monel
FEATURES		
Available custom-built?	Yes	
Performance guarantees	Yes	
Cleaning	Chemical or manual	
Design codes applicable	European and overseas	
EXTRAS		
FURTHER COMMENTS		

MANUFACTURER	*E.J. BOWMAN (BIRMINGHAM) LTD.*
MODEL NAME AND NUMBER	PK600-3350-6 (largest size)
TYPE OF HEAT EXCHANGER	Shell and tube
TYPICAL APPLICATIONS	Photographic processing machines; laundries; glassworks; bottle-washing plants
OPERATING TEMPERATURE RANGE (°C)	Max. 100
LIQUID FLOW RATE (l/h)	Shell-side: 40 000; tube-side: 40 000
PRESSURE DROPS (mm w.g.)	Shell-side: 4 000; tube-side: 4 000
MAXIMUM WORKING PRESSURE (kN/m^2)	Shell-side: 700; tube-side: 700
DUTY (kW)	1400
EFFICIENCY (%)	60 (single unit); 84 (2 units in series); 94 (3 units in series)
TYPICAL PAYBACK PERIOD (years)	Less than 1
CONSTRUCTION	
Materials	Cupro-nickel tube stack: 70/30 brass shell; bronze end-covers
FEATURES	
Available custom-built?	No
Performance guarantees	Yes
Cleaning	Removable end-covers and tube stack
Design codes applicable	None
EXTRAS	None
FURTHER COMMENTS	Designed for recovering heat from waste contaminated hot process water and preheating incoming cold water.

MANUFACTURER	*BRONSWERK HEAT TRANSFER BV*
MODEL NAME AND NUMBER	Liquid-to-Liquid Heat Exchanger
TYPE OF HEAT EXCHANGER	Shell and tube
TYPICAL APPLICATIONS	Exchange liquid-to-liquid as: boiler feedwater; pre-heaters; heat recovery of water of textile firms condensate coolers
OPERATING TEMPERATURE RANGE (^{o}C)	In counter flow: from 25-300
LIQUID FLOW RATE (l/h)	Unlimited
PRESSURE DROPS	Min. 0.05-0.002 bar
MAXIMUM WORKING PRESSURE (kN/m^2)	400 bar
DUTY (kW)	Unlimited
EFFICIENCY (%)	Depends on temperature difference
TYPICAL PAYBACK PERIOD (years)	Depends on sort of problem
CONSTRUCTION	
Materials	Steel, copper, copper alloys, titanium, stainless steel
FEATURES	
Available custom-built?	Only custom-built
Performance guarantees	Always given
Cleaning	Automatic cleaning or manual cleaning
Design codes applicable	All codes of: Germany; The Netherlands; UK; France; Scandinavia; USA
EXTRAS	Heat recovery systems with pumps; storage tanks; heat exchangers and cleaning systems
FURTHER COMMENTS	

MANUFACTURER	*BRONSWERK HEAT TRANSFER BV*
MODEL NAME AND NUMBER	Tubular Heat Exchangers
TYPE OF HEAT EXCHANGER	Tube
TYPICAL APPLICATIONS	Fresh water to foul water
OPERATING TEMPERATURE RANGE (°C)	0-300
LIQUID FLOW RATE (l/h)	Max. approx. 300.000
PRESSURE DROPS (mm w.g.)	3000-10 000
MAXIMUM WORKING PRESSURE (kN/m^2)	350
DUTY (kW)	30 000 000
EFFICIENCY (%)	Up to 99.9
TYPICAL PAYBACK PERIOD (years)	Custom-built
CONSTRUCTION	
Materials	All metals
FEATURES	
Available custom-built?	Only custom-built
Performance guarantees	BHT
Cleaning	Automatically; intermittent by hand
Design codes applicable	ASME/TEMA; AD multiclatter; BS; Stoonwernen; Swedish code; Lloyds
EXTRAS	Control equipment; autocleaning
FURTHER COMMENTS	

MANUFACTURER	*CHEMOKOMPLEX*	
MODEL NAME AND NUMBER	Shell Floating-head Water Coolers, liquid-to-liquid	
TYPE OF HEAT EXCHANGERS	Shell and tube; usually more heat exchanger elements are connected	
TYPICAL APPLICATIONS	For cooling power plant or other industrial condensates	
OPERATING TEMPERATURE RANGE (°C)	-20-200	
LIQUID FLOW RATE (l/h)	1500-200 000 l/h shell-side: 1.2 m/s max. ;	500-50 000 l/h tube-side 2 m/s max.
PRESSURE DROPS (mm w.g.)	Shell-side: 100-3000 ;	tube-side: 100-6000
MAXIMUM WORKING PRESSURE (kN/m^2)	Both shell and tube: 1200	
DUTY (kW)	4-2000 (for 1 heat exchanger element)	
EFFICIENCY (%)	99	
TYPICAL PAYBACK PERIOD (years)	1-2	
CONSTRUCTION		
Materials	Shell-side: stainless or carbon steel tube or plate Tube-side: carbon or stainless steel tubes; brass or copper tubes	
FEATURES		
Available custom-built?	Yes	
Performance guarantees	Suitable types are chosen by manufacturer according to specified data; guarantees given up to certain contamination level	
Cleaning	Tube bundle can be pulled out and cleaned mechanically	
Design codes applicable	MSZ 10425-70; MSZ 13822-66; MSZ 13823-66; MSZ 13824-66; MSZ 13833-71; MSZ 13834-68; DIN 28 180; DIN 28 182; DIN 28 183	
EXTRAS	None	
FURTHER COMMENTS	The above heat exchangers are custom-built.	

MANUFACTURER	*CRANE LTD.*
MODEL NAME AND NUMBER	Condensate Booster Drainage Control System (CBA 1-5)
TYPE OF HEAT EXCHANGER	
TYPICAL APPLICATIONS	Used where required to remove condensate and non-condensible gases at optimum rate to improve efficiency of system; typical processes benefiting from CBA installation include: heating; cooking; evaporating; drying; pressing; curing; typical industries include: food; brewing; lumber; plywood; paper; vender; tobacco; chemicals; plastics; textiles; rendering
OPERATING TEMPERATURE RANGE (°C)	Max. 207
LIQUID FLOW RATE (l/h)	27 000
PRESSURE DROPS (mm w.g.)	Unit increases pressure of condensate to boiler working pressure
MAXIMUM WORKING PRESSURE (kN/m^2)	2067
DUTY (kW)	
EFFICIENCY (%)	
TYPICAL PAYBACK PERIOD (years)	Less than 1
CONSTRUCTION	
Materials	Cast-iron casing; gunmetal impeller; stainless-steel jet nozzle; bronze jet venturi tube
FEATURES	
Available custom-built?	Each CBA individually sized for specific application
Performance guarantees	All crane pumps guaranteed for 12 months from date of dispatch; tests meet requirements of BS Spec. 599
Cleaning	Regular service and maintenance contract available for all crane pumps
Design codes applicable	
EXTRAS	Flash fitting
	Speed increasing gear with extended base plate
FURTHER COMMENTS	CBA system is designed to economize in steam plant by recovering condensate at same temperature and pressure at which it is formed without need to trap atmosphere; this leads to considerable fuel savings and other benefits, including: improved process line; efficiency; quality.

MANUFACTURER	*CTC HEAT*	
MODEL NAME AND NUMBER	SKR (0.5R; 1.0R; 1.5R)	SKR-X series
TYPE OF HEAT EXCHANGER	Shell and tube (with finned tubes)	Shell and tube (with cross-spiral indented tube surfaces)
TYPICAL APPLICATIONS	Water to water waste heat recovery	Liquid to liquid waste heat recovery
OPERATING TEMPERATURE RANGE (°C)	Up to 200	150 standard
LIQUID FLOW RATE /l/h)	Various (shell volume up to 1020 l)	Various
PRESSURE DROPS (mm w.g.)	Depends on flow rate →	
MAXIMUM WORKING PRESSURE (kN/m^2)	Shell-side: 600-1600; tube: 1300-2100	Shell-side: 1600; tube: 1600
DUTY (kW)	Various →	
EFFICIENCY (%)	Dependent on duty →	
TYPICAL PAYBACK PERIOD (years)	N/a →	
CONSTRUCTION		
Materials	Copper tubes; mild-steel shell	Stainless-steel tubes, shell
FEATURES		
Available custom-built?	Yes, stainless-steel shell and tube units available (SKR-X series)	
Performance guarantees	Yes →	
Cleaning		
Design codes applicable	Shell to SIS 1330 →	
EXTRAS	Thermal insulation of shell available on some SKR and all SKR-X units; flanged or welded connections available	
FURTHER COMMENTS		

MANUFACTURER	*CURWEN & NEWBERY LTD.*
MODEL NAME AND NUMBER	CN Continuous Blowdown Heat Recovery Unit (Sizes CB1-CB6)
TYPE OF HEAT EXCHANGER	Shell and tube
TYPICAL APPLICATIONS	On continuous blowdown heat recovery units designed for recovering waste heat contained in continuous and intermittent blowdown from steam boilers; heat is normally used for heating boiler feedwater or other hot-water requirements
OPERATING TEMPERATURE RANGE (°C)	Normally between temperature of cold water and blowdown: e.g. 10-212
LIQUID FLOW RATE (l/h)	300-2750
PRESSURE DROPS (mm w.g.)	Shell-side: 7000 Tube: 3000
MAXIMUM WORKING PRESSURE (kN/m^2)	Shell-side: 700 Tube: 2400
DUTY (kW)	Up to 300
EFFICIENCY (%)	60
TYPICAL PAYBACK PERIOD (years)	0.5-2
CONSTRUCTION	
Materials	Shell-side: mild steel Tube-side: copper
FEATURES	
Available custom-built?	CN continuous blowdown heat recovery unit specifically designed for incorporation with standard range of CN blowdown tanks
Performance guarantees	Units to perform as specified within measuring tolerances
Cleaning	Back-flushing of heat recovery unit recommended to remove sludge, etc.; can be achieved by using MCW at 200 kN/m^2
Design codes applicable	Shell generally constructed to BS 1500
EXTRAS	CN blowdown tank; aftercooler; thermostatic control valves; non-return valves and strainers
FURTHER COMMENTS	

MANUFACTURER	*DU PONT DE NEMOURS INTERNATIONAL SA*	
MODEL NAME AND NUMBER	Du Pont Shell and Tube Heat Exchanger - Model 525C-3	Model 105C-3
TYPE OF HEAT EXCHANGER	Shell and tube →	
TYPICAL APPLICATIONS	Heat recovery from corrosive liquids →	
OPERATING TEMPERATURE RANGE (°C)	10-150 →	
LIQUID FLOW RATE (l/h)	Shell-side: 50 000	
PRESSURE DROPS (mm w.g.)		
MAXIMUM WORKING PRESSURE (kN/m^2)	Internal tube: 1000 (10°C)-260 (150°C) External to tube 440 (10°C)-130 (150°C)	As for 525C-3
DUTY (kW)	Dependent on flow rates →	
EFFICIENCY (%)		
TYPICAL PAYBACK PERIOD (years)	N/a →	
CONSTRUCTION		
Materials	Shell-carbon steel; fibreglass-reinforced epoxy Tubes: Teflon fluorocarbon resin	As for 525C-3
FEATURES		
Available custom-built?	Yes →	
Performance guarantees		
Cleaning	Strainers required for particles 25% of tube diameter	Back flush or introduce compressed air to promote turbulence
Design codes applicable	ASME (metal shells) →	
EXTRAS		
FURTHER COMMENTS		

MANUFACTURER	DU PONT DE NEMOURS INTERNATIONAL SA	
MODEL NAME AND NUMBER	Shell and Tube (model 525-ct-5)	Tankcoils (model D 2500-I-5-L)
TYPE OF HEAT EXCHANGER	Shell and tube	Immersion coil
TYPICAL APPLICATIONS	Shell-side: 98% H_2SO_4; temperature 125 ^oC-60 ^oC Tube-side: 37% H_2SO_4; temperature 35 ^oC-55 ^oC Acid reconcentration system: steam generation	Tank-side: 98% H_2SO_4 temperature 120 ^oC-90 ^oC Tube-side: water; 20 ^oC-90 ^oC
OPERATING TEMPERATURE RANGE (^oC)	-30-150 →	→
LIQUID FLOW RATE (l/h)	Shell side: 50 000 kg/h per shell Tube side: 50 000 kg/h	Tank-side: unlimited Tube-side: 50 000-200 000. kg/h
PRESSURE DROPS (mm w.g.)	Depending flow rates very low, average: 0.1 kg/cm^2	Depending flow rates very low, average tube side: 1.0 kg/cm^2
MAXIMUM WORKING PRESSURE (kN/m^2)	1100 , depending on temperature	11 kg/cm^2, depending on temperature
DUTY (kW)	Dependent on flow rates	Dependent on flow rates
EFFICIENCY (%)	No fouling →	→
TYPICAL PAYBACK PERIOD (years)	12 months →	→
CONSTRUCTION		
Materials	Shell-side: carbon steel; enamelled; FEP Teflon-lined; stainless steel Tubes: FEP Teflon	Tank-side: Teflon-lined; brick-lined Tubes: FEP Teflon
FEATURES		
Available custom-built?	Yes →	→
Performance guarantees	1 year →	→
Cleaning	Strainers required for particles 25% of tube diameter	Back-flush; or introduce compressed air to promote turbulence
Design codes applicable	All American and European codes: TUV; Stoomwezen; BS; BR: Boiler code; - ASME, etc.	
EXTRAS	Engineering support after sales service →	→

MANUFACTURER	*ELKSTROM & SON*
MODEL NAME AND NUMBER	
TYPE OF HEAT EXCHANGER	Shell and tube
TYPICAL APPLICATIONS	Chemical and petrochemical industries; food industry
OPERATING TEMPERATURE RANGE (°C)	Up to 500
LIQUID FLOW RATE (l/h)	Up to 4 000 000
PRESSURE DROPS (mm w.g.)	
MAXIMUM WORKING PRESSURE (kN/m^2)	30 000
DUTY (kW)	
EFFICIENCY (%)	
TYPICAL PAYBACK PERIOD (years)	
CONSTRUCTION	
Materials	Carbon steel; stainless steel; high nickel alloys; titanium
FEATURES	
Available custom-built?	Yes
Performance guarantees	Yes
Cleaning	
Design codes applicable	ASME/TEMA; BS; AD-Merkblatt (TUV); Swedish code
EXTRAS	
FURTHER COMMENTS	

MANUFACTURER	*GESTRA (UK) LTD.*
MODEL NAME AND NUMBER	Residual Blowdown Heat Exchangers Type HE, HE-S (16 in range)
TYPE OF HEAT EXCHANGER	Shell and tube
TYPICAL APPLICATIONS	Heat recovery from residual boiler blowdown
OPERATING TEMPERATURE RANGE (°C)	Blowdown inlet typically 105
LIQUID FLOW RATE (l/h)	455-9205
PRESSURE DROPS (mm w.g.)	
MAXIMUM WORKING PRESSURE (kN/m^2)	Shell- and tube-side: 800
DUTY (kW)	Up to 650
EFFICIENCY (%)	Greater than 75 when combined with flash vessel
TYPICAL PAYBACK PERIOD (years)	Less than 1
CONSTRUCTION	
Materials	Steel (shell); copper (tube) on HE model; stainless steel (tube) on HE-S model
FEATURES	
Available custom-built?	Normally from standard range
Performance guarantees	Yes
Cleaning	Normal cleaning techniques: access provided by isolating valves on circuit
Design codes applicable	BS
EXTRAS	
FURTHER COMMENTS	

MANUFACTURER	*HARRIS THERMAL TRANSFER PRODUCTS, INC.*
MODEL NAME AND NUMBER	
TYPE OF HEAT EXCHANGER	Shell and tube
TYPICAL APPLICATIONS	Paper mill
OPERATING TEMPERATURE RANGE (^{o}C)	46/102
LIQUID FLOW RATE (l/h)	Shell-side: 454 000 tube-side: 567 500
PRESSURE DROPS (mm w.g.)	Shell-side: 14 060 tube-side: 14 060
MAXIMUM WORKING PRESSURE (kN/m^2)	Shell-side: 414 Tube-side: 414
DUTY (kW)	29 300
EFFICIENCY (%)	87
TYPICAL PAYBACK PERIOD (years)	3
CONSTRUCTION	
Materials	Carbon steel: (shell) stainless steel (tube)
FEATURES	
Available custom-built?	Custom-built
Performance guarantees	Yes
Cleaning	Chemical
Design codes applicable	ASME section VIII, div. 1
EXTRAS	
FURTHER COMMENTS	

MANUFACTURER	*HEAT ENERGY RECOVERY SERVICES*
MODEL NAME AND NUMBER	Addon Heat Recovery Unit (models 100, 150, 200)
TYPE OF HEAT EXCHANGER	Storage heat exchanger formed from cylindrical water cylinder (copper) with heat exchanger constructed in cylinder walls.
TYPICAL APPLICATIONS	For storing heat energy recovered from refrigeration plant up to 5 kW; operating mode: water-cooled condenser
OPERATING TEMPERATURE RANGE (°C)	Up to 71°C (because of unique feature of construction, but performance related to condensing temperature: 49°C
LIQUID FLOW RATE (l/h)	Storage capacity: 100; 150; 200; (22; 33; 44 gal)
PRESSURE DROPS (mm w.g.)	Refrigeration coil designed and rated for equivalent condensing temperature drop: 2 degC or 5 degC, when installed as a desuperheater
MAXIMUM WORKING PRESSURE (kN/m^2)	Normal refrigeration limits
DUTY (kW)	0.5-6
EFFICIENCY (%)	N/a; unit recovers heat normally wasted by air-cooled condensers
TYPICAL PAYBACK PERIOD (years)	2-5 years (commercial; farm installation: 2-3)
CONSTRUCTION	
Materials	Heavy-gauge copper for cylinder and heat exchanger; thermal insulation, cork and foam; outer casing, welded rust-proofed steel; stone enamel finish or natural
FEATURES	
Available custom-built?	Available as standard units; can be custom-built to suit specific requirements
Performance guarantees	Directly related to original refrigeration system and working load
Cleaning	Not normally needed, but inspection boss and plug provided; anode protection standard to all units
Design codes applicable	Three basic modes: Addon 100 (designed for series connection); 150; 200
EXTRAS	Standard: 2-3 kW immersion heater and thermostats can be fitted; thermometer N.B.: Cold-water feed incorporates drain cock (chromium-plated)
FURTHER COMMENTS	Units are free-standing.

MANUFACTURER	*HIRAKAWA IRON WORKS LTD.*
MODEL NAME AND NUMBER	MP
TYPE OF HEAT EXCHANGER	Shell and tube
TYPICAL APPLICATIONS	Industrial process steam cooling, water on tube side
OPERATING TEMPERATURE RANGE (°C)	
LIQUID FLOW RATE (l/h)	Shell-side: 3.1-9.7 Tube-side: 4.7-18.5
PRESSURE DROPS (mm w.g.)	Shell-side: 10 000-500 000 Tube-side: 10 000-50 000
MAXIMUM WORKING PRESSURE (kN/m^2)	Shell-side: 500-1000 Tube-side: 500-1000
DUTY (kW)	99-465
EFFICIENCY (%)	
TYPICAL PAYBACK PERIOD (years)	Approx: 1
CONSTRUCTION	
Materials	Shell-side: SS (rolled steel for general structure) Tube-side: STB (carbon steel boiler and exchanger tubes)
FEATURES	
Available custom-built?	Yes
Performance guarantees	1 year
Cleaning	
Design codes applicable	Boiler Construction code
EXTRAS	
FURTHER COMMENTS	

MANUFACTURER	*HOLDEN & BROOKE LTD.*
MODEL NAME AND NUMBER	Shell and Tube Heat Exchangers types B; F; H; J; L; R; P
TYPE OF HEAT EXCHANGER	Shell and tube
TYPICAL APPLICATIONS	Heat recovery from effluent streams (liquid or condensing vapours)
OPERATING TEMPERATURE RANGE (°C)	-180-650 (1)
LIQUID FLOW RATE (l/h)	Dependent on requirements
PRESSURE DROPS (mm w.g.)	Dependent on duty
MAXIMUM WORKING PRESSURE (kN/m^2)	38000 (tube-side); 5000 (shell-side) (1)
DUTY (kW)	Dependent on requirements
EFFICIENCY (%)	Dependent on duty
TYPICAL PAYBACK PERIOD (years)	Dependent on duty
CONSTRUCTION	
Materials	Carbon steel; stainless steel; copper and alloys; monel; inconel- incoloy; titanium; bimetal (1)
FEATURES	
Available custom-built?	Yes
Performance guarantees	Yes
Cleaning	Conventional cleaning techniques
Design codes applicable	ASME; TEMA; BS; and other European codes
EXTRAS	
FURTHER COMMENTS	
NOTE	(1) These figures relate to the general product range of Holden and Brooke, which includes: shell and tube gas coolers; calorifiers.

MANUFACTURER	*IMI RANGE LTD.*
MODEL NAME AND NUMBER	Hercules Heat Recovery Water Heating Units
TYPE OF HEAT EXCHANGER	Plain and extended surface tube
TYPICAL APPLICATIONS	Heat recovery in industrial; commercial; and domestic installations, utilizing waste heat from discharge water or other fluids in process operations, laundrettes, dishwashers, refrigeration processes, etc.
OPERATING TEMPERATURE RANGE (°C)	50-100
LIQUID FLOW RATE (l/h)	
PRESSURE DROPS (mm w.g.)	
MAXIMUM WORKING PRESSURE (kN/m^2)	Shell-side: 500; tube-side: 1000
DUTY (kW)	1-100
EFFICIENCY (%)	
TYPICAL PAYBACK PERIOD (years)	2
CONSTRUCTION	
Materials	Copper; copper alloy
FEATURES	
Available custom-built?	Yes
Performance guarantees	By agreement with customer
Cleaning	
Design codes applicable	
EXTRAS	
FURTHER COMMENTS	Units normally designed to suit customer's individual applications.

MANUFACTURER	*JAEGGI LTD.*
MODEL NAME AND NUMBER	
TYPE OF HEAT EXCHANGER	Shell and tube
TYPICAL APPLICATIONS	All liquid-to-liquid heat exchange
OPERATING TEMPERATURE RANGE (^{o}C)	900
LIQUID FLOW RATE (l/h)	As required
PRESSURE DROPS (mm w.g.)	
MAXIMUM WORKING PRESSURE (kN/m^2)	10 000
DUTY (kW)	To suit customer's requirements
EFFICIENCY (%)	
TYPICAL PAYBACK PERIOD (years)	
CONSTRUCTION	
Materials	Mild steel; stainless steel; high-temperature resistance steel; titanium
FEATURES	
Available custom-built?	Yes
Performance guarantees	
Cleaning	Inside and/or outside tubes, dependent on design
Design codes applicable	SVDB (Switzerland); Services des Mines (France); Lloyds; ASME; TUEV (Germany)
EXTRAS	According to customer's specifications
FURTHER COMMENTS	

MANUFACTURER	*PAUL MUELLER CO.*
MODEL NAME AND NUMBER	Mueller Temp-Plate Heat Transfer Surface
TYPE OF HEAT EXCHANGER	Single and double embossed panel-type heat exchanger
TYPICAL APPLICATIONS	Immersion sections; clamp-on sections; vaporizers; water chillers; cryogenic shrouds; fluidized bed dryers; effluent water coolers; tank shells and heads
OPERATING TEMPERATURE RANGE ($^{\circ}C$)	-160-350
LIQUID FLOW RATE (l/h)	480-72000
PRESSURE DROPS (mm w.g.)	Designed to meet customer's requirements
MAXIMUM WORKING PRESSURE (kN/m^2)	Up to 80 bar
DUTY (kW)	
EFFICIENCY (%)	50-95 depends on application
TYPICAL PAYBACK PERIOD (years)	4 months to 3 years, depends on application
CONSTRUCTION	
Materials	Stainless steel; all chrome-nickel alloys, carbon steel
FEATURES	
Available custom-built?	Designed to customer's specifications; all shapes,sizes, etc.
Performance guarantees	Heat transfer calculations included
Cleaning	Simple cleaning: rinse; brush; etc.
Design codes applicable	ASME
EXTRAS	Applications are unlimited: from single units to banks of many units; totally enclosed shell and plate units; falling film cryogenic vaporizers; many energy-saving applications; jacketed shells for ammonia, steam, water, hot oil, Freon, etc.
FURTHER COMMENTS	Send for Mueller Temp-Plate Idea Book, TP-107, and Heat Transfer Catalogue, TP-101.

MANUFACTURER	*NORTHVALE (DIVISION OF BSS LTD.)*
MODEL NAME AND NUMBER	Northvale UW; FW series
TYPE OF HEAT EXCHANGER	Shell and tube
TYPICAL APPLICATIONS	Dye works; laundry; keg-washing plant; condensate cooling; effluent cooling
OPERATING TEMPERATURE RANGE (°C)	0-200

LIQUID FLOW RATE (l/h)	All units are custom-designed and built; figures given are typical only	300 to 100 000 (shell-and tube-side)
PRESSURE DROPS (mm w.g.)		500 to 3000 (shell-and tube-side)
MAXIMUM WORKING PRESSURE (kN/m^2)		2000 (shell-and tube-side)
DUTY (kW)		25 to 3000
EFFICIENCY (%)	Thermal effectiveness	70-95
TYPICAL PAYBACK PERIOD (years)		0.5-5

CONSTRUCTION	
Materials	Carbon steel; stainless steel; copper alloys
FEATURES	
Available custom-built?	Yes
Performance guarantees	Yes, subject to full and correct information being provided by the customer
Cleaning	Provision for mechanical or chemical cleaning
Design codes applicable	BS 5500; BS 1500; BS 3274; ASME section VIII; TEMA
EXTRAS	Control valves and instrumentation; relief valves; bursting discs; strainers; etc.
FURTHER COMMENTS	Units can be supplied fully packaged with controls and instrumentation; alternative units can be supplied to recover heat from flash steam.

MANUFACTURER	*OSLO SVEISEBEDRIFT*
MODEL NAME AND NUMBER	Series 705 and 715
TYPE OF HEAT EXCHANGER	Shell and tube
TYPICAL APPLICATIONS	Industry; ships; schools; hospitals; swimming-pools
OPERATING TEMPERATURE RANGE (^{o}C)	Max. steam pressure: 1000 kN/m^2; normally: in, 90-70 out, 5-55; (or in, 11-90; out, 60-80)
LIQUID FLOW RATE (l/h)	20-700 l/min
PRESSURE DROPS (mm w.g.)	0.5-1.5
MAXIMUM WORKING PRESSURE (kN/m^2)	Shell: 500 Tube (copper): 1000
DUTY	25-1800 Mcal/h
EFFICIENCY (%)	
TYPICAL PAYBACK PERIOD (years)	3
CONSTRUCTION	
Materials	Shell: steel (st 37) or stainless steel; heat exchanger: copper tube
FEATURES	
Available custom-built?	All diameters in 580 mm
Performance guarantees	2 years
Cleaning	
Design codes applicable	
EXTRAS	
FURTHER COMMENTS	

MANUFACTURER	*RYCROFT (CALORIFIERS) LTD.*
MODEL NAME AND NUMBER	NSSZF
TYPE OF HEAT EXCHANGER	Two - shell and tube heat exchangers in series
TYPICAL APPLICATIONS	Heat extraction from dye effluent and other process liquids
OPERATING TEMPERATURE RANGE (°C)	115-45
LIQUID FLOW RATE (l/h)	Shell-side: 11000; tube-side: 18000
PRESSURE DROPS (mm w.g.)	Shell-side: 1500; tube-side: 3000
MAXIMUM WORKING PRESSURE (kN/m^2)	Shell-side: 1000; tube-side: 1000
DUTY (kW)	915
EFFICIENCY (%)	
TYPICAL PAYBACK PERIOD (years)	
CONSTRUCTION	
Materials	Stainless steel throughout
FEATURES	
Available custom-built?	Yes
Performance guarantees	Normal
Cleaning	Chemical
Design codes applicable	BS 3274
EXTRAS	
FURTHER COMMENTS	Custom-built calorifiers designed to raise temperature of domestic hot water or industrial hot water using hot effluents discharged to waste; storage or non-storage recovery systems.

MANUFACTURER	*FIRMA SCHIFF & STERN KG*
MODEL NAME AND NUMBER	Schiffstern-Thermex (Size 0-13 (108-650mm); Length 1-4 (1300mm-3500mm overall)
TYPE OF HEAT EXCHANGER	U-tube counter-flow heat exchanger with extractable nest of tubes
TYPICAL APPLICATIONS	For district heating systems; warm-water preparation; industrial appliance (heat recovery, cooling and heating of process water)
OPERATING TEMPERATURE RANGE ($^{\circ}$C)	Up to 350
LIQUID FLOW RATE (l/h)	Tube-side: 300 (size 0)-300.000 (size 13); shell-side: 1500 (size 0)-360.000 (size 13)
PRESSURE DROPS (mm w.g.)	Tube-side: up to 5000; shell-side: up to 7000 (as a function of flow velocity)
MAXIMUM WORKING PRESSURE (kN/m^2)	Tube-side: up to 4000; shell-side: up to 2500
DUTY (kW)	Function of inlet and outlet temperature tube-side and shell-side
EFFICIENCY (%)	99
TYPICAL PAYBACK PERIOD (years)	
CONSTRUCTION	
Materials	Carbon steel; stainless steel; combinations other tube and tube plate materials available
FEATURES	
Available custom-built?	Many sizes ex-stock; others within 2-4 months
Performance guarantees	1 year from date of delivery
Cleaning	Easy: bundle extractable; cleaning with cheap chemicals
Design codes applicable	AD Merkblatter
EXTRAS	Special design to customer's requirements
FURTHER COMMENTS	Design is back by three Austrian patents; applications for patents abroad pending.

MANUFACTURER	*FIRMA SCHIFF & STERN KG*
MODEL NAME AND NUMBER	Spiral Heat Exchanger (in various sizes: shell diameter: 115-1240mm; 1200-4000mm long)
TYPE OF HEAT EXCHANGER	Counterflow HX, with specially wound tubes and extractable nest of tubes; wide range of working pressure and material
TYPICAL APPLICATIONS	For district heating systems; warm-water preparation; industrial appliance (heat recovery, cooling and heating or process water)
OPERATING TEMPERATURE RANGE (°C)	Up to 350
LIQUID FLOW RATE (l/h)	Tube-side: 200-500 000; shell-side: 1500-80 000 (depending on size)
PRESSURE DROPS (mm w.g.)	Tube-side: up to 6000; shell-side: up to 500 depending on the flow velocity)
MAXIMUM WORKING PRESSURE (kN/m^2)	Tube-side: up to 4000; shell-side: up to 2500
DUTY (kW)	10-10 000, depending on working conditions
EFFICIENCY (%)	99
TYPICAL PAYBACK PERIOD (years)	
CONSTRUCTION	
Materials	Shell: carbon steel, plain or coated, and stainless steel; tubes and tube plates: steel, stainless steel; nickel-bronze; copper
FEATURES	
Available custom-built?	Many sizes ex-stock; others within 2-4 months
Performance guarantees	1 year from date of delivery
Cleaning	Easy: bundle extractable; cleaning with cheap chemicals
Design codes applicable	AD Merkblatter
EXTRAS	Special design to customer's requirements
FURTHER COMMENTS	Main feature: upright installation; extreme small space required; many international patents granted.

MANUFACTURER	*W. SCHMIDT KG*
MODEL NAME AND NUMBER	Sigma
TYPE OF HEAT EXCHANGER	Sigma plate heat exchangers for liquid-to-liquid applications; also with dry substance with particles smaller than 1mm; spiral heat exchangers for liquid-to-liquid applications; also dry substances with particles sized up to 20mm
TYPICAL APPLICATIONS	Pasteurizer in food industry for milk, beer and fruit juices; in the chemical industry for all pumpable media
OPERATING TEMPERATURE RANGE (°C)	When using soft gaskets (nitrile-rubber, ethylene-propylene rubber, viton, silicon) up to 150; for higher temperatures, hard gaskets (klingerite, oilit) are used
LIQUID FLOW RATE (l/h)	Up to 1 500 per heat exchanger
PRESSURE DROPS (mm w.g.)	Depending on the grouping of the plates or on height of gaps (spiral heat exchangers)
MAXIMUM WORKING PRESSURE (kN/m^2)	1600
DUTY (kW)	Up to 50 000 per exchanger
EFFICIENCY (%)	Plate heat exchangers: up to 93; spiral heat exchangers: up to approx. 85
TYPICAL PAYBACK PERIOD (years)	
CONSTRUCTION	
Materials	Stainless steels: 4301/4541, 4401/4571, 4449; titanium; titanium-palladium; hastelloy C; hastelloy B; incolloy
FEATURES	
Available custom-built?	Yes
Performance guarantees	Indicated temperatures and pressures as well as the materials are guaranteed
Cleaning	CIP by means of circulation of cleaning solution or mechanical cleaning by opening the heat exchanger
Design codes applicable	Construction according to the German AD Merkblatter
EXTRAS	Diameter of connections according to customer's requirements and to flow rate; different kinds of supports, closure by means of screws or central closure; stainless steel lagging of support and closing lid is possible
FURTHER COMMENTS	

MANUFACTURER	*SENIOR PLATECOIL LTD.*
MODEL NAME AND NUMBER	Series H Plate Coils (series P for special duties)
TYPE OF HEAT EXCHANGER	Plate
TYPICAL APPLICATIONS	Process liquid-to-liquid heat recovery, ideally suited to process vessel heating
OPERATING TEMPERATURE RANGE (°C)	Up to 400°C (for most heat transfer fluids)
LIQUID FLOW RATE (l/h)	To customer's requirements
PRESSURE DROPS (mm w.g.)	To customer's requirements
MAXIMUM WORKING PRESSURE (kN/m^2)	Plate: 2400 (higher with seriesP)
DUTY (kW)	To customer's requirements
EFFICIENCY (%)	Dependent on duty
TYPICAL PAYBACK PERIOD (years)	Dependent on duty
CONSTRUCTION	
Materials	Stainless steel or titanium (as standard)
FEATURES	
Available custom-built?	Yes
Performance guarantees	
Cleaning	Flushing out, external spraying or brushing
Design codes applicable	Safety factor: 5:1
EXTRAS	Fabricated in other materials; optional connections and distribution systems (internal)
FURTHER COMMENTS	

MANUFACTURER	*SERCK HEAT TRANSFER*
MODEL NAME AND NUMBER	AA 12; 13 Tubular Heat Exchangers
TYPE OF HEAT EXCHANGER	Shell and tube
TYPICAL APPLICATIONS	Low- and high-grade heat recovery
OPERATING TEMPERATURE RANGE (°C)	Up to 250
LIQUID FLOW RATE (l/h)	To customer's requirements
PRESSURE DROPS (mm w.g.)	To customer's requirements
MAXIMUM WORKING PRESSURE (kN/m^2)	1000 (but can be increased with design modifications)
DUTY (kW)	To customer's requirements
EFFICIENCY (%)	Dependent on duty
TYPICAL PAYBACK PERIOD (years)	Dependent on duty
CONSTRUCTION	
Materials	Shell-side: brass; copper nickel; stainless steel; titanium (tubes, tube plates and baffles); Shell-side: mild steel or cast iron (shell); Tube-side: mild steel; cast iron or gunmetal (channels)
FEATURES	
Available custom-built?	Yes, either assembled from standard parts or designed and built for specific applications
Performance guarantees	Thermal and pressure performance guaranteed, providing fluid flows, temperatures and pressures maintained at design levels
Cleaning	In-situ by chemical processes or by tube stack removal (shell-side); brushing through, utilizing removable end-covers on channels (tube-side)
Design codes applicable	Most national codes as required
EXTRAS	Temperature controls (if required); changeover valves for multiple modules
FURTHER COMMENTS	Materials are chosen for specific fluids being handled.

MANUFACTURER	*SERCK HEAT TRANSFER*
MODEL NAME AND NUMBER	BR30 Plate Heat Exchangers
TYPE OF HEAT EXCHANGER	Plate
TYPICAL APPLICATIONS	Low-grade heat recovery
OPERATING TEMPERATURE RANGE (°C)	Up to 150
LIQUID FLOW RATE (l/h)	To customer's requirements
PRESSURE DROPS (mm w.g.)	To customer's requirements
MAXIMUM WORKING PRESSURE (kN/m^2)	1000
DUTY (kW)	To customer's requirements
EFFICIENCY (%)	Depends on duty
TYPICAL PAYBACK PERIOD (years)	Depends on duty
CONSTRUCTION	
Materials	Plate: stainless steel; titanium (mild steel frames)
FEATURES	
Available custom-built?	Yes, assembled from standard units
Performance guarantees	Thermal and pressure performance guaranteed, providing fluid flows, temperatures and pressures maintained at design levels
Cleaning	CIP; or simple plate removal
Design codes applicable	None
EXTRAS	Up to 2 sets of plates can be accommodated within a single frame (depending on performance requirements) for compactness
FURTHER COMMENTS	Seals provided for any fluids being handled.

MANUFACTURER	*STRUTHERS WELLS CORP.*
MODEL NAME AND NUMBER	Equipment custom-designed; standard models not available
TYPE OF HEAT EXCHANGER	Shell and tube
TYPICAL APPLICATIONS	Boiler feedwater preheat by process-plant effluent streams
OPERATING TEMPERATURE RANGE (°C)	Shell-side; tube-side: 200-600
LIQUID FLOW RATE (l/h)	Shell-side; tube-side: 50 000-1 500 000
PRESSURE DROPS (mm w.g.)	Shell-side; tube-side: 500-7500
MAXIMUM WORKING PRESSURE (kN/m^2)	Shell-side; tube-side: 2450
DUTY (kW)	Shell-side; tube-side: 40 000-1 000 000
EFFICIENCY (%)	85-90
TYPICAL PAYBACK PERIOD (years)	2-6
CONSTRUCTION	
Materials	Full range
FEATURES	
Available custom-built?	Yes
Performance guarantees	Yes
Cleaning	Chemical or mechanical
Design codes applicable	ASME section I; ASME section VIII
EXTRAS	Full range of controls; bypass arrangements; special tube-end welding techniques
FURTHER COMMENTS	

MANUFACTURER	*AB SVENSKA MASKINVERKEN*
MODEL NAME AND NUMBER	Bendek B-BR (with built-in drain cooler)
TYPE OF HEAT EXCHANGER	Shell and tube with spirally winded tubes
TYPICAL APPLICATIONS	Cargo heating in acid and product ships; heavy oil heating
OPERATING TEMPERATURE RANGE (^{o}C)	Up to 400
LIQUID FLOW RATE (l/h)	5000-100 000
PRESSURE DROPS (mm w.g.)	5000-10 000
MAXIMUM WORKING PRESSURE (kN/m^2)	250 000
DUTY (kW)	50-1000
EFFICIENCY (%)	98
TYPICAL PAYBACK PERIOD (years)	2
CONSTRUCTION	
Materials	Carbon and stainless steel
FEATURES	
Available custom-built?	Standard
Performance guarantees	2 years
Cleaning	Self-cleaning
Design codes applicable	DIN; LR; BV; DNV; ABS
EXTRAS	The heaters are provided with built-in drain cooler which gives a reduced steam consumption of up to 15%
FURTHER COMMENTS	

MANUFACTURER	*AB SVENSKA MASKINVERKEN*
MODEL NAME AND NUMBER	Sunrod CAA-FAA (with built-in drain cooler)
TYPE OF HEAT EXCHANGER	Shell and tube with extended surface
TYPICAL APPLICATIONS	Oil heating in storage tanks and cargo heating in ships; oil heating for steam generators
OPERATING TEMPERATURE RANGE (°C)	Up to 225
LIQUID FLOW RATE (l/h)	5000-2 500 000
PRESSURE DROPS (mm w.g.)	65 000
MAXIMUM WORKING PRESSURE (kN/m^2)	50-5000
DUTY (kW)	
EFFICIENCY (%)	98
TYPICAL PAYBACK PERIOD (years)	2
CONSTRUCTION	
Materials	Carbon steel
FEATURES	
Available custom-built?	Standard
Performance guarantees	2 years
Cleaning	Self-cleaning
Design codes applicable	ASME; DIN; LR; DNV; ABS; BV; NK; GL
EXTRAS	The heaters are provided with built-in drain cooler which gives a reduced steam consumption of up to 15%
FURTHER COMMENTS	Sunrod type units also available for heat recovery from all types of prime movers (exhaust up to 650°C). Steel tubing (DIN 35.8) is used, and sootblowers or water nozzles are provided for cooling. Silencers are available as extras and units may be used for steam-raising.

MANUFACTURER	*TRANTER INC.*
MODEL NAME AND NUMBER	Platecoil
TYPE OF HEAT EXCHANGER	Platecoil immersion coils for heating and cooling
TYPICAL APPLICATIONS	Heating and cooling plating solutions and numerous metal-cleaning solutions; heating chemical and petroleum storage tanks; heating and cooling chemical reactors
OPEATING TEMPERATURE RANGE (°C)	-20-340
LIQUID FLOW RATE (l/h)	Up to 11340
PRESSURE DROPS (mm w.g.)	Various
MAXIMUM WORKING PRESSURE (kN/m^2)	Up to 2750 (heavy-duty plates)
DUTY (kW)	Various
EFFICIENCY (%)	Dependent on duty
TYPICAL PAYBACK PERIOD (years)	Various
CONSTRUCTION	
Materials	Carbon steel; stainless steels; titanium; hastelloy; monel; nickel
FEATURES	
Available custom-built?	Yes
Performance guarantees	In certain cases
Cleaning	Chemical cleaning
Design codes applicable	ASME
EXTRAS	Fabrication into banks; suction heaters; storage-tank heaters; vessels- drum warmers; boxes and numerous other assemblies
FURTHER COMMENTS	While this type of heat exchanger is available as a standard 'plate', it may be fabricated into many shapes.

MANUFACTURER	*URANUS SA*
MODEL NAME AND NUMBER	Uranus UP - US
TYPE OF HEAT EXCHANGER	Plate
TYPICAL APPLICATIONS	Swimming-pool, hot domestic water, heat recovery equipment
OPERATING TEMPERATURE RANGE (^{o}C)	Up to 130
LIQUID FLOW RATE (l/h)	100.000
PRESSURE DROPS (mm w.g.)	10 bar
MAXIMUM WORKING PRESSURE (kN/m^{2})	1500
DUTY (kW)	5000
EFFICIENCY (%)	98
TYPICAL PAYBACK PERIOD (years)	5
CONSTRUCTION	
Materials	Stainless-steel; titanium
FEATURES	
Available custom-built?	Normally built up from standard plates to any size
Performance guarantees	Yes
Cleaning	On-line flushing, plates can be split for access
Design codes applicable	
EXTRAS	
FURTHER COMMENTS	

MANUFACTURER	*VICARB SA*
MODEL NAME AND NUMBER	ECM (4 types); V (8 types)
TYPE OF HEAT EXCHANGER	ECM block-type graphite heat exchanger; V plate heat exchanger
TYPICAL APPLICATIONS	ECM heat recovery from corrosive liquid to corrosive liquid; V, many kinds of heat recovery
OPERATING TEMPERATURE RANGE (°C)	ECM: 0-200; V: 0-150
LIQUID FLOW RATE (l/h)	ECM: up to 500 000; V: Up to 3 500 000
PRESSURE DROPS (mm w.g.)	
MAXIMUM WORKING PRESSURE (kN/m^2)	ECM: 700; V: 1500
DUTY (kW)	
EFFICIENCY (%)	
TYPICAL PAYBACK PERIOD (years)	
CONSTRUCTION	
Materials	ECM: graphite; V: stainless steel; titanium; nickel; hastelloy; monel; titanium
FEATURES	
Available custom-built?	Yes
Performance guarantees	Yes
Cleaning	Chemical or manual
Design codes applicable	European and other foreign
EXTRAS	
FURTHER COMMENTS	

11 Heat recovery from prime movers – equipment

Data has already been presented in Chapters 8 and 9 on gas–gas and gas–liquid heat recovery devices, several of which can be used to recover heat from prime movers. This is particularly true of the waste heat boilers detailed in Chapter 9. However, a number of manufacturers either specialize solely in heat recovery from prime movers, or include such systems, giving them a separate identity, in their product range. In addition, although comparatively small in number in both categories, some manufacturers of prime movers offer 'packages' with waste heat recovery equipment, and prime themselves, particularly steam turbines, are available as users of waste heat for power generation.

The recovery of heat from reciprocating engines and gas turbines (and in some cases, steam turbines) is necessary and desirable in order to make up for the low thermal efficiency of the units, and can be of benefit in many industrial processes, providing high-grade heat. (One particular application of prime mover heat recovery, gas engine-driven heat pumps – is discussed in Chapter 6.) For a detailed discussion of the techniques which may be used, and the benefits and applications, the reader is recommended to consult Chapter 4.

11.1 Data presentation

The equipment covered by this chapter includes the following:

Waste heat boilers;
Reboilers;
Recuperators;
Regenerators;
Thermal fluid heaters;
Gas turbines;
Steam turbines;
Economizers.

Obviously many of the above types of heat recovery equipment will be found, in slightly different forms, in other data chapters, and detailed descriptions will be found in the corresponding introductory chapters at the beginning of this Directory.

Because of the wide spread of devices for heat recovery from prime movers, the difficulty in presenting consistent data may be evident to the reader. Many systems are custom-built, however, and it is hoped that this chapter will at least provide the potential user with an idea of the types of equipment available and their relative efficiencies and suitabilities for different types of prime movers.

11.1.1 *Format of data*

The data on prime mover waste heat recovery is set out in the following manner:

(1) *Equipment manufacturer*: The prime manufacturer or subsidiary company is given by name. Addresses, telephone number and product categories of each manufacturer are listed in Appendix 1, together with agents, where known. In addition, the manufacturer was asked to identify in this section whether the manufactures the prime mover and/or the heat recovery device.
(2) *Model number and name*: The model name and, where applicable, model number are given for ease of identification. In many cases, systems are custom-built, and while these may be given a proprietary-name, model numbers are generally omitted. Where it aids identification, the type of heat recovery unit is named.
(3) *Type of prime mover*: The type of prime mover to which the heat recovery equipment may be applied is given.
(4) *Flow volumes*: The flow volumes of both the prime mover exhaust and the supply-side are given. The fluid on the supply-side may be a gas, liquid or vapour, and in the case of a turbine used to recover waste heat, only the exhaust-side flow is relevant, as the heat is converted by the turbine into electrical or mechanical energy.
(5) *Pressure drop*: Most prime movers have maximum permissible pressure drops on the exhaust gas-side, and for continuing satisfactory performance, a waste heat unit so applied must meet this requirement. Where appropriate, pressure drops are given for the supply-side fluid.
(6) *Operating temperature range*: The operating temperature range, normally given in terms of the maximum exhaust gas temperature accepted, is given. In some cases the operating temperature may be dictated by the choice of fluid used on the supply-side of the heat exchanger. For example, units for raising steam will have a different operating temperature range, perhaps, than those used to heat a high-temperature organic fluid.
(7) *Duty*: The heat recovered, or the power output (in the case of turbines) is given. Unless stated otherwise, this should be regarded as a maximum output figure.
(8) *Proportion of exhaust heat recovered*: The amount of heat which the heat recovery system will take out of the exhaust gas stream is given, normally expressed as a percentage. Where a turbine itself is used as the energy recovery system, the duty is given.

(9) *Typical payback period*: The return on the investment, expressed in terms of the installed cost of the system related to the energy or fuel savings obtained, is given. This is normally 1 to 3 years, but can be considerably longer for this type of equipment.
(10) *Construction*: Materials of construction and, where it can be quantified, the weight range of the equipment are given. In the vast majority of prime mover heat recovery applications (excluding those where a turbine may be used on the end of an industrial-process exhaust), the waste gas conditions are accurately known and provided that the dew point is not reached, materials suitable for such heat recovery devices are well proven in this role.
(11) *Features*: In common with the other types of heat recovery equipment described in this book, there are numerous other parameters of interest to a potential purchaser, and some of the more important ones are listed so that manufacturers may comment on their applicability to their product range:

(i) Available custom-built: manufacturers fall into three categories here. In the case of heat recovery from reciprocating engine exhaust gases, more are offering standard items of equipment designed to match the principal manufacturers' products (e.g. Waukesha, Caterpillar and Cummings engines), and a manufacturer is commonly able to provide performance charts for each exchanger, appropriate to each range of engine and fuel type available. This is particularly true of manufacturers in the USA. However, a number of manufacturers offer custom-built systems. A similar situation exists with gas turbine waste heat recovery; for example, Eclipse Lookout produce a packaged system in a number of size ranges. In this area, custom-built systems still predominate. A number of manufacturers in the prime mover heat recovery sector build only custom-designed units.

(ii) Performance guarantees: performance guarantees are given in almost all cases. As in other instances, these rest upon the accuracy of the conditions specified to the manufacturer concerning the environment into which the equipment will be placed. In the case of prime movers, most well-established manufacturers will have sufficient experience to know such conditions very accurately.

(iii) Installation method: unless the equipment is of such a size as to necessitate construction onsite, a packaged unit ready for connection to a suitably prepared exhaust duct may be appropriate in the majority of cases. Some manufacturers incorporate all controls in the 'package', others may offer these as optional extras, possibly necessitating more extensive sitework.

(iv) Installer: installation may be carried out by the manufacturer, the manufacturer of the prime mover, a contractor nominated by any of the parties or the user himself. Some manufacturers leave the initiative in this respect to the customer, offering assistance if requested.

(v) Cleaning methods: techniques for cleaning the heat exchanger surfaces are identified. These may only be recommended by the manufacturer, or provision may be made for them in the heat recovery 'package' or they may be available as optional extras.

(vi) Corrosion resistance: as discussed above (materials of construction), the corrosion resistance of these units is generally predictable, and within limits laid down by the user/manufacturer.

(vii) Condensate collection: provision may be made for draining-off condensate, but operation of most of the heat recovery systems described here relies for long life on the fact that condensation should be avoided.

(viii) Performance control and turndown range: the method of performance control, if required, and the turndown range, necessary to match reduced load on the prime mover, are specified where possible.

(ix) Silencers: heat recovery units on reciprocating engines are available integral with a silencer. Where appropriate, manufacturers indicate whether a silencer is integral or an extra unit.

(12) *Extras available*: Optional extras covering cleaning, control, etc. are commonly available, and manufacturers were given an opportunity to list these. They include sootblowers, control systems, superheaters, auxiliary firing equipment, economizers and silencers.

(13) *Further comments*: A space was provided on the questionnaire for manufacturers to add any comments which they felt could be of value in equipment selection. These are reproduced here.

Manufacturers' data sheets relating to equipment for heat recovery from prime movers follow on the next page.

MANUFACTURER	*ABCO INDUSTRIES INC.*
MODEL NAME AND NUMBER	Natural-circulation heat exchangers NWSB or Forced-circulation FWSB models 1-50
TYPE OF PRIME MOVER	Gas turbine; reciprocating gas; or diesel engine
FLOW VOLUMES (m^3/h)	
Exhaust-side	180 000 (typical)
Supply-side	360 000
PRESSURE DROP (mm w.g.)	
Exhaust-side	65 (ave.)
Supply-side	Negligible in natural circulation type; up to 1000 mm in forced circulation
OPERATING TEMPERATURE RANGE (°C)	225 (steam); 370 (circulating thermal fluids)
DUTY (kW)	14 500
PROPORTION OF EXHAUST HEAT RECOVERED (%)	65 (typical)
TYPICAL PAYBACK PERIOD (years)	1 or less
CONSTRUCTION	
Materials	Carbon steel
Weight range (kg)	To 25 000
FEATURES	
Available custom-built?	All units
Performance guarantee	All units
Installation method	Others
Installed by?	Contractor
Cleaning	Manual or steam/air sootblower
Corrosion resistance	Per specification/application
Condensate collection	As required
Performance control and turndown range	
Silencer	None
EXTRAS	Sootblowers; controls; pumping equipment; superheaters; auxiliary firing equipment
FURTHER COMMENTS	

MANUFACTURER	*SA BABCOCK BELGIUM NV*	
MODEL NAME AND NUMBER	Diesel Engine Heat Exchanger	Gas Turbine Heat Exchanger
TYPE OF PRIME MOVER	Diesel engines, static or marine	Gas turbine
FLOW VOLUMES (m^3/h)		
Exhaust-side	20 000–200 000	100 000–1 200 000 or above
Supply-side		
PRESSURE DROP (mm w.g.)		
Exhaust-side	100	50
Supply-side		
OPERATING TEMPERATURE RANGE (°C)	At inlet: 250–400; at outlet: 200–300	At inlet: 400–700; at outlet: 200
DUTY (kW)	700–7000	10 000–200 000
PROPORTION OF EXHAUST HEAT RECOVERED (%)	20–60	60–75
TYPICAL PAYBACK PERIOD (years)	1 ⟶	
CONSTRUCTION		
Materials	Mild steel ⟶	
Weight range (kg)	8000	Up to 500 000
FEATURES		
Available custom-built?	Custom-built or standard ⟶	
Performance guarantee	Steam production	Steam production; temperature gas-outlet temperature
Installation method	Package ⟶	
Installed by?	Manufacturer ⟶	
Cleaning	No ⟶	
Corrosion resistance	No risk due to choice of metal temperature ⟶	
Condensate collection	No ⟶	
Performance control and turndown range	From 0–100 ⟶	
Silencer	Boiler acts as a silencer ⟶	
EXTRAS		
FURTHER COMMENTS	Material proposed above: La Mont waste heat recovery boilers ⟶	
		We can also propose natural circulation boilers.
	Above ranges are typical and not considered as limits. ⟶	

MANUFACTURER	*BEVERLEY CHEMICAL ENGINEERING LTD.*
MODEL NAME AND NUMBER	Custom-built
TYPE OF PRIME MOVER	Gas turbine or reciprocating diesel
FLOW VOLUMES (m^3/h)	
Exhaust-side	Up to 80 000 Nm^3/h
Supply-side	Up to 300 m^3/h liquid; 80 000 Nm^3/h air
PRESSURE DROP (mm w.g.)	
Exhaust-side	0-500
Supply-side	0-40000 (liquid), 0-500 (air)
OPERATING TEMPERATURE RANGE (°C)	Dependent on prime mover, normally 600°C max. inlet down to 200°C outlet
DUTY (kW)	To suit prime mover, 15 000
PROPORTION OF EXHAUST HEAT RECOVERED (%)	Up to 65
TYPICAL PAYBACK PERIOD (years)	1-2
CONSTRUCTION	
Materials	Carbon or stainless steel
Weight range (kg)	Depends on duty
FEATURES	
Available custom-built?	Yes
Performance guarantee	Available
Installation method	Supplied as package
Installed by?	Manufacturer
Cleaning	Facility built in, if required
Corrosion resistance	In design
Condensate collection	
Performance control and turndown range	Incorporated to requirements
Silencer	
EXTRAS	
FURTHER COMMENTS	

MANUFACTURER	*BONO GMBH*
MODEL NAME AND NUMBER	RA Steam Boilers
TYPE OF PRIME MOVER	Reciprocating diesel engine
FLOW VOLUMES (m^3/h)	
Exhaust-side	10 000-100 000 Nm^3/h
Supply-side	
PRESSURE DROP (mm w.g.)	
Exhaust-side	50-130
Supply-side	
OPERATING TEMPERATURE RANGE (^{o}C)	Inlet temperature: 310-420
DUTY (kW)	
PROPORTION OF EXHAUST HEAT RECOVERED (%)	50-60
TYPICAL PAYBACK PERIOD (years)	
CONSTRUCTION	
Materials	Tubes: ASTM 106B
Weight range (kg)	6000-15 000
FEATURES	
Available custom-built?	
Performance guarantee	
Installation method	
Installed by?	Contractor
Cleaning	
Corrosion resistance	
Condensate collection	If required
Performance control and turndown range	
Silencer	
EXTRAS	
FURTHER COMMENTS	Steam boilers of horizontal-finned tubes, forced-circulation type, with steam drum incorporated or separate; with or without gas bypass duct.

MANUFACTURER	*BRONSWERK HEAT TRANSFER BV*
MODEL NAME AND NUMBER	Heat Exchanger
TYPE OF PRIME MOVER	Flue gases of diesel engines and gas engines
FLOW VOLUMES (m^3/h)	
Exhaust-side	Up to 100 000
Supply-side	Water-air; thermo-oil 30 000-1 000 000 kg
PRESSURE DROP (mm w.g.)	
Exhaust-side	Approx. 200
Supply-side	Air: 1000 max.; fluids: 10 000 max.
OPERATING TEMPERATURE RANGE (°C)	Up to 250
DUTY (kW)	300 000
PROPORTION OF EXHAUST HEAT RECOVERED (%)	20-50
TYPICAL PAYBACK PERIOD (years)	1
CONSTRUCTION	
Materials	Metals, normally stainless on exhaust-side
Weight range (kg)	Units up to 20 tonne
FEATURES	
Available custom-built?	Yes
Performance guarantee	Yes
Installation method	Connected to suitable flanged fittings
Installed by?	Contractor
Cleaning	Manual
Corrosion resistance	Appropriate to materials
Condensate collection	Not normally required
Performance control and turndown range	By flow modulation
Silencer	Available
EXTRAS	
FURTHER COMMENTS	

MANUFACTURER	*COALTECH*
MODEL NAME AND NUMBER	Gas Turbine Regenerator
TYPE OF PRIME MOVER	Gas turbine
FLOW VOLUMES (m^3/h)	
Exhaust-side	156 530
Supply-side	146 810
PRESSURE DROP (mm w.g.)	
Exhaust-side	350
Supply-side	2380
OPERATING TEMPERATURE RANGE (°C)	480-540
DUTY (kW)	5480
PROPORTION OF EXHAUST HEAT RECOVERED (%)	87
TYPICAL PAYBACK PERIOD (years)	Less than 1
CONSTRUCTION	
Materials	Shell (carbon steel); tubes and manifolds (300 stainless steel)
Weight range (kg)	31 800
FEATURES	
Available custom-built?	Yes
Performance guarantee	Full factory warranty for 1 year
Installation method	Factory-assembled components
Installed by?	Contractor
Cleaning	Not required
Corrosion resistance	Yes
Condensate collection	
Performance control and turndown range	Custom-built
Silencer	By others
EXTRAS	
FURTHER COMMENTS	All designs customed-engineered to suit installations.

MANUFACTURER	*CONSECO*
MODEL NAME AND NUMBER	Conseco Heat Recovery Boilers
TYPE OF PRIME MOVER	Gas turbine and other internal combustion engines
FLOW VOLUMES (m^3/h)	
Exhaust-side	41 231 kg/h at 530°C (1)
Supply-side	6864 kg/h (steam) at 121°C
PRESSURE DROP (mm w.g.)	
Exhaust-side	152 (1)
Supply-side	0.703 kg/cm^2
OPERATING TEMPERATURE RANGE (°C)	Up to 540 (930 in special applications) (2)
DUTY (kW)	750-5400
PROPORTION OF EXHAUST HEAT RECOVERED (%)	65 (hot water and low pressure steam); 75 (high pressure steam raising)
TYPICAL PAYBACK PERIOD (years)	Approx. 3 (taking into account operating/maintenance costs)
CONSTRUCTION	
Materials	
Weight range (kg)	
FEATURES	
Available custom-built?	Yes
Performance guarantee	Yes, each unit tested at factory under onsite conditions
Installation method	Supplied as package for connecting to services
Installed by?	Contractor
Cleaning	Ease of accessibility for cleaning internally
Corrosion resistance	Yes, and feedwater treatment available for boilers
Condensate collection	N/a
Performance control and turndown range	Boilers can be fully automatic to produce varying quantities of fixed-quality steam, depending on turbine output
Silencer	N/a
EXTRAS	Economizers; superheaters; supplementary firing
FURTHER COMMENTS	Gas bypass control system incorporated; system can be used to heat water; raise steam; or heat a high-temperature heat transfer fluid.
NOTES	(1) Example quoted for Conseco boiler on Ruston TA 1750 gas turbine.
	(2) Can be applied to raise steam from incinerator waste heat, gases exhausting from the incinerator at temperatures approaching 950°C .

MANUFACTURER	*DANKS OF NETHERTON LTD.*
MODEL NAME AND NUMBER	Custom-built
TYPE OF PRIME MOVER	Gas turbine and reciprocating diesel
FLOW VOLUMES (m^3/h)	
Exhaust-side	
Supply-side	75 000 stp (typical)
PRESSURE DROP (mm w.g.)	
Exhaust-side	
Supply-side	100
OPERATING TEMPERATURE RANGE (^{o}C)	1100, from prime mover
DUTY (kW)	6700 (typical)
PROPORTION OF EXHAUST HEAT RECOVERED (%)	55 (typical)
TYPICAL PAYBACK PERIOD (years)	Less than 1
CONSTRUCTION	
Materials	To shell and water tube boiler codes
Weight range (kg)	7500 (typical)
FEATURES	
Available custom-built?	Yes
Performance guarantee	Yes
Installation method	As package unit
Installed by?	Danks Technical Services Ltd.
Cleaning	Yes
Corrosion resistance	
Condensate collection	
Performance control and turndown range	Custom-built
Silencer	
EXTRAS	Duct and other burners for supplementary firing
FURTHER COMMENTS	

MANUFACTURER	*ECLIPSE LOOKOUT CO.*
MODEL NAME AND NUMBER	Packaged Thermal Fluid Heat Recovery Heaters (6PTL-75PTL)
TYPE OF PRIME MOVER	Gas turbine
FLOW VOLUMES (m^3/h)	
Exhaust-side	12 000 (6 PTL)-150 000 (75 PTL)
Supply-side	Thermal fluid to suit (exit temperature, typically 315°C)
PRESSURE DROP (mm w.g.)	
Exhaust-side	N/a
Supply-side	To user's requirements
OPERATING TEMPERATURE RANGE (°C)	Suit all gas turbine exhausts: e.g. 500
DUTY (kW)	439 (6PTL)-5272(75PTL) (based on turbine exhaust, 427°C; exit from heater, 315°C)
PROPORTION OF EXHAUST HEAT RECOVERED (%)	Dependent on user's requirements
TYPICAL PAYBACK PERIOD (years)	1-2
CONSTRUCTION	
Materials	Carbon-steel finned tubes; stainless-steel tube sheets; carbon-steel ducts
Weight range (kg)	Up to 18000
FEATURES	
Available custom-built?	Normally standard package
Performance guarantee	Yes
Installation method	Supplied as preassembled package
Installed by?	Contractor
Cleaning	Standard techniques
Corrosion resistance	Normal material considerations
Condensate collection	N/a
Performance control and turndown range	To match gas turbine: full control/monitoring equipment in package
Silencer	N/a
EXTRAS	System design assistance; fluid-side accessories
FURTHER COMMENTS	

MANUFACTURER	*ELKSTROM & SON*
MODEL NAME AND NUMBER	
TYPE OF PRIME MOVER	Gas turbine / Diesel engine } exhaust
FLOW VOLUMES (m^3/h)	
Exhaust-side	Very much varied
Supply-side	Very much varied
PRESSURE DROP (mm w.g.)	
Exhaust-side	Very much varied
Supply-side	Very much varied
OPERATING TEMPERATURE RANGE (°C)	Very much varied
DUTY (kW)	Very much varied
PROPORTION OF EXHAUST HEAT RECOVERED (%)	Very much varied
TYPICAL PAYBACK PERIOD (years)	Very much varied
CONSTRUCTION	
Materials	Stainless steel; carbon steel
Weight range (kg)	
FEATURES	
Available custom-built?	Yes
Performance guarantees	Yes
Installation method	Connected to suitable flanged fittings
Installed by?	Contractor
Cleaning	Manual
Corrosion resistance	Appropriate to materials
Condensate collection	Not normally required
Performance control and turndown range	
Silencer	
EXTRAS	
FURTHER COMMENTS	

MANUFACTURER	*HAMWORTHY ENGINEERING CO. LTD.*
MODEL NAME AND NUMBER	
TYPE OF PRIME MOVER	Gas Turbine or Diesel
FLOW VOLUMES (m^3/h)	
Exhaust-side	Up to 200 000
Supply-side	
PRESSURE DROP (mm w.g.)	
Exhaust-side	As required
Supply-side	
OPERATING TEMPERATURE RANGE (^{o}C)	Max. 650
DUTY (kW)	
PROPORTION OF EXHAUST HEAT RECOVERED (%)	
TYPICAL PAYBACK PERIOD (years)	
CONSTRUCTION	
Materials	Steel
Weight range (kg)	
FEATURES	
Available custom-built?	Yes
Performance guarantee	Yes
Installation method	
Installed by?	Customer
Cleaning	Water wash or steam sootblower
Corrosion resistance	
Condensate collection	
Performance control and turndown range	
Silencer	
EXTRAS	
FURTHER COMMENTS	

MANUFACTURER	*HAMWORTHY ENGINEERING CO. LTD.*
MODEL NAME AND NUMBER	HR type
TYPE OF PRIME MOVER	Diesel
FLOW VOLUMES (m^3/h)	
Exhaust-side	Up to 6000 x 1000 modules
Supply-side	
PRESSURE DROP (mm w.g.)	
Exhaust-side	As required
Supply-side	
OPERATING TEMPERATURE RANGE (°C)	Max. 650
DUTY (kW)	
PROPORTION OF EXHAUST HEAT RECOVERED (%)	
TYPICAL PAYBACK PERIOD (years)	
CONSTRUCTION	
Materials	Cast iron
Weight range (kg)	
FEATURES	
Available custom-built?	No
Performance guarantee	Yes
Installation method	
Installed by?	Customer
Cleaning	Brush
Corrosion resistance	
Condensate collection	
Performance control and turndown range	
Silencer	
EXTRAS	
FURTHER COMMENTS	

MANUFACTURER	*HAPPEL KG*
MODEL NAME AND NUMBER	Happel/GEA Regenerators
TYPE OF PRIME MOVER	Gas turbines
FLOW VOLUMES (m^3/h)	
Exhaust-side	Dependent on requirements of gas turbine
Supply-side	Dependent on requirements of gas turbine
PRESSURE DROP (mm w.g.)	
Exhaust-side	Dependent on requirements of gas turbine
Supply-side	Dependent on requirements of gas turbine
OPERATING TEMPERATURE RANGE (°C)	Suit all gas turbine exhausts; e.g. 500
DUTY (kW)	Dependent on gas turbine size; e.g. 40 000
PROPORTION OF EXHAUST HEAT RECOVERED (%)	80 (saving 25% gas turbine fuel)
TYPICAL PAYBACK PERIOD (years)	1-6 (dependent on gas turbine utilization)
CONSTRUCTION	
Materials	Steel (shell and tube type unit)
Weight range (kg)	
FEATURES	
Available custom-built?	Yes
Performance guarantee	Yes
Installation method	Flanged to exhaust duct
Installed by?	Manufacturer or contractor
Cleaning	Access for cleaning inside tubes of heat exchanger (exhaust gas-side)
Corrosion resistance	Good
Condensate collection	N/a
Performance control and turndown range	Follows prime mover performance
Silencer	N/a
EXTRAS	
FURTHER COMMENTS	

MANUFACTURER	*HITACHI ZOSEN (HITACHI SHIPBUILDING & ENGINEERING CO. LTD.)*	
MODEL NAME AND NUMBER	OC-10	OC-20
TYPE OF PRIME MOVER	Overhang condensing-type geared steam turbine	
FLOW VOLUMES (m^3/h)		
Exhaust-side	4-10 tonne/h (steam)	10-16 tonne/h (steam)
Supply-side		
PRESSURE DROP (mm w.g.)		
Exhaust-side	Vac. 650 mm w.g. (86.7 kPa) - 722 mm w.g. (96.3 kPa)	Vac. 670 mm w.g. (89.3 kPa) - 700 mm w.g. (93.3 kPa)
Supply-side	1.5 k.g.f./cm^2 (gauge) (0.147 MPa)-5 k.g.f/cm^2 (gauge) (0.490 MPa)	4 k.g.f./cm^2 (gauge) (0.392 MPa)-9 k.g.f./cm^2 (gauge) (0.883 MPa)
OPERATING TEMPERATURE RANGE (°C)	Saturated temperature - 300 ⟶	
DUTY (kW)	250-1000 kW	900 kW-1500 kW
PROPORTION OF EXHAUST HEAT RECOVERED (%)		
TYPICAL PAYBACK PERIOD (years)	3-4 (including waste heat boiler; generator; and the other all necessary auxiliary equipment)	
CONSTRUCTION		
Materials		
Weight range (kg)	14 000-23 000 kg (including turbine proper; reduction gear; generator; condenser; common-bed; all LO equipment which include LO cooler; oil pumps; oil filter; etc., gland packing seal equipment.	20 000-27 000 kg
FEATURES		
Available custom-built?	Custom-built ⟶	
Performance guarantee	Steam rate (steam flow, kg/output, kW) and max. continuous output	
Installation method	Packaged type ⟶	
Installed by?	Manufacturer ⟶	
Cleaning		
Corrosion resistance		
Condensate collection	Condenser hot well ⟶	
Performance control and turndown range	Speed governor and/or initial pressure regulator; 0, full load turndown range	
Silencer		
EXTRAS		
FURTHER COMMENTS		

MANUFACTURER	*HITACHI ZOSEN (HITACHI SHIPBUILDING & ENGINEERING CO. LTD.)*	
MODEL NAME AND NUMBER	Blast Furnace Top-Pressure Recovery Gas Turbine Type SHR	
TYPE OF PRIME MOVER	Axial double-flow reaction-type gas expander	Axial single-flow reaction-type gas expander
FLOW VOLUMES (m^3/h)		
Exhaust-side	220 000-800 000 Nm^3/h as wet gas	80 000-430 000 Nm^3/h as wet gas
Supply-side		
PRESSURE DROP (mm w.g.)		
Exhaust-side	Approx. 1000 mm w.g. (9.81 kPa)	
Supply-side	0.5 k.g.f./cm^2 (gauge (0.049 MPa)-3.0 k.g.f./cm^2 (gauge) (0.294 MPa)	
OPERATING TEMPERATURE RANGE (°C)	30-120	
DUTY (kW)	2000-23 000 kW	1500-13 000
PROPORTION OF EXHAUST HEAT RECOVERED (%)		
TYPICAL PAYBACK PERIOD (years)	Approx. 1 (including all necessary auxiliary equipments such as LO unit; sealing-gas unit; etc.)	
CONSTRUCTION		
Materials		
Weight range (kg)		
FEATURES		
Available custom-built?	Custom-built	
Performance guarantee	Max. continuous output	
Installation method		
Installed by?	Manufacturer	
Cleaning		
Corrosion resistance		
Condensate collection		
Performance control and turndown range	Initial pressure regulator; speed governor and output control; 0, full load of turndown range	
Silencer		
EXTRAS		
FURTHER COMMENTS		

MANUFACTURER	*NORTHERN ENGINEERING INDUSTRIES (POWER PLANT DIVISION)*
MODEL NAME AND NUMBER	Waste Heat Boiler
TYPE OF PRIME MOVER	Gas turbine in power station or industrial process
FLOW VOLUMES	
Exhaust-side	Up to 1.3 x 10^6 kg/h
Supply-side	Up to 150 000 kg/h (evaporation rate)
PRESSURE DROP (mm w.g.)	
Exhaust-side	Up to 254
Supply-side	To customer's requirements
OPERATING TEMPERATURE RANGE (^{o}C)	Up to 540 (depending on gas turbine type)
DUTY (kW)	20 000-100 000
PROPORTION OF EXHAUST HEAT RECOVERED (%)	Typically 80-92% (giving overall 40-45% efficiency on a combined-cycle plant)
TYPICAL PAYBACK PERIOD (years)	Related to total-plant cost
CONSTRUCTION	
Materials	Superheater, low chrome steel; evaporator, carbon steel
Weight range (kg)	
FEATURES	
Available custom-built?	Yes
Performance guarantee	Yes
Installation method	Site-assembly of modules
Installed by?	Manufacturer or contractor
Cleaning	Offload water washing (for some fuels, onload systems available)
Corrosion resistance	Yes, design above dewpoint
Condensate collection	Yes
Performance control and turndown range	Follow gas turbine performance, normally down to 70% full load
Silencer	Yes, on safety devices
EXTRAS	
FURTHER COMMENTS	Minimum of sitework needed, because of modular construction.

MANUFACTURER	*POTT INDUSTRIES INC. ENGINEERING CONTROLS DIVISION*
MODEL NAME AND NUMBER	ECX, VP, ECWT
TYPE OF PRIME MOVER	Turbines and reciprocating engines
FLOW VOLUMES (m^3/h)	
Exhaust-side	4064-1 820 800
Supply-side	
PRESSURE DROP (mm w.g.)	
Exhaust-side	3.74-11.2
Supply-side	
OPERATING TEMPERATURE RANGE	315-649°F (160-350°C)
DUTY (kW)	100-15 000
PROPORTION OF EXHAUST HEAT RECOVERED (%)	40-60
TYPICAL PAYBACK PERIOD (years)	
CONSTRUCTION	
Materials	Steel (to ASME Code section VIII)
Weight range (kg)	453-27 215
FEATURES	
Available custom-built?	All custom-built
Performance guarantee	Yes
Installation method	Contractor
Cleaning	Flue-gas brushes and/or sootblowers
Corrosion resistance	Normal corrosion allowances
Condensate collection	
Performance control and turndown range	Exhaust gas bypass valve
Silencer	Included
EXTRAS	
FURTHER COMMENTS	Vaporphase packaged jacket water and exhaust waste heat recovery silencer recovers both jacket water and exhaust in the form of 15 p.s.i.g. steam for total recovery, while cooling the reciprocating engine.

MANUFACTURER	*RILEY-BEAIRD INC.*
MODEL NAME AND NUMBER	Maxim BVS Heat Recovery Silencers BVS 25-3-BVS 5550-30 (16 models)
TYPE OF PRIME MOVER	Reciprocating engines
FLOW VOLUMES	
Exhaust-side	270-55000 kg/h
Supply-side	
PRESSURE DROP (mm w.g.)	
Exhaust-side	100-360
Supply-side	
OPERATING TEMPERATURE RANGE (°C)	260-600 (cooling exhaust to not less than 177)
DUTY (kW)	7-5000
PROPORTION OF EXHAUST HEAT RECOVERED (%)	Up to 72
TYPICAL PAYBACK PERIOD (years)	
CONSTRUCTION	
Materials	Steel (to section VIII, division 1, ASME Pressure Vessel Code)
Weight range (kg)	419-30675 (dry weight)
FEATURES	
Available custom-built?	Normally standard units
Performance guarantee	Yes
Installation method	Connected to flanged exhaust and water-supply ducts
Installed by?	Customer or contractor
Cleaning	Removable cover plates give access to gas flow passages and heat transfer surface
Corrosion resistance	Good optional low-alloy weathering steel used in dewage gas-fuelled engine units
Condensate collection	Operates above dew point
Performance control and turndown range	Level control systems; high and low alarm switches; safety valves
Silencer	Combined with heat recovery unit
EXTRAS	Presented controls; factory applied insulation
FURTHER COMMENTS	

MANUFACTURER	*RILEY-BEAIRD INC.*
MODEL NAME AND NUMBER	Heat exchanger GTW; HSS; MFT; TRP; BVS; WVS; SOH; PCB
TYPE OF PRIME MOVER	All engine types (also incinerators and high-temperature flue gases)
FLOW VOLUMES (m^3/h)	
Exhaust-side	250-100 000 Nm^3/h
Supply-side	To suit customer's requirements (waste heat boiler)
PRESSURE DROP (mm w.g.)	
Exhaust-side	To suit customer's requirements
Supply-side	
OPERATING TEMPERATURE RANGE (°C)	260-1100
DUTY (kW)	7-5000
PROPORTION OF EXHAUST HEAT RECOVERED (%)	Up to 72
TYPICAL PAYBACK PERIOD (years)	Dependent on cost of fuel
CONSTRUCTION	
Materials	Steel; stainless steel; alonized steel: section I, section IV, section VII, division 1 to ASME Pressure Vessel Code
Weight range (kg)	100-50 000
FEATURES	
Available custom-built?	Yes
Performance guarantee	Yes
Installation method	Connected to flanged exhaust and liquid-supply piping
Installed by?	Customer or contractor
Cleaning	Varies with model: chemical cleaning required
Corrosion resistance	Good optional low-alloy weathering steel used in sewage gas-fuelled engine units; alonizing used in high-temperature service
Condensate collection	Operates above dew point
Performance control and turndown range	Level control system; high and low alarm switches; safety valves; gas bypass on control coolers
Silencer	Combined with heat recovery unit or added in series, if required
EXTRAS	Preselected controls; factory applied insulation
FURTHER COMMENTS	Most experienced company in engine heat recovery in the USA; no European manufacture.

MANUFACTURER	*RUSTON GAS TURBINES LTD.*	
MODEL NAME AND NUMBER	TA 1750	TB 5000
TYPE OF PRIME MOVER	Gas turbine →	
FLOW VOLUMES (m^3/h)		
Exhaust-side		
Supply-side		
PRESSURE DROP (mm w.g.)		
Exhaust-side	Variable →	
Supply-side		
OPERATING TEMPERATURE RANGE (°C)		
DUTY (kW)	1300	3500
PROPORTION OF EXHAUST HEAT RECOVERED (%)	Up to 70 →	
TYPICAL PAYBACK PERIOD (years)	Variable →	
CONSTRUCTION		
Materials		
Weight range (kg)		
FEATURES		
Available custom-built?	Yes →	
Performance guarantees	Yes →	
Installation method	Supervision by RGT →	
Installed by?	Supervision by RGT →	
Cleaning		
Corrosion resistance		
Condensate collection		
Performance control and turndown range		
Silencer	Yes →	
EXTRAS		
FURTHER COMMENTS		

MANUFACTURER	*SENIOR ECONOMIZERS LTD.*
MODEL NAME AND NUMBER	Steel-finned or Cast-iron Finned Surface
TYPE OF PRIME MOVER	Economizer for use with gas turbine or diesel engine
FLOW VOLUMES (m^3/h)	
Exhaust-side	No limit on volumes
Supply-side	No limit on volumes
PRESSURE DROP (mm w.g.)	
Exhaust-side	
Supply-side	To suit user's requirements
OPERATING TEMPERATURE RANGE (^{o}C)	Standard range up to 600
DUTY (kW)	No limit, to suit heat required or heat available
PROPORTION OF EXHAUST HEAT RECOVERED (%)	Dependent on application
TYPICAL PAYBACK PERIOD (years)	Normally find clients require 2-3-year period
CONSTRUCTION	
Materials	Carbon steel; cast iron
Weight range (kg)	
FEATURES	
Available custom-built?	Yes
Performance guarantee	Yes
Installation method	Shop-assembled or site-erected
Installed by?	Manufacturer
Cleaning	Steam sootblowing; compressed air and water washing
Corrosion resistance	Within limits
Condensate collection	Where required
Performance control and turndown range	Control to approx. 25% of full load
Silencer	Yes
EXTRAS	
FURTHER COMMENTS	

MANUFACTURER	*STRUTHERS WELLS CORP.*
MODEL NAME AND NUMBER	Custom-designed
TYPE OF PRIME MOVER	Gas turbine
FLOW VOLUMES (m^3/h)	
Exhaust-side	18 000-600 000
Supply-side	
PRESSURE DROP (mm w.g.)	
Exhaust-side	15-30
Supply-side	
OPERATING TEMPERATURE RANGE (°C)	400-600
DUTY (kW)	60 000-1 200 000
PROPORTION OF EXHAUST HEAT RECOVERED (%)	85-92
TYPICAL PAYBACK PERIOD (years)	2-5
CONSTRUCTION	
Materials	Carbon steel; refractory
Weight range (kg)	200 000-1 000 000
FEATURES	
Available custom-built?	Yes
Performance guarantee	Yes
Installation method	Field assembly
Installed by?	Manufacturer or customer
Cleaning	Normally by sootblowers on gas-side; chemical cleaning of steam-side
Corrosion resistance	Yes
Condensate collection	No
Performance control and turndown range	Fully modulating control; turndown range infinite with dump stack
Silencer	Yes, when required
EXTRAS	Full range of controls; dump stack; steam superheater; auxiliary firing
FURTHER COMMENTS	

MANUFACTURER	*TERRY CORP.*
MODEL NAME AND NUMBER	Terry LP Turbine Prime Mover, types F; GF; GAF; GA; GHF
TYPE OF PRIME MOVER	Steam turbine used to generate power from high pressure or waste; low pressure steam for driving generators, compressors, pumps
FLOW VOLUMES (m^3/h)	
Exhaust-side	Exhaust waste steam: 1.5-11 bar
Supply-side	N/a
STEAM PRESSURE DROP (mm w.g.)	
Exhaust-side	Depends on steam pressure, condensing to 400 p.s.i.g.
Supply-side	5-1200 p.s.i.g.
OPERATING TEMPERATURE RANGE	Appropriate to steam conditions up to 1000°F
DUTY (kW)	Up to 12000
PROPORTION OF EXHAUST HEAT RECOVERED (%)	Depends on steam conditions
TYPICAL PAYBACK PERIOD (years)	Less than 1 in some cases
CONSTRUCTION	
Materials	Steel; iron
Weight range (kg)	
FEATURES	
Available custom-built?	Yes
Performance guarantee	Yes
Installation method	As for turbine practice
Installed by?	Manufacturer, customer or contractor
Cleaning	N/a
Corrosion resistance	Conventional turbine considerations: steam path is stainless steel
Condensate collection	Steam/water separator needed to prevent condensate entering turbine at start of machine upstream
Performance control and turndown range	Built in variable speed control with turndown of up to 10:1 available; allows optimizing speed of driven unit for best efficiency
Silencer	N/a
EXTRAS	Surface and barometric condenser
FURTHER COMMENTS	Steam turbines can be used in place of pressure-reducing valves to more efficiently utilize high-pressure steam energy.

12 Incinerators with heat recovery equipment

In Chapters 8 and 9, data on gas-to-gas and gas-to-liquid heat recovery equipment is given, and several manufacturers in both areas indicate that incineration plant is a potential application for their heat exchangers. However, because of the growing importance of this sector, especially in countries where pollution-control legislation is necessitating the addition of incineration plant (in particular, fume incinerators and oxidizers) downstream of industrial processes, it was believed that a section devoted to manufacturers of incineration plant with heat recovery and specialist suppliers of equipment for heat recovery from such plant would be of value. This chapter, together with the brief description of some combined incineration/heat recovery plant in Chapter 5, brings together this data.

12.1 Data presentation

The equipment covered by this chapter includes:

Solid waste incinerators;
Liquid waste incinerators;
Fume incinerators;
Sludge and surry incineration;
Heat recovery equipment associated with these incinerators.

In the majority of the cases, the manufacturer of the incineration plant has provided data on the type of heat recovery equipment used. However, some heat recovery equipment manufacturers produce units specifically designed for incinerators, and the equipment data is also given by these organizations.

Other data on heat recovery equipment having a more general field of application, including incineration plant, will be found in Chapters 8 and 9, and these are referenced in Appendix 1, where the product category of each manufacturer is listed.

12.1.1 *Format of data*

The data on incinerators and their associated heat recovery equipment is listed in the following manner:

(1) *Equipment manufacturer*: The prime manufacturer or licensee of the incineration equipment and/or heat recovery equipment is given. Addresses, country of origin, telephone number and product categories of each manufacturer are listed in Appendix 1. Where known, agents or overseas licensees for the prime manufacturer are also given, supplemented by the name of the prime manufacturer for cross-reference.
(2) *Model name and number*: The name of the incinerator/heat recovery unit is given, and where appropriate the model number or range of models. In many cases the equipmen is custom-built, and it is not possible to give a specific name or identification number.
(3) *Type*: The type of incinerator to which the heat recovery equipment may be applied is indicated. This is normally a solid, liquid or fume incinerator.
(4) *Applications*: Incinerators may be used in industry, or for disposal of municipal refuse (including hospitals, etc.). This section indicates the broad application areas for the incinerators.
(5) *Incinerator capacity*: Sizing of the incinerator is of prime importance. Many companies offer a range of capacities, and these are given in this section, normally in terms of throughput per day.
(6) *Incinerator thermal data*: A number of items of data of interest to potential users are specified:

- (i) Temperature: where known, the incineration temperature is given. This may be affected by the use of afterburners or preheating, and this is indicated in some cases.
- (ii) Fuel type: fuel is used either to initiate combustion or in some cases to keep combustion at a satisfactory level during incineration. The type(s) of fuel used are given.
- (iii) Fuel-utilization rate: the rate of which the incinerator uses this fuel is given , where applicable.
- (iv) Supplementary firing: additional fuel may be required with some waste types, or supplementary firing may be installed in the flue for fume treatment.

(7) *Flue gas data*: Data on the flue gases downstream of the incineration chamber are given. These indicate the temperature, normally a maximum, which the heat recover unit is likely to see, and the maximum flow rate of gases, for sizing of the heat exchanger unit.
(8) *Heat recovery data*: The efficiency of heat recovery from the incinerator flue gases is given, together with a brief description of the type(s) of waste heat recovery equipment which may be used on the incinerator. These include waste heat boilers (with economizers and superheaters), gas–gas heat recovery units, thermal fluid heaters, high-pressure hot-water heaters, rotary dryers, fluidized bed systems and

economizers operating alone.

(9) *Pollution-control system*: In some cases, the incinerator itself fulfils the requirements of pollution control. However, scrubbers and other systems such as electrostatic precipitators may be used. Manufacturers indicate in this section which type of pollution-control equipment is fitted.

(10) *Extras available and further comments*: The manufacturer is given the opportunity to list optional equipment and add any other comments which may be of value to potential users. Options include full-control systems, total systems, pollution-control and heat recovery options, automatic stoking and filters.

Manufacturers' data sheets relating to incinerators with heat recovery equipment and specialist equipment for heat recovery from incineration plant follow on the next page.

MANUFACTURER	*SA BABCOCK BELGIUM NV*	
MODEL NAME AND NUMBER	Waste Heat Recovery After-Incinerator	
TYPE	Liquid waste	Solid waste
APPLICATIONS	Industrial	Industrial; municipal
CAPACITY	Depending on nature of liquid	500 tonne/day
INCINERATOR		
Temperature (°C)	According to waste fuel ⟶	
Fuel type		
Fuel utilization rate		
Supplementary firing		
FLUE GAS		
Temperature (°C)	800	900
Flow rate (m^3/h)	280 000 Nm^3/h	Up to 180 000 Nm^3/h
HEAT RECOVERY		
Efficiency (%)	80	85
Equipment type	Boiler with superheater and economizer ⟶	
POLLUTION-CONTROL SYSTEM		
EXTRAS AVAILABLE		
FURTHER COMMENTS	Above ranges are typical and not considered as limits.	

MANUFACTURER	*HENRY BALFOUR & CO. LTD.*
MODEL NAME AND NUMBER	NICE Process
TYPE	Liquid waste (normally designed to recover salts)
APPLICATIONS	Industrial, principally chemical effluents
CAPACITY	Typically 0.7-700 tonne/day
INCINERATOR	
Temperature (°C)	800-1100
Fuel type	Natural gas; fuel oil; distillation residues
Fuel utilization rate	Dependent on unit size
Supplementary firing	No
FLUE GAS	
Temperature (°C)	Up to 1500
Flow rate	Heat releases 0.5-17 MW (standard); specials to 60 MW
HEAT RECOVERY	
Efficiency (%)	High efficiency, if used for preheating/evaporation by direct contact
Equipment type	Evaporators; incineration liquid preheaters; and concentrators
POLLUTION-CONTROL SYSTEM	Venturi scrubber; quenching of combustion gases
EXTRAS AVAILABLE	

MANUFACTURER	*BEVERLEY CHEMICAL ENGINEERING LTD.*
MODEL NAME AND NUMBER	Custom-Built
TYPE	Fume; liquid; solid waste
APPLICATIONS	Industrial; municipal
CAPACITY	Up to 80 000 Nm^3/h; 100 000 l/day
INCINERATOR	
Temperature (^oC)	Up to 1200
Fuel type	Waste
Fuel utilization rate	As required
Supplementary firing	Gas or oil
FLUE GAS	
Temperature (^oC)	Up to 1200
Flow rate (m^3/h)	Up to 80 000 Nm^3/h
HEAT RECOVERY	
Efficiency (%)	Up to 60
Equipment type	Gas-to-gas (air preheater) Gas-to-liquid (boiler or thermal fluid heater)
POLLUTION-CONTROL SYSTEM	Incinerator with or without scrubber
EXTRAS AVAILABLE	

MANUFACTURER	*BRONSWERK HEAT TRANSFER BV*
MODEL NAME AND NUMBER	BANOdour
TYPE	Air and flue gases with bad odours
APPLICATIONS	Industrial
CAPACITY	Up to 100 000 m^3/h
INCINERATOR	
Temperature (°C)	Max. 1000
Fuel type	Natural gas/oil
Fuel utilization rate	70%
Supplementary firing	
FLUE GAS	
Temperature (°C)	400-200
Flow rate (m^3/h)	100 000
HEAT RECOVERY	
Efficiency (%)	70
Equipment type	Burner/heat exchanger
POLLUTION-CONTROL SYSTEM	Odour-burning with heat recovery by preheating
EXTRAS AVAILABLE	Controls; filters; cleaning equipment

MANUFACTURER	*BURKE THERMAL ENGINEERING LTD.*
MODEL NAME AND NUMBER	BTE 700
TYPE	Fume
APPLICATIONS	Industrial
CAPACITY	2 million m^3/day
INCINERATOR	
Temperature (^oC)	750
Fuel type	Natural gas
Fuel utilization rate	
Supplementary firing	
FLUE GAS	
Temperature (^oC)	750
Flow rate (m^3/h)	11 000
HEAT RECOVERY	
Efficiency (%)	65
Equipment type	Air preheater plus heat pipes
POLLUTION-CONTROL SYSTEM	Oxidation
EXTRAS AVAILABLE	

MANUFACTURER	*CEA COMBUSTION INC.*	
MODEL NAME AND NUMBER	Thermal oxidizer in fibre plant	Thermal Oxidizer in Resin Impregnation plant
TYPE	Fume; liquid; solid waste	Fumes
APPLICATIONS	Industrial ⟶	
CAPACITY	75 000 m^3/h fumes; 12 000 l/day liquid;60 tonne/day solid	14 400 m^3/h fumes
INCINERATOR		
Temperature (°C)	815 ⟶	
Fuel type	Natural gas; fuel oil ⟶	
Fuel utilization rate	Normally uses combustion waste	
Supplementary firing	Not needed ⟶	
FLUE GAS		
Temperature (°C)		
Flow rate (m^3/h)		
HEAT RECOVERY		
Efficiency (%)	Thermal fluids heated to 230 °C 72.5 tonne/h dry saturated steam o.p.	80% reduction in HPHW generator fuel requirements
Equipment type	Steam generators and thermal fluid heater	High-pressure hot-water heater and fume pre-heater
POLLUTION-CONTROL SYSTEM	Primary incineration meets requirements	Incoming fumes preheated using oxidizer exhausts; and primary incineration then meets requirements
EXTRAS AVAILABLE	Other heat recovery equipment available includes: gas- air heaters; asphalt heaters; CEA. Combustion products are tailored to customer requirements, two examples of installations are given.	

MANUFACTURER	*C-E AIR PREHEATER CO. INC.*
MODEL NAME AND NUMBER	Direct-flame Fume Incinerator
TYPE	Fume
APPLICATIONS	Industrial
CAPACITY	1 672 800 m^3/day (max.); 96 000 m^3/day (min.)
INCINERATOR	
Temperature (^oC)	815
Fuel type	Raw gas; residual oils; propane
Fuel utilization rate	
Supplementary firing	
FLUE GAS	
Temperature (^oC)	Variable
Flow rate (m^3/h)	
HEAT RECOVERY	
Efficiency (%)	35-70 (air heater)
Equipment type	Air heater
POLLUTION-CONTROL SYSTEM	
EXTRAS AVAILABLE	Boilers or economizers by Deltak

MANUFACTURER	*CLEERBURN LTD.*
MODEL NAME AND NUMBER	Cleerburn CL range; CL10, 20, 30, 50; with various specials depending on waste type
TYPE	Solid waste
APPLICATIONS	Schools; offices; timber; and other industrial wastes
CAPACITY	6-20 m^3/day (general guidance only)
INCINERATOR	
Temperature (°C)	1000
Fuel type	Oil (gas optional)
Fuel utilization rate	Dependent on waste type and dryness
Supplementary firing	Oil (CL Mk IIB range)
FLUE GAS	
Temperature (°C)	
Flow rate (m^3/h)	
HEAT RECOVERY	
Efficiency (%)	
Equipment type	Air heaters; water heaters
POLLUTION-CONTROL SYSTEMS	Varies with model after-burners; reheat chambers; fly-ash traps; grit arrestors
EXTRAS AVAILABLE	

MANUFACTURER	*CLEERBURN LTD.*
MODEL NAME AND NUMBER	Cleerburn RL and RL (PR) series, RL/20-RL/120, RL/20(PR)-RL/120(PR) (10 models)
TYPE	Difficult solid wastes (PR models for plastics)
APPLICATIONS	Industrial wastes
CAPACITY	1.25-4 tonne/day (general guidance only)
INCINERATOR	
Temperature (°C)	Max. 1400
Fuel type	32 s oil (waste oils and gas optional)
Fuel utilization rate	Dependent on waste type and dryness
Supplementary firing	Oil as standard
FLUE GAS	
Temperature (°C)	
Flow rate (m^3/h)	
HEAT RECOVERY	
Efficiency (%)	65-70
Equipment type	Water heaters; steam-raising system as separate units
POLLUTION-CONTROL SYSTEM	Two-stage smoke combustion (sealed-flame principle on RL(PR) models)
EXTRAS AVAILABLE	

MANUFACTURER	*COMBUSTION POWER CO.*	
MODEL NAME AND NUMBER	Custom-Design	
TYPE	Liquid; solid waste	
APPLICATIONS	Industrial	
CAPACITY	Largest unit is rated at an energy output of 36 MW (natural-gas equivalent)	
INCINERATOR		
Temperature (°C)	760-1100	
Fuel type	Industrial wastes; woodyard debris- fly-ash char; clarifier sludge	
Fuel itilization rate	Max. 25 tonne/h, as required, high moisture, high inert	
Supplementary firing		
FLUE GAS		
Temperature (°C)		
Flow rate (m^3/h)		
HEAT RECOVERY		
Efficiency (%)	Dependent on maximum operating temperature and on moisture content of fuel	
Equipment type	Boiler for 1 system;	Rotary dryex for second system
POLLUTION-CONTROL SYSTEM	Multiple cyclones	3 cyclones
EXTRAS AVAILABLE		

MANUFACTURER	*COMTRO EQUIPMENT CORP.*		
MODEL NAME AND NUMBER	A35 Crittenton	A45 Moore	A48 Knoll
TYPE	Solid waste →		
APPLICATIONS	Industrial →		
CAPACITY	3 ton	10 ton	20 ton
INCINERATOR			
Temperature (°C)	982°C →		
Fuel type	Gas; oil	Oil	Oil
Fuel utilization rate	800 000 Btu/ton →		
Supplementary firing	No →		
FLUE GAS			
Temperature (°C)	982°C →		
Flow rate (m^3/h)	2290 m^3/h	5189 m^3/h	7592 m^3/h
HEAT RECOVERY			
Efficiency (%)	62	65	63
Equipment type	Fire tube boiler (steam) →		
POLLUTION-CONTROL SYSTEM	None (secondary afterburner)	Cyclone (afterburner)	None (afterburner)
EXTRAS AVAILABLE			

MANUFACTURER	*DANKS OF NETHERTON LTD.*
MODEL NAME AND NUMBER	Custom-built
TYPE	Solid waste
APPLICATIONS	Industrial
CAPACITY	70 tonne/day
INCINERATOR	
Temperature (°C)	1400
Fuel type	Wood; agricultural waste products
Fuel utilization rate	3 tonne/h
Supplementary firing	Oil; gas
FLUE GAS	
Temperature (°C)	250-300
Flow rate (m^3/h)	21 000 stp
HEAT RECOVERY	
Efficiency (%)	70
Equipment type	Boiler
POLLUTION-CONTROL SYSTEM	Custom-built
EXTRAS AVAILABLE	Automatic stoker or fluidized bed firing.

MANUFACTURER	*ECLIPSE LOOKOUT CO.*
MODEL NAME AND NUMBER	Eclipse Fume Incinerators
TYPE	Fume
APPLICATIONS	Industrial
CAPACITY	8490 m^3/min gas flow
INCINERATOR	
Temperature (^{o}C)	Up to 1000
Fuel type	Gas, oil, or both
Fuel utilization rate	N/a
Supplementary firing	N/a
FLUE GAS	
Temperature (^{o}C)	Up to 1000
Flow rate (m^3/h)	Up to 8490 m^3/min
HEAT RECOVERY	
Efficiency (%)	Up to 85
Equipment type	Air-to-air; air-to-steam or hot water; air-to-liquid (see appropriate Chapters for details)
POLLUTION-CONTROL SYSTEM	Most designed to meet LA Code 66

MANUFACTURER	*ELBOMA SPRL*
MODEL NAME AND NUMBER	FreeCALOR Special Boilers and Furnaces for waste incineration with recovery of the calories
TYPE	Solid waste
APPLICATIONS	Industrial
CAPACITY	Smallest type: 150 000 kcal/h Biggest type : 10 000 000 kcal/h
INCINERATOR	
Temperature (°C)	Dependent on medium: hot water; superheated water; steam and thermal fluid
Fuel type	Self-combustion
Fuel utilization rate	
Supplementary firing	
FLUE GAS	
Temperature (°C)	Approx. 80 deg. C above the medium
Flow rate (m^3/h)	Dependent on capacity of apparatus
HEAT RECOVERY	
Efficiency (%)	Approx. 80
Equipment type	(i.e. clean air act)
POLLUTION-CONTROL SYSTEM	If necessary (i.e. clean air act) the Multi-Cyclone dedusting system is installed
EXTRAS AVAILABLE	Operating temperatures: warm water max. 110°C; superheated water max. 200°C; steam max. 20 bar; thermal fluid 360°C.

MANUFACTURER	*GOTAVERKEN ANGTEKNIK AB*
MODEL NAME AND NUMBER	EGB
TYPE	Exhaust gas boilers
APPLICATIONS	Marine incinerators
CAPACITY (steam)	0.5-10 tonne/h
INCINERATOR	
Temperature (^{o}C)	
Fuel type	
Fuel-utilization rate	
Supplementary firing	
FLUE GAS	
Temperature (^{o}C)	320°C-360°C
Flow rate (m^3/h)	
HEAT RECOVERY	
Efficiency (%)	
Equipment type	
POLLUTION-CONTROL SYSTEM	
EXTRAS AVAILABLE	

MANUFACTURER	*GRANCO EQUIPMENT INC.*
MODEL NAME AND NUMBER	Incin-o-Wheel
TYPE	Fume incinerator
APPLICATIONS	Industrial
CAPACITY	Single units up to 17 000 Nm^3/h
INCINERATOR	
Temperature (°C)	370
Fuel type	Gas, light oil
Fuel utilization rate	Dependent on size and solvent concentration
Supplementary firing	
FLUE GAS	
Temperature (°C)	
Flow rate (m^3/h)	
HEAT RECOVERY	
Efficiency (%)	70-75 at rated flow
Equipment type	Ceramic heat wheel regenerator
POLLUTION-CONTROL SYSTEM	Complete systems can be furnished
EXTRAS AVAILABLE	

MANUFACTURER	*HARRIS THERMAL TRANSFER PRODUCTS*
MODEL NAME AND NUMBER	
TYPE	Solid waste
APPLICATIONS	Industrial
CAPACITY	Unknown
INCINERATOR	
Temperature (^{o}C)	1315
Fuel type	Paper; wood; plastic
Fuel utilization rate	Unknown
Supplementary firing	Oil
FLUE GAS	
Temperature (^{o}C)	760 inlet
Flow rate (m^3/h)	141 000
HEAT RECOVERY	
Efficiency (%)	88
Equipment type	Finned unit, thermal fluid heater
POLLUTION-CONTROL SYSTEM	
EXTRAS AVAILABLE	

MANUFACTURER	*HEENAN ENVIRONMENTAL SYSTEMS LTD.*
MODEL NAME AND NUMBER	Custom-built
TYPE	Fume, liquid, solid, sludges, and slurries
APPLICATIONS	Mainly industrial
CAPACITY	Fume: 4000-4 x 10^6 m^3/day Liquid: 2000-100 000 l/day Solid: 2-100 tonne/day
INCINERATOR	
Temperature (°C)	700-2300
Fuel type	Oil; gas
Fuel utilization rate	Variable
Supplementary firing	Dependent on CV of waste
FLUE GAS	
Temperature (°C)	700-2000
Flow rate (m^3/h)	200-200 000
HEAT RECOVERY	
Efficiency (%)	40-85 dependent on process
Equipment type	Boiler; thermal fluid; air preheater (various suppliers used)
POLLUTION-CONTROL SYSTEM	Wet-gas scrubber; bag filter; cyclone (various suppliers used)
EXTRAS AVAILABLE	Turnkey project capability

MANUFACTURER	*ROBERT JENKINS SYSTEMS LTD.*
MODEL NAME AND NUMBER	Consumat Incineration/Energy recovery systems
TYPE	Primarily solid waste, can be adapted for liquid
APPLICATIONS	Industrial, municipal
CAPACITY	0.6-27 tonne/day (units may also work in parallel)
INCINERATOR	
Temperature (°C)	800
Fuel type	Gas; oil
Fuel utilization rate	Dependent on size
Supplementary firing	Yes
FLUE GAS	
Temperature (°C)	1000-1200
Flow rate (m^3/h)	Depends on refuse, excess air, etc.
HEAT RECOVERY	
Efficiency (%)	E.g. unit processing 350 kg/h hospital waste generates 1250 kg/h steam at medium pressure
Equipment type	Waste heat boiler or hot-water boiler
POLLUTION-CONTROL SYSTEM	Supplementary burner for smoke oxidation
EXTRAS AVAILABLE	Automatic ash removal; liquid waste adaptor; automatic feed; loading device; air curtain; ash collector; waste heat recovery unit.

MANUFACTURER	*KLEINEWFERS GMBH*
MODEL NAME AND NUMBER	Combustor Units
TYPE	For fumes; waste gases; liquids
APPLICATIONS	Industrial
CAPACITY	3.000-200.000 Nm^3/h
INCINERATOR	
Temperature (°C)	Up to 1500
Fuel type	All fuels
Fuel utilization rate	
Supplementary firing	All fuels
FLUE GAS	
Temperature (°C)	Up to 1500
Flow rate (m^3/h)	Up to 200 000 Nm^3/h
HEAT RECOVERY	
Efficiency (%)	Up to 90
Equipment type	Boilers; thermal fluid heaters; gas-to-gas heat exchangers
POLLUTION-CONTROL SYSTEMS	Yes, FID, etc.
EXTRAS AVAILABLE	Complete units with all necessary instruments and parts.

MANUFACTURER	*KRAUSS-MAFFEI INDUSTRIEANLAGEN GMBH*
MODEL NAME AND NUMBER	Custom-built incineration plant with waste heat recovery
TYPE	Liquid; solid waste
APPLICATIONS	Industrial; special
CAPACITY	Example: solid and liquid waste unit: 10 tonne/h solid; 6.2 tonne/h paste; 4.8 tonne/h liquid
INCINERATOR	
Temperature (^{o}C)	Combustion at 1000°C
Fuel type	
Fuel utilization rate	105 MJ
Supplementary firing	Yes
FLUE GAS	
Temperature (^{o}C)	
Flow rate	
HEAT RECOVERY	
Efficiency (%)	34 tonne/day steam at 250°C, to 1450 kW turbogenerator; condensate return to plant
Equipment type	Waste heat boiler raising steam for turbogenerators
POLLUTION-CONTROL SYSTEM	Electrostatic precipitation; scubbers
EXTRAS AVAILABLE	Fluid bed furnaces; rotary kiln systems also available.

MANUFACTURER	*LYON & PYE LTD.*
MODEL NAME AND NUMBER	Portico P2-P8 Rotair Incinerators
TYPE	Liquid; solid waste
APPLICATIONS	Industrial; municipal; hospital; etc.
CAPACITY	From 0.25-300 ton/day
INCINERATOR	
Temperature (^{o}C)	Surface: 50
Fuel type	Oil (all grades), gas (natural, LPG, town)
Fuel utilization rate	Various
Supplementary firing	Various
FUEL GAS	
Temperature (^{o}C)	600-1000
Flow rate (m^3/h)	Various
HEAT RECOVERY	
Efficiency (%)	Approx. 60
Equipment type	Boiler/direct water jacket
POLLUTION-CONTROL SYSTEM	Cyclones; Venturi jet scrubbers; simple jet scrubbers; cyclonic scrubbers; etc.
EXTRAS AVAILABLE	Boilers; thermal fluid; air heaters; handling plants; pueumatic conveying autofeeders; free-standing chimneys; acoustic booths; etc.

MANUFACTURER	*METALLURGICAL ENGINEERS LTD.*
MODEL NAME AND NUMBER	Exchangers for incinerators only
TYPE	Liquid; solid waste
APPLICATIONS	Industrial; municipal
CAPACITY	
INCINERATOR	
Temperature (°C)	
Fuel type	
Fuel utilization rate	
Supplementary firing	
FLUE GAS	
Temperature (°C)	Up to 1150
Flow rate (m^3/h)	Up to 25 000 Nm^3/h
HEAT RECOVERY	
Efficiency (%)	
Equipment type	Combustion air preheater; boiler; thermal fluid heater; special combustion air preheaters for fluidized bed incinerators
POLLUTION-CONTROL SYSTEM	
EXTRAS AVAILABLE	

MANUFACTURER	*MOLLER & JOCHUMSEN (ENGINEERING) AS*
MODEL NAME AND NUMBER	Thermal Combustion Plant type FT1; FT2
TYPE	Fume (aircontaining organics or solid particles)
APPLICATIONS	Industrial (foodstuff; chemical; rubber)
CAPACITY	3000-12000 Nm^3/h
INCINERATOR	
Temperature (^oC)	Inlet up to 300, raised to 800-900
Fuel type	Natural gas (or oil)
Fuel utilization rate	(FT1) 17-27 kg/1000 Nm^3 (gas)
Supplementary firing	Yes
FLUE GAS	
Temperature (^oC)	800-900
Flow rate (m^3/h)	See above
HEAT RECOVERY	
Efficiency (%)	Dependent on user's requirement
Equipment type	Air or liquid heater for FT1; FT2 has heat exchanger for combustion air preheating
POLLUTION-CONTROL SYSTEM	Fume incineration
EXTRAS AVAILABLE	A catalytic system which may not need supplementary firing is also available.

MANUFACTURER	*MOTHERWELL BRIDGE TACOL LTD.*	
MODEL NAME AND NUMBER	Type 1 Cascade Rotary Drum	Type 2 Rotary Kiln
TYPE	Solid waste	Liquid; pasty; solid as admix
APPLICATIONS	Municipal	Industrial
CAPACITY	40-2000 tonne/day	4-100 tonne/day
INCINERATOR		
Temperature (°C)	1000 →	
Fuel type	None used →	
Fuel utilization rate	Normally self-sustaining →	
Supplementary firing	Oil; gas, as required →	
FLUE GAS		
Temperature (°C)	300 →	
Flow rate (m^3/h)	Variable →	
HEAT RECOVERY		
Efficiency (%)	Variable dependent on CV of wastes burnt →	
Equipment type	Integral water tube or waste heat	Waste heat
POLLUTION-CONTROL SYSTEM	Electrostatic precipitators	High efficiency cyclone or scrubbers
EXTRAS AVAILABLE		

MANUFACTURER	*POTT INDUSTRIES INC., ENGINEERING CONTROLS DIVISION*
MODEL NAME AND NUMBER	ECX series
TYPE	Fume; liquid; solid waste
APPLICATIONS	Industrial; municipal
CAPACITY	
INCINERATOR	
Temperature (°C)	760-871
Fuel type	Gas
Fuel utilization rate	
Supplementary firing	
FLUE GAS	
Temperature (°C)	760-871
Flow rate (m^3/h)	30 000-200 000
HEAT RECOVERY	
Efficiency (%)	40-60
Equipment type	Fire tube boiler; thermal fluid heater; air preheater
POLLUTION-CONTROL SYSTEM	
EXTRAS AVAILABLE	

MANUFACTURER	*SMITH ENGINEERING CO.*	
MODEL NAME AND NUMBER	Custom-built, sized to air flow	
TYPE	Fumes off paint-curing oven	
APPLICATIONS	Can plant (Industrial) →	
CAPACITY	N/a →	
INCINERATOR		
Temperature (°C)	760 →	
Fuel type	Natural gas →	
Fuel utilization rate	5000 s.c.f.h.	262 lb/h
Supplementary firing		
FLUE GAS		
Temperature (°C)	132	177
Flow rate	48 000 lb/h	45 035 lb/h
HEAT RECOVERY		
Efficiency (%)	1st=50; 2nd=20; 3rd=20	1st=70; 2nd=50
Equipment type	Air-to-air; air-to-water; air	Air-to-air
POLLUTION-CONTROL SYSTEM	Yes →	
EXTRAS AVAILABLE		

MANUFACTURER	*TRANE THERMAL CO.*	
MODEL NAME AND NUMBER		
TYPE	Fume	Liquid waste
APPLICATIONS	Industrial ⟶	
CAPACITY	15 000-90 000 m^3/h	Varies
INCINERATOR		
Temperature (°C)	500-1000	1000
Fuel type	All types, including waste ⟶	
Fuel utilization rate		
Supplementary firing	Yes, natural gas or fuel oil	Yes, natural gas or fuel oil
FLUE GAS		
Temperature (°C)		
Flow rate (m^3/h)		
HEAT RECOVERY		
Efficiency (%)	70-80 ⟶	
Equipment type	Recuperation; waste heat boiler	Waste heat boiler
POLLUTION-CONTROL SYSTEM	If required	Scrubbers; mist eliminators
EXTRAS AVAILABLE		

MANUFACTURER	*VYNCKE WT*
MODEL NAME AND NUMBER	VIF 150 to 6000
TYPE	Solid waste: wood; board; paper; plastics; textile; sisal; coconut fibre; shells; etc.
APPLICATIONS	Industrial
CAPACITY	150: 150 000 kcal/h = 175 kW) (m^3/day dependent on the 600: 60 000 kcal/h = 7000 kW) calorific value of the burnt waste)
INCINERATOR	
Temperature (°C)	800-1200
Fuel type	See Type
Fuel utilization rate	100%
Supplementary firing	Only for moist waste when not predryed
FLUE GAS	
Temperature (°C)	250
Flow rate (m^3/h)	(calculated at 250°C) 900-40 000
HEAT RECOVERY	
Efficiency (%)	Dependent on type: 76-84
Equipment type	Three draught fire tube boilers (water and steam); thermal fluid heaters (350°C)
POLLUTION-CONTROL SYSTEM	Cyclone (dry type) (dedusting to 150 mg/Nm^3)
EXTRAS AVAILABLE	Automatic storing; metering; transporting; feeding systems of the waste

13 Heat pumps

The heat pump is unique among the equipment described in this Directory in that it is able to produce heat at a higher temperature than the heat source from which the heat is being recovered. As explained in Chapter 6, this is achieved only by the expenditure of some energy – in the vapour-compression cycle the energy is used to drive the refrigerant gas compressor.

Heat pumps have been used for many years in HVAC applications, where they are commonly regarded not so much as heat recovery devices, but as means for effectively providing heating and cooling in a building by functioning as a refrigerator in one section and a heater in another. Nevertheless, this is a most effective energy-conservation technique. As discussed in Chapter 6, heat pumps may also use ambient air (or water) as the heat source, a method for domestic heating now being actively promoted in Europe (having been used routinely in the USA for decades).

Application of heat pumps to industrial-process heat recovery has, on the other hand, proceeded much more slowly, and few manufacturers offer systems for heat recovery from, say, process liquid effluents. More use has been made in the industrial sector of the combined heating and refrigeration capabilities of heat pumps, and a classic example of this in the UK is cited in Chapter 6. Similarly, little or no reference is made in manufacturers' literature to the use of diesel or gas engines to drive heat pump compressors, electric motors being the common drive mechanism. A number of large industrial concerns are active in this area throughout the world, however, and the heat pump scene as far as process industries are concerned is likely to be transformed over the next decade, both in Europe and North America, not to mention Japan. Potential (and existing) applications in industry are fully documented in Chapter 6.

13.1 Data presentation

Heat pump equipment is normally categorized in terms of the cycle – all those detailed below are operating on the vapour-compression cycle – and by the heat source and heat sink:

Air-to-air heat pumps;
Water-to-air heat pumps;
Air-to-water heat pumps;
Liquid–liquid heat pumps.

The heat pump industry in Europe relies heavily, at present, on licences from companies in the USA, but there are a number of notable exceptions. Also, HVAC applications predominate, but some very interesting process heat pump 'packages' are now being offered. The data is presented in a format similar to that in other chapters, but the properties of heat pumps lead to considerable differences in the type of data given.

13.1.1 *Format of data*

Data on commercially available heat pump systems is arranged in the following manner:

(1) *Equipment manufacturer*: The prime manufacturer or subsidiary company is given by name. Addresses, country of origin, telephone number and agents in other countries (or manufacturing licensees) are listed in Appendix 1. Where the addresses of agents are given, their supplier is also noted, so that reference may be made back to the product data in this chapter.
(2) *Model name and number*: Most heat pumps systems have a model name and an identification number, and these are given as appropriate. In some cases, where a number of models exist in one particular range, the range is indicated by the model numbers of the smallest and largest units, and the number of heat pumps available spanning that range.
(3) *Type*: The type of heat pump, normally expressed in terms of the type of heat source and sink used (see start of Section 13.1) is given.
(4) *Application*: Application areas for the heat pump system are listed. With regard to HVAC use, the particular types of buildings – e.g. schools, hospitals, commercial premises or swimming-pools, – are frequently given. Process use is also stated where appropriate, but for a more detailed analysis of this, the reader is recommended to consult Chapter 6.
(5) *Compressor data*: The main component of a heat pump is the compressor (assuming a vapour-compression cycle unit) and its drive. This section describes the main parameters of the heat pump in this respect, as follows:

(i) Compressor type: normally this will be either a reciprocating, centrifugal or screw type, depending upon the duty and the type of prime mover, the operating temperature range and the application (see Chapter 6).
(ii) Drive: the type of drive in this case is identified by the terms 'hermetic' or 'open'. Open-drive compressors, as their name implies, have a mechanical linkage between the drive and the compressor, as in a diesel driving a generator set. At the other extreme, a hermetic unit consists of drive and compressor

in an integrated unit, completely sealed. Such a system is comparatively cheap to manufacture and eliminates the need for rotating seals. It is only used for electric-motor drives.

(iii) Drive power source: in cases where an open compressor is used, the drive power source is listed. This may be any one of a number of prime movers.

(iv) Power input – cooling: the heat transport capability of a heat pump varies, depending upon whether it is being used in a heating or cooling duty (HVAC systems are normally reversible, and may be used to cool a building in summer, as well as providing space heating in winter). This also affects the power needed to drive the compressor, and it is normal practice on reversible heat pump systems to quote a power input in both the heating and cooling modes. This power input is often given at standard conditions of evaporating and condensing temperatures (ARI) or at temperatures specified in the manufacturer's literature.

(v) Power input – heating: see (iv) above.

(6) *Refrigerant used*: The choice of refrigerant (or working fluid) in the heat pump is determined largely by the operating temperature range, which fixes the pressure ratio. R22 is the most common for HVAC applications, others such as R114 being used at higher temperatures. These are discussed in more detail in Chapter 6.
(7) *Heat pump capacity*: As in the case of the compressor input power, the heat pump capacity, i.e. the amount of heating or cooling it will carry out, varies depending upon the mode in which it is operating. This may be determined to the ARI standard, or be given at specific temperatures by the manufacturer.
(8) *Operating temperature*: The operating temperature range of many heat pumps is limited by compressor/refrigerant characteristics, and this applies particularly when heat pumps are required for industrial-process applications, a condensing temperature little in excess of 100°C being currently the maximum commercially attainable. Where possible, in this section, the minimum evaporating temperature and maximum condensing temperature are given. Readers are, however, encouraged to consult the manufacturers, who may in some cases be able to increase the condensing temperature of a particular unit by using a different working fluid.
(9) *Typical payback periods*: It is difficult to obtain data on the payback periods for heat pump systems in air conditioning in many cases, because it is often difficult to define, and can only be compared to systems which would be used instead of a heat pump, generally having a lower efficiency. However, where heat pumps may be retrofitted to a process, the return on the investment may be considered in the same way as any other piece of heat recovery equipment.
(10) *Features*: As with more conventional heat recovery equipment, there are a number of features of heat pumps, in addition to their overall performance characteristics, which are of considerable interest to a potential purchaser. Some of these are listed for each manufacturer, as follows:

(i) Available custom-built: some manufacturers are able to 'custom-build' heat pump 'packages' to meet the requirements of process applications. In HVAC installations, the size of the unit is generally fixed within the range offered by the manufacturer, but the number of these standard modules may be varied to suit the needs of each building, room size and number of rooms.

(ii) Available in 'package': most heat pumps are available in packaged units, including necessary evaporating, condensing and intercooling heat exchangers.

(iii) Performance guarantees: guarantees are normally given on performance, assuming that the data given to the manufacturer for his assessment is accurate. Components of the heat pump system also regularly carry some form of guarantee – some manufacturers, for example, offer a one-year guarantee on all parts, with an additional four-years' guarantee on the compressor.

(iv) Installer: the manufacturer, an authorized dealer, or a contractor nominated by the purchaser or consulting engineers, may be responsible for installation of the heat pump system. In many countries outside the USA, contractors with wide experience of heat pump systems are scarce, and it is wise to consult the manufacturer in cases of doubt, who may insist that one of his factory-trained dealer-network members carries out the work.

(v) Performance control and turndown range: Heat pumps can have both automatic control and manual over-ride facilities, applying both to normal heating or cooling duty, and reversing of the cycle. Control of supplementary heating may also be automatic. A number of control features on most heat pump systems are offered as optional extras. Heat pumps used in process applications can have good turndown ratios, but if driven by a fossil-fuelled prime mover, the user must remember that this component will also have an optimum running speed and load characteristic for maximum fuel economy.

(vi) Defrosting: because the evaporator coil cools down the ambient air on heat pumps used in HVAC and refrigeration duties, frost can form on this coil. Defrost mechanisms are normally used to overcome this – a common technique is to reverse the heat pump at regular intervals (an energy-consuming method), or to reverse only when a sensor indicates frost on the coil. Supplementary coil heating may also be used. Process-plant use normally does not require such precautions.

(vii) Reversibility: heat pumps used in applications other than industrial processes may be reversible, both for ease of defrosting of the evaporator coil and for functioning both as a heater in winter and a cooling system in summer.

(11) *Extras available*: Heat pump manufacturers in general offer a wide range of optional equipment with their systems. This includes remote thermostats, automatic changeover from heating to cooling, night setback (to save energy when buildings are unoccupied), air filtration, fresh-air control systems, supplementary electric heaters, heat-storage tanks, sound absorbers and 'specials'.

(12) *Further comments:* Provision is made for each manufacturer to provide any

additional data which he feels may be of value to the potential user.

Manufacturers' data sheets relating to heat pump equipment follow on the next page.

MANUFACTURER	*AAF LTD.*
MODEL NAME AND NUMBER	Enercon HV; CW; VW; and RCRM units
TYPE	Water-to-air
APPLICATIONS	Office buildings; particularly suitable for multiroom buildings where a high degree of flexibility is required
COMPRESSOR	
Type	
Drive	Hermetic; semihermetic
Drive power source (open only)	
Power input (cooling)	From 0.69 up to 52.0 kW
Power input (heating)	From 0.72 up to 5.6 kW
REFRIGERANTS USED	R22
HEAT PUMP CAPACITY	
Cooling	From 1.6 up to 165 kW
Heating	From 1.8 up to 153 kW
TEMPERATURE (°C)	
Operating range (min. evaporating to max. condensing temperature)	Information not available for general release; however, all Enercon products are suitable for supply-water temperatures between 16-33°C
TYPICAL PAYBACK PERIODS (years)	
Airconditioning	
Process applications	
FEATURES	
Available custom-built?	Under certain circumstances equipment can be modified to suit customer requirements
Available in package	Yes
Performance guarantee	Yes, ARI rated; underwriters laboratory approved
Installed by?	Main HV contractor
Performance control and turndown range	
Defrosting	Not necessary
Reversability	Completely reversible
EXTRAS AVAILABLE	Manual or automatic changeover thermostats; wall- or unit-mounted washable, permanent or throwaway, filters; random start relays available; manual or automatic fresh-air dampers; flexible hose connections
FURTHER COMMENTS	CW; HW; and VW units carry a 5-year refrigerant warranty, and a 1-year component parts warranty; RCRM units have a 1-year warranty.

MANUFACTURER	*L'AIR CONDITIONNE-AIRWELL*					
MODEL NAME AND NUMBER	Heat Pump RO			Self-contained (Reverse-Cycle System)		
	2.5;	3;	4;	3;	5;	7.5;
TYPE	Air-to-water			Air-to-air		
APPLICATIONS	Residential: office; swimming-pool			Residential: office		
COMPRESSOR						
Type	Reciprocating →					
Drive	Hermetic →					
Drive power source (open only)						
Power input (cooling)	No cooling →			3440	5540	6995
Power input (heating)	3190W	3890W	4850W	3440W	5540W	6995W
REFRIGERANTS USED	R22	R22	R22	R22	R22	R22
HEAT PUMP CAPACITY						
Cooling	No cooling →			8240W	14100W	19030W
Heating (1)	8400W	10700W	13000W	10200W	16350W	23450W (2)
TEMPERATURE (°C)						
Operating range (min. evaporating to max. condensing temperature)	Min. evaporating: -4 Max. condensing: 68			Min. evaporating: -11 Max. condensing: 68		
TYPICAL PAYBACK PERIOD (years)						
Airconditioning						
Process applications						
FEATURES						
Available custom-built?	No →					
Available in package	No →					
Performance guarantee						
Installed by?						
Performance control and turndown range	Turndown range about + 5°C (3)			Turndown range about + 5°C (4)		
Defrosting	No →			Yes →		
Reversability	No →			Yes →		
EXTRAS AVAILABLE	Electrical heating (W)			9000	10500	18000
FURTHER COMMENTS						
NOTES						

NOTES

(1) Outside wet bulb, 7°C; temperature of water, 53°C.
(2) Outside dry bulb, 7°C inside temperatures, 20°C.
(3) For oil burner.
(4) For electrical heating.

MANUFACTURER	*ROBERT BOSCH GMBH*					
MODEL NAME AND NUMBER	Industrial Heat Pump	PIW 12;	PIW 17;	PIW 23;	PIW 33;	PIW 43;
TYPE		Water-to-water →				
APPLICATIONS	Heat sources: cooling circuits or waste water; heat users: warm water (e.g. for showers)					
COMPRESSOR						
Type		Reciprocating →				
Drive		Semihermetic →				
Drive power source (open only)		Electric →				
Power input (cooling)	At $t_o = 10^oC, t_c = 50^oC$	2.4kW	3.5kW	4.9kW	7.0kW	9.0kW
Power input (heating)	See further comments	2.4kW	3.5kW	4.9kW	7.0kW	9.0kW
REFRIGERANTS USED		R12 →				
HEAT PUMP CAPACITY						
Cooling	At $t_o = 10^oC, t_c = 50^oC$	7.2kW	10.2kW	14.3kW	20.4kW	26.3kW
Heating	At $t_o = 10^oC, t_c = 50^oC$	9.5kW	13.5kW	18.9kW	27.0kW	34.8kW
TEMPERATURE						
Operating range (min. evaporating to max. condensing temperature)		From $t_o = 0^oC$		to $t_c = 60^oC$ →		
TYPICAL PAYBACK PERIOD (years)						
Airconditioning						
Process applications		2-7 years →				
FEATURES						
Available custom-built		No →				
Available in package		Yes →				
Performance guarantee		Yes →				
Installed by?		Plumber or industrial user →				
Performance control and turndown range		No →				
Defrosting		No →				
Reversability		No →				
EXTRAS AVAILABLE		Air-to-water heat pumps (on request) →				
FURTHER COMMENTS	These heat pumps are not reversible, but they can be used for cooling (by evaporator) and heating (by condenser) simultaneously; therefore, the same power input is valid for cooling and heating.					

MANUFACTURER	*CARRIER INTERNATIONAL CORP.*
MODEL NAME AND NUMBER	30GR 008, 010, 015
TYPE	Water-to-air
APPLICATIONS	
COMPRESSOR	
Type	Reciprocating
Drive	Semi-hermetic
Drive power source (open only)	
Power input (cooling)	
Power input (heating)	
REFRIGERANTS USED	R22
HEAT PUMP CAPACITY	
Cooling	13.2-53.6 kW
Heating	7.9-53.9 kW
TEMPERATURE (°C)	
Operating range (min. evaporating to max. condensing temperature)	Cooling, 25-45°C; heating, -15-15°C
TYPICAL PAYBACK PERIOD (years)	
Airconditioning	
Process applications	
FEATURES	
Available custom-built	No
Available in package	Yes
Performance guarantee	Yes ± 5% on capacity as per US Standard (A.R.I.)
Installed by?	Contractor
Performance control and turndown range	Electrothermostatic control
Defrosting	Automatic defrost cycle
Reversability	Yes
EXTRAS AVAILABLE	Safety cut-offs Supplementary electric heaters 'Motormaster' control system
FURTHER COMMENTS	

MANUFACTURER	*CARRIER INTERNATIONAL CORP.*
MODEL NAME AND NUMBER	38CQ Split System/40 AQ-FS; 38AC Split System/40RT; 50RQ Single Package; 38RQ Split System/40 AQ-FS; 50MQ Single Package
TYPE	Air-air (reverse cycle)
APPLICATIONS	Air conditioning
COMPRESSOR	
Type	Reciprocating
Drive	Hermetic or semihermetic
Drive power source (open only)	
Power input (cooling	
Power input (heating)	
REFRIGERANTS USED	R500, R22
HEAT PUMP CAPACITY	
Cooling	
Heating	38CQ - 8.3-11.9kW; 38RQ - 10.1-15.9kW; 38AC - 28.2-100.5kW; 50MQ - 8.6-14.4kW; 50DQ - 50kW; 50RQ - 15.2-28kW;
TEMPERATURE (^{o}C)	
Operating range (min. evaporating to max. condensing temperature)	-24 to +22°C (heating) 8 to 50°C (Cooling)
TYPICAL PAYBACK PERIOD (years)	
Airconditioning	Depends on energy costs; 3-7 years in USA
Process applications	N/a
FEATURES	
Available custom built	No
Available in package	Yes
Performance guarantee	Yes $\pm$ 5% on capacity as per US Standard (A.R.I.)
Installed by?	Contractor
Performance control and turndown range	Indoor thermostat
Defrosting	Defrost cycle incorporated
Reversability	Yes
EXTRAS AVAILABLE	
FURTHER COMMENTS	
Air Volumes (m^3/h)	40 FS/AQ Indoor Unit 680-3830; 40 RT Indoor Unit 5760-2550 50 MQ Single package 1330-3340; 50 DQ 7510-10520 50 RQ 3170-5670

MANUFACTURER	*CLIREF*					
MODEL NAME AND NUMBER	PLC A C;	RLA C;	DRLA DH;	EVE;	RL;	DRL
TYPE	Water-to-air →			Water-to-water →		
APPLICATIONS						
COMPRESSOR						
Type						
Drive	Hermetic	Semi-hermetic	Semi-hermetic	Hermetic	Semi-hermetic	Semi-hermetic
Drive power source (open only)	N/a					
Power input (cooling)	3-13 ton	10-30 ton	30-72 ton	2-10 ton	6-72 ton	80-200 ton
Power input (heating)	5-18 ton →			3-13 ton →		
REFRIGERANTS USED	R22 →					
HEAT PUMP CAPACITY						
Cooling						
Heating						
TEMPERATURE (°C)						
Operating range (min. evaporating to max. condensing temperature)	-15-60	-25-60	-25-60	-15-60	-25-60	-25-60
TYPICAL PAYBACK PERIOD (years)						
Airconditioning	3-9 →					
Process applications	2-5 →					

MANUFACTURER	*COMMAND-AIRE CORP.*
MODEL NAME AND NUMBER	Console units SWPC (½-1½ ton) Rooftop units RWP/RUWP (2-45 ton) Vertical indoor units and Horizontal Ceiling Units SWP/H (0.75-25 tons)
TYPE	Water-to-air
APPLICATIONS	Comfort airconditioning: residential, commercial; industrial
COMPRESSOR	
Type	Reciprocating
Drive	Hermetic
Drive power source (open only)	
Power input (cooling)	1310 ton for 2½ ton unit (typical) at ARI conditions
Power input (heating)	986 ton for 2½ ton unit (typical) at ARI conditions
REFRIGERANTS USED	R22
HEAT PUMP CAPACITY	
Cooling	2 kW
Heating	2.6 kW
TEMPERATURE (°C)	
Operating range (min. evaporating to max. condensing temperature)	Min. evaporating temperature 2 Max. condensing temperature 55
TYPICAL PAYBACK PERIODS (years)	
Airconditioning	1-3
Process applications	N/a
FEATURES	
Available custom-built	No
Available in package	Yes
Performance guarantee	To meet catalogue ratings
Installed by?	Mechanical contractors
Performance control and turndown range	1 step up to 5 ton; 2 or 3 step above 5 ton
Defrosting	Not required
Reversability	Automatic or manual by electric valve
EXTRAS AVAILABLE	Day-night control; outside air economizer cycle; special air filtering
FURTHER COMMENTS	High coefficient of performance.

MANUFACTURER	*CORBRIDGE SERVICES LTD. (METRO OF DENMARK)*			
MODEL NAME AND NUMBER	3000 MV	3000 MJ	2003 MV	2004 MJ
TYPE	Air-to-water	Water-to-water	Air-to-water	Water-to-water
APPLICATIONS	Domestic hot water →			
COMPRESSOR				
Type	Centrifugal →			
Drive	Hermetic →			
Drive power source (open only)				
Power input (cooling)	0.560 W →			
Power input (heating)				
REFRIGERANTS USED	Freon →			
HEAT PUMP CAPACITY				
Cooling				
Heating	Max. 1.5 kW →			
TEMPERATURE (°C)				
Operating range (min. evaporating to max. condensing temperature)	52 →			
TYPICAL PAYBACK PERIODS (years)				
Airconditioning	As little as 2, dependent on ambient input →			
Process applications	Resaurant hot water from kitchen hoods →			
FEATURES				
Available custom-built	No →			
Available in package	Yes →			
Performance guarantee	Yes →			
Installed by?	Local installers →			
Performance control and turndown range	Yes →			
Defrosting	Built-in →			
Reversability	No →			
EXTRAS AVAILABLE	Built-in 280 l tank, →			
FURTHER COMMENTS	2 kW and 6 kW input models available.			

MANUFACTURER	*DUNHAM-BUSH LTD.*
MODEL NAME AND NUMBER	From PCW 025H-HRS (25 h.p.)-PCW 100H-HRD (100 h.p.) Reciprocating (-R22) From PCX 120 H5-PCX 750-05 (-R22) PCX 121-OHR-PCX 351-OHR (-R500)
TYPE	Liquid-to-liquid (heat source, water; heating medium, water)
APPLICATIONS	Office buildings; banks; stores; etc.
COMPRESSOR	
Type	Reciprocating 25-100 h.p.; screw 100-750 h.p.
Drive	Generally available in semi-hermetic and open configuration
Drive power source (open only)	Electric motor
Power input (cooling)	From 20 kW-600 kW, according to model
Power input (heating)	
REFRIGERANTS USED	R22, R500
HEAT PUMP CAPACITY	
Cooling	73 kW-11968 kW
Heating	88 kW-2372 kW
TEMPERATURE (°C)	
Operating range (min. evaporating to max. condensing temperature)	Min. evaporating = 1°C; Max. condensing R22 = 51°C; R500 = 71°C
TYPICAL PAYBACK PERIODS (years)	
Airconditioning	Totally dependent upon the individual application
Process applications	
FEATURES	
Available custom-built	Yes
Available in package	Yes
Performance guarantee	Yes
Installed by?	Approved supplier and installer
Performance control and turndown range	100-10% of full-load duty
Defrosting	Not applicable
Reversability	No
EXTRAS AVAILABLE	Hot-gas injection for lower than 10% capacity control; Double-bundle condensers for reclaim circuits
FURTHER COMMENTS	

MANUFACTURER	*FGF EQUIPMENT*
MODEL NAME AND NUMBER	PCDF 1-20 (8 models) VMCRO 3-30 (7 models)
TYPE	PCDF: Air-to-air associated with heat pipes VMCRO: Air-to-water
APPLICATIONS	Heat recovery on exhaust air for buildings, swimming-pools, hospitals, etc.
COMPRESSOR	
Type	Reciprocating
Drive	Hermetic
Drive power source (open only)	
Power input (cooling)	2 kW
Power input (heating)	0.9-13.6 kW for PCDF with heat pipes; 2.8-31.8 kW for VMCRO
REFRIGERANTS USED	R22
HEAT PUMP CAPACITY	
Cooling	3100-47 000 kW/h for PCDF with heat pipe; VMCRO not used in cooling
Heating	7.8-148 kW for PCDF with heat pipes; 8.9-5.6 for VMCRO
TEMPERATURE (°C)	
Operating range (min. evaporating to max. condensing temperature)	-5-15°C evaporating 20-65°C condensing
TYPICAL PAYBACK PERIODS (years)	
Airconditioning	
Process applications	
FEATURES	
Available custom built	
Available in package	
Performance guarantee	
Installed by?	
Performance control and turndown range	
Defrosting	
Reversability	An option on PCDF only
EXTRAS AVAILABLE	The PCDF heat pump is associated with heat pipes; performances in low temperatures (when needed for energy of heating) are doubled because of the heat pipe; the coefficient of efficiency may be more than 10
FURTHER COMMENTS	

MANUFACTURER	*AB SVENSKA FLAKTFABRIKEN*
MODEL NAME AND NUMBER	VKBB-02-5-39; VKBB-02-6-39 (6 sizes)
TYPE	Air-to-air
APPLICATIONS	
COMPRESSOR	
Type	Reciprocating
Drive	Semihermetic
Drive power source (open only)	
Power input (cooling)	Max. 14; 19; 27; and 20; 27; 31 kW (6 sizes)
Power input (heating)	Max. 16; 21; 30; and 21; 30; 37 kW (6 sizes)
REFRIGERANTS USED	
HEAT PUMP CAPACITY	
Cooling (kW)	Max. 52; 69; 90; and 74; 92; 116 (6 sizes)
Heating (kW)	Max. 61; 78; 100; and 84; 105; 135 (6 sizes) (at 15°C outdoor air temperature)
TEMPERATURE (°C)	
Operating range (min. evaporating to max. condensing temperature)	Cooling 0-50; Heating -30-50
TYPICAL PAYBACK PERIODS (years)	
Airconditioning	
Process applications	
FEATURES	
Available custom-built	No
Available in package	Yes
Performance guarantee	See catalogue
Installed by?	Contractor
Performance control and turndown range	Yes
Defrosting	Defrost timer/defrost thermostat; defrosts every 90 min, if required
Reversability	By a 4-way valve
EXTRAS AVAILABLE	Special versions for exhaust air only; sound absorbers
FURTHER COMMENTS	Complete air-handling unit comprising: supply-air fan; exhaust-air fan; filters G80/F45/F85; Cooling heat pump unit; reheating system with controls (water or electric); fully equipped control system.

MANUFACTURER	*FRIEDRICH AIR CONDITIONING & REFRIGERATION CO.*
MODEL NAME AND NUMBER	PYA Packaged Heat Pumps PYA-024-PYA-060
TYPE	Air-to-air
APPLICATIONS	Offices and room heating cooling, using ambient air source
COMPRESSOR	
Type	Centrifugal
Drive	Hermetic
Drive power source (open only)	
Power input (cooling)	3.8-8.9 kW (ARI 240)
Power input (heating)	3.5 kW-7.3 kW (ARI 240)
REFRIGERANTS USED	R22
HEAT PUMP CAPACITY	
Cooling	6.74-17 kW (ARI 240)
Heating	7.3-18 kW (ARI 240)
TEMPERATURE (^{o}C)	
Operating range (min. evaporating to max. condensing temperature)	Appropriate for air-source space heating only
TYPICAL PAYBACK PERIODS (years)	
Airconditioning	N/a
Process applications	N/a
FEATURES	
Available custom-built	No
Available in package	Yes
Performance guarantee	Yes
Installed by?	Contractor
Performance control and turndown range	Automatic control system; thermostat plus control by optional supplementary heating
Defrosting	Automatic defrost cycle
Reversability	Yes
EXTRAS AVAILABLE	Supplementary electric heaters
FURTHER COMMENTS	

MANUFACTURER	*FRIEDRICH AIR CONDITIONING & REFRIGERATION CO.*
MODEL NAME AND NUMBER	Series 800 Heat Recovery Systems (Console models)
TYPE	Water-to-air, decentralized system
APPLICATIONS	Offices; schools; hospitals
COMPRESSOR	
Type	Centrifugal
Drive	Hermetic
Drive power source (open only)	
Power input (cooling)	(kW to ARI Rating) 0.82 (800-7 model)-2.45 (800-16 model)
Power input (heating)	(kW to ARI Rating) 0.75 (800-7 model)-2.30 (800-16 model)
REFRIGERANTS USED	R22
HEAT PUMP CAPACITY	
Cooling	(kW to ARI Rating) 2.05-5.3
Heating	(kW to ARI Rating) 2.05-5.6
TEMPERATURE (°C)	
Operating range (min. evaporating to max. condensing temperature)	15-46
TYPICAL PAYBACK PERIODS (years)	
Airconditioning	
Process applications	N/a
FEATURES	
Available custom-built	Yes, from standard modules
Available in package	Yes
Performance guarantee	Yes
Installed by?	Contractor
Performance control and turndown range	Manual changeover
Defrosting	N/a
Reversability	Yes
EXTRAS AVAILABLE	Automatic changeover; remote thermostat; fresh-air control; night setback; time delay
FURTHER COMMENTS	Can be linked to solar collectors.

MANUFACTURER	*FRIEDRICH AIR CONDITIONING & REFRIGERATION CO.*
MODEL NAME AND NUMBER	V/H Heat Recovery Systems (decentralized heat pump system), H/V10-V240
TYPE	Water-to-air, and supplementary water heating, if required
APPLICATIONS	Offices; schools; hospitals
COMPRESSOR	
Type	Centrifugal
Drive	Hermetic
Drive power source (open only)	
Power input (cooling)	(kW to ARI 240-67): 1.35-24.8
Power input (heating)	(kW to ARI 240-67): 1.30-19.4
REFRIGERANTS USED	R22
HEAT PUMP CAPACITY	
Cooling	(kW to ARI 240-67): 2.95-70.5
Heating	(kW to ARI 240-67): 3.23-60.0
TEMPERATURE (°C)	
Operating range (min. evaporating to max. condensing temperature)	15.6-40.5
TYPICAL PAYBACK PERIODS (years)	
Airconditioning	
Process applications	
FEATURES	
Available custom-built	Yes, from standard modules
Available in package	Yes
Performance guarantee	Yes
Installed by?	Contractor
Performance control and turndown range	Time delay; remote thermostat; night setbacks
Defrosting	N/a
Reversability	Yes
EXTRAS AVAILABLE	Controls/thermostat for automatic changeover from cooling to heating and vice versa; automatic pump alternators; no-flow relay channels
FURTHER COMMENTS	Systems available as vertical (stored concealed in mechanical equipment rooms, etc.); or horizontal (above ceiling); largest units available in vertical mode only.

MANUFACTURER	*GENERAL ELECTRIC CO.*
MODEL NAME AND NUMBER	IGETRIC Weathertron, models WB818; 824; WR030 050; WA075-200 (9 sizes)
TYPE	Air-to-air split system
APPLICATIONS	Commercial and industrial-building space heating/ cooling
COMPRESSOR	
Type	Centrifugal
Drive	Hermetic
Drive power source (open only)	
Power input (cooling)	1.8 (WB818)-22.4 kW (WA200)
Power input (heating)	2.1 (WB818)-19.7 kW (WA200)
REFRIGERANTS USED	R22
HEAT PUMP CAPACITY	
Cooling	4.5 (WB818)-57.1 kW (WA200); Rated at 35°C ambient d.b. 27°C inside d.b., 20°C w.b.
Heating	5.1 (WB818)-58.6 kW (WA200); Rated at 8°C d.b., 6°C w.b. ambient; 21°C d.b. inside
TEMPERATURE (°C)	
Operating range (min. evaporating to max. condensing temperature)	-20-24 outdoor to indoor d.b.
TYPICAL PAYBACK PERIODS (years)	
Airconditioning	
Process applications	
FEATURES	
Available custom-built?	No
Available in package	Yes
Performance guarantee	1 year on all parts; 4 years additional on compressor
Installed by?	Manufacturer's agent or contractor
Performance control and turndown range	Control by automatic thermostat
Defrosting	Yes, defrosts only when coil partially blocked by ice
Reversability	Yes, automatic thermostat and electrically operated 4-way switchover valve
EXTRAS AVAILABLE	Accessory pressure control for cooling to -18°C; steam and hot-water coils; return air plenums; and air distributors (on some models); cooling only duty if required.

MANUFACTURER	*GENERAL ELECTRIC CO.*
MODEL NAME AND NUMBER	IGETRIC Weathertron, model WC 024-WC100 (7 sizes)
TYPE	Air-to-air (single package)
APPLICATIONS	Office-building space heating/cooling
COMPRESSOR	
Type	Centrifugal
Drive	Hermetic
Drive power source (open only)	
Power input (cooling)	2.5 (WC024)-10.3 kW (WC100)
Power input (heating)	2.2 (WC024)-8.1 kW (WC100)
REFRIGERANTS USED	R22
HEAT PUMP CAPACITY	
Cooling	5.84 (WC024)-28.6 kW(WC100); rated at 35°C ambient d.b. 27°C inside d.b., 20°C w.b.
Heating	6.35 (WC024)-29.3 kW (WC100); rated at 8°C d.b., 6°C w.b., ambient; 21°C d.b. inside
TEMPERATURE (°C)	
Operating range (min. evaporating to max. condensing temperature)	-20-24 outdoor to indoor d.b.
TYPICAL PAYBACK PERIODS (years)	
Airconditioning	
Process applications	
FEATURES	
Available custom-built?	Yes
Available in package	Yes
Performance guarantee	1 year on all parts; 4 years additional on compressor
Installed by?	Manufacturer's agent or contractor
Performance control and turndown range	Via automatic thermostat, aided by supplementary heaters as necessary
Defrosting	Yes, defrosts only when coil partially blocked by ice
Reversability	Yes, automatic thermostat
EXTRAS AVAILABLE	High static-pressure indoor fans; seasonal selection thermostats (instead of automatic); low ambient cooling operation (to 13°C); and evaporator defrosts control on indoor coil (cooling to -1°C)
FURTHER COMMENTS	

MANUFACTURER	*GRENCO BV*
MODEL NAME AND NUMBER	
TYPE	Ground-to-air; air-to-air; water-to-air; Ground-to-water; air-to-water; water-to-water
APPLICATIONS	Comfort heating Process heating
COMPRESSOR	
Type	RC9/RC11/AC80 (reciprocating) MS10 (screw)
Drive	SAC 80 (semihermetic) RC9/RC11/AC80/MS10 (open)
Drive power source (open only)	Electric motor Diesel/gas engine
Power input (cooling)	$t_o = 10^oC$ $t_c = 25^oC$ $Q_o/P = 5$ $t_o / t_c = {-5}/{+35}$ $Q_o/P = 4$
Power input (heating)	$t_o = 5$ $t_c = 40$ AC COP = 5.5; MS COP = 6.5
REFRIGERANTS USED	NH3/R12/R22
HEAT PUMP CAPACITY	
Cooling	$t_o/t_c = -10/25$ 80-1700 kW
Heating	$t_o/t_c = 5/40$ 160-3500 kW
TEMPERATURE (^oC)	
Operating range (min. evaporating to max. condensing temperature)	R22 t_o min. = -5 t_c max. = 55 R12 t_o min. = -5 t_c max. = 75
TYPICAL PAYBACK PERIODS (years)	
Airconditioning	Dependent on local energy prices; and
Process applications	yearly operating hours
FEATURES	
Available custom-built	Yes
Available in package	Yes
Performance guarantee	Yes
Installed by?	Manufacturer
Performance control and turndown range	25-100% at fixed rev/min
Defrosting	Yes
Reversability	Yes
EXTRAS AVAILABLE	
FURTHER COMMENTS	

MANUFACTURER	*HITACHI LTD.*
MODEL NAME AND NUMBER	RHU-303A; RHU-503A and RHU-753A Heat Pump Water Chiller
TYPE	Air-to-water
APPLICATIONS	Cooling/heating for offices; shops; stores; public facilities; industrial spaces
COMPRESSOR	
Type	Reciprocating
Drive	Hermetic
Drive power source (open only)	
Power input (cooling)	RHU-303A: 3.0; RHU-503A: 5.1; RHU-753A: 7.6 kW
Power input (heating)	RHU-303A: 2.7; RHU-503A: 4.9; RHU-753A: 6.7 kW
REFRIGERANTS USED	R22
HEAT PUMP CAPACITY	
Cooling	RHU-303A: 6.4; RHU-503A: 11.6; RHU-753A: 17.1 kW
Heating	RHU-303A: 8.1; RHU-503A: 14.5; RHU-753A: 21.3 kW
TEMPERATURE (°C) Operating range	(see below)

Min./Max. ambient and water temperature	Cooling	Ambient Temperature	20-40	Heating	Ambient Temperature	Wet-bulb-6.5-24 Dry-bulb-7.0-18
		Chilled-water temperature	5-15		Hot-Water temperature	57 max.

TYPICAL PAYBACK PERIODS (years)	
Airconditioning	Approx. 2 (based on 2.5 P/kW)
Process applications	
FEATURES	
Available custom-built	No
Available in package	Supplied in standard package form
Performance guarantee	2 years, labour/parts
Installed by?	Climate Equipment Ltd., and approved installers
Performance control and turndown range	Cooling: 5°C-15°C; Heating: 30°C-57°C
Defrosting	Thermostat; pressure switch; timer
Reversability	Yes, automatic heating or cooling
EXTRAS AVAILABLE	Refrigerant head pressure; fan-speed control
FURTHER COMMENTS	Hitachi Compressor Protection System (Hicompro System) comprises compressor/motor internal thermostat and mercury overcurrnet relays; compressor crankcase oil heater; fan-motor thermal overcurrent relays; timer relays to prevent short-cycling; discharge gas thermostat; refrigerant liquid-line fusible plug; and automatic deforst system.

MANUFACTURER	*HITACHI LTD.*
MODEL NAME AND NUMBER	RUA-512AH; RUA-762AH Self-contained Heat Pump Roof-top type Airconditioner
TYPE	Air-to-air
APPLICATIONS	Cooling/heating for offices; shops; stores; public facilities, industrial spaces
COMPRESSOR	
Type	Reciprocating
Drive	Hermetic
Drive power source (open only)	Electric motor
Power input (cooling)	RUA-512AH 5.4 kW RUA-762AH 7.8 kW
Power input (heating)	RUA-512AH 4.5 kW RUA-762AH 6.3 kW
REFRIGERANTS USED	Refrigerant 22
HEAT PUMP CAPACITY	
Cooling	RUA-512AH 14.2 kW RUA-762AH 21.0 kW
Heating	RUA-512AH 15.1 kW RUA-762AH 22.0 kW
TEMPERATURE (°C)	Indoor Temperature: / Outdoor Temperature:
Operating range (min. air to max. air dry bulb/ wet bulb temperatures)	Indoor: Cooling, 35 d.b./21 w.b. max.; 19.5 d.b./14 w.b. min. Heating, 27 d.b. max.; 15 d.b. min. Outdoor: Cooling, 50 d.b. max.; - 15 d.b. min. Heating, 24 d.b. max.; 20 d.b. min.
TYPICAL PAYBACK PERIODS (years)	
Airconditioning	Approx. 1.5 (based on 2.5p/kW)
Process applications	
FEATURES	
Available custom-built	No
Available in package	Supplied in standard package form
Performance guarantee	2 years, labour and parts
Installed by?	Climate Equipment Ltd., and approved installers
Performance control and turndown range	Room thermostat: range 18°C-30°C
Defrosting	Thermostat; pressure switch; timer
Reversability	Yes, automatic heating or cooling
EXTRAS AVAILABLE	Electric auxiliary air heater; air filter; refrigerant head pressure/fan-speed control
FURTHER COMMENTS	Hitachi compressor protection system (Hicompro System) comprises compressor/motor internal thermostat and mercury overcurrent relays; fan-motor thermal overcurrent relays; timer relays to prevent short-cycling; refrigerant discharge-gas thermostat; indoor-coil; freeze-protection thermostat; refrigerant dual pressure-switch and compressor crankcase oil heater; and automatic defrost system.

MANUFACTURER	*HITACHI ZOSEN (HITACHI SHIPBUILDING & ENGINEERING CO. LTD.)*
MODEL NAME AND NUMBER	Hi-Thermo 1700; 3400; 7000; 10000; 14000; 17000; 20000
TYPE	Air-to-air
APPLICATIONS	Office building; factory; supermarket
COMPRESSOR	
Type	Reciprocating
Drive	Hermetic
Drive power source (open only)	
Power input (cooling)	4-42 kW
Power input (heating)	3-31 kW
REFRIGERANTS USED	R22
HEAT PUMP CAPACITY	
Cooling	6100-66000 kcal/h at 35°C of the outdoor air temperature and 27°C of the indoor air temperature
Heating	7000-75000 kcal/h at 7°C of the outdoor air temperature and 21°C of the indoor air temperature
TEMPERATURE (°C)	
Operating range (min. evaporating to max. condensing temperature)	Min. evaporating temperature: -15°C; Max. condensing temperature: 55°C
TYPICAL PAYBACK PERIODS (years)	
Airconditioning	
Process applications	
FEATURES	
Available custom-built	Custom-built
Available in package	
Performance guarantee	Thermal output
Installed by?	Manufacturer
Performance control and turndown range	Performance control/step control turn range: 50-17%
Defrosting	Reverse operation
Reversability	Possible
EXTRAS AVAILABLE	
FURTHER COMMENTS	

MANUFACTURER	*INTERNATIONAL RESEARCH & DEVELOPMENT CO. LTD.*
MODEL NAME AND NUMBER	Specials only
TYPE	Liquid-to-liquid
APPLICATIONS	Industrial process plant
COMPRESSOR	
Type	Reciprocating
Drive	Open
Drive power source (open only)	Gas engine
Power input (cooling)	N/a
Power input (heating)	75 kW
REFRIGERANTS USED	R114
HEAT PUMP CAPACITY	
Cooling	N/a
Heating	250-350 kW
TEMPERATURE (°C)	
Operating range (min. evaporating to max. condensing temperature)	30-120
TYPICAL PAYBACK PERIODS (years)	
Airconditioning	N/a
Process applications	2-4
FEATURES	
Available custom-built	Yes
Available in package	Yes
Performance guarantee	Yes
Installed by?	Manufacturer or contractor
Performance control and turndown range	By engine or compressor modulation
Defrosting	N/a
Reversability	N/a
EXTRAS AVAILABLE	Prime mover heat recovery systems
FURTHER COMMENTS	Specials built only for demonstrations; evaluation; etc.

MANUFACTURER	*SINGER (CLIMATE CONTROL DIVISION)*
MODEL NAME AND NUMBER	WMB; CC; FV
TYPE	(Ceiling or wall-mounted) water-to-air
APPLICATIONS	Office blocks; hotels
COMPRESSOR	
Type	
Drive	Hermetic, semihermetic
Drive power source (open only)	
Power input (cooling)	
Power input (heating)	
REFRIGERANTS USED	R22
HEAT PUMP CAPACITY	
Cooling	1.96 kW-9.46 kW; and to 67.3 kW
Heating	3.16 kW-18 kW; and to 67.8 kW
TEMPERATURE (°C)	
Operating range (min. evaporating to max. condensing temperature)	
TYPICAL PAYBACK PERIODS (years)	
Airconditioning	
Process applications	
FEATURES	
Available custom-built	
Available in package	Yes
Performance guarantee	Yes
Installed by?	Mechanical contractors
Performance control and turndown range	
Defrosting	
Reversability	
EXTRAS AVAILABLE	
FURTHER COMMENTS	

MANUFACTURER	*SULZER BROS*
MODEL NAME AND NUMBER	Sulzer Escher Wyss Liquifrigor or Unitop (also specials)
TYPE	Ground; water; air; solar-heated water; serving air- or water-cooled condensers
APPLICATIONS	Chemical, food, and other industries; airconditioning; ice-rinks; swimming-ppols
COMPRESSOR	
Type	Normally reciprocating or turbocompressors
Drive	Semihermetic/open available
Drive power source (open only)	Normally electric motor; other drives under consideration
Power input (cooling)	
Power input (heating)	e.g. Swimming-pool heating: coefficient of performance, 4-8
REFRIGERANTS USED	Dependent on operating temperature range
HEAT PUMP CAPACITY	
Cooling	
Heating	Liquifrigor up to 582 kW; Unitop up to 1740 kW
TEMPERATURE (°C)	
Operating range (min. evaporating to max. condensing temperature)	Produces hot water at up to 50°C; ground water at 10°C can provide heat for condenser water at 40°C plus
TYPICAL PAYBACK PERIODS (years)	
Airconditioning	Dependent on application
Process applications	
FEATURES	
Available custom-built	Yes
Available in package	Yes
Performance guarantee	Yes
Installed by?	Manufacturer or contractor
Performance control and turndown range	Automatic control and modulation available
Defrosting	If required by application
Reversability	If required by application
EXTRAS AVAILABLE	
FURTHER COMMENTS	Special refrigerants; heat exchangers; and custom-built heat pump systems are available; some heat pumps can condense at a sufficiently high temperature to produce hot water above 70°C.

MANUFACTURER	*TECHNIBEL SA*
MODEL NAME AND NUMBER	PSA 308; PSA 411; PSA 514
TYPE	Air-to-air
APPLICATIONS	Industrial applications, office buildings
COMPRESSOR	
Type	Centrifugal
Drive	Hermetic
Drive power source (open only)	
Power input (cooling)	PSA 308: 3.9 kW PSA 411: 4.8 kW PSA 514: 6.0 kW
Power input (heating)	PSA 308: 3.5 kW PSA 411: 4.3 kW PSA 514: 5.6 kW
REFRIGERANTS USED	R22
HEAT PUMP CAPACITY	
Cooling	PSA 308: 8720W PSA 411: 11630W PSA 514: 13950W
Heating	PSA 308: 9420W PSA 411: 12300W PSA 514: 15000W
TEMPERATURE (°C)	
Operating range (min. evaporating to max. condensing temperature)	(Outside temperature: -30°C up to 45°C)
TYPICAL PAYBACK PERIODS (years)	
Airconditioning	1 if airconditioning used anyway
Process applications	5-6
FEATURES	
Available custom-built	No
Available in package	Yes
Performance guarantee	Yes
Installed by?	Authorized dealers only - lists available
Performance control and turndown range	Good. On-off, but expansion valve controlled
Defrosting	Yes
Reversability	Yes
EXTRAS AVAILABLE	Supplementary heaters; remote control; external thermostat to limit use of supplementary heaters; hot-water coils; humidifiers; low air flow control
FURTHER COMMENTS	Special outdoor coil design gives 30% faster defrost time and frosting occurs one-seventh less often than conventional fin-tube coils.

MANUFACTURER	*TRANE*
MODEL NAME AND NUMBER	Trane Heat Recovery Cold Generators (Water- or air-cooled, dual condensers) Trane Heat Recovery CenTraVac (Water cooled, dual condenser or double-bundle condenser)
TYPE	Liquid-to-liquid; water-to-air (reciprocating compressors); water-to-water (centrifugal compressors)
APPLICATIONS	Airconditioning and process
COMPRESSOR	
Type	Reciprocating compressor on cold generators; centrifugal compressors on CentraVac
Drive	Semihermetic reciprocating compressors standard; open reciprocating compressors as special design for water-cooled split systems; hermetic or open centrifugal compressors
Drive power source (open only)	Open reciprocating compressors normally electric motor; open centrifugals normally electric motor, but steam turbines; gas/diesel engine options available
Power input (cooling)	Variable from 34-3200 kW, dependent on application
Power input (heating)	Variable from 34-3200 kW, dependent on application
REFRIGERANTS USED	Reciprocating compressors: R22; R12; R502; Centrifugal compressors: R113; R11; R12; R22
HEAT PUMP CAPACITY	
Cooling	Variable from 75-11300 kW, dependent on application
Heating	Variable from 109-14500 kW, dependent on application
TEMPERATURE (°C)	
Operating range (min. evaporating to max. condensing temperature)	Reciprocating compressors: min. evaporating temperature, dependent on type of heat source, fluid, down to -20°C max. condensing temperature, dependent on actual evaporating temperature, up to 63°C Centrifugal compressor: min. evaporating temperature, dependent on type of heat source fluid, down to -12°C max. condensing temperature, dependent on actual evaporating temperature, up to 50°C
TYPICAL PAYBACK PERIODS (years)	
Airconditioning	From 1 year, but variable dependent on application
Process applications	From 1 year, but variable dependent on application
FEATURES	
Available custom-built	Yes
Available in package	As packaged refrigeration/heat recovery units only
Performance guarantee	Yes
Installed by?	Mechanical contractor
Performance control and turndown range	Reciprocating compressors: 4-step/8-step control; centrifugal compressors: full modulation down to 10%
Defrosting	Not necessary
Reversability	No
EXTRAS AVAILABLE	Units available for external location using air-cooled heat-rejection condenser and water-cooled heat recovery condenser: reciprocating compressors only
FURTHER COMMENTS	Standard line products may be used in applications to produce heat from cool-temperature energy sources; single condenser units air- or water-cooled with reciprocating compressors.

MANUFACTURER	*WESTAIR DYNAMICS LTD.*			
MODEL NAME AND NUMBER	Westair Dehumidifier HTS 2	Westair Dehumidifier TS 2	Westair Dehumidifier TS 5 Mk III	Westair Dehumidifier TSM
TYPE	Air-to-air →	→	→	→
APPLICATIONS	Process drying →	→	→	→
COMPRESSOR				
Type	Reciprocating →	→	→	→
Drive	Hermetic →	→	→	→
Drive power source (open only)				
Power input (cooling)	3 kW	3.5 kW	8.5 kW →	→
Power input (heating)	3 kW →	→	9 kW →	→
REFRIGERANTS USED	R22 →	→	→	→
HEAT PUMP CAPACITY				
Cooling				
Heating				
TEMPERATURE (°C)				
Operating range (min. evaporating to max. condensing temperature)	20°C-50°C	30°C-50°C →	→	→
TYPICAL PAYBACK PERIODS (years)				
Airconditioning				
Process applications	1-4 →	→	→	→
FEATURES				
Available custom-built	No →	→	→	Yes
Available in package	Yes →	→	→	→
Performance guarantee	No →	→	→	→
Installed by?	Customer or supplier →	→	→	→
Performance control and turndown range	N/a →	→	→	1-10
Defrosting	Yes	No →	→	→
Reversability	No →	→	→	→
EXTRAS AVAILABLE	Electric controls remote controls →	→	→	→
FURTHER COMMENTS				

MANUFACTURER	*WESTINGHOUSE*
MODEL NAME AND NUMBER	Templifier Heat Pump Water Heater Model TPB 020; 025; 030; 045; 055; 060 (series A, B, C, D)
TYPE	Water-to-water
APPLICATIONS	Process water heating (or glycol heating) in industry; hospitals; offices; etc.
COMPRESSOR	
Type	Centrifugal (type CH)
Drive	Hermetic
Drive power source (open only)	
Power input (cooling)	N/a
Power input (heating)	17.3-57.6 kW (series A); 17.5-58.3 kW (series B); 13.5-45.4 kW (series C); 11.5-38.2 kW (series D)
REFRIGERANTS USED	Series A: R500; series B: R12; series C: R21-114; series D: R114
HEAT PUMP CAPACITY	
Cooling	N/a
Heating	76-252 kW (series A); 70-234 kW (series B);[1] 49-163 kW (series C); 31-103 kW (series D)
TEMPERATURE (°C)	
Operating range (min. evaporating to max. condensing temperature)	16 (evaporating) to Series A 57 (condensing), series B 66 (condensing), series C 82 (condensing), series D 104 (condensing) (max.)
TYPICAL PAYBACK PERIODS (years)	
Airconditioning	N/a
Process applications	1.5-4 (typical)
FEATURES	
Available custom-built	Generally not, although equipment selections for temperature ranges outside tabulated data may be made by Westinghouse
Available in package	Yes
Performance guarantee	Yes
Installed by?	Manufacturer
Performance control and turndown range	Thermostats and time-delay switch; cut-out modulation
Defrosting	Normally not required
Reversability	No
EXTRAS AVAILABLE	
FURTHER COMMENTS	
NOTE	(1) Based on water-source leaving temperature of 29.5°C (series A); 35°C (series B); 40.6 °C (series C); 51.7°C (series D); heating to maximum appropriate condensing temperatures.

MANUFACTURER	*WESTINGHOUSE*
MODEL NAME AND NUMBER	Templifier Model TPE 050; 063; 079; 100
TYPE	Water-to-water
APPLICATIONS	Industrial process heat recovery
COMPRESSOR	
Type	Centrifugal (Westinghouse CE type)
Drive	Hermetic
Drive power source (open only)	
Power input (cooling)	N/a
Power input (heating)	65.5 kW (TPE 050); 142 kW (TPE 063); 216 kW (TPE 079); 338 kW (TPE 100)
REFRIGERANTS USED	R12, R114 (for higher temperatures)
HEAT PUMP CAPACITY	
Cooling	N/a
Heating	284 kW (TPE 050); 625 kW (TPE 063); (1) 997 kW (TPE 079); 1536 kW (TPE 100)
TEMPERATURE (°C)	
Operating range (min. evaporating to max. condensing temperature)	15 (evaporating); 110 (condensing)
TYPICAL PAYBACK PERIODS (years)	
Airconditioning	N/a
Process applications	2-4
FEATURES	
Available custom-built	No, but can be engineered into systems; production components may be arranged to suit
Available in package	Yes, package sizes of up to 3000 kW output available
Performance guarantee	Yes
Installed by?	Manufacturer
Performance control and turndown range	Thermostat for cycling compressor and capacity control
Defrosting	N/a
Reversability	N/a (normally not required for process waste heat recovery applications)
EXTRAS AVAILABLE	
FURTHER COMMENTS	Protective controls standard.
NOTE	(1) Based on water source at 35°C; delivery temperature 65.5°C.

MANUFACTURER	*ZOHAR*
MODEL NAME AND NUMBER	Zohar Airconditioners, Models 250-B, 300-B, 370-B, 450-B
TYPE	Air-to-air
APPLICATIONS	Space heating and cooling, plus dehumidification
COMPRESSOR	
Type	
Drive	Hermetic
Drive power source (open only)	
Power input (cooling)	1.22 kW-2.95 kW
Power input (heating)	0.98 kW-2.6 kW
REFRIGERANTS USED	R22
HEAT PUMP CAPACITY	
Cooling	2.6 kW-5.9 kW
Heating	2.84 kW-5.12 kW
TEMPERATURE (°C)	
Operating range (min. evaporating to max. condensing temperature)	Appropriate to airconditioning applications only
TYPICAL PAYBACK PERIODS (years)	
Airconditioning	Unknown
Process applications	N/a
FEATURES	
Available custom-built	Yes
Available in package	Yes
Performance guarantee	Yes
Installed by?	As required by client
Performance control and turndown range	Air dampers and continuously variable thermostat
Defrosting	Yes, supplementary heaters incorporated when heat pump reversed
Reversability	Yes
EXTRAS AVAILABLE	
FURTHER COMMENTS	Cooling only (non-reversible) system also available. All machines function as dehumidifiers.

14 Heat recovery from light fittings

The use of air and water-cooled luminaires is described in Chapter 7. This type of system is impossible to categorize in a similar manner to that adopted for other heat recovery systems, but for completeness manufacturers' data is given in this chapter.

14.1 Data presentation

The data presented is generally in not so precise a form as that for the other heat recovery equipment. Most systems are custom-built, so potential users will liase closely with manufacturers when drawing up equipment specifications. However, it is hoped that the brief notes given on manufacturers will be of value.

There is one system included in the replies to the survey which is not associated directly with light fittings, but could use heat from these fittings. This is the 'Protectasol' air conditioning window.

14.1.1 *Format of data*

The data on luminaires is arranged in the following manner:

(1) *Equipment manufacturer*: The prime manufacturer or subsidiary company is given by name. Addresses, country of origin, telephone number and product categories of each manufacturer are listed in Appendix 1.
(2) *Model name and number*: The name of the equipment and, where known, the model number are given. Most models in this category are custom-built.
(3) *Custom-built units*: Manufacturers were asked to provide data, where possible, on the size range and performance of typical custom-built units.
(4) *Standard units*: For manufacturers offering a standard range of heat recovery luminaires, data in sizes is given.
(5) *Further comments*: Any additional comments are given, which the manufacturer

feels may assist potential users.

Manufacturers' data sheets relating to equipment for heat recovery from light fittings follow on the next page

MANUFACTURER	*BBI LIGHTING*
MODEL NAME AND NUMBER	Visolite Therm Control
CUSTOM-BUILT UNITS	
Size and performance details on typical examples	Performance will be to client's specification
STANDARD UNITS	
Size and performance details	
FURTHER COMMENTS	BBI Visolite air-handling luminaires can be supplied with 100-mm diam. spigots for fully ducted air-extract and/or air-input systems, or with slots for exhaust air only into negative plenum; the patented 'Aerostat' acts as a stabilizer and ensures a uniform air-volume intake.

MANUFACTURER	*COURTNEY, POPE LIGHTING LTD.*
MODEL NAME AND NUMBER	F1477; F1016
CUSTOM-BUILT UNITS	
Size and performance details on typical examples	
STANDARD UNITS	
Size and performance details	F1477: 1200, 1500 and 1800 mm uses conventional fake ceilings; lamp wattages, 40, 65 or 85; air flow, 45 l/s; suitable for use with air-supply diffusers F1016: air flow, 50 l/s (not suitable for use with air-supply diffusers)
FURTHER COMMENTS	F2573; 2016; 2060 also available: air flows, 40-55 l/s (none suitable for use with air-supply diffusers).

MANUFACTURER	*AB SVENSKA FLAKTFABRIKEN*
MODEL NAME AND NUMBER	Combivent
CUSTOM-BUILT UNITS	
Size and performance details on typical examples	
STANDARD UNITS	
Size and performance details	600 mm long air supply and exhaust section; with light-fitting section 40 or 65 W with 2,3 or 4 fluorescent tubes: 1310 mm long for 40 W unit, 1610 mm long for 65 W unit; air supply and exhaust sections fitted with air distributors, control units and optional reheaters
FURTHER COMMENTS	Modular system; computer-aided design service available.

MANUFACTURER	*MOORLITE ELECTRICAL LTD.*
MODEL NAME AND NUMBER	Air extract: LBR REG 4000 range MFA BAT; Air supply MFS
CUSTOM-BUILT UNITS	
Size and performance details on typical examples	Sized to suit luminaire and ceiling system; typical data for special 1 x 6 85 lamp 300 x 1800; luminaire shown on LAB/AH 009/2
STANDARD UNITS	
Size and performance details	1200 x 300; 1800 x 300; 2400 x 300; 1200 x 600; 1800 x 600; 2400 x 600; 600 x 600; Data for 1200 x 600 K4000 type shown on LAB/AH/010
FURTHER COMMENTS	Supply-air facilities both for flange-mounted header boot with capacitory of 8 l/s per 300 mm length, also integrated curved-blade diffuser from Barber Colman/Roy Air, are available.

MANUFACTURER	*OSRAM-GEC*
MODEL NAME AND NUMBER	Recessed Comfort and Module Luminaires
CUSTOM-BUILT UNITS	
Size and performance details on typical examples	Custom-built service available
STANDARD UNITS	
Size and performance details	Lighting levels up to 40 000 lumens; total input energy range: 109-445 W; air flow range: 18-170 m^3/h; heat transfer capability: 69-75%
FURTHER COMMENTS	Optional extras include 'lay-in' type for suspended ceilings with no side trim; damper control; spigot for ducted airconditioning systems; slotted flange for supply-air plenum boxes; and linking pieces for continuous mounting.

MANUFACTURER	*PROTECTA-SOL AG*
MODEL NAME AND NUMBER	Airconditioning Window/Facade (1)
CUSTOM-BUILT UNITS	
Size and Performance details	Designed to fit any windows in an airconditioned building, by passing room air at 20°C (typically) between a facade (interior) and the window; the internal surface of the glass is maintained at 18-20°C, even in external ambients of -10°C. With a K value of 0.3-0.6 kcal/m^2h°C, considerably lower than that of double-glazing; effective use of internal heat gains can result in fuel/energy savings.
STANDARD UNITS	
Size and performance details	
FURTHER COMMENTS	
NOTE	(1) Not an air-handling luminaire, but system has potential as a use for heat from luminaires, as well as other heat sources.

MANUFACTURER	*THORN LIGHTING LTD.*
MODEL NAME AND NUMBER	Arena (Programme 1)
CUSTOM-BUILT UNITS	
Size and performance details	
STANDARD UNITS	
Size and performance details	300 mm wide x 1200 mm, 1500mm, 1800 mm or 2400 mm long; ducted or plenum exhaust air flow rates of 5-200 l/s with heat extraction 65-80%
FURTHER COMMENTS	It is a linear integrated ceiling system using fluorescent tubes of 40W, 65W, 75W and 85W ratings twin- or triple-tubed; uses a very wide range of light controllers.

MANUFACTURER	*THORN LIGHTING LTD.*
MODEL NAME AND NUMBER	Programme 2
CUSTOM-BUILT UNITS	
Size and performance details on typical examples	
STANDARD UNITS	
Size and performance details	1200 mm and 1500 mm square ducted or plenum exhaust air flow rates of 7-40 l/s with heat extraction of 58-80%
FURTHER COMMENTS	It is modular integrated ceiling system using new Format or Kolorformat or purpose-made luminaires.

MANUFACTURER	*THORN LIGHTING LTD.*
MODEL NAME AND NUMBER	New format FTRB; FTRF; FTRX
CUSTOM-BUILT UNITS	
Size and performance details on typical examples	
STANDARD UNITS	
Size and performance details	600 mm wide x 600 mm, 1200 mm or 1800 mm long ducted or plenum exhaust at flow rates 7-40 l/s and heat extraction of 58%-80%
FURTHER COMMENTS	Air-handling fluorescent-tube luminaire taking two, three or four 40 W or 75 W tubes or two or three 40 W U-tubes; the FTRF controller is prismatic; FTRX controller is specular reflector.

MANUFACTURER	*THORN LIGHTING LTD.*
MODEL NAME AND NUMBER	Kolorformat DKC 250/400; FTRF or DKK 125/250/400; FTRF
CUSTOM-BUILT UNITS	
Size and performance details on typical examples	
STANDARD UNITS	
Size and performance details	600 mm x 600 mm ducted or plenum exhaust of air flow rates of 7-40 l/s with heat extraction of 60-75%
FURTHER COMMENTS	Air-handling, high-pressure mercury halide lamp luminaire taking 125 W, 250 W or 400 W lamps.

Appendix 1
Heat recovery equipment manufacturers

In Chapters 8–14, comprehensive data is given on a wide range of heat recovery equipment, listed alphabetically under the name of the manufacturer or licensee.

The alphabetical list here details the addresses, country of origin and telephone numbers of these manufacturers, and also lists some agents and licensees outside the manufacturers' countries of origin.

The data is presented in the following format, the manufacturers being listed alphabetically:

Manufacturer
Address		Product category
Country
Telephone number (including area code where known)

The product category (e.g. gas–gas, heat pumps) describes the type of equipment manufactured, in terms of the Chapter headings 8–14.

In cases where a manufacturer who has completed, for example, a data sheet on gas–liquid heat recovery equipment, but has indicated incinerators as an application area, a reference is made to this, suggesting that the reader study the appropriate data sheet (in this case, in Chapter 9).

Where an agent, subsidiary, or licencee is given, the name and country of origin of the primary supplier/manufacturer or parent company is given.

Manufacturers

AAF Ltd. Bassington Industrial Estate Cramlington Northumberland NE23 8AF UK 067 071 3477	Gas–Gas Heat Pumps UK Subsidiary of American Air Filter (USA)
ABCO Industries Inc. PO Box 268 Abilene Texas 79604 USA 915/677-2011	Gas–Liquid Prime Mover heat recovery Incineration plant (see Chapter 9)
Acoustics and Envirometrics Ltd. Ruxley Towers Claygate Surrey KT10 OUF UK 0372 67281	Gas–Gas Agent for Munters Econovent (Sweden)
Aero-Plast Azay-le-Brule 79400 Saint-Maixent-L'École France (49) 26-15-37	Gas–Gas Heat Pumps
Aerotherm Wärmetechnik GmbH 6501 Stadecken-Elsheim Mainz West Germany (061 36) 5081	Gas–Gas Licensee of Wing Co. (USA)
AGA-CTC Vavmevaxlave AB Box 60 37201 Ronneby Sweden 0457 12890	Liquid–Liquid Gas–Liquid Prime Mover heat recovery (see Chapter 9)
Eduard Ahlborn 32 Hildesheim Luntzelstrasse 22 Postfach 530 West Germany (0 51 21) 5-32-71	Liquid–Liquid

Air Energy Recovery Company Inc.
3900 Lemmon Avenue — Gas–Gas
Dallas
Texas 75219
USA
(214) 522-4090

Air Froehlich AG für Energierückgewinnung
Berglistrasse 2 — Gas–Gas
CH-9320 Arbon — Gas–Liquid
Switzerland
071/46-55-25

Airscrew Fans Ltd.
Special Applications Division — Gas–Gas
Weybridge — UK Distributor for Isothermics (USA)
Surrey
UK
0932 45599

Airwell
78400 Chatou — Heat Pumps
France
976 30 30

Alfa-Laval Co. Ltd.
Great West Road — Liquid–Liquid
Brentford — Gas–Liquid
Middlesex TW8 9BT
UK
01-560 1221

Allied Air Products Co.
315 East Franklin Street — Gas–Gas
Newberg
Oregon 97132
USA

American Air Filter Co. Inc. (AAF)
215 Central Avenue — Gas–Gas
Louisville — Heat Pumps
Kentucky 40201
USA

APV Co. Ltd.
PO Box No. 4
Manor Royal
Crawley
West Sussex RH10 2QB
UK
(0293) 27777

Liquid–Liquid

AREX
Box 119
S-61500 Valdemarsvik
Sweden
0123/120-50

Gas–Liquid
Gas–Gas (see Chapter 9)

Armca Specialities Ltd.
Armca House
102 Beehive Lane
Ilford
Essex IG4 5EQ
UK
01-551-0037

Gas–Gas
UK Agent for Wing (USA) and Aerotherm (Germany)

Armstrong Engineering Associates Inc.
PO Box 566
West Chester
Pennsylvania 19380
USA
(215) 436-6080

Gas–Liquid

ASET
2 Rue de Bourgogne
69800 Saint-Priest
France
(78) 20 16 16

Gas–Gas
Liquid–Liquid
Gas–Liquid

Atlas Industrial MFG Co.
81 Somerset Place
Clifton
New Jersey 07012
USA

Liquid–Liquid

Babcock Product Engineering Ltd.
11 The Boulevard
Crawley
Sussex RH10 1UK
UK
0293 28755

Gas–Liquid

Fives-Cail Babcock 7 Rue Montalivet 75383 Paris France (1) 266-35-25	Gas–Gas Gas–Liquid Incinerator heat recovery (see Chapter 9)
SA Babcock Belgium NV B4330 Grace-Hollogne Belgium 041/6388-04	Gas–Gas Gas–Liquid Association with Prime Movers (see Chapter 9) Incineration and heat recovery Prime Mover heat recovery
Babcock-Krauss-Maffei Industriean Lagen GmbH D-8000 Munchen 50 Tannenberg 4 West Germany 089/8-11-10-51	Incineration and heat recovery
Bahco Ventilation S-199 01 Enkoping Sweden	Gas–Gas
Henry Balfour and Co. Ltd. Leven Fife KY8 4RW Scotland UK 0333 23020	Incineration and heat recovery
Barriquand 9 to 13 Rue Saint-Claude 42300 Roanne France (77) 71-68-13	Gas–Gas Liquid–Liquid
BBI Lighting Morison House Rankine Road Daneshill West Basingstoke Hampshire RG24 OPH UK 0256 61181	Air handling luminaires

Beltran and Cooper Ltd.
20 Altrincham Road
Wilmslow
Cheshire SK9 5ND
UK
(0625) 532783

Gas–Gas
Gas–Liquid
Incinerator heat recovery (see Chapter 9)

Berner International Co. Ltd.
Nihon Jitensha Kaikan
9-15, 1-Chome Akasaka
Minato-Ku
Tokyo 107
Japan
585 6421

Gas–Gas
Agent for Munters Econovent (Sweden)

Beverley Chemical Engineering Ltd.
Billingshurst
Sussex RH14 9SA
UK
040 381 2091

Gas–Gas
Gas–Liquid
Incinerators with heat recovery

Bono GmbH
Rudolf-Winkel-Strasse 20
Göttingen 3400
West Germany

Prime Mover heat recovery

Robert Bosch GmbH
Geschäftsbereich Junkers
D7314 Wernau
West Germany

Heat Pumps

E. J. Bowman (Birmingham) Ltd.
Aston Brook Street East
Birmingham B6 4AP
UK
021-359 5401

Liquid–Liquid

Bronswerk Heat Transfer BV
Stationsweg 22
Postbus 92
3860 V.B.-Nykevk
The Netherlands
(03494) 2444

Gas–Gas
Liquid–Liquid
Gas–Liquid
Prime Mover heat recovery
Incinerators with heat recovery

Peter Brotherhood Ltd. Peterborough PE4 6AB UK 0733 71321	Gas–Liquid
Brunnschweiler SPA Castella Postale 129 1-34170 Gorizia Italy (0481) 81981/2	Gas–Gas
Bry-Air Inc. PO Box 795 Route 37 Sunbury Ohio 43074 USA 614 965 2974	Gas–Gas
Burke Thermal Engineering Ltd. Nucleus Works Mill Lane Alton Hampshire UK	Gas–Gas Incinerator heat recovery UK Agent for Q-Dot Corp. (USA)
Cargocaire Engineering Corp. 6 Chestnut St Amesbury Massachusetts 01913 USA 617 388 0600	Gas–Gas Owned by Munters Econovent (Sweden)
Carlyle Air Conditioning Co. Ltd. Clifton House Uxbridge Road London W5 5SX UK 01-579 7101	Heat Pumps
Carrier International Corporation Carrier Parkway Syracuse NY 13201 USA (315) 432 6000	Heat Pumps

Casinghini and Figlio
Via Larga, 16
Milano
Codice Postale 20122
Italy

Gas–Liquid
Associated with Prime Movers

CEA Combustion Inc.
61 Taylor Reed Place
Stamford
Connecticut 06906
USA
(203) 359-1320

Incineration and heat recovery

C–E Air Preheater
Combustion Engineering Inc.
Andover Road
Wellsville
New York 14895
USA
716/593-2700

Gas–Gas
Incineration and heat recovery

Chemokomplex
Budapest 62
Nepkoztarsasag Utja 60
PO Box 141
Hungary
128 430

Liquid–Liquid

Cleerburn Ltd.
1 High Street
Slough
SL1 1DY
UK
0753 26764

Incineration and heat recovery

Climate Equipment Ltd.
18 Central Chambers
The Broadway
Ealing
London W5 2NR
UK
01-567 7258

Heat Pumps
UK Agents for Hitachi (Japan)

Coaltech
See Farrier Products

CLIREF Z. L. 'Les Meurières' 69780 Mions France (78) 20-95-48	Heat Pumps
Combustion Power Co. Inc. 1346 Willow Road Menlo Park California 94025 USA (415) 324-4744	Incineration and heat recovery
Command-Aire Corp. 3221 Speight Avenue Box 7916 Waco Texas USA (817) 753-3601	Heat Pumps
Conseco 611 North Road Medford Wisconsin 54451 USA (715) 748-5888	Gas–Liquid Prime Mover heat recovery
Contapol Ltd. 36 Crescent Road Worthing West Sussex UK 0903 208211	Gas–Liquid Prime Mover heat recovery Incineration and heat recovery UK Distributor for Conseco (USA), Eclipse Lookout (USA), Kraftenlagen (Germany)
Cool Heat Environment Control 176 High Road London N2 9AS UK 01-883 0007	Heat Pumps UK Distributor for Zohar (Israel)
Corbridge Services Post Office Buildings Hill Street Corbridge-on-Tyne Northumberland NE45 5AA UK 043 471 3000	Heat Pumps UK Agent for Metro

Corning Glass International
1a Cumberland House
Kensington Court
London W8 5NP
UK
01-937 1795

Gas–Gas
UK Office of Corning (USA)

Corning Glass Works
Corning
New York 14830
USA
607 974 9000

Gas–Gas

Cornwell Products Ltd.
57–61 Mortimer Street
London W1N 8QN
UK
01-580 5025

Incineration and heat recovery
UK Agent for Moller and Jochumsen (Denmark)

Courtney, Pope Lighting Ltd.
Amhurst Park Works
341 Seven Sisters Road
Tottenham
London N15 6RB
UK
01-800 1270

Air handling luminaires

Crane Ltd.
Pumps Division
Furnival Street
Reddish
Stockport
Greater Manchester
UK

Liquid–Liquid

CTC Heat (London) Ltd.
286b Station Road
Harrow
Middlesex HA1 2DX
UK
01-427 8454

Liquid–Liquid

Curwen and Newbery Ltd.
Westcroft House
Alfred Street
Westbury
Wiltshire
UK
(0373) 823646

Gas–Gas
Gas–Liquid
Liquid–Liquid
UK Licensee, Sun-Econ Inc. (USA)

Danks of Netherton Ltd.
257 Halesowen Road
Dudley
West Midlands DY2 9PG
UK
0384 66417

Gas–Liquid
Prime Mover heat recovery
Incinerator heat recovery

Deltak Corp.
13330 12th Avenue North
(Xenium at 12th Avenue)
Minneapolis
MN55441
USA
(612) 544-3371

Gas–Liquid

DES Ltd.
Commerce Way
Croydon
Surrey
UK
01-681 0771

Heat Pumps
UK Distributor for Carlyle

DesChamps Laboratories Inc.
PO Box 348
5–B Merry Lane
East Hanover
New Jersey 07936
USA
(201) 884-1460

Gas–Gas

Dravo Corporation
Fabricated Products Division
PO Box 9305
Pittsburgh
Pennsylvania 15225
USA
(412) 771-1200

Gas–Gas

Dunham-Bush International
175 South Street
West Hartford
Connecticut 06110
USA
203 249 8671

Heat Pumps

Dunham-Bush Ltd.
Fitzherbert Road
Portsmouth PO6 1RR
UK
070 18 70961

Heat Pumps
UK Subsidiary of Dunhan-Bush (USA)

Du Pont de Nemours International SA
82 Route des Acacias
CH-1211
Geneva 24
Switzerland
(022) 27-87-07

Liquid–Liquid

Eclipse Lookout Co.
PO Box 4756
Chattanooga
Tennessee 37405
USA
615/265-3441

Gas–Gas
Prime Mover heat recovery
Incinerator heat recovery (see also Chapter 9)
Gas–Liquid

Elkström and Son
PO Box 3007
S-291 03 Kristianstad
Sweden
044/12-32-20

Liquid–Liquid
Prime Mover heat recovery

Elboma
Fratersplein 10-11
B-9000 Gent
Belgium
091/25-77-45

Incineration and heat recovery

Elliott Turbomachinery Ltd.
40 Medina Road
Cowes
Isle of Wight
PO31 7DA
UK
098 382 4111

Liquid–Liquid

Energy Recovery Co.
PO Box 158
N112 W18700 Mequon Road
Germantown
Wisconsin 53022
USA
(414) 255-5210

Gas–Gas

Energy Systems Inc. 5 Lawrence Avenue Bloomfield New Jersey 07003 USA	Gas–Gas
Farrier Products Church Road and Derry Court PO Box 1602 York Pennsylvania 17405 USA (717) 767-6717	Prime Mover heat recovery
FGF Equipment 62 Rue du Faubourg Poissonnière 75010 Paris France 246 6012	Gas–Gas Heat Pumps
Fläkt (UK) Ltd. Staines House 158 High Street Staines Middlesex TW18 4AR UK 0784 57221	Heat Pumps Gas–Gas UK Head Office of Fläkt Group (Sweden)
AB Svenska Fläktfabriken S-551 84 Jonkoping Sweden 036 118500	Gas–Gas
AB Svenska Fläktfabriken Fack 104 60 Stockholm Sweden 23 83 00	Heat Pumps Gas–Gas Air handling luminaires
Friedrich Air Conditioning and Refrigeration Co. 4200 North PanAm Expressway PO Box 1540 San Antonio Texas 78295 USA 512 225 2000	Heat Pumps

Fuel Furnaces Ltd.
Shady Lane
Birmingham
B44 9EX
UK
021-360 7000

Gas–Liquid
Prime Mover heat recovery (see Chapter 9)

Gaylord Industries Inc.
9600 South West Seely Avenue
PO Box 558
Wilsonville
Oregon 97070
USA
503/682-3801

Gas–Gas

GEA Airexchangers Inc.
411 Gordon Avenue
Thomasville
Georgia
USA

Gas–Gas
Heat Recovery from Incineration
US Agents for GEA and Happel

GEA Airexchangers Ltd.
Leicester Square Chambers
42 Leicester Square
London WC2H 7LL
UK

Gas–Gas
Heat Recovery from Incineration
UK Agents for GEA and Happel

GEA Kuhlturmbau und Luftkondensation GmbH
Konigsallee 43–47
4630 Bochum 1
West Germany
(0234) 338-1

Prime Mover heat recovery

General Electic Co.
Central Air Conditioning Dept.
Appliance Park, Building 6
Louisville
Kentucky 40225
USA

Heat Pumps

Generator AB
PO Box 45
Partville
Sweden
(31) 44-09-20

Gas–Liquid

Gestra (UK) Ltd.
9–11 Bancroft Court
Hitchin
Herts SG5 1PH
UK
0462 50026

Liquid–Liquid

Gibbons Bros Ltd.
PO Box 20
Lenches Bridge
Brierley Hill
West Midlands DY 6 8XA
UK
0384 55491

Gas–Gas
UK Representatives of Holcroft (USA)

Girdwood Halton (Air Conditioning) Ltd.
18–20 Thorpe Road
Norwich
Norfolk NR1 1RY
UK
0603 611314

Heat Pumps
UK Distributors for Singer (USA)

AB Gotaverken
S-402 70 Goteborg
Sweden
031/22-83-00

Incinerator and heat recovery

Graham Manufacturing Ltd.
Gloucester Road
Cheltenham GL51 8NF
UK
0242 32361

Granco Equipment Inc.
1425 Burlingame, South West
Grand Rapids
Michigan 49509
USA
(616) 241-5603

Gas–Gas
Incineration and heat recovery

E. Green and Son Ltd.
PO Box
Calder Vale Road
Wakefield
West Yorkshire WF1 5PF
UK
0924 71171

Gas–Gas
Gas–Liquid
Incineration and heat recovery (see Chapter 9)
Heat recovery from Prime Movers (see Chapter 9)

Grenco Bedrijfskoeling BV Postbus 205, S'Hertogenbosch Holland	Heat Pumps
Hamworthy Engineering Co. Ltd. Fleets Corner Poole Dorset BH17 7LA UK 020 13 5123	Prime Mover with heat recovery
Happel Kg PO Box 2880 4690 Herne 2 West Germany	Prime Mover heat recovery
Harris Thermal Transfer Products Inc. 19830 South West 102nd Street Tualatin Oregon 97062 USA 503 638 7626	Gas–Gas Gas–Liquid Liquid–Liquid Incinerators and heat recovery
Hawthorn Leslie (Engineers) Ltd. St Peters Works Newcastle upon Tyne NE99 1PD UK 0632 656043	Gas–Liquid Associated with Prime Movers (see Chapter 9)
Heat Energy Recovery Services Lodge Hill St John's Road Mortimer Berkshire RG7 1TR UK 0732 332151	Liquid–Liquid
Heat-Frig Ltd. Torrie Lodge Portsmouth Road Esher Surrey UK 0372 62131	Heat Pumps UK Distributor for Westinghouse (USA)

Heenan Environmental Systems Ltd. PO Box 14 Shrub Hill Road Worcester WR4 9HA UK 0905 23461	Incineration and heat recovery
Hirakawa Iron Works Ltd. 11, Kita 1-chome Oyodo-cho, Oyodo-ku, Osaka-shi 531 Japan	Gas–Liquid Liquid–Liquid
Hirt Combustion Engineers 931 South Maple Avenue Montebello California 90640 USA (213) 723-8294	Gas–Gas Gas–Liquid Incinerator heat recovery (see Chapters 8,9)
Hitachi Ltd. 6–2, 2-Chome Otemachi Chiyoda-ku Tokyo Japan	Heat Pumps
Hitachi Zosen **Hitachai Shipbuilding and Engineering Co. Ltd.** 6–14 Edobori 1-Chome, Nishi-Ku Osaka 550 Japan Osaka (443) 8051	Gas–Gas Gas–Liquid Incinerator heat recovery (see Chapter 9) Heat Pumps Prime Mover heat recovery
Hitachi Zosen International SA Winchester House 77 London Wall London EC2N 1BQ UK 01-588 3531	Gas–Gas Gas–Liquid Incinerator heat recovery (see Chapter 9) Prime Mover heat recovery UK Agent for Hitachi Zosen (Japan)

Hitachi Zosen International SA
345 Park Avenue
New York
NY 10022
USA
212 355 5650

Gas–Gas
Gas–Liquid
Incineration heat recovery (see Chapter 9)
Prime Mover heat recovery
UK Agents for Hitachi Zosen (Japan)

Holcroft
12068 Market Street
Livonia
Michigan 48150
USA
(313) 591-1000

Gas–Gas

Holden and Brooke Ltd.
Sirius Works
Haverford Street
Manchester M12 5JL
UK
061-273 8262

Liquid–Liquid

Hotwork Development Ltd.
Little Royd Mills
Low Road, Earlsheaton
Dewsbury
West Yorkshire WF12 8BY
UK
0924 463325

Gas–Gas

James Howden and Co. Ltd.
195 Scotland Street
Glasgow G5 8PJ
UK
041-429 2131

Gas–Gas

Hygrotherm Engineering Ltd.
Botanical House
1 Botanical Gardens
Talbot Road
Manchester M16 0HL
UK
061-872 6861

Gas–Gas
Gas–Liquid
Incinerator heat recovery (see Chapters 8, 9)
UK Licensee of Hirt Combustion
Engineers (USA)

IMI Range Ltd.
PO Box 1
Bridge Street, Stalybridge
Cheshire SK15 1PQ
UK
061-338 3353

Liquid–Liquid

International Research and Development Co. Ltd.
Fossway
Newcastle upon Tyne
NE6 2YD
UK
0632 650451

Gas–Gas
Heat Pumps
Gas–Liquid

Ionics Inc.
PO Box 99
Bridgeville
Pennsylvania 15017
USA
412 343 1040

Gas–Gas

Isoterix Ltd.
1 Bank House
Balcombe Road
Horley
Surrey
UK
029 34 2099

Gas–Gas

Isothermics Inc.
PO Box 86
Augusta
New Jersey 07822
USA
(201) 383-3500

Gas–Gas

ITT Reznor
Barnfield Road
Park Farm Industrial Estate
Folkstone
Kent CT19 5DR
UK
0303 56581

Gas–Gas
UK Subsidiary of ITT Reznor (USA)

ITT Reznor
Mercer
Pennsylvania 16137
USA

Gas–Gas

Jaeggi Ltd.
Morgen Strasse 89
CH-3018
Berne
Switzerland
031 55 13 55

Gas–Gas
Gas–Liquid
Liquid–Liquid

Robert Jenkins Systems Ltd.
Wortley Road
Rotherham
Yorkshire S61 1LT
UK
Rotherham 79140

Incineration and heat recovery

Johnson Construction Company AB
PO Box 1259
Uretenvagen 13
171 24 Solna
Sweden
08 289400

Gas–Gas

Junkers GmbH
Heat Pumps (see Robert Bosh GmbH)

Kleinewefers GmbH
Kleinewefers-Kaland erstrasse
Postfach 1560
4150 Krefeld 1
West Germany
(02151) 89-61

Gas–Gas
Gas–Liquid
Incinerators and heat recovery

Kraftanlagen AG
6900 Heidelberg 1
Postfach 10 34 20
Im Breitspiel 7
West Germany
06221/39-41

Gas–Gas
Agent for Munters Econovent (Sweden)

Laminaire Products
272–274 Soho Road
Birmingham B21 9LX
UK
021-551 1851

Gas–Gas

Lumitron Ltd. Chandos Road London NW10 6PA UK 01-965 0211	Air-handling luminaires
Lyon and Pye Ltd. 543 Prescot Road St Helens Lancashire WA10 3BZ UK St. Helens 20925	Incineration and heat recovery
Metallurgical Engineers Ltd. Boundary House Boston Road London W7 2QQ UK 01-567 9612-3-4	Gas–Gas Gas–Liquid Incinerators and heat recovery
Metro Bymosevej 1-3 DK-3200 Helsinge Denmark 03 29 62 11	Heat Pumps
Møller and Jochumsen AS DK 8700 Horsens Denmark (05) 62-23-00	Incineration and heat recovery
Moorlite Electrical Ltd. Burlington Street Ashton-under-Lyne Lancashire OL7 OAX UK 061-330 6811	Air-handling luminaires
Motherwell Bridge Tacol Ltd. Green Dragon House 64–70 High Street Croydon CRO 1NA UK 01-686 6917	Incineration and heat recovery

Paul Mueller Co.
PO Box 828
Springfield
Missouri 65801
USA
(417) 865-2831

Liquid–Liquid

Munters Econovent AB
Industrivagan 1
Fack
S-19120 Sollentuna
Sweden
08 960140

Gas–Gas

NEI (Overseas) Ltd.
PO Box 1NT
Cuthbert House
All Saints
Newcastle upon Tyne NE99 1NT
UK
0632 24013

Gas–Liquid
Prime Mover heat recovery
Heat Pumps
(Address for overseas inquiries)

NEI Projects (Process Engineering) Ltd.
Victoria Works
Gateshead NE8 3HS
UK
0632 772271

Prime Mover heat recovery
Heat Pumps
UK Agent for Westinghouse (USA)

NEI Thompson Cochrane Ltd.
Carntyre Works
300 Myreside Street
Glasgow G32 6BS
UK
041-556 5252

Gas–Liquid

D. J. Neil Ltd.
PO Box 31
Macclesfield
Cheshire SK10 2EX
UK
0625 613666

Liquid–Liquid
Gas–Liquid
UK Distributor for Schiff and Stern KG (Austria); and Generator AB (Sweden)

Noren Products
3511 Haven Avenue
Menlo Park
California 94025
USA

Gas–Gas

Northern Engineering Industries Ltd.
PO Box 1NT
Cuthbert House
All Saints
Newcastle upon Tyne NE99 1NT
UK
0632 24013

Gas–Liquid
Prime Mover heat recovery
Heat Pumps
(See Operating Division Addresses)
Heat Pumps
UK Agent for Westinghouse (USA)

Northvale (Division of BSS) Ltd.
Uxbridge Road
Leicester
UK
0533 659 12

Liquid–Liquid

Novenco Ltd.
Tundry Way
Chainbridge Road
Blaydon-on-Tyne
Tyne and Wear NE21 5SN
UK

Gas–Gas
UK Subsidiary of Nordisk Ventilator AS (Norway)

Oslo Sveisebedrift
PO Box 15
3301 Hokksund
Norway

Liquid–Liquid

OSO-Hokksund
3300 Hokksund
Norway
(02) 85-17-80

Liquid–Liquid

Osram
PO Box 17
East Lane
Wembley HA9 7PG
UK
01-904 4321

Air-handling luminaires

Pott Industries Inc.
Engineering Controls Division
611 East Marceau Street
St Louis
Missouri 63111
USA
(314) 638-4000

Prime Mover heat recovery
Incinerators and heat recovery

Protecta-Sol AG Postfach 55 CH-8712 Stafa Switzerland 01 928 61 11	Air-conditioned window
Q-Dot International Corp. 151 Regal Row Dallas Texas 75247 USA (214) 630-1224	Gas–Gas Gas–Liquid
Recuperator Spa 20020 Lainate (Mi) Via Mantova 4 Italy (02) 9371686/7	Gas–Gas
Riley-Beaird Inc. PO Box 31115 Shreveport Louisiana 71130 USA (318) 868-4441	Prime Mover heat recovery
Ruston Gas Turbines Ltd. PO Box 1 Lincoln UK 0522 25212	Prime Mover heat recovery
Rycroft (Calorifiers) Ltd. Ryco Works 129 Sunbridge Road Bradford Yorkshire BD1 2NL UK 0274 31563	Liquid–Liquid
Schiff and Stern KG 1 Haidequerstrasse 3 1110 Vienna Austria (0222) 76-15-15	Liquid–Liquid

W. Schmidt KG Kuhlerwerke 7518 Bretten/Baden Postfach 95 West Germany (72 52) 1051	Liquid–Liquid
Senior Economizers Ltd. Otterspool Way Watford Bypass Watford Hertfordshire WD2 8HX UK 0923 26091	Gas–Liquid Prime Mover heat recovery (see also Chapter 9) Incinerators and heat recovery (see Chapter 9)
Senior Platecoil Ltd. Otterspool Way Watford Bypass Watford Hertfordshire WD2 8HX UK 0923 26091	
Serck GmbH 2 Hamburg 70 Tilsiterstrasse 90 West Germany 040/6-93-50-11	Liquid–Liquid (Manufacturing division of Serck Heat Transfer (UK)
Serck Heat Transfer PO Box 598B Warwick Road Birmingham B11 2QY UK 021-772 4353	Liquid–Liquid
Serck Heat Transfer Airport Office Park Cincinnati Ohio 45275 USA	(Office for Serck Heat Transfer (UK)
Singer Co. Climate Control Division 1300 Federal Boulevard Carteret New Jersey 07008 USA (201) 636-3300	Heat Pumps

Smith Engineering Co.
PO Box 3696
El Monte
California 91733
USA
(213) 443-0214

Gas–Gas
Incinerators and heat recovery

Smith Environmental Corp.
PO Box 3696
El Monte
California 91733
USA
(213) 686-2155

Incineration and heat recovery

Sorbent Drying Ltd.
23 Tolworth Park Road
Surbiton
Surrey KT6 7RN
UK
01-390 0166

Gas–Gas
UK Agents for Van Swaay Installaties
(The Netherlands)

Standard and Pochin Ltd.
Evington Valley Road
Leicester LE5 5LS
UK
(0533) 736114

Gas–Gas

Stein Atkinson Stordy Ltd.
Ounsdale Road
Wombourne
West Midlands WV5 8BY
UK
090 77 4171

Gas–Gas

Stierle Hochdruck Economiser KG
68 Manheim
Karl-Ludwig-Strasse 14
West Germany
0621/40-90-19

Gas–Liquid

Stone-Platt Crawley Ltd.
PO Box 5
Gatwick Road
Crawley
West Sussex RH10 2RN
UK
0293 27711

Gas–Liquid
Incinerator heat recovery
(see Chapter 9)
Prime Mover heat recovery (see Chapter 9)

Struthers Wells Corp.
PO Box 8
Warren
Pennsylvania 16365
USA
814/726-1000

Gas–Liquid
Prime Movers with heat recovery (see also Chapters 8, 9)
Liquid–Liquid

Studsvik Energiteknik AB
S-611 01 Nykoping 1
Sweden

Gas–Liquid
Gas–Gas (see Chapter 9)
(See also Arex)

Sulzer Bros Ltd.
CH-8401 Winterthur
Switzerland
052 81 1122

Heat Pumps

Sulzer Bros (UK) Ltd.
Farnborough
Hampshire
UK
0252 44311

Heat Pumps

Sulzer Bros (South Africa) Ltd.
Johannesburg 2000
PO Box 930
South Africa
24 89 61

Heat Pumps
Agents for Sulzer Bros. (Switzerland)

Sulzer-EscherWyss GmbH
D-8990 Lindau-Bodensee
Postfach 1380
West Germany
083 82 761

Heat Pumps

Sunbeam Equipment Corp.
Comtro Division
Suite 510
Century Plaza Building
Lansdale
Pennsylvania 19446
USA
(215) 699-4421

Incineration and heat recovery

Company	Products
AB Svenska Maskinverken Fack S-175-02 Jarfalla Sweden	Gas–Liquid Prime Mover heat recovery (see Chapter 9) Liquid–Liquid
Syrec 21 Rue Paul Doumer 781100 Le Vesinet France 906 54 58	Gas–Gas Heat Pumps (see product data for FGF and Aero-Plast (France)
Technibel SA Route Départmentale 28 Reyrieux 01600 Trevoux France	Heat Pumps
Temperature Ltd. 192-206 York Road London SW11 3SS UK 01-223 0511	Heat Pumps
Terry Corp. PO Box 1200 Hartford Connecticut 06101 USA 203 688 6211	Prime Mover heat recovery
Thermal Efficiency Ltd. Otterspool Way Watford Bypass Watford Hertfordshire WD2 8HX UK 0923 26091, and 0923 35571	Gas–Gas
Thorn Lighting Ltd. Great Cambridge Road Enfield Middlesex EN1 1UL UK 01-363 5353	Air handling luminaires

Tjernlund Customair Inc.
2140 Kasota Avenue
St Paul
Minnesota 55108
USA
(612) 645-4507

Gas–Gas
Air handling luminaires

Torin Corporation
Applied Products Division
Heatbank Dept
436 West Willard Street
Kalamazoo
Michigan 49006
USA
(616) 349-6681

Gas–Gas

Trane
Commercial Air Conditioning Division
La Crosse
WI 54601
USA
(608) 782-8000

Gas–Gas
Heat Pumps

Trane Air Conditioning
Peloquin Chambers
18 St Augustines Parade
Bristol
UK

Gas–Gas
UK Office of Trane (USA)

Trane Thermal Co.
Brook Road
Conshohocken
Pennsylvania 19428
USA
(215) 828-5400

Gas–Gas
Incineration and heat recovery
(see also Chapter 8)

Tranter Inc.
735 E. Hazel Street
Lansing
Michigan 48909
USA

Liquid–Liquid

United Air Specialists (UK) Ltd.
15 Waterloo Place
Leanington Spa
Warwickshire CV32 5LA UK
0926 311621

Gas–Gas

United Air Specialists Inc.
6665 Creek Road
Cincinnati
Ohio 45242
USA
(513) 891-0400

Gas–Gas

Uranus SA
24 Rue Saint Michel
Lyon 69007
France
(78) 58-43-20

Liquid–Liquid

Valmet Oy
Pansio Works
20240 Turku 24
Finland
921 402 300

Gas–Gas
Gas–Liquid

Van Den Bergh and Partners Ltd.
7 Fairacres
Dedworth Road
Windsor
Berkshire SL4 4LE
UK
07535 69561

Liquid–Liquid
UK Agents for W. Schmidt (West Germany)

Van Swaay Installaties BV
Bredewater 24
Zoetermeer
The Netherlands
079-219363

Gas–Gas

Ventilation Equipment and Conditioning Ltd.
Vequip Works
320–322 Latimer Road
London W10 6QR
UK
01-969 7553

Gas–Gas

Vicarb SA
24 Avenue Marcel-Cachin
38400 St Martin D'Heres
France

Liquid–Liquid

Vyncke WT Gentsteenweg 224 8730 Harelbeke Belgium	Incinerators and heat recovery
Wanson Company Ltd. 7 Elstree Way Borehamwood Hertfordshire WD6 1SA UK 01-953 7111	Gas–Liquid
Weltemp Ltd. 6 Ark Road Kingston upon Thames Surrey KT2 6EF UK 01-549 0106	Gas–Gas Heat Pumps UK Distributors for Technibel (France)
Westair Dynamics Ltd. Thames Works Central Avenue East Molesey Surrey KT8 0QZ UK 01-979 9031	Heat Pumps
Westinghouse Commercial-Industrial Templifier Department Staunton Virginia 24401 USA	Heat Pumps
Westinghouse Air Conditioning International Westinghouse Building Gateway Center Pittsburgh Pennsylvania 15222 USA	Heat Pumps (International Office)

Westinghouse Electric
1 The Curfew Yard
Thames Street
Windsor
Berkshire
UK
075 35 54711

Heat Pumps
UK Office of Westinghouse (USA)

Lee Wilson Engineering Co. Inc.
20005 Lake Road
Cleveland
Ohio 44116
USA
(216) 331-6600

Gas–Gas

Wing Co.
Division of Wing Industries Inc.
125 Moen Avenue
Cranford NJ 07016
USA
(201) 272-3600

Gas–Gas

Zenit
Association for Foreign Trade Agencies Department
PO Box 88
15000 Praha 5
Czechoslovakia
02/53-03-11

Gas–Gas
Agent for Munters Econovent (Sweden)

Zohar Engineering Co. Ltd.
13 Nakhalat-Benjamin Street
65-161 Tel Aviv
Israel
03 623334

Heat Pumps

Appendix 2
Financial analyses for assessing economics of waste heat recovery equipment

A2.1 Introduction

The phrases 'payback period' and 'rate of return on investment' have appeared regularly in the main text, with only brief descriptions of their meaning in the context of waste heat recovery equipment. One of the primary objectives of private companies is to achieve maximum profits, and investment in heat recovery equipment should be shown to contribute towards this aim. While the desire to save energy on a national, or international, scale is important, and the policy of central governments and their agencies, most organizations see the financial implications as the over-riding consideration. Thus, the expected profitability of investment in waste heat recovery systems is a critical factor in determining if companies will adopt the systems.

The benefits which may accrue to a company investing in waste heat recovery equipment could be several:

(i) fuel savings;
(ii) reduced size, hence capital cost of heating/refrigeration plant;
(iii) reduced maintenance costs for existing equipment;
(iv) reduced costs of labour;
(v) reduced pollution;
(vi) reduced costs of effluent treatment;
(vii) improvements in the product;
(viii) possible revenues from the sale of 'exported' heat or electricity.

A reduction in fuel utilization is generally the most important benefit resulting from investment in heat recovery equipment. The other benefits may also occur, but are generally restricted to particular applications, and are highly unlikely to occur all together. A reduction in the cost of heating or refrigeration plant normally only applies when heat recovery equipment may be designed into a building at an early stage in its conception. Reduced maintenance costs can result from reductions in stack gas temperatures. Labour cost reductions are normally associated with savings in time associated

Table A2.1 Potential costs due to investment in heat recovery equipment

Type of costs	Example of costs
(1) Pre-engineering and planning costs	Engineering consultant's inhouse manpower and materials to determine type, size and location of heat exchanger
(2) Acquisition costs of heat recovery equipment	Purchase and installation costs of the system
(3) Acquisition costs of necessary additions to existing equipment	Purchase and installation costs of new controls, burners, stack dampers and fans to protect the furnace and recuperator from higher temperatures entering the furnace due to preheating of combustion air
(4) Replacement costs	Cost of replacing the inner shell of the recuperator in *n* years, net of the salvage value of the existing shell
(5) Costs of modification and repair of existing equipment	Cost of repairing furnace doors to overcome greater heat loss resulting from increased pressure due to preheating of combustion air
(6) Space costs	Cost of useful floor space occupied by waste heat stream generator, cost of useful overhead space occupied by evaporator
(7) Costs of production downtime during installation	Loss of output for a week, net of the associated savings in operating costs
(8) Costs of adjustments	Lower production; labour costs of commissioning
(9) Maintenance cost of equipment	Costs of servicing the heat exchanger
(10) Property and/or equipment taxes of heat recovery equipment	Additional property tax incurred on capitalized value of recuperator (USA)
(11) Change in insurance or hazards costs	Higher insurance rates due to greater fire risks; increased cost of accidents due to more hot spots within a tighter space

Note: In addition, attention should be given to the length of intended use, expected lives of related equipment and the flexibility of alternative equipment to future modification and expansion.

with the startup of equipment. Pollution reduction is indirectly a benefit. Gases may have to be cooled before entering electrostatic precipitators. More important are the cost benefits due to heat recovery downstream of pollution control equipment (see Chapter 5). Effluent treatment costs may vary as a function of the liquid effluent discharge temperature. Some authorities may argue that higher temperature effluents are more difficult to treat because the conditions would assist growth of organisms. Heat recovery from these effluents prior to discharge could help.

The benefits from the 'export' of heat or power is most likely to result from investing in total energy plant. However, in Chapter 2 an example is given of the transport of steam between two factory sites adjacent to one another.

Against the benefits, one has to weigh the costs of the investment. These are summarized in Table A2.1. A number of examples of these costs are also given.

A2.2 Techniques of financial analysis

A2.2.1 *Payback period*

A first estimate of the effectiveness of heat recovery equipment may be obtained by calculating either the payback period, or the return on the investment. The information needed to carry out the calculations is very basic, and generally easy to obtain. The following data is required:

(a) Capital cost of energy-conservation equipment, including installation cost.
(b) Annual operating cost.
(c) Annual fuel savings.
(d) Fuel price.
(e) Equipment life.

The fuel price should ideally take into account increases which will occur over the life of the equipment. Current fuel prices may be used, but this will result in a pessimistic figure for the payback period.

Using the above data, the payback period, which is the ratio of the capital cost of the installation to the net annual savings, may be calculated:

$$\text{Payback period} = \frac{A}{(C \times D) - B}.$$

The payback period is then compared to the equipment life E. In some cases a payback period of less than ½E is considered profitable where E is less than 10 years. In the author's experience, industry is generally interested in payback periods of 2 years or less, although this may be influenced by the level of overall investment at any one time.

The disadvantages of the payback-period method, which is unlikely to be used by most companies as the final detailed analysis, are that in its simplest form it does not permit consideration of cash flows beyond the payback period, which may be only a few months. Also, the technique neglects to discount costs occurring at different times to a common base for comparison.

This could result in being unable to discriminate between two systems, having identical payback times, but having different annual returns. One may save £5000 in the first year of operation, the other £7000. A reversal of savings in the second year would make the payback equal, but the benefits in terms of a return on earnings would be greater for the unit recovering £7000 in the first year.

There are situations in which the simple calculation of payback period can be of value. For example, a rapid payback may be the prime criterion when the investor has funds available for only a short time. A speculator will also be interested in a quick return, and this technique, in spite of its shortcomings, will be valid here. Third, if the expected life of the assets is very difficult to predict, the calculation of the payback period is helpful in assessing the likelihood of achieving a successful investment.

A more accurate method for determining the payback period, overcoming some of the drawbacks of failing to discount costs, is to use the following equation, which enables one to determine the number of years R, for which the expression equates to zero:

$$A = \sum_{j=1}^{R} \frac{C_j - B_j}{(1+i)^j}$$

where A = initial investment cost, C_j = benefits in year j, B_j = costs in year j, R = break-even number of years, and i = discount rate.

Where yearly net benefits are uneven, an iterative process can be used to determine the solution. If, on the other hand, yearly net benefits are expected to be about uniform, the following formula can be used.

$$R = \frac{-\log\left(1 - i\frac{A}{C}\right)}{\log\ (1+i)}$$

where R = breakeven number years, C = yearly net benefits, A = initial investment cost, and i = discount rate.

A2.3 Return on investment

The equipment depreciation may be taken into account by calculation of the return on the investment:

$$\text{Return on investment} = \frac{[(C \times D) - B] - \text{Depreciation charge}}{A}$$

where A = total cost of equipment, B = annual operating cost, C = annual savings, and D = fuel price.

A return on the investment of 25 per cent or more is desirable. This technique also does not take into account the timing of the cash flows, and is based on the concept of 'original book value' which generally does not include all costs. It, therefore, results in only a rough approximation of an investment's value.

A2.4 Other factors

There are a number of more sophisticated accounting techniques which are routinely used by companies investing in capital plant, and it is beyond the scope of this book to deal with them. For treatments, the reader is advised to consult Harker (1978), NBS Handbook 115 (1974) or Ruegg (1976).

Other factors to be taken into account include (depending upon government policies), investment grants, loans, etc. (Some of these are specifically tailored to heat recovery equipment, or energy conservation in general.) Where pollution-control legislation is in force, or proposed, it may also be argued that companies should be compensated for the extra investment required, particularly if they benefit all by recovering the heat used for, say, fume incineration.

References

Harker, J. H. (1978), 'Economic balancing of heat exchangers', *Processing,* June, 85.

NBS (1974), *Energy Conservation Program Guide for Industry and Commerce,* NBS Handbook 115, US Bureau of Standards, Washington, DC, September.

Ruegg, R. (1977), 'Economics of waste heat recovery', chapter 3 in: NBS Report PB-264959 *Waste Heat Management Guidebook,* Federal Energy Administration, Washington, DC.

Appendix 3
Useful conversion factors

Length	1 ft	= 0.3048 m
Area	1 ft^2	= 0.0929 m^2
	1 in^2	= 6.451 cm^2
Volume	1 ft^3	= 0.0283 m^3
	1 Imperial gallon (gal)	= 4.546 l
	1 US gallon	= 0.833 Imperial gallons
Volume rate of flow	1 gal/min	= 0.0758 l/s
Mass	1 lb	= 0.4536 kg
	1 ton	= 1.016 tonnes (t)
	1 t	= 1000 kg
Mass flow rate	1 lb/h	= 1.259×10^{-4} kg/s
Density	1 lb/ft^3	= 16.019 kg/m^3
Force	1 lbf	= 4.448N
Pressure	1 lbf/in^2	= 6.894 kN/m^2 (kPa)
	1 bar	= 10^5 N/m^2 (Pa)
	1 atm	= 101.325 kN/m^2 (kPa)
Dynamic viscosity	1 poise (P)	= 0.1 N s/m^2 (Pa s)
	1 lbf s/ft^2	= 0.047 N s/m^2 (Pa s)
Energy	1 kW h	= 3.6×10^6 J
	1 hp h	= 2.684×10^6 J
	1 Btu	= 1.055 kJ
	1 Btu	= 0.251 kcal

Power	1 hp	= 0.745 kW
	1 hp	= 1.013 metric hp
Temperature	(°F − 32) x 5/9	= °C (K)
Quantity of heat	1 Btu	= 1.055 kJ
	1 kcal	= 4.186 kJ
Heat flow rate	1 Btu/h	= 0.293 W
	1 kcal/h	= 1.163 W
Density of heat flow	1 Btu/ft^2 h	= 3.154 W/m^2
Thermal conductivity	1 Btu/ft h °F	= 1.730 W/m °C (W/m K)
Coefficient of heat transfer	1 Btu/ft^2 h °F	= 5.678 W/m^2 °C (W/m^2 K)
Specific heat capacity	1 Btu/lb °F	= 4.186 x 10^3 J/kg °C (J/kg K)
Enthalpy	1 Btu/lb	= 2.326 J/g
	1 kcal/kg	= 4.186 J/g
Calorific value	1 Btu/ft^3	= 0.037 J/cm^3
(Volume basis)	1 therm/gal	= 2.32 x 10^4 J/cm^3
Light-illumination	1 lux	= 1 lumen/m^2
	1 foot candle	= 1 lumen/ft^2

Energy equivalents

Calorific value of fuel oil	177000 Btu/gal
	186 MJ /gal
	41 MJ/l
Average efficiency of steam raising	125 lb/steam/gal fuel oil
	12.47 kg/l fuel oil
Heating value of steam	1050 Btu/lb
	1108 MJ/lb
	2443 MJ/kg
One tonne coal equivalent	27.3 GJ (heat supplied)
Primary fuel equivalent conversion factors	
Fuel oil	46 MJ/l
Electricity (UK)	14.40 MJ/kW h
Gas	146 MJ/therm
Derv	43 MJ/l
Other liquid fuels	46 MJ/l
Coal	29.8 GJ/t

Bibliography

Heat exchanger types and application

Anon. (1974), Spiral heat exchangers , *Heating Air Conditioning J.* 43, 510
Anon. (1975), Waste heat recovery systems , Processing report, *Processing* (UK), May, 46.
Anon. (1976), Energy conservation in building design, PSA Library Service Bibliography, London.
Anon. (1976), Systems and systems control in buildings, PSA Library Service Bibliography, London.
Anon. (1977), Plant heating using incinerator waste heat, *Industrial Heating,* vol. 44, pt 12, 14.
Anon. (1977), Ceramic wheel heat recovery saves energy for process heating, *Prec. Metal,* vol. 35, pt 12, 33.
Anon. (1977), Waste heat recycling system in finishing of coil steel, *Ind. Heating,* vol. 44, pt 7, 10–11.
Anon. (1977), 'Big paint booth recovers exhaust heat , *Ind. Finish,* vol. 53, no. 10, 36–39.
Anon. (1977), Heat recuperator: improved heat transfer efficiency , *Ind. Gas,* vol. 57, no.3, 14–16.
Anon. (1978), Cracking power recovery from gas turbines, *Processing,* February, 34–35.
Anon. (1978), Extended surfaces take the heat out of exchanger tubing, *Processing* June, 59–61.
Ambrose, E. R. (1975), Heat reclaiming systems – a review, *Heating, Piping Air Conditioning,* 47, 55–58.
ASHRAE (1975), Energy conservation in new building design, ASHRAE Standard 90–75, New York.
Balmer, I. R. (1975), Effective use of fuels and resources in the metals industries , *Metallurgia Metal Forming,* September, 319–321.
Berg, C. A. (1973), *Energy Conservation through Effective Utilisation,* National Bureau of Standards NBSIR 73 102, Washington DC, February.
Bernstein, H. M. and McCarthy, P. M. (1975), Analysis of factors related to energy use in the commercial sector, *Proc. 57th Annual Conference, Planning, 1975: Innovation and Action, San Antonio, Texas, USA, 1975,* American Institute of Planners.
Botterill, J. S. M. (1975), *Fluid-Bed Heat Transfer,* Academic Press, London and New York.
Bowlen, K. L. (1974), Energy recovery from exhaust air for year round environmental control, *Trans. ASHRAE,* vol. 1, 314–321.
Briley, G. C. (1976), Conserve energy – refrigerate with waste heat, *Hydrocarbon Process,* vol. 355, no. 5, 173–174.
Bunt, B. P. (1975), The energy used by packaging and its minimisation, *J. Soc. Dairy Technol.,* vol. 28, no. 3.

Burke, B. (1977), Energy saving and air pollution control, *HV News,* vol. 20, pt 6, 30–31, 33, 36.

Castrell, C. J. (1976), *Waste Energy Conservation Techniques,* ASME Paper 76-Pet-21, New York.

Challand, T. B. (1976), Computer systems help boost heat recovery, *Oil Gas J.,* September 6, 105–113.

Challand, T. B. and Yang, S. F. (1977), Optimum refinery energy use via computer programs, *Proc. Computer Conf., New Orleans, USA, 1973,* Report CC-77-80, National Petroleum Refiners' Association.

Clay, P. E. (1968), Ins and outs of heat recovery equipment, *Air Conditioning, Heating Ventilating,* vol. 65, no. 1.

Cook, C. S. (1974), *Evaluation of a Fossil Fuel fired Ceramic Regenerative Heat Exchanger,* Report PB-236 346, General Electric Co., Philadelphia, USA, October.

Day, C. E. (1969), How we specify fuels for our plants, *Trans. ASME,* Paper 69-FU-7, New York.

Drake, D. E. (1977), Putting low temperature waste heat to work, *Plant Eng.,* vol. 31, no. 17, 18 August, 135–136.

Dutch Patent 7403678 (1974), *Heat pumps in bottle washing plant,* Milpro NV Brussels, 19 March.

Edwards, J. V. (1971), Computation of transient temperatures in regenerators, *Int. J. Heat Mass Transfer,* vol. 14, 1175–1202.

Electricity Council (1976), *Heat Recovery for Industry,* UK Electricity Council Publication EC 3484, Technical Information Series no. 18.

FEA (1974a), *Industrial Energy Study of Selected Food Industries,* FEA Report EI 1652, PB-237 316, July.

FEA (1974b), *Fuel and Energy Conservation in the Coal Industries,* Hittman Associates Inc., Columbia, USA, FEA Report EI 1659, PB-237 151, May.

Federal Energy Authority (1975), Industrial energy study of the petroleum refining industry, Report PB-238671/2, FEA/EI-1656 (USA).

Fleming, J. (1976), Recover energy with exchangers, *Hydrocarbon Process,* vol. 55, no. 7, 101–104.

Fuchs, W., James, G. R. and Stokes, K. J. (1977), Economics of flue gas heat recovery, *Chem. Engng. Proc.,* vol. 73, no. 11, 65–70.

Fulton, R. and Strindehag, O. (1976), *Ecoterm – A Method of Waste Heat Recovery,* Fläkt/SF Air Treatment Ltd., Staines.

Gentry, C. B. (1976), The ceramic heat wheel, *Light Metal Age,* June.

Gray, P. M. J. (1975), Conservation in primary extraction processing, *Metallurgist Materials Technol.,* vol. 7, no. 2, February.

Homfeld, E. W. (1970), Utilising gas engine waste heat in a cracking plant, *Gas Warme International,* vol. 19, no. 8 (in German).

Howard, J. R. and Sanderson, P. R. (1977), Towards more versatile fluidised-bed heat exchangers, *Applied Energy,* vol. 3, 115–125.

Huang, F. and Elshout, R. (1976), Crude preheat can save fuel, *Oil Gas J.,* vol. 76, no. 5, 65–68.

Johnson, L. W. (1977), Guidelines for incineration and fuel conservation, *Ind. Heating,* vol. 44, pt 3, 9–12.

Kern, W. I. (1975), Increasing heat exchanger efficiency through continuous mechanical tube maintenance, *Combustion* (USA), August, 18–27.

Lloyd, S. and Starling, C. (1976), *Heat Recovery from Buildings: An annoted bibliography with a survey of available products and their suppliers,* BSRIA Bibliography 104/76, April.

Lock, J. (1978), Ploughing back the energy, *Processing,* June, 67.

Marriott, J. (1971), Where and how to use plate heat exchangers, *Chemical Engineering,* 127–133, 5 April.

Miller, R. R. (1974), Trends in in-plant environment, *Trans. ASHRAE,* vol. 1, 211–215.

Mortimer, J. (1973), Exchanging heat to save fuel, *Engineer,* 2–9 August.

Pearson, S. G. (1977), Industrial plant heat recovery, 5th Annual ASME Ind. Power Conf., Paper 77-IPC-PWR-8.

Pettman, M. J. (1975), Energy in plant, *Hydrocarbon Processing* (USA), vol. 54, no. 1.
Quartulli, O. J. (1975), Stop wastes: re-use process condensate, *Hydrocarbon Processing*, vol. 54, no. 10, 94–99.
Queen, D. M. (1974), *Industrial Energy Study of the Hydraulic Cement Industry*, FEA Report EI 1665, PB-237, August, 142.
Reay, D. A. (1978), Heat recovery from exhaust gases, *Energy Management*, January, 2–3.
Ricco, L. J. (1978), CPI niche for heat pipes, *Chem. Eng.*, vol. 85, no. 2, 84–86.
Richardson, B. L. (1977), Today's dryers keep dollars out of the stack, *Textile World*, vol. 127, no. 10, 97–106.
Ricken, J. and Schwarz, P. (1977), Heat recovery in supermarkets, *Linde Rep. Sci. Technol.*, no. 25, 2–15.
Robson, B. G. (1970), Heat recovery from exhaust air with rotary heat exchangers, *Australian J. Refrig.*, vol. 24, no. 3, 16–20.
Sander, U. (1977), Waste heat recovery in sulphuric acid plants, *Chem. Engng. Prog.*, vol. 73, pt 3, 61–64.
Santoleri, J. J. (1975), Energy recovery from low heating value industrial waste, ASME Paper 75-IPWR-13, New York.
Schoenberger, P. K. (1975), Energy saving techniques for existing buildings, *Heating, Piping Air Conditioning*, January.
Schultz, G. V. (1977), Reclaiming dirty exhaust heat, *Factory Manager*, vol. 10, no. 2, 15–17.
Schwindt, H. J. (1974), Utilisation of waste heat from inductive melting installations, *Electrowarme International*, vol. 32, no. B2.
Svensson, J. (1976), Waste heat recovery from incineration, *Shipbuilding Marine Engng. Int.*, vol. 99, pt 1204, 674, 7, 80, 4.
Szabo, B. S. (1967), The economics of heat recovery systems, *Air Conditioning Heating Vent.*, June, 59–64.
Thielbahr, W. H. (1976), Heat exchanger technology needs for conservation research and technology, US Naval Weapons Center Technical Memorandum 2930.
Van den Hoogen, B. (1973), Designing gas turbine heat recovery boilers, *Gas Turbine Int.*, November–December, 32–35.
Waterland, A. F. (1973), Energy conservation in an industrial plant, *Trans. ASME Power Division*, Paper 73-IPWR-9.
White, W. C. (1974), *Energy problems and challenges in fertilizer production, Proc. Fertilizer Institute Round Table Meeting, Washington DC, 4 December 1974.*
Woodall, L.C. and Godshall, E.F. (1976), Energy economics in a dye house, *Textile Ind.*, vol. 140, no. 10, 90–107.
Woods, S. E. (1974), Heat conservation in zinc and lead extraction and refining, *Metals Materials*, vol. 8, no. 3, March.
Wright, R. (1968), Finned tubes for waste heat recovery, *Chem. Proc. Engineering*, Heat Transfer Survey.

Energy systems and general publications

Anon. (1963), *Waste heat recovery, Proc. Inst. Fuel Conf., Bournemouth, UK, 1961*, Chapman and Hall, London.
Anon. (1974), *Technology of Efficient Energy Utilisation*, Report of NATO Science Committee Conference, Les Arcs, France, October 1973, (published by Scientific Affairs Division, NATO, Brussels).

Anon. (1974), Energy conservation in commercial, residential and industrial buildings, *Proc. Conf. Ohio State University, 5–7 May 1974* (published by National Science Foundation, NSF RA N74 123, PB-240 306, May).
Anon. (1975), Efficient use of fuels in process and manufacturing industries, *Proc. Inst. Gas Technol. Symposium, Chicago, USA, 16–19 April 1974* (published IGT, Chicago).
Anon. (1975), Efficient use of fuels in the metallurgical industries, *Proc. Inst. Gas Technol. Symposium, Chicago, USA, 9–13 December 1974* (published IGT, Chicago).
Anon. (1975), Energy use and conservation in the metals industry, *Proc. AIME Energy Symposium, New York, 1975.*
Anon. (1975), Waste heat utilisation, *Proc. National Conf. Gatlinburg, Tennessee, USA, 27–29 October 1971,* Oak Ridge National Laboratory, USA.
Anon. (1975), *Energy R and D – Problems and Perspectives,* OECD, Paris.
Anon. (1975), Energy recovery in process plate, *Proc. Conf. Inst. Mech. Engrs, London, 29–31 January 1975.*
Anon. (1978), *World Energy Resources, 1985–2020,* IPC, London.
Anon. (1978), *Energy for Industry and Commerce,* Cambridge Information and Research Services, Cambridge.
Anderson, L. L. and Tillman, D. A. (1978), *Fuels from Waste,* Academic Press, London.
Boyen, J. L. (1975), *Practical Heat Recovery,* John Wiley, New York.
Burberry, P. (1977), *Building for Energy Conservation,* Architectural Press, London.
Chapman, P. (1975), *Fuel's Paradise,* Penguin Books, Harmondsworth.
Crawley, G. M. (1975), *Energy,* Collier-Macmillan, London.
Department of Energy (1975), Energy saving: the fuel industries and some large firms, UK Departments of Energy Paper no. 5, HMSO, London.
Dryden, I. G. C. (ed.) (1975), *The Efficient Use of Energy,* Institute of Fuel/IPC, Guildford.
Esso. (1975), *Energy Saving in Industry,* Esso Petroleum Co. Publications (UK).
Gurney, J. D. (1978), Industrial combined heat and power: a case history, *Proc. Nat. Energy Managers' Conference, Birmingham, UK, 10–11 October 1978.*
Gyftopoulos, E. P. and Lazaridis, L. J. (1975), *Potential Fuel Effectiveness in Industry,* Ballinger New York, USA.
Harker, J. H. (1978), Economic balancing of heat exchangers, *Processing, June, 85.*
Ion, D. C. (1975), *Availability of World Energy Resources,* Graham and Trotman, London.
Kaya, A. (1978), Industrial energy control: the computer takes charge, *IEEE Spectrum,* July, 48–53.
Kell, J. R. and Martin, P. L. (1978), *Heating and Air Conditioning of Buildings,* Architectural Press, London.
Kelly's 1979 Guide to Energy Equipment and Services, Kelly's Directories, London.
Kreider, K. G. and McNeil, M. B. (1977), *Waste Heat Management Guidebook,* US National Bureau of Standards Handbook 121, Washington DC.
Leach, G. (1976), *Energy and Food Production,* IPC, London.
Lindberg, L. (1977), *The Energy Syndrome,* Lexington, New York.
Loftness, R. L. (1977), *Energy Handbook,* Van Nostrand Rheinhold, New York.
McKay (1977), World energy resources, *Proc. Conf. Moral Re-Armament, Caux, Switzerland, July, 1977* UK AERE Report R-8856.
Meador, R. (1978), *Future Energy Alternatives,* Ann Arbor Science Inc. Michigan.
O'Callaghan, P. W. (1978), *Building for Energy Conservation,* Pergamon Press, Oxford.
Payne, G. A. (1978), *The Energy Manager's Handbook,* IPC, London
Reay, D. A. (1979), *Industrial Energy Conservation – a Handbook for Engineers and Managers* 2nd ed., Pergamon Press, Oxford.
Richardson, H. W. (1975), *Economic Aspects of the Energy Crisis,* Lexington Books, Hampshire.
Simon, A. L. (1975), *Energy Resources,* Pergamon Press, Oxford.

Smith, C. B. (1976), *Efficient Electricity Use,* Pergamon Press, Oxford.
Tyldesley, J. R. (1977), *Applied Thermodynamics and Energy Conversion,* Longman, London.
Veziroglu, T. N. and Seifritz, W. (1978), Hydrogen Energy System, *Proc. 2nd World Hydrogen Energy Conference, Zurich 1978,* Pergamon Press, Oxford.
Wong, H. V. (1977), *Essential Formulae and Data on Heat Transfer for Engineers,* Longman, London.
Xlapedes, D. N. (1976), *Encyclopaedia of Energy,* McGraw-Hill, New York.

Heat Pumps

Anon. (1975), Aktuelle Wege zu verbesserter Energie anwendung, *VDI-Berichte 250* (published by VDI-Verlag GmbH, Düsseldorf).
Anon. (1975), Industrial heat pump cuts the fuel bill, *Electrical Review,* 28 March–4 April, 404–405.
Anon. (1976), Heat pump refinements, *New Scientist,* 22 January, 180.
Anon. (1978), Higher temperatures from heat pump heating packages, *Processing,* September, 49.
Blakeley, R. E., Ng., D. N. and Treece, R. J. (1978), The design and development of an absorption cycle heat pump optimised for the achievement of maximum coefficient of performance, *Proc. EEC Contractors' Meeting, Brussels, 27–28 September 1978.*
Bowen, J. L. (1975), Energy conservation in the seventies, *Refrig. Air Conditioning,* April.
Bridgers, F. H. (1975), How new technology may save energy in existing buildings, *Heating, Piping Air Conditioning,* vol. 47, no. 9, 50–55.
BSRIA (1975), *BSRIA Bibliography 103 (Heat Pumps),* BSRIA, Bracknell.
Dawes, J. (1978), Heat recovery and heat pumps for swimming pools, *Heating Air Conditioning J.,* October, 24–28.
Juttemann, H. (1974), *Heat Pumps in Large Buildings,* OA-Trans.-939, Electricity Council, London.
Leidenfrost, W. and Eisele, E. H. (1975), Rotating heat exchangers and the optimisation of a heat pump, *IEE Trans. Industry Application,* 1A-8, 3.
Macadam, J. A. (1974), *Heat pumps – the British experience,* Building Research Establishment Note N117/74.
Macmichael, D. B. A. and Reay, D. A. (1979), *Heat Pumps,* Pergamon Press, Oxford.
Miller, W. (1977), Energy conservation in timber-drying kilns by vapour recompression, *Forest. Prod. J.,* vol. 27, pt 9, 54–58.
Nicolich, M. J. (1977), Residential heat pump use: saving electrical energy, *ASHRAE J.,* vol. 19, pt 12, 22–23.
Perry, E. J. (1977), The role of refrigeration in energy conservation, *ASHRAE J.,* vol. 19, pt 12, 17–21.
Perry, E. J. (1978), Heat pumps – the future, *Heating Air Cond. J.,* September, 24–31.
Petersen, B. (1978), Diesel heat pump for district heating plants and the heating of large housing blocks, *Proc. EEC Contractors' Meeting, Brussels, 27–28 September 1978.*
Reay, D. A., Macmichael D. B. A. and Searle, N. K. (1978), Feasibility and design study of a gas engine-driven high temperature industrial heat pump, *Proc. EEC Contractors' Meeting, Brussels, 27–28 September 1978.*
Ross, P. N. (1975), The Templifier for process heat, *Proc. Annual Conf. EEI Conservation and Energy Management Division, Atlanta, Georgia, USA, 16–18 March 1975.*
Trenkowitz, G. (1972), Use of heat pumps for heat recovery, *Elektrowarme Int.,* vol. 30, no. 4, A180-A187.
Trenkowitz, G. (1974), Energy saving by using heat pumps, Ki Klima/Kalte-Ingenieur 4/74, 155–162.
Villaume, M. (1973), Centralised heat pumps and waste heat recovery, *Chaud. Froid. Plomb.,* vol. 27, no. 73-113-119, January, February, 320–321, (in French).

Westbrook, N. J. (1978), Evaporation advances in the chemical industry, *Processing*, September, 39–41.

Yanagimachi, M. (1965), Air source heat pump/heat system of Hiroshima regional station of Japan Broadcasting Corporation, *Trans. SHASE, Japan*, vol. 3, 23–30.

Prime movers

Butler, P. (1975), A showpiece to save money by promoting waste heat recovery, *Engineer*, 23 October.

Ford, E. (1971), Prospects for natural gas fuelled total energy systems, *Steam Heating Engng.*, vol. 40, 472.

Fraize, W. E. and Kinney, C. (1978), Effects of steam injection on the performance of gas turbine power cycles, ASME Paper 78-GT-11.

Klein, S. (1970), Total on-site power, *Machine Design*, vol. 42, no. 2.

McDonald, C. F. (1978), Role of the recuperator in high performance gas turbine applications, ASME Paper 78-GT-46.

Mottram, A. W. T. (1971), The gas turbine – recent improvements and their effect on the range of applications, *Proc. 8th World Energy Conference, Bucharest, 18 June–2 July, 1971.*

Stansell, J. (1975), Powerful future for gas turbines, *Electrical Review*, 28 November.

Wittner, B. R. and Culp, R. E. (1973), Turbine total energy for offshore rigs, *Gas Turbine International*, November-December.

Subject index

Index to manufacturers